CAMBRIDGE MONOGRAPHS
ON MECHANICS AND
APPLIED MATHEMATICS

GENERAL EDITORS

G. K. BATCHELOR, F.R.S.
Professor of Applied Mathematics at the University of Cambridge

J. W. MILES
Professor of Applied Mechanics and Geophysics, University of California, La Jolla

HYDRODYNAMIC STABILITY

HYDRODYNAMIC STABILITY

P. G. DRAZIN

Professor of Applied Mathematics, University of Bristol

W. H. REID

Professor of Applied Mathematics, University of Chicago

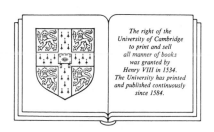

The right of the
University of Cambridge
to print and sell
all manner of books
was granted by
Henry VIII in 1534.
The University has printed
and published continuously
since 1584.

CAMBRIDGE UNIVERSITY PRESS

CAMBRIDGE

LONDON NEW YORK NEW ROCHELLE
MELBOURNE SYDNEY

Published by the Press Syndicate of the University of Cambridge
The Pitt Building, Trumpington Street, Cambridge CB2 1RP
32 East 57th Street, New York, NY 10022, USA
10 Stamford Road, Oakleigh, Melbourne 3166, Australia

First published 1981
First paperback edition 1982
Reprinted 1984, 1985

Printed in the United States of America

British Library cataloguing in publication data
Drazin, P G
Hydrodynamic stability.–
(Cambridge monographs on mechanics and applied mathematics).
1. Hydrodynamics 2. Stability
I. Title II. Reid, William Hill III. Series
532'.51 QC151 80-40273
ISBN 0 521 22798 4 hard covers
ISBN 0 521 28980 7 paperback

TO OUR PARENTS

Isaac and Leah Drazin

William and Edna Reid

CONTENTS

PREFACE

For nearly a century now, hydrodynamic stability has been recognized as one of the central problems of fluid mechanics. It is concerned with when and how laminar flows break down, their subsequent development, and their eventual transition to turbulence. It has many applications in engineering, in meteorology and oceanography, and in astrophysics and geophysics. Some of these applications are mentioned, but the book is written from the point of view intrinsic to fluid mechanics and applied mathematics. Thus, although we have emphasized the analytical aspects of the theory, we have also tried, wherever possible, to relate the theory to experimental and numerical results.

Our aim in writing this book has been twofold. Firstly, in Chapters 1–4, to describe the fundamental ideas, methods, and results in three major areas of the subject: thermal convection, rotating and curved flows, and parallel shear flows. Secondly, to provide an introduction to some aspects of the subject which are of current research interest. These include some of the more recent developments in the asymptotic theory of the Orr–Sommerfeld equation in Chapter 5, some applications of the linear stability theory in Chapter 6 and finally, in Chapter 7, a discussion of some of the fundamental ideas involved in current work on the nonlinear theory of hydrodynamic stability.

Each chapter ends with a number of problems which often extend or supplement the main text as well as provide exercises to help the reader understand the topics. An asterisk is used to indicate those problems which we judge to be relatively long or difficult. Some hints and references are given to help in the solution of many of the problems. We have also prepared answers to the problems which may be obtained from either author for a nominal charge to cover reproduction and postage costs.

Thus this is a textbook suitable for a graduate course on the fundamental ideas and methods and on the major applications of the theory of hydrodynamic stability. It also leads the reader up to the frontiers of research on selected topics. In general we have assumed that the reader is familiar with whatever mathematical methods are needed, notably in the theories of ordinary and of partial differential equations and in the theory of functions of a complex variable. But we have explained some specialized and modern mathematical points at length where it seems that they are likely to be unfamiliar to most readers.

We are grateful to our many colleagues throughout the world who have responded so generously to our various inquiries. In particular, we thank A. Davey, T. H. Hughes, and L. M. Mack for providing new or unpublished numerical results, R. J. Donnelly, E. L. Koschmieder and S. A. Thorpe for providing photographs, J. P. Cleave for advice on some mathematical points, L. C. Woods for advice on the presentation of the material, B. S. Ng for detailed comments on Chapters 1–5, and A. Davey and J. T. Stuart for constructive criticisms of a draft of Chapter 7. For help with the typing of the manuscript we also thank N. Thorp in Bristol and M. Bowie, F. Flowers, L. Henley, and M. Newman in Chicago. We are especially indebted to S. Chandrasekhar and C. C. Lin, who have contributed so much to the theory of hydrodynamic stability; through their papers and books, and through our personal contacts with them, they have greatly influenced our work on the subject. One of us (W.H.R.) also wishes to acknowledge with thanks the generous support provided over the years by the U.S. National Science Foundation, most recently under grant no. MCS 78–01249.

And, finally, we should like to thank G. K. Batchelor not only for his help as editor of this series but also for his kindness during an early stage in our careers when it was our good fortune to be associated with him.

Bristol   P.G.D.
Chicago   W.H.R.
August 1979

For this new impression we have corrected some misprints and other errors, and we have inserted Problem 1.12 on p. 31 and an addendum on pp. 479–80.
March 1982   P.G.D.
                                                                   W.H.R.

CHAPTER 1

INTRODUCTION

Yet not every solution of the equations of motion, even if it is exact,
can actually occur in Nature. The flows that occur in Nature must not
only obey the equations of fluid dynamics, but also be stable.

– L. D. Landau & E. M. Lifshitz (1959)

1 Introduction

The essential problems of hydrodynamic stability were recognized
and formulated in the nineteenth century, notably by Helmholtz,
Kelvin, Rayleigh and Reynolds. It is difficult to introduce these
problems more clearly than in Osborne Reynolds's (1883) own
description of his classic series of experiments on the instability of
flow in a pipe.

The ... experiments were made on three tubes .... The diameters of
these were nearly 1 inch, $\frac{1}{2}$ inch and $\frac{1}{4}$ inch. They were all ... fitted with
trumpet mouthpieces, so that the water might enter without disturbance.
The water was drawn through the tubes out of a large glass tank, in which
the tubes were immersed, arrangements being made so that a streak or
streaks of highly coloured water entered the tubes with the clear water.
The general results were as follows:–
(1) When the velocities were sufficiently low, the streak of colour
extended in a beautiful straight line through the tube, Fig. 1.1(a).
(2) If the water in the tank had not quite settled to rest, at sufficiently low
velocities, the streak would shift about the tube, but there was no
appearance of sinuosity.
(3) As the velocity was increased by small stages, at some point in the
tube, always at a considerable distance from the trumpet or intake, the
colour band would all at once mix up with the surrounding water, and fill
the rest of the tube with a mass of coloured water, as in Fig. 1.1(b). Any
increase in the velocity caused the point of break down to approach the
trumpet, but with no velocities that were tried did it reach this. On viewing
the tube by the light of an electric spark, the mass of colour resolved itself
into a mass of more or less distinct curls, showing eddies, as in Fig. 1.1(c).

Reynolds went on to show that the *laminar flow*, the smooth flow
he described in paragraph (1), breaks down when Va/ν exceeds a
certain critical value, V being the maximum velocity of the water in
the tube, a the radius of the tube, and ν the kinematic viscosity of

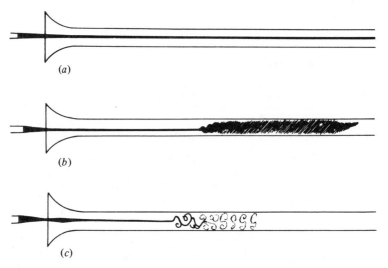

Fig. 1.1. (*a*) Laminar flow in a pipe. (*b*) Transition to turbulent flow in a pipe. (*c*) Transition to turbulent flow as seen when illuminated by a spark. (From Reynolds 1883.)

water at the appropriate temperature. This dimensionless number Va/v, now called the *Reynolds number*, specifies any class of dynamically similar flows through a pipe; here we shall denote the number by R. The series of experiments gave the critical value R_c of the Reynolds number as nearly 13 000. However,

the critical velocity was very sensitive to disturbance in the water before entering the tubes . . . . This at once suggested the idea that the condition might be one of instability for disturbance of a certain magnitude and stable for smaller disturbances.

At the critical velocity

another phenomenon . . . was the intermittent character of the disturbance. The disturbance would suddenly come on through a certain length of tube and pass away and then come on again, giving the appearance of flashes, and these flashes would often commence successively at one point in the pipe. The appearance when the flashes succeeded each other rapidly was as shown . . .

in Fig. 1.2. Such 'flashes' are now called *turbulent spots* or *turbulent bursts*. Below the critical Reynolds number there was laminar Poiseuille flow with a parabolic velocity profile, the resistance of the

Fig. 1.2. Turbulent spots in a pipe. (From Reynolds 1883.)

pipe to the flow of water being proportional to the mean velocity. As the velocity increased above its critical value, Reynolds found that the flow became *turbulent*, with a chaotic motion that strongly diffused the dye throughout the water in the tube. The resistance of the pipe to turbulent flow grew in proportion to the square of the mean velocity.

Later experimentalists have introduced disturbances of finite amplitude at the intake or used tubes with roughened walls to find R_c as low as 2000, and have used such regular flows and such smooth-walled tubes that R_c was 40 000 or even more.

Reynolds's description illustrates the aims of the study of hydrodynamic stability: to find whether a given laminar flow is unstable and, if so, to find how it breaks down into turbulence or some other laminar flow.

Methods of analysing the stability of flows were formulated in Reynolds's time. The method of normal modes for studying the oscillations and instability of a dynamical system of particles and rigid bodies was already highly developed. A known solution of Newton's or Lagrange's equations of motion for the system was perturbed. The equations were linearized by neglecting products of the perturbations. It was further assumed that the perturbation of each quantity could be resolved into independent components or modes varying with time t like e^{st} for some constant s, which is in general complex. The values of s for the modes were calculated from the linearized equations. If the real part of s was found to be positive for any mode, the system was deemed unstable because a general initial small perturbation of the system would grow exponentially in time until it was no longer small. Stokes, Kelvin and Rayleigh adapted this method of normal modes to fluid dynamics. The essential mathematical difference between fluid and particle dynamics is that the equations of motion are partial rather than ordinary differential equations. This difference leads to many technical difficulties in hydrodynamic stability, which, to this day, have been overcome for only a few classes of flows with very simple configurations. Indeed, Reynolds's experiment itself is still imperfectly understood. However, we can explain qualitatively the transition from laminar flow to turbulence with some confidence. Poiseuille flow with a parabolic profile is stable to infinitesimal disturbances at all Reynolds numbers. Some way below the observed critical Reynolds number a finite disturbance may grow if it is not too small. Above the critical Reynolds number quite small disturbances, perhaps introduced at the inlet or by an irregularity of the wall of the tube, grow rapidly with a sinuous motion. Soon they grow so much that nonlinearity becomes strong and large eddies (Fig. 1.1(c)) or turbulent spots (Fig. 1.2) form. At high Reynolds numbers turbulence ensues at once and the flow becomes random and strongly nonlinear everywhere.† This instability of Poiseuille flow may be contrasted with that of plane Poiseuille flow, which is unstable to infinitesimal disturbances at sufficiently large values of the Reynolds number. This explanation is supported by the treatment of the theory of the linear stability of Poiseuille flow in § 31 and of the nonlinear stability in § 49.1.

The physical mechanisms of Reynolds's experiments on instability of Poiseuille flow in a pipe are vividly illustrated by a film loop made by Stewart (FL 1968) for the Educational Development Center. This loop consists of edited excerpts from his longer film on *Turbulence* (Stewart F 1968). Details of these and other motion pictures on hydrodynamic stability may be found at the end of the bibliography.

2 Mechanisms of instability

Few laminar flows correspond to known solutions of the nonlinear equations of motion. Fewer are simple enough to allow detailed analysis of their instability. Consequently research on hydrodynamic stability has been deep but narrow. We are forced to study

† Many of the features of the transition from laminar to turbulent flow can easily be seen by observing the smoke from a cigarette. Light the cigarette, point the burning tip upwards in still air, and observe the smoke as it rises from rest at the hot tip.

the stability of a few classes of simple laminar flows, mostly with planar, axial or spherical symmetry. Unfortunately, their simplicity obscures some general aspects of instability, especially three-dimensional ones, such as flow with stretching of vortex lines. To gain a wider understanding of hydrodynamic stability and to put these simple flows in perspective as prototypes, it is helpful to sketch the important physical mechanisms of instability.

Broadly speaking, one may say that instability occurs because there is some disturbance of the equilibrium of the external forces, inertia and viscous stresses of a fluid. We shall discuss the external forces first. External forces of interest are buoyancy in a fluid of variable density, surface tension, magnetohydrodynamic forces, etc. It is also convenient to regard centrifugal and Coriolis forces as external forces when there is rotation of the whole system in which the fluid moves. If heavy fluid rests above light fluid it is clear that the fluids tend to overturn under the action of gravity. A similar instability occurs on the free surface of a container of liquid when it is moved downwards at a uniform acceleration greater than the gravitational acceleration. There is in fact a close analogy between the problem of instability of a fluid of variable density, namely Bénard convection bounded by horizontal planes, and the problem of instability of axisymmetric swirling flow of homogeneous fluid, namely Taylor vortices bounded by two coaxial rotating cylinders. The analogue of the density turns out to be the square of the circulation or swirl. If in Couette flow the circulation around the inner cylinder is greater in magnitude than that around the outer, the centrifugal force tends to throw out the fluid near the inner cylinder as an overturning instability. This centrifugal instability may occur also in flows along a rigid curved surface such as a concave wall of a channel. Surface tension resists the increase of area of a surface and so exerts a stabilizing influence, particularly on disturbances of small length scale. A magnetic field can inhibit the motion of an electrically conducting fluid across the magnetic lines of force and thereby usually tends to stabilize flows.

In the absence of any external force or of viscosity, a fluid moves according to the equilibrium between its inertia and internal stresses of pressure. A small disturbance may upset this equilibrium. The tendency of fluid to move down pressure gradients can be seen to amplify disturbances of certain flows and thereby create instability. This instability can be depicted more reliably in terms of interactions of the vortex lines, which are convected and stretched by the motion of the fluid.

An obvious effect of viscosity is to dissipate the energy of any disturbance and thereby stabilize a flow. Indeed, for this reason any bounded flow is stable if the viscosity is large enough. So, by and large, viscosity has a stabilizing influence. Viscosity has also the more complicated effect of diffusing momentum. This can make some flows, notably parallel shear flows, *un*stable although the same flows of an inviscid fluid are stable.

Thermal conductivity, or molecular diffusion of heat, has some effects similar to those of viscosity, or molecular diffusion of momentum. It tends to smooth out the temperature differences of a disturbance and is usually a stabilizing influence.

It is natural to consider the stability of primarily steady flows, but unsteady ones are also of some practical importance. The acceleration of a laminar flow plays an identifiable role in its stability. Analysis is difficult in general, but it emerges that acceleration of a laminar flow has a stabilizing and deceleration a destabilizing tendency. Flows that oscillate in time, such as Poiseuille flow driven through a circular pipe by an oscillatory pressure gradient, have intricate stability characteristics. Parametric stability or instability may occur, whereby the free oscillations of disturbances of the mean flow resonate with the forced oscillations of the flow.

Finally, the boundaries of a flow are an important factor. They constrain the development of a disturbance and usually the closer they are together the more stable the flow. However, they sometimes give rise to strong shear in boundary layers which is diffused outwards by viscosity and so leads to instability of the flow.

In a typical flow more than one of these mechanisms may act. For example, in plane Poiseuille flow the dual effects of viscosity, the inertia and the boundaries all influence the instability. Plane Poiseuille flow of inviscid fluid is stable. At large but finite Reynolds numbers the diffusion of momentum from thin shear layers near the walls leads to instability. At small Reynolds numbers the dissipative role of viscosity is dominant and there is stability. This leads to a critical value of the dimensionless number R representing the ratio of the magnitudes of the destabilizing forces of shear and stabilizing viscous forces for which their effects may be said to balance.

This summary of mechanisms of hydrodynamic stability will be given substance by detailed problems in this and later chapters. But the necessary details of the instability of any particular flow should not obscure these general mechanisms, whose recognition helps to classify as well as to understand problems. Instability arising from an upset of the equilibrium between an external force and dissipative effects is usually simpler than inertial instability. Prototype problems of this simpler instability will be analysed first, the linear instability of a fluid heated from below in Chapter 2 and centrifugal instability in Chapter 3. If fluid lies at rest between two horizontal plates, the lower one being hotter than the upper, then we have light fluid below heavy fluid. The buoyancy tends to overturn the fluid. This tendency is countered by the dissipative and diffusive effects of viscosity and thermal conductivity. The dimensionless number typifying the ratio of the destabilizing buoyancy to the stabilizing diffusive forces is called the *Rayleigh number*; its critical value is calculated and related to experiments. In this flow, viscosity and thermal conductivity have only stabilizing effects. The instability of Couette flow between rotating cylinders is analogous, viscosity tending only to stabilize the centrifugal instability. The linear instability of parallel shear flows is treated in Chapter 4, where it is shown that viscosity typically plays the dual roles of stabilizer and destabilizer. There is an imbalance between the inertia and both the dissipative and diffusive effects of viscosity. The physical mechanism of this instability and its mathematical description is more difficult to understand than that of instability due to an external force, and will be explained in detail in Chapter 4. The most difficult mathematical topics, involving the asymptotic theory of the solutions of the Orr–Sommerfeld equation which governs the problem, will be elaborated in Chapter 5. Chapter 5 may be ignored by the less mathematically inclined reader who is interested in stability characteristics rather than in deducing them. More complicated flows whose stability is governed by more than two of the mechanisms will be presented in Chapter 6. Their mathematical difficulties are so formidable that only a few illustrative examples will be given. Also a few other problems will be discussed, in particular the instability of unsteady basic flows. Chapters 2–6 treat almost entirely linear theory, which is now well formed and understood. The important topic of nonlinear instability is introduced in Chapter 7. Nonlinear theory has not reached a definitive stage yet, knowledge of the transition from laminar flow to turbulence being far from complete. Research has intensified in the last two decades, however, and has led to an increased understanding of transition beyond the stage of linear instability.

To prepare for this course of detailed analysis, this chapter ends with an account of the concepts of stability and of the methods of linear stability, including some classic problems as illustrations.

3 Fundamental concepts of hydrodynamic stability

To analyse the stability of any laminar flow one must first find the velocity $U(x, t)$ and other fields, such as pressure $P(x, t)$ and temperature $\Theta(x, t)$, needed to specify the laminar flow at each point x and time t. These fields define the *basic flow*. The fields may be steady or unsteady, and should satisfy the appropriate equations of motion and boundary conditions. Choice of suitable equations to model an observed flow and solution of the equations are often difficult tasks, but we shall suppose here that the equations and their solution are completely known, even though minor features of the observed flow may be neglected or only an approximate solution found.

Physically we want to know whether the basic flow can be observed or not. If it is disturbed slightly, the disturbance may either die away, persist as a disturbance of similar magnitude or grow so much that the basic flow becomes a different laminar or a turbulent flow. Broadly speaking, we call such disturbances (*asymptotically*) *stable*, *neutrally stable* or *unstable* respectively. All possible slight disturbances are likely to be excited in some degree by small irregularities or vibrations of the basic flow in practice, so it will persist only if it is stable to *all* slight disturbances. In seeking more precise definitions of stability we may be guided by the considerable mathematical literature of stability of solutions of ordinary and partial differential equations, but must frame the definitions to further our physical understanding. The choice of useful definitions of 'disturbed slightly', 'die away' and 'disturbance of similar magnitude' is usually clear unless the basic flow is unsteady or nonlinearity is significant. Definitions for these two cases are still controversial, and will be discussed in § 48 and Chapter 7. But at the outset we should recognize that the important thing is to understand how disturbances evolve in time rather than to argue about the definition of their stability.

It may help the mathematically inclined reader to formalize these definitions. However, the physically inclined reader may ignore this paragraph because in most applications common sense makes a formal definition unnecessary. Following the theory of stability of systems of ordinary differential equations, we say a basic flow is *stable* (*in the sense of Liapounov*) if, for any $\varepsilon > 0$, there exists some positive number δ (depending upon ε) such that if

$$\|\mathbf{u}(\mathbf{x}, 0) - \mathbf{U}(\mathbf{x}, 0)\|, \|p(\mathbf{x}, 0) - P(\mathbf{x}, 0)\|, \text{ etc.} < \delta,$$

then

$$\|\mathbf{u}(\mathbf{x}, t) - \mathbf{U}(\mathbf{x}, t)\|, \|p(\mathbf{x}, t) - P(\mathbf{x}, t)\|, \text{ etc.} < \varepsilon \quad \text{for all } t \geqslant 0,$$

where $\mathbf{u}$ is the velocity field and p is the pressure field, which satisfy the equations of motion and the boundary conditions. This definition means that the flow is stable if the perturbation is small for all time provided it is small initially, or, in yet other words, if the solution is uniformly continuous for all time with respect to the initial conditions. The precise meaning of 'small' or 'continuous' perturbation has to be assigned by definition of the positive definite norm. There is some freedom here, but it is physically useful to choose

$$\|\mathbf{u}(\mathbf{x}, t) - \mathbf{U}(\mathbf{x}, t)\| = \max_{\mathbf{x} \in \mathcal{V}} |\mathbf{u}(\mathbf{x}, t) - \mathbf{U}(\mathbf{x}, t)|$$

or

$$\left\{ \int_{\mathcal{V}} \tfrac{1}{2}\rho(\mathbf{u} - \mathbf{U})^2 \, d\mathbf{x} \right\}^{1/2} \quad \text{or} \quad \left\{ \int_{\mathcal{V}} [\{\mathcal{V}^{1/3}\boldsymbol{\nabla}(\mathbf{u} - \mathbf{U})\}^2 + (\mathbf{u} - \mathbf{U})^2] \, d\mathbf{x} \right\}^{1/2},$$

for example, where $\mathcal{V}$ is the domain of flow and ρ is the density of the fluid. Similarly, we may say the basic flow is *asymptotically stable* (*in the sense of Liapounov*) if moreover

$$\|\mathbf{u}(\mathbf{x}, t) - \mathbf{U}(\mathbf{x}, t)\|, \text{ etc.} \to 0 \quad \text{as} \quad t \to +\infty.$$

These definitions may be unsatisfactory when the norm of the basic flow itself decreases or increases substantially in time. Then a time-dependent norm may have to be carefully chosen to represent what the experimentalist or observer means intuitively by stability.

Other perturbations that might lead to instability arise from small changes in the boundary conditions due to irregularities in Nature or imperfections of laboratory equipment. The mathematical treatment of these perturbations is closely related to that of a small initial disturbance of the basic flow.

Also, it must be recognized that an unstable basic flow free of any disturbance cannot instantaneously be set up in the laboratory or arise in Nature. Rather a stable basic flow evolves in space or time until it becomes unstable, and the nature of the instability may be affected by the means of evolution.

Here we consider steady basic flows and *assume* that the equations of motion and the boundary conditions may be linearized for sufficiently small disturbances. Linearization is straightforward in principle and practice: products of the increments $\mathbf{u}'(\mathbf{x}, t) = \mathbf{u} - \mathbf{U}$, $p'(\mathbf{x}, t) = p - P$, etc., that is, of the total velocity $\mathbf{u}(\mathbf{x}, t)$ and pressure $p(\mathbf{x}, t)$, etc. of the disturbed flow less their respective values for the basic flow, are neglected. Thereby a linear homogeneous system of partial differential equations and boundary conditions is obtained. These have coefficients that may vary in space but not time because the basic flow is steady. Experience with the method of separation of variables and with Laplace transforms suggests that in general the solutions of the system can be expressed as the real parts of integrals of components, each component varying with time like e^{st} for some complex number $s = \sigma + i\omega$. The linear system will determine the values of s and the spatial variation of corresponding components as eigenvalues and eigenfunctions.

If the basic flow has some simple symmetry, the linear system may be transformed with respect to some of the space variables as well as time. For example, Poiseuille flow has basic velocity and pressure respectively given by

$$\mathbf{U} = V(1 - r^2/a^2)\mathbf{i},$$

$$P = p_0 - 4\rho\nu Vx/a^2 \quad \text{for } 0 \le r \le a, \, 0 \le \theta < 2\pi, \, -\infty < x < \infty,$$

where ρ is the density of the fluid, cylindrical polar coordinates (x, r, θ) are used, and $\mathbf{i}$ is the unit vector parallel to the x-axis. The axial symmetry of this flow is such that the coefficients of the linearized system of differential equations and boundary conditions for $\mathbf{u}'$, p' are functions of r alone. So the Laplace transform of the system with respect to t, a Fourier series in θ and the Fourier transform with respect to x may be taken to express the velocity perturbation in the form

$$\mathbf{u}'(x, r, \theta, t) = \text{Re} \int_{-\infty}^{\infty} dk \sum_{n=-\infty}^{\infty} \int ds\, \hat{\mathbf{u}}(r, k, n, s)\, e^{st+i(kx+n\theta)},$$

where the contour of the s-integral is the Bromwich contour for the inversion of the Laplace transform. Here $\hat{\mathbf{u}}$ is to be found from the initial velocity and pressure fields and the transformed system of *ordinary* differential equations in r and of the boundary conditions. This system gives an eigenvalue relation of the form

$$\mathcal{F}(s, k, n, V, a, \nu) = 0,$$

and eigenfunctions $\hat{\mathbf{u}}, \hat{p}$ except for an arbitrary multiplicative function of k, n, s. The multiplicative function is specified by the initial conditions. This is the *method of normal modes*, whereby small disturbances are resolved into modes, which may be treated separately because each satisfies the linear system. Its success depends on finding a complete set of normal modes to represent the development of an arbitrary initial disturbance.

If $\sigma > 0$ for a mode, then the corresponding disturbance will be amplified, growing exponentially with time until it is so large that nonlinearity becomes significant. If $\sigma = 0$ the mode is said to be *neutrally stable*, and if $\sigma < 0$ *asymptotically stable* or *stable*. Thus a mode is *unstable* if $\sigma > 0$, and stable if $\sigma \leq 0$ because then it remains small for all time. A small disturbance of the basic flow will in general excite all modes, so that if $\sigma > 0$ for at least one mode then the flow is unstable. Conversely, if $\sigma \leq 0$ for *all* of a complete set of modes then the flow is stable. A mode is *marginally stable* if $\sigma = 0$ for critical values of the parameters on which the eigenvalue s depends but $\sigma > 0$ for some neighbouring values of the parameters. Plane Poiseuille flow, with basic velocity $\mathbf{U} = V(1 - z^2/a^2)\mathbf{i}$ between rigid walls at $z = \pm a$, provides an example of this. The eigenvalue relation has the form,

$$\mathcal{F}(s, k, l, V, a, \nu) = 0,$$

where k and l are the wavenumbers of the mode in the x-direction and y-direction. This gives a mode with $\sigma = 0$ for certain values of k, l, $R = Va/\nu$ but $\sigma > 0$ for slightly larger values of R and the same k, l. The values of the parameters for marginal stability are often sought to give a criterion of stability, though it should be remembered that neutral stability is not necessarily marginal stability. The critical relationship between the parameters is the equation of the *curve* (or *surface*) of *marginal stability* or, more shortly, the *marginal curve* (or *surface*). (On a *neutral curve*, $\sigma = 0$ but σ is not necessarily positive for any neighbouring values of the parameters.) For example, the solution $\mathscr{F} = 0$ for plane Poiseuille flow gives $\sigma = \sigma(ak, al, R)$. The equation $\sigma = 0$ then would give the neutral curve relating k and R for each value of l. (In general, there may be no, one or many branches of the neutral curve for each mode.) If $\sigma > 0$ for any neighbouring values of k, l, R this neutral curve is also a marginal curve. The minimum value of R on all the marginal curves for all k, l, would then be called the *critical Reynolds number* R_c at which instability arises, because there would be instability for some values of R just greater than R_c and none with $R < R_c$. One would intuitively expect instability for all values of R with $R > R_c$, and this intuition is correct for plane Poiseuille flow, though some flows are exceptions to this rule. If $\omega \neq 0$ as $\sigma \downarrow 0$ for a disturbance, *oscillatory instability* sets in. This is sometimes called *overstability*, a term coined by Eddington (1926, p. 201). However, if $s = 0$ at marginal stability, i.e., $\sigma = \omega = 0$, there is said to be *exchange of stabilities*. This definition has been traced back to Poincaré (1885), p. 270 but seems to have been used first in the present sense by Jeffreys (1926), p. 833. Then instability sets in as a steady secondary flow, such as in the case of the convection cells that arise when a fluid is heated from below. Exchange of stabilities is typical of non-dissipative flow of an inviscid fluid, for which s^2 is real, as it is for normal modes of a conservative dynamical system, and for which a mode may travel as a wave with unchanged form at a uniform velocity. This *principle of exchange of stabilities* is valid also for some dissipative flows.

Reynolds's discovery that the criterion for stability of Poiseuille flow depends upon V, a, ν only through the combination $R = Va/\nu$ led to the widespread recognition of the importance of dimensional analysis in fluid dynamics. Indeed, we have just implicitly assumed that the eigenvalue relation $\mathcal{F}(s, k, l, V, a, \nu) = 0$ for plane Poiseuille flow gives solutions $s = s(ak, al, R)$. Naturally, we shall often use dimensionless variables in order to understand hydrodynamic stability better. We shall use an asterisk as a subscript to the dimensional form and omit the asterisk for the dimensionless form where it is desirable to use the two forms of the same physical quantities. Sometimes we shall need only the dimensional form and so shall not use the asterisks. Thus we may henceforth write $\mathbf{u}_*(\mathbf{x}_*, t_*)$ for the dimensional and $\mathbf{u}(\mathbf{x}, t)$ for the dimensionless total velocity of a disturbed flow. For the example of plane Poiseuille flow we should accordingly take $\mathbf{u} = \mathbf{u}_*/V$, $\mathbf{x} = \mathbf{x}_*/a$, $t = Vt_*/a$, $p = p_*/\rho V^2$, $\mathbf{U} = \mathbf{U}_*/V$, etc.

A normal mode depends on time exponentially with a complex exponent. It is the real part of a complex solution, whose real and imaginary parts separately are solutions because the system is linear. So, by convention, we shall drop explicit mention of taking the real part. Thus, for the example of Poiseuille flow, we shall write the real velocity perturbation as simply

$$\mathbf{u}' = \hat{\mathbf{u}}\, e^{st+i(kx+n\theta)},$$

instead of the proper expression

$$\mathbf{u}' = \mathrm{Re}\,\{\hat{\mathbf{u}}\, e^{st+i(kx+n\theta)}\}.$$

However, for initial-value problems and nonlinear theory we shall write out the variables explicitly to avoid any ambiguity.

We must not forget that a small disturbance is a superposition of normal modes, not a single one, in practice. So, if any basic flow is unstable, a localized initial disturbance not only will grow but also may move and spread, each unstable component growing at its own rate and moving at its own phase velocity. There is said to be *convected instability* when no unstable mode has group velocity equal to zero, because then the disturbance will remain small at any *fixed* point although it will grow as its centre moves far upstream or downstream. There is said to be *absolute instability* if some unstable mode has group velocity equal to zero, because then the disturbance will grow at some fixed points. These ideas are developed for parallel shear flows in §§ 24 and 47.

Also we must not forget that real disturbances are not infinitesimal, so that the concepts introduced in this section must be refined and extended. This is done in Chapter 7. This section is necessarily rather general, but it is well illustrated by the film *Flow instabilities* (Mollo-Christensen F 1968). Indeed, the whole subject of hydrodynamic instability is vividly introduced by the film. Excerpts of the film, on an ingenious experiment to illustrate the marginal curve, make up a film loop (Mollo-Christensen FL 1968b).

4 Kelvin–Helmholtz instability

To illustrate some of the mechanisms and concepts of stability we shall now work through a classic problem that demands little mathematics. Consider the basic flow of incompressible inviscid fluids in two horizontal parallel infinite streams of different velocities and densities, one stream above the other. Then the basic flow is given by

$$\mathbf{U} = \begin{cases} U_2 \mathbf{i} \\ U_1 \mathbf{i}, \end{cases} \quad \rho = \begin{cases} \rho_2 \\ \rho_1, \end{cases} \quad P = \begin{cases} p_0 - g\rho_2 z & (z > 0) \\ p_0 - g\rho_1 z & (z < 0), \end{cases} \quad (4.1)$$

say, where U_1, U_2 are the velocities of the two streams, ρ_1, ρ_2 are the densities, p_0 is a constant pressure, z is the height and g is the acceleration due to gravity. Helmholtz (1868) remarked that 'every perfect geometrically sharp edge by which a fluid flows must tear it asunder and establish a surface of separation, however slowly the rest of the fluid may move,' thereby recognizing the basic flow, but the problem of instability was first posed and solved by Kelvin (1871); it is now called *Kelvin–Helmholtz instability*.

The physical mechanism of Kelvin–Helmholtz instability has been described by Batchelor (1967), pp. 515–16 in terms of the vorticity dynamics. For simplicity of description, let us suppose for the moment that $U_1 = -V$, $U_2 = V > 0$, and that $\rho_2 = \rho_1$, so that we consider the special case of a vortex sheet in a homogeneous fluid. Thus buoyancy is ignored. Then consider an initial disturbance which slightly displaces the sheet so that its elevation is sinusoidal. Again for simplicity we suppose that the flow is two-dimensional in the xz-plane, so that the elevation of the sheet is given by $z = \zeta(x, t)$

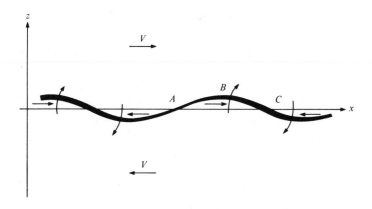

Fig. 1.3. Growth of a sinusoidal disturbance of a vortex sheet with positive vorticity normal to the paper. The local strength of the sheet is represented by the thickness of the sheet. The curved arrows indicate the directions of the self-induced movement of the vorticity in the sheet, and show (a) the accumulation of vorticity at points like C and (b) the general rotation about points like C, which together lead to exponential growth of the disturbance. (After Batchelor 1967, Fig. 7.1.3.)

at subsequent times. Batchelor traces the vorticity dynamics as follows, using the fundamental properties that each vortex line in an inviscid fluid is carried with the fluid and induces a rotating flow with circulation equal to the strength of the vortex line. The vorticity $\partial u/\partial z - \partial w/\partial x$ of the sheet is positive for $V > 0$. So positive vorticity is swept away from points like A (in Fig. 1.3) where $\zeta = 0$, $\partial \zeta/\partial x > 0$ and towards points like C where $\zeta = 0$, $\partial \zeta/\partial x < 0$, because vorticity in parts of the sheet displaced downwards (or upwards) induces a velocity with a positive (or negative) x-component at any part of the sheet where $z > 0$ (or < 0). In particular, the induced velocity at points like B due to each part of the displaced vortex sheet has positive x-component. Now the positive vorticity accumulating at points like C will induce clockwise velocities around such points and thereby amplify the sinusoidal displacement of the vortex sheet. These processes of accumulation of vorticity at points like C and of rotation of neighbouring points of the sheet will continue together, leading to exponential growth of the disturbance without change of the spatial form of the disturbance so long as the disturbance is small enough not to significantly change the basic state. We shall next substantiate this physical description with some mathematical details.

Kelvin assumed that the disturbed flow was irrotational on each side of the vortex sheet. This follows if the initial disturbance of the flow is irrotational, because irrotational flow of an inviscid fluid persists. However, initial rotational disturbances are also possible. To simplify the mathematics we shall adopt Kelvin's restrictive assumption, remembering that it allows a proof of instability but not stability because it gives no information about rotational disturbances. In fact, as we shall see in Problem 1.6 and Chapter 4, rotational disturbances are no more unstable and so Kelvin did find a necessary as well as a sufficient condition for instability. Thus we assume the existence of a velocity potential ϕ on each side of the interface between the two streams with $\mathbf{u} = \nabla\phi$, where

$$\phi = \begin{cases} \phi_2 & (z > \zeta) \\ \phi_1 & (z < \zeta), \end{cases} \tag{4.2}$$

the interface having elevation

$$z = \zeta(x, y, t) \tag{4.3}$$

when the flow is disturbed. Then the equations of continuity and incompressibility give $\nabla \cdot \mathbf{u} = 0$ and therefore the Laplacians of the potentials vanish, i.e.

$$\Delta\phi_2 = 0 \ (z > \zeta), \quad \Delta\phi_1 = 0 \ (z < \zeta). \tag{4.4}$$

Note that Euler's equations of motion have been used only implicitly in taking the irrotational flow as persistent.

The boundary conditions are as follows:

(a) The initial disturbance may be supposed to occur in a finite region so that for all time

$$\nabla\phi \to \mathbf{U} \quad \text{as} \quad z \to \pm\infty. \tag{4.5}$$

(b) The fluid particles at the interface must move with the interface without the two fluids occupying the same point at the same time and without a cavity forming between the fluids. Therefore the vertical velocity at the interface is given by

$$\frac{\partial\phi}{\partial z} = \frac{D\zeta}{Dt} = \frac{\partial\zeta}{\partial t} + \frac{\partial\phi}{\partial x}\frac{\partial\zeta}{\partial x} + \frac{\partial\phi}{\partial y}\frac{\partial\zeta}{\partial y} \quad (z = \zeta),$$

the material derivative of the surface elevation (cf. Lamb 1932, p. 7). This kinematic condition is the same as that for surface gravity waves, which may be regarded as a special form of this instability. There is a discontinuity of tangential velocity at the interface, and it leads to the two equations,

$$\frac{\partial \phi_i}{\partial z} = \frac{\partial \zeta}{\partial t} + \frac{\partial \phi_i}{\partial x} \frac{\partial \zeta}{\partial x} + \frac{\partial \phi_i}{\partial y} \frac{\partial \zeta}{\partial y} \quad (z = \zeta, i = 1, 2). \tag{4.6}$$

(c) The normal stress of the fluid is continuous at the interface. For an inviscid fluid, this gives the dynamical condition that the pressure is continuous. Therefore

$$\rho_1 \{ C_1 - \tfrac{1}{2} (\nabla \phi_1)^2 - \partial \phi_1 / \partial t - gz \}$$
$$= \rho_2 \{ C_2 - \tfrac{1}{2} (\nabla \phi_2)^2 - \partial \phi_2 / \partial t - gz \} \quad (z = \zeta), \tag{4.7}$$

by Bernoulli's theorem for irrotational flow, which is valid on each side of the vortex sheet, $z = \zeta$. In order that the basic flow satisfies this condition, the constants C_1, C_2 must be related so that

$$\rho_1 (C_1 - \tfrac{1}{2} U_1^2) = \rho_2 (C_2 - \tfrac{1}{2} U_2^2). \tag{4.8}$$

Equations (4.2)–(4.8) pose the nonlinear problem for instability of the basic flow (4.1). For linear stability we first put

$$\phi_2 = U_2 x + \phi_2' \ (z > \zeta), \quad \phi_1 = U_1 x + \phi_1' \ (z < \zeta) \tag{4.9}$$

and neglect products of the small increments ϕ_1', ϕ_2', ζ. There is no length scale in the basic velocity so it is far from clear how small ζ must be in order that the linearization is valid. But we can plausibly justify the linearization if the surface displacement and its slopes are small, i.e. $\partial \zeta / \partial x$, $\partial \zeta / \partial y \ll 1$ and $g\zeta \ll U_1^2$, U_2^2. If this is granted, linearization of equations (4.4)–(4.7) is straightforward, giving

$$\Delta \phi_2' = 0 \ (z > 0), \quad \Delta \phi_1' = 0 \ (z < 0); \tag{4.10}$$

$$\nabla \phi' \to 0 \quad \text{as} \quad z \to \pm \infty, \tag{4.11}$$

$$\partial \phi_i' / \partial z = \partial \zeta / \partial t + U_i \partial \zeta / \partial x \ (z = 0, i = 1, 2), \tag{4.12}$$

$$\rho_1 (U_1 \partial \phi_1' / \partial x + \partial \phi_1' / \partial t + g\zeta)$$
$$= \rho_2 (U_2 \partial \phi_2' / \partial x + \partial \phi_2' / \partial t + g\zeta) \ (z = 0). \tag{4.13}$$

It can be seen that all coefficients of this linear partial differential system are constants and that the boundaries are horizontal. So we use the method of normal modes, assuming that an arbitrary disturbance may be resolved into independent modes of the form

$$(\zeta, \phi_1', \phi_2') = (\hat{\zeta}, \hat{\phi}_1, \hat{\phi}_2) \, e^{i(kx+ly)+st}. \tag{4.14}$$

Equations (4.10) now give

$$\hat{\phi}_2 = A_2 \, e^{-\tilde{k}z} + B_2 \, e^{\tilde{k}z}, \tag{4.15}$$

where A_2 and B_2 are arbitrary constants and $\tilde{k} = (k^2 + l^2)^{1/2}$ is the total wavenumber. The boundary condition (4.11) at infinity implies that $B_2 = 0$ and therefore

$$\hat{\phi}_2 = A_2 \, e^{-\tilde{k}z}. \tag{4.16}$$

Similarly, we find

$$\hat{\phi}_1 = A_1 \, e^{\tilde{k}z}. \tag{4.17}$$

Now equations (4.12), (4.13) give three homogeneous linear algebraic equations for the three unknown constants $\hat{\zeta}, A_1, A_2$. Equations (4.12) give

$$A_2 = -(s + ikU_2)\hat{\zeta}/\tilde{k}, \quad A_1 = (s + ikU_1)\hat{\zeta}/\tilde{k} \tag{4.18}$$

and thence give the eigenfunctions (4.14) except for an arbitrary multiplicative constant. Then equation (4.13) gives the eigenvalue relation,

$$\rho_1\{\tilde{k}g + (s + ikU_1)^2\} = \rho_2\{\tilde{k}g - (s + ikU_2)^2\}. \tag{4.19}$$

The solution of this quadratic equation gives two modes with

$$s = -ik\frac{\rho_1 U_1 + \rho_2 U_2}{\rho_1 + \rho_2} \pm \left\{ \frac{k^2 \rho_1 \rho_2 (U_1 - U_2)^2}{(\rho_1 + \rho_2)^2} - \frac{\tilde{k}g(\rho_1 - \rho_2)}{\rho_1 + \rho_2} \right\}^{1/2}. \tag{4.20}$$

Both are neutrally stable if

$$\tilde{k}g(\rho_1^2 - \rho_2^2) \geqslant k^2 \rho_1 \rho_2 (U_1 - U_2)^2, \tag{4.21}$$

the equality giving marginal stability. One mode is asymptotically stable but the other unstable if

$$\tilde{k}g(\rho_1^2 - \rho_2^2) < k^2 \rho_1 \rho_2 (U_1 - U_2)^2. \tag{4.22}$$

This is accordingly a necessary and sufficient condition for instability of the mode with wavenumbers k, l. Thus the flow is always unstable (to short waves) if $U_1 \neq U_2$. (for $\rho_1 = \rho_2$)

Surface gravity waves. To interpret this result it is simplest to consider special cases separately. When $\rho_2 = 0$ and $U_1 = U_2 = 0$ we have the model of surface gravity waves on deep water. They are stable with phase-velocity

$$c = is/\tilde{k} = \pm(g/\tilde{k})^{1/2}, \qquad (4.23)$$

as is well known. This illustrates the identity of waves and oscillatory stable normal modes. It is often helpful to regard waves as a special case of hydrodynamic stability.

Internal gravity waves. When the basic flow is at rest ($U_1 = U_2 = 0$) we find

$$s = \pm\{\tilde{k}g(\rho_2 - \rho_1)/(\rho_1 + \rho_2)\}^{1/2}. \qquad (4.24)$$

There is instability if and only if $\rho_1 < \rho_2$, that is heavy fluid rests above light fluid. The stable waves have phase-velocity given by

$$c = \pm\{g(\rho_1 - \rho_2)/\tilde{k}(\rho_1 + \rho_2)\}^{1/2}. \qquad (4.25)$$

The eigenfunctions (4.14) for the velocity potential die away exponentially with distance from the interface, as in all cases of Kelvin–Helmholtz instability, and thus the motion is confined to the vicinity of the interface between the two fluids. These waves are a special case of internal gravity waves, which may propagate in the interior of a stratified fluid far from any boundary (see § 44.2). They can be observed between layers of fresh and salt water that occur in estuaries; the upper surface of the fresh water may be very smooth while strong internal gravity waves occur at the interface of the salt water a metre or two below, because for fresh and salt water $(\rho_1 - \rho_2)/(\rho_1 + \rho_2) \approx 10^{-2} \ll 1$, and so equations (4.18) give relatively small fluid velocities for given amplitude of interfacial elevation.

Rayleigh–Taylor instability. The eigenvalues (4.24) can be interpreted differently if the whole fluid system has an upward vertical acceleration f. Then, by the principle of equivalence in dynamics, or by solution of the problem of normal modes,

$$s = \pm\{\tilde{k}g'(\rho_2 - \rho_1)/(\rho_1 + \rho_2)\}^{1/2}, \qquad (4.26)$$

where $g' = f + g$ is the apparent gravitational or net vertical acceleration of the system. It follows that there is instability if and only if $g' < 0$, i.e. the net acceleration is directed from the lighter towards the heavier fluid. This is called Rayleigh–Taylor instability after Rayleigh's (1883b) theory of the stability of a stratified fluid at rest under the influence of gravity and Taylor's (1950) recognition of the significance of accelerations other than gravity. Rayleigh–Taylor instability can be simply observed by rapidly accelerating a beaker of water downwards (and standing clear!). Quantitative observations have been made by Lewis (1950) and by others. *Instability due to shear.* When there is a vortex sheet in a homogeneous fluid ($\rho_1 = \rho_2$, $U_1 \neq U_2$), equation (4.20) gives

$$s = -\tfrac{1}{2}\mathrm{i}k(U_1 + U_2) \pm \tfrac{1}{2}k(U_1 - U_2). \qquad (4.27)$$

This flow is always unstable, the waves moving with phase-velocity $c = \tfrac{1}{2}k(U_1 + U_2)/\tilde{k}$, the average velocity of the basic flow resolved in the direction $(k/\tilde{k}, l/\tilde{k}, 0)$ of propagation. Waves of all lengths are unstable, there being no length scale of the basic flow. For a real shear layer of finite thickness, we shall show (Chapter 4) that short waves are stable. However, the above result shows the instability of a shear layer to waves whose lengths are much greater than the thickness of the layer. It can also be seen that the wave of given length $\lambda = 2\pi/\tilde{k}$ which grows most rapidly is the one which propagates in the direction of the basic flow ($k = \tilde{k}$). So, after some time, waves in the direction of the basic flow will be dominant.

The full condition (4.22) of Kelvin–Helmholtz instability represents an imbalance of the destabilizing effect of inertia over the stabilizing effect of buoyancy when the heavy fluid is below. Kelvin (1871) used this theory as a model of the generation of ocean waves by wind. Helmholtz (1890) applied the theory to billow clouds, whose presence in regular lines marks instability of winds with strong shear. A striking photograph of billow clouds is shown in Fig. 1.4(a), the photographer having had the good fortune to view a regular train of clouds in a direction perpendicular to that of wave propagation. Reynolds's (1883) paper on pipe flow also describes some experiments on Kelvin–Helmholtz instability, though he did not recognize them as such. Reynolds filled a tube with water above carbon disulphide and tilted it. The ensuing relative motion of the two fluids led to instability at their interface. Thorpe (1969) has recently refined this experiment and clearly identified the Kelvin–Helmholtz instability. Some of his results are illustrated in Fig. 1.4(b) and (c).

Surface waves are illustrated on film by Bryson (FL 1967, containing excerpts from F 1967) and internal gravity waves by

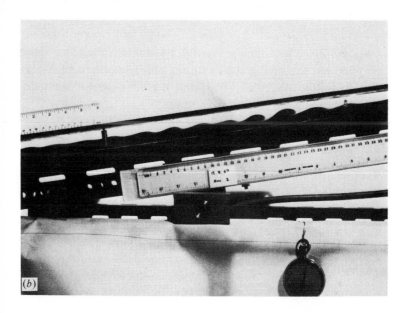

Fig. 1.4. Kelvin–Helmholtz instability. (a) Billow clouds near Denver, Colorado, photographed by Paul E. Branstine. For the meteorological details see Colson (1954). (b) Development of instability at the interface of two fluids of equal depth in relative acceleration owing to the tilt of the channel. (From Thorpe 1968.) (c) The same run of Thorpe's experiment about half a second later.

Long (F 1968, a few sequences only). Kelvin–Helmholtz instability modified by viscosity is shown in one short sequence by Mollo-Christensen & Wille (FL 1968) and Rayleigh–Taylor instability modified by surface tension in another.

5 Break-up of a liquid jet in air

Another illustration of the methods of hydrodynamic stability is Rayleigh's (1879) (see also Rayleigh 1894, pp. 351–62) classic theory of the break-up of a liquid jet in air, such as the formation of drops by a thin jet of water from a hose. Plateau had previously shown that capillarity would lead to instability of a round jet, because an axisymmetric deformation could decrease the surface area of the jet and thereby release surface energy. Rayleigh analysed in detail the instability of a uniform basic flow of incompressible inviscid liquid within the cylinder $r = a$, the liquid having a free surface governed by its surface tension γ. Without loss of generality a Galilean transformation may be made to reduce this uniform flow to rest, so that we take the basic state

$$\mathbf{U} = 0, \quad P = p_\infty + \gamma/a \quad (0 \leqslant r \leqslant a), \qquad (5.1)$$

where p_∞ is the pressure outside the jet.
Euler's equation of motion for an inviscid fluid gives

$$\rho(\partial \mathbf{u}/\partial t + \mathbf{u} \cdot \nabla \mathbf{u}) = -\nabla p \qquad (5.2)$$

and the equations of continuity and incompressibility give

$$\nabla \cdot \mathbf{u} = 0. \qquad (5.3)$$

It should be recalled that the pressure decreases outward across the liquid surface by $\gamma(R_1^{-1} + R_2^{-1})$ and that differential geometry gives the sum of the principal curvatures of the surface $R_1^{-1} + R_2^{-1} = \nabla \cdot \mathbf{n}$, where $\mathbf{n}$ is the unit outward normal from the liquid. If the surface of the jet is disturbed slightly, such that its equation becomes $r = \zeta(x, \theta, t)$ in terms of cylindrical polar coordinates (x, r, θ), then the dynamic boundary condition for the free surface is

$$p = p_\infty + \gamma \nabla \cdot \mathbf{n} \quad (r = \zeta), \qquad (5.4)$$

where

$$\mathbf{n} = (-\partial \zeta/\partial x, 1, -\partial \zeta/r\partial \theta)\{(\partial \zeta/\partial x)^2 + 1 + (\partial \zeta/r\partial \theta)^2\}^{-1/2}. \qquad (5.5)$$

The kinematic condition that each particle on the surface remains there gives

$$\left\{ \text{from } \frac{DF}{Dt} = 0 \quad F = r - \zeta(x, \theta, t) = 0 \right\}$$

$$u_r = D\zeta/Dt \quad (r = \zeta). \qquad (5.6)$$

Linearization of these equations and boundary conditions by neglecting products of the increments $\mathbf{u}' = \mathbf{u}$, $p' = p - P$, $\zeta' = \zeta - a$ gives

$$\rho \partial \mathbf{u}'/\partial t = -\nabla p', \qquad (5.7)$$

$$\nabla \cdot \mathbf{u}' = 0, \qquad (5.8)$$

$$p' = -\gamma(\zeta'/a^2 + \partial^2 \zeta'/\partial x^2 + \partial^2 \zeta'/a^2 \partial \theta^2) \quad (r = a), \qquad (5.9)$$

$$u_r' = \partial \zeta'/\partial t \quad (r = a). \qquad (5.10)$$

By the method of normal modes we consider only a typical wave component with

$$(\mathbf{u}', p', \zeta') = (\hat{\mathbf{u}}(r), \hat{p}(r), \hat{\zeta}) \, \mathrm{e}^{st + \mathrm{i}(kx + n\theta)}, \qquad (5.11)$$

where k is a real axial wavenumber and n an integer. Now equations (5.7) and (5.8) give

$$\Delta p' = -\rho \partial (\nabla \cdot \mathbf{u}') / \partial t = 0,$$

where the Laplacian is given by

$$\Delta = \partial^2 / \partial r^2 + \partial / r \partial r + \partial^2 / r^2 \partial \theta^2 + \partial^2 / \partial x^2.$$

Therefore

$$\frac{d^2 \hat{p}}{dr^2} + \frac{d\hat{p}}{r dr} - (k^2 + n^2 / r^2) \hat{p} = 0. \tag{5.12}$$

This is the modified Bessel equation of order n for the functions $I_n(kr)$, $K_n(kr)$. It is an equation symmetric in $\pm n$, so we may take $n \geq 0$ without loss of generality. Now $I_n(kr) \sim (\frac{1}{2}kr)^n / n!$ and $K_n(kr)$ is unbounded as $r \to 0$. Therefore, on invoking the condition that the pressure is bounded at the centre of the jet, we have

$$\hat{p} = A I_n(kr) \tag{5.13}$$

for some constant A. Then equation (5.7) gives

$$\hat{\mathbf{u}} = -A(\rho s)^{-1} (ik I_n(kr), k I_n'(kr), inr^{-1} I_n(kr)). \tag{5.14}$$

Finally, boundary conditions (5.9) and (5.10) give

$$A I_n(\alpha) = -\gamma (1 - \alpha^2 - n^2) \hat{\zeta} / a^2, \tag{5.15}$$

and

$$-A(a\rho s)^{-1} \alpha I_n'(\alpha) = s\hat{\zeta}, \tag{5.16}$$

respectively, where the dimensionless wavenumber is given by $\alpha = ak$. These two homogeneous equations for the unknown constants A and $\hat{\zeta}$ give the eigenvalue relation

$$s^2 = \frac{\gamma}{a^3 \rho} \frac{\alpha I_n'(\alpha)}{I_n(\alpha)} (1 - \alpha^2 - n^2). \tag{5.17}$$

To find the stability characteristics from this relation, we must use some properties of the modified Bessel function. In fact $\alpha I_n'(\alpha) / I_n(\alpha)$ is positive for each non-zero real value of α. It follows that $s^2 < 0$ for all α if $n \neq 0$, but that $s^2 > 0$ for $-1 < \alpha < 1$ and $s^2 \leq 0$ for $\alpha \geq 1$ or $\alpha \leq -1$ if $n = 0$. Therefore, the jet is stable to all non-axisymmetric modes, but is unstable to axisymmetric modes whose wavelength $\lambda = 2\pi/k$ is greater than the circumference $2\pi a$ of the jet.

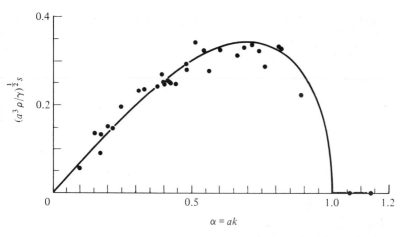

Fig. 1.5. The growth rates of unstable capillary modes of a uniform jet of fluid. (The solid line is based on the solution of equation (5.17) for an inviscid fluid by Chandrasekhar (1961), p.538 and the experimental points are from Donnelly & Glaberson (1966).)

Rayleigh evaluated the right-hand side of equation (5.17) numerically to find the greatest value of s for $n = 0$. He found that the greatest value of the dimensionless logarithmic growth rate $(a^3\rho/\gamma)^{1/2}s$ was 0.3433 for $\alpha = 0.6970$, giving the wavelength $\lambda = 9.016a$ of greatest instability. Rayleigh (1879) then put forward for the first time the idea that the fastest growing mode would become the dominant (or 'most dangerous') one and would be the mode observed in practice, because all modes would have comparable amplitude initially. (This idea is not necessarily correct in view of initial conditions and of nonlinearity but is often a good working rule.) For the break-up of a jet of water of diameter 10 mm, $(a^3\rho/\gamma)^{1/2}s = 0.3433$ gives a disturbance growing by the factor e in about an eighth of a second. For thinner jets this disintegration is faster.

In practice non-uniformity of the basic velocity of the jet, interaction with the ambient air, viscosity and nonlinearity are also significant. Chandrasekhar (1961), § 111 has given a full linear viscous theory. Bohr (1909) wrote a classic memoir on the nonlinear theory, more recent work being done by Wang (1968) and Yuen (1968). However, if a jet is not so thick that there is shear at the nozzle and if the liquid is not too viscous, Rayleigh's theory agrees

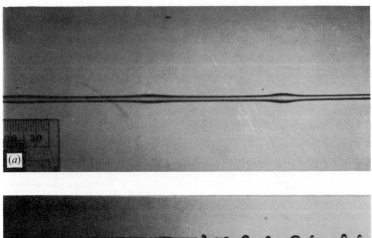

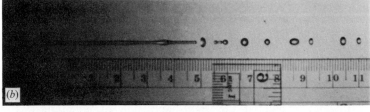

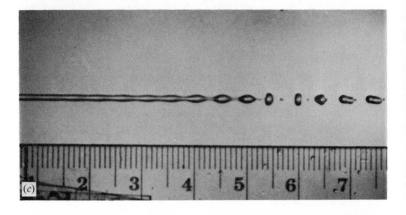

well with experiments, as shown in Fig. 1.5. Some photographs of
this instability are shown in Fig. 1.6. Rayleigh also found 'that there
is a recognized, if not a very pleasant, word' *varicosity* to describe
the symmetric instability photographed. This is sometimes called
sausage instability now, no more pleasant a description.

   You can see the break-up of a jet in your own kitchen after gently
turning on the water tap. It is better illustrated, with the aid of

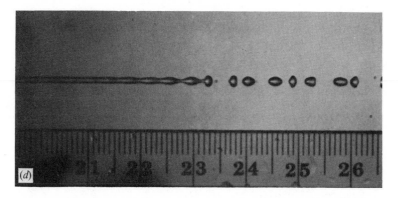

Fig. 1.6. The break-up of a jet of water, cf. Fig. 1.5. (a) $\alpha = 0.148$. (b) $\alpha = 0.262$. (c) $\alpha = 0.678$. (d) $\alpha = 0.805$. (From Donnelly & Glaberson 1966.)

high-speed photography and carefully controlled lighting, by the motion picture of Trefethen (FL 1965, edited from F 1965). Further work on drop formation in a circular liquid jet is reviewed by Bogy (1979).

Problems for chapter 1

1.1. *Kelvin's minimum energy principle and the stability of potential flow.* Prove that, of all flows of an inviscid incompressible fluid in a simply connected domain with prescribed normal flux at the surface, the irrotational flow has the least kinetic energy.

Discuss why a basic irrotational flow, although the one of minimum energy, is not necessarily stable by considering two-dimensional perturbations of the uniform basic flow $\mathbf{U} = U\mathbf{i}$ of an inviscid incompressible fluid in the square $-L < x, y < L$. Show that the vorticity-perturbation equation is satisfied by $\omega'(x, y, t) = f(x - Ut, y)$ for any differentiable function f, and hence that the problem for the velocity perturbation $\mathbf{u}'(x, y, t)$ is not well posed if the boundary conditions are merely that the normal component of $\mathbf{u}'$ vanishes on the sides of the square. If further $\omega'(x, y, 0) = g(x, y)$ in the square and $\omega'(-L, y, t) = h(y, t)$ for $t > 0$, where $U > 0$ and g and h are given functions, find ω' inside the square for $t > 0$. Hence show that ω' need not be small as $t \to \infty$, even if $g \equiv 0$. [Lamb (1932), § 45; $\partial\omega'/\partial t + U\partial\omega'/\partial x = 0$, $\omega' = g(x - Ut, y)$ for $0 < t < (x + L)/U$ and $\omega' = h(y, t - x/U)$ for $t > (x + L)/U$.]

1.2. *The stability of uniform rotation of a fluid.* Consider the stability of a uniform rotation of a viscous incompressible fluid within a rigid container.

First show that the Navier–Stokes equations referred to a frame rotating with constant angular velocity Ω are

$$\frac{\partial \mathbf{u}}{\partial t} + \mathbf{u} \cdot \nabla \mathbf{u} + 2\Omega \times \mathbf{u} = -\nabla \left\{ \frac{p}{\rho} + \tfrac{1}{2}(\Omega \times \mathbf{x})^2 \right\} + \nu \Delta \mathbf{u}$$

and

$$\nabla \cdot \mathbf{u} = 0.$$

Then the basic flow may be taken to have velocity $\mathbf{U} = 0$ and pressure $P = -\tfrac{1}{2}\rho(\Omega \times \mathbf{x})^2$ within a stationary domain $\mathcal{V}$ relative to the frame. Using Gauss's divergence theorem and the boundary condition that $\mathbf{u} = 0$ on $\partial\mathcal{V}$ (i.e. the container), prove that

$$\frac{d}{dt} \int_{\mathcal{V}} \tfrac{1}{2} u_i^2 \, d\mathbf{x} = -\nu \int_{\mathcal{V}} (\partial u_i / \partial x_j)^2 \, d\mathbf{x}.$$

Hence deduce that the flow is stable. [§ 53.1; Sorokin (1961), p. 372.]

*1.3. *The stability of a vortex street.* A vortex street in the wake of a long cylinder is modelled by two-dimensional irrotational flow of an inviscid incompressible fluid about two long staggered rows of line vortices, each vortex of one row having equal but opposite circulation to each vortex of the other row. Taking line vortices of circulation κ at points $(ma + Vt, \tfrac{1}{2}b)$ and $-\kappa$ at $((n + \tfrac{1}{2})a + Vt, -\tfrac{1}{2}b)$ in the xy-plane for m, $n = 0, \pm 1, \pm 2, \ldots$, show that $V = (\kappa/2a) \tanh(\pi b/a)$. Then find the infinite ordinary differential system governing the perturbed motion of the line vortices, linearize it, apply the method of normal modes, and prove Kármán's result that such an unsymmetrical vortex sheet is stable. [Lamb (1932), § 156.]

1.4. *The effect of surface tension on Kelvin–Helmholtz instability.* If there is surface tension γ between the two fluids, show that equation (4.20) becomes

$$s = -ik\frac{\rho_1 U_1 + \rho_2 U_2}{\rho_1 + \rho_2} \pm \left\{ \frac{k^2 \rho_1 \rho_2 (U_1 - U_2)^2}{(\rho_1 + \rho_2)^2} - \frac{\tilde{k}^2}{\rho_1 + \rho_2} \left[\frac{g(\rho_1 - \rho_2)}{\tilde{k}} + \tilde{k}\gamma \right] \right\}^{1/2}.$$

Deduce that the flow is stable if and only if

$$(U_1 - U_2)^2 \leqslant 2(\rho_1 + \rho_2)\{g\gamma(\rho_1 - \rho_2)\}^{1/2}/\rho_1\rho_2,$$

the 'cut-off' wavelength being $\lambda = 2\pi/\tilde{k} = 2\pi\{\gamma/g(\rho_1 - \rho_2)\}^{1/2}$.

Show that the wind generates waves on the sea if the difference of the basic water and air speeds is such that

$$|U_1 - U_2| > 6.6 \text{ m s}^{-1},$$

and that the cut-off wave has length $\lambda = 0.017$ m and speed 0.008 m s^{-1} relative to the sea. [Use $\rho_1 = 1020$ kg m^{-3}, $\rho_2 = 1.25$ kg m^{-3}, $g = 9.8$ m s^{-2}, $\gamma = 0.074$ N m^{-1}. Kelvin (1871), Chandrasekhar (1961), § 101.]

1.5. *An initial-value problem for a vortex sheet.* Suppose that the basic flow is a vortex sheet in a homogeneous incompressible inviscid fluid, taking equation (4.1) with $U_2 = V > 0$, $U_1 = -V$, $\rho_2 = \rho_1$. Consider an irrotational disturbance for which the interface is released from rest with

$$\zeta = H \exp(-x^2/2L^2) \quad \text{at } t = 0.$$

Deduce that

$$\zeta = H \exp\{(V^2t^2 - x^2)/2L^2\} \cos(Vxt/L^2) \quad \text{for } t > 0,$$

and find ϕ_1' and ϕ_2'.

$$[\zeta(x, 0) = (2\pi)^{-1/2}HL \int_{-\infty}^{\infty} \exp(-\tfrac{1}{2}k^2L^2 + ikx) \, dk.$$

Therefore

$$\zeta(x, t) = (2\pi)^{-1/2}HL \int_{-\infty}^{\infty} \exp(-\tfrac{1}{2}k^2L^2 + ikx) \cosh(kVt) \, dk,$$

because $s = \pm kV$ and $\partial\zeta/\partial t = 0$ at $t = 0$.

$$\phi_1' = 2(2\pi)^{-1/2}HLV$$
$$\times \int_0^{\infty} \exp(-\tfrac{1}{2}k^2L^2 + kz)(\sin kx \cosh kVt + \cos kx \sinh kVt) \, dk,$$

$$\phi_2' = 2(2\pi)^{-1/2}HLV$$
$$\times \int_0^{\infty} \exp(-\tfrac{1}{2}k^2L^2 - kz)(\sin kx \cosh kVt - \cos kx \sinh kVt) \, dk.$$

Note that $\nabla\phi_1'$, $\nabla\phi_2' \neq 0$ at $t = 0$.]

1.6. *The effect of rotational disturbances on Kelvin–Helmholtz instability.* Show that the linearized Euler equations of motion give

$$\partial\mathbf{u}'/\partial t + U_2 \partial\mathbf{u}'/\partial x = -\rho_2^{-1}\nabla p'$$

for $z > \zeta$. Deduce that $\Delta p' = 0$ for an incompressible fluid and that the perturbation of vorticity satisfies the equation

$$\boldsymbol{\omega}'(\mathbf{x}, t) = \boldsymbol{\omega}'(x - U_2t, y, z).$$

Hence argue plausibly that the presence of vorticity in the initial disturbance does not affect the criterion of instability (4.22). [See § 24 for a deeper discussion.]

1.7. *The instability of a jet.* Consider the two-dimensional instability of a 'top-hat' jet of thickness $2L$ in a homogeneous incompressible fluid, with basic velocity $\mathbf{U} = U\mathbf{i}$, where $U = 0$ for $|z| > L$ and $U = V$ for $|z| < L$. If the disturbance has velocity potential $\phi' \propto \exp(st + ikx)$ show that $(s + ikU)^{-1}\partial\phi'/\partial z$ and $(s + ikU)\phi'$ are continuous at the interfaces $z = \pm L$.

Assuming that ϕ' is an odd function of z (in which case u' is odd, w' is even, and the mode is called *sinuous*), obtain the eigenvalue relation,

$$(s+ikV)^2 \tanh(|k|L) + s^2 = 0.$$

Deduce that this jet is unstable to waves of all lengths. Assuming that ϕ' is even (varicose mode), obtain the eigenvalue relation,

$$(s+ikV)^2 \coth(|k|L) + s^2 = 0.$$

[Rayleigh (1894), pp. 380–1, Lamb (1932), pp. 374–5.]

1.8. *The instability of a gas jet*. For a gas jet in liquid, with the liquid at rest in the region $r > a$ *outside* the free surface, prove that equation (5.17) is replaced by

$$a^3 \rho s^2 / \gamma = -\frac{\alpha K_n'(\alpha)}{K_n(\alpha)}(1 - \alpha^2 - n^2).$$

Hence show that there is instability only for the varicose mode ($n = 0$) with $-1 < \alpha < 1$, and that $(a^3 \rho / \gamma)^{1/2} s$ attains its maximum value 0.820 for $\alpha = 0.484$. [Rayleigh (1892b).]

1.9. *The role of vorticity in capillary instability*. Equation (5.7) implies that the vorticity $\omega' = \nabla \times u'$ is independent of time. Discuss the significance of this in the initial-value problem of capillary instability.

1.10. *The capillary stability of a plane jet*. A liquid of density ρ and surface tension γ is at rest between the free surfaces $z = \pm a$. Show that the stability of the liquid is governed by the equation

$$\Delta p' = 0 \quad \text{for} \quad -\infty < x, y < \infty, \; -a \leqslant z \leqslant a$$

and boundary conditions

$$\rho \frac{\partial^2 p'}{\partial t^2} = \pm \gamma \left(\frac{\partial^2}{\partial x^2} + \frac{\partial^2}{\partial y^2} \right) \frac{\partial p'}{\partial z} \quad \text{at } z = \pm a.$$

Taking normal modes with $p' \propto \exp\{st + i(kx + ly)\}$, deduce that

$$a^3 \rho s^2 / \gamma = -\tilde{\alpha}^3 \tanh \tilde{\alpha} \quad \text{or} \quad -\tilde{\alpha}^3 \coth \tilde{\alpha},$$

where $\tilde{\alpha} = a(k^2 + l^2)^{1/2}$. Deduce that this plane jet is stable, unlike the round jet, which is unstable to axisymmetric disturbances. [$\rho \partial u'/\partial t = -\nabla p', \nabla \cdot u' = 0; p' = \gamma \nabla \cdot n, w' = \partial \zeta'/\partial t, n = \pm(-\partial \zeta'/\partial x, -\partial \zeta'/\partial y, 1)$ at $z = \pm a$. Note that $s \sim -\gamma a^3 \tilde{\alpha}^3 / \rho$ as $\tilde{\alpha} \to \infty$ for short capillary waves at a free surface, in common with the round jet.]

1.11. *Rayleigh–Taylor instability in a porous medium*. Consider the stability of a basic flow in which two incompressible fluids move with a horizontal interface and uniform vertical velocity in a uniform porous medium. You are given that motion of a fluid in a porous medium is governed by Darcy's law, namely that $\mathbf{u} = \nabla\phi$, where $\phi = -k(p + g\rho z)/\mu$, ρ is the density and μ the dynamic viscosity of the fluid, and k is the permeability of the medium to the fluid.

Let the lower fluid have density ρ_1 and viscosity μ_1 and the upper fluid ρ_2 and μ_2; let the medium have permeability k_1 to the lower and k_2 to the upper fluid; and let the basic velocity be $W\mathbf{k}$. Then show that the flow is stable if and only if

$$\left(\frac{\mu_1}{k_1} - \frac{\mu_2}{k_2}\right) W + g(\rho_1 - \rho_2) \geq 0.$$

[Take the equation of the mean interface to be $z = 0$ instantaneously and of the disturbed interface to be $z = \zeta$, where $\zeta = \hat{\zeta} \exp(st + ikx)$. (There need be no danger of confusion of the different ks.) Show that $\Delta\phi = 0$ if $z \neq \zeta$. Use continuity of normal velocity and pressure at $z = \zeta$ to show on linearization that

$$\frac{s}{k}\left(\frac{\mu_1}{k_1} + \frac{\mu_2}{k_2}\right) = -g(\rho_1 - \rho_2) - \left(\frac{\mu_1}{k_1} - \frac{\mu_2}{k_2}\right) W.$$

See Saffman & Taylor (1958).]

1.12. *Rayleigh–Taylor stability of superposed fluids confined by a vertical cylinder.* Consider an inviscid incompressible fluid of density ρ_1 at rest beneath a similar fluid of density ρ_2, the fluids being confined by a long vertical rigid cylinder with equation $r = a$ and there being surface tension γ at the interface with equation $z = 0$. Here cylindrical polar coordinates (r, θ, z) are used and Oz is the upward vertical. Then show, much as in Problem 1.4, that small irrotational disturbances of the state of rest may be found in terms of the normal modes of the form

$$\phi' \propto \cos n\theta \, J_n(kr) \, e^{-k|z| + ist} \quad \text{and} \quad \zeta \propto \cos n\theta \, J_n(kr) \, e^{ist}$$

for $n = 0, 1, 2, \ldots$, where

$$s^2 = \{g(\rho_2 - \rho_1)k - \gamma k^3\}/(\rho_1 + \rho_2)$$

and ka equals the mth positive zero $j'_{n,m}$ of the derivative J'_n of the Bessel function for $m = 1, 2, \ldots$. Deduce that there is stability if

$$a^2 g(\rho_2 - \rho_1) < \gamma j'^2_{1,1}.$$

This shows that a cut-off of long waves by a horizontal boundary together with a cut-off of short waves by surface tension may render a heavy fluid above a light one stable. This physical point is illustrated by an experiment (for which viscosity also damps short waves) of Mollo-Christensen (F1968). [Maxwell, J. C. (1876). Capillary action. *Encyclopaedia Britannica* (9th edn), vol. V, p. 59. Also *Scientific papers* (1890), vol. II, p. 587. Cambridge University Press.]

CHAPTER 2

THERMAL INSTABILITY

Science is nothing without generalisations. Detached and ill-assorted facts are only raw material, and in the absence of a theoretical solvent, have but little nutritive value.　— Lord Rayleigh (1884)

6 Introduction

Thermal instability often arises when a fluid is heated from below. The classic example of this, described in this chapter, is a horizontal layer of fluid with its lower side hotter than its upper. The basic state is then one of rest with light fluid below heavy fluid. When the temperature difference across the layer is great enough, the stabilizing effects of viscosity and thermal conductivity are overcome by the destabilizing buoyancy, and an overturning instability ensues as thermal convection. This convective instability may be distinguished from free convection, such as that due to a hot *vertical* plate, for which hydrostatic equilibrium is impossible. Again, a basic flow of free convection may itself be unstable. Our concern here with convection is only with thermal instability. Convective instability seems to have been first described by James Thomson (1882), the elder brother of Lord Kelvin, but the first quantitative experiments were made by Bénard (1900).

Rayleigh (1916a) wrote that

Bénard worked with very thin layers, only about 1 mm deep, standing on a levelled metallic plate which was maintained at a uniform temperature. The upper surface was usually free, and being in contact with the air was at a lower temperature. Various liquids were employed – some, indeed, which would be solids under ordinary conditions. The layer rapidly resolves itself into a number of *cells*, the motion being an ascension in the middle of a cell and a descension at the common boundary between a cell and its neighbours . . . .

The cells acquire

surfaces *nearly* identical, their forms being nearly regular convex polygons of, in general, 4 to 7 sides. The boundaries are vertical . . . .

Fig. 2.1 shows a plan of the convection cells in a silicone oil, with regular hexagons as the predominant polygons. Bénard convection

Fig. 2.1. Bénard cells under an air surface. (From Koschmieder & Pallas 1974.)

can also be seen in a short film sequence of Mollo-Christensen (FL 1968a).

Stimulated by Bénard's experiments, Rayleigh (1916a) formulated the theory of convective instability of a layer of fluid between horizontal planes. He chose equations of motion and boundary conditions to model the experiments, and derived the linear equations for normal modes. He then showed that instability would occur only when the adverse temperature gradient was so large that the dimensionless parameter $g\alpha\beta d^{4}/\kappa\nu$ exceeded a certain critical value. Here g is the acceleration due to gravity, α the coefficient of thermal expansion of the fluid, $\beta = -d\Theta/dz$ the magnitude of the vertical temperature gradient of the basic state of rest, d the depth of the layer of the fluid, κ its thermal diffusivity and ν its kinematic viscosity. This parameter is now called the *Rayleigh number*. We shall denote it by R in this chapter, where there is no danger of confusing it with the Reynolds number, but by Ra where confusion may occur. The Rayleigh number is a characteristic ratio of the destabilizing effect of buoyancy to the stabilizing effects of diffusion and dissipation. Rayleigh's paper is the foundation of scores of papers on thermal convection. His and many of the early theoretical papers are readily available in a recent anthology (Saltzman 1962).

Block (1956) showed physically and Pearson (1958) analytically that most of the motions observed by Bénard, being in very thin layers with a free surface, were driven by the variation of surface tension with temperature, *not* by thermal instability of light fluid below heavy fluid. However, Rayleigh's model is in accord with experiments on layers of fluid with rigid boundaries and on thicker layers with a free surface, because the importance of the variation of surface tension relative to that of buoyancy diminishes as the thickness of the layer increases. In spite of this ironic discovery, the convection in a horizontal layer of fluid heated from below is still called *Bénard convection*.

7 The equations of motion

7.1 *The exact equations*

The equations of motion of a heat-conducting viscous fluid under the action of gravity can be found in textbooks (e.g. Batchelor 1967). In the notation of Cartesian tensors with position vector $\mathbf{x} = x_j$ and velocity $\mathbf{u} = u_j$ ($j = 1, 2, 3$), the equations are as follows. The equation of continuity is

$$\frac{\partial \rho}{\partial t} + \frac{\partial(\rho u_j)}{\partial x_j} = 0. \tag{7.1}$$

The equations of motion are the Navier–Stokes equations,

$$\rho \frac{Du_i}{Dt} = -g\rho\delta_{i3} + \frac{\partial \sigma_{ij}}{\partial x_j}, \tag{7.2}$$

where $D/Dt = \partial/\partial t + \mathbf{u} \cdot \nabla$, the x_3-axis is the upward vertical, the stress tensor is given by

$$\sigma_{ij} = -p\delta_{ij} + \mu \left(\frac{\partial u_i}{\partial x_j} + \frac{\partial u_j}{\partial x_i} - \frac{2}{3} \frac{\partial u_k}{\partial x_k} \delta_{ij} \right) + \lambda \frac{\partial u_k}{\partial x_k} \delta_{ij}, \qquad (7.3)$$

μ is the coefficient of dynamic viscosity of the fluid, and λ is that of bulk viscosity (or second viscosity). The equation of energy, or of heat conduction, is

$$\rho \frac{DE}{Dt} = \frac{\partial}{\partial x_j} \left(k \frac{\partial \theta}{\partial x_j} \right) - p \frac{\partial u_j}{\partial x_j} + \Phi, \qquad (7.4)$$

where E is the internal energy per unit mass of the fluid, k is the thermal conductivity, θ is the temperature, and the rate of viscous dissipation per unit volume of fluid is given by

$$\Phi = \tfrac{1}{2}\mu \left(\frac{\partial u_i}{\partial x_j} + \frac{\partial u_j}{\partial x_i} \right)^2 + (\lambda - \tfrac{2}{3}\mu) \left(\frac{\partial u_k}{\partial x_k} \right)^2. \qquad (7.5)$$

For a calorically perfect gas $E = c_v \theta$, and for a liquid $E = c\theta$, where c_v is the specific heat at constant volume and c the specific heat.

In general, the equations of state for a fluid specify ρ, μ, λ, k, c and E as functions of p and θ. For layers of real fluid in which the pressure does not vary much, these functions are almost independent of p.

7.2 The Boussinesq equations

To these equations of motion, Rayleigh (1916a) applied the *Boussinesq approximation*, due independently to Oberbeck (1879) and Boussinesq (1903). The basis of this approximation is that there are flows in which the temperature varies little, and therefore the density varies little, yet in which the buoyancy drives the motion. Then the variation of density is neglected everywhere except in the buoyancy. On the basis of this approximation for small temperature difference between the bottom and top of the layer of fluid,

$$\rho = \rho_0 \{1 - \alpha(\theta - \theta_0)\}, \qquad (7.6)$$

where ρ_0 is the density of the fluid at the temperature θ_0 of the bottom of the layer and α is the constant coefficient of cubical expansion. For a perfect gas, $\alpha = 1/\theta_0 \approx 3 \times 10^{-3} \, \mathrm{K}^{-1}$, and for a typical liquid used in experiments $\alpha \approx 5 \times 10^{-4} \, \mathrm{K}^{-1}$. If $\theta_0 - \theta \leqslant 10 \, \mathrm{K}$, then $(\rho - \rho_0)/\rho_0 = \alpha(\theta_0 - \theta) \ll 1$, but nevertheless the buoyancy $g(\rho - \rho_0)$ is of the same order of magnitude as the inertia, acceleration or viscous stresses of the fluid and so is not negligible. For most real fluids $d\mu/\mu$ $d\theta$, dk/k $d\theta$, dc/c $d\theta \lesssim \alpha$, so that μ, k and c, or c_v, are treated as constants in the Boussinesq approximation. (The coefficient of bulk viscosity λ is neglected, because it only arises as a factor of $\partial u_i/\partial x_j$, which is of order α.) In short, one approximates the thermodynamic variables as constants except for the pressure and temperature and except for the density when multiplied by g. This approximation works well for flows with temperature differences of a few degrees or less, such as those in Bénard's experiments, and can be formally justified by dimensional analysis (Spiegel & Veronis 1960, Mihaljan 1962). Here we shall give only a partial justification.

The differentials of density in the continuity equation (7.1) are of order α, so the approximation gives

$$\frac{\partial u_j}{\partial x_j} = 0, \tag{7.7}$$

as for an incompressible fluid. Then the stress tensor is given by

$$\sigma_{ij} = -p\delta_{ij} + \mu\left(\frac{\partial u_i}{\partial x_j} + \frac{\partial u_j}{\partial x_i}\right). \tag{7.8}$$

Again, on treating ρ and μ as constants in each term other than the buoyancy, the Navier–Stokes equations become

$$\frac{Du_i}{Dt} = -\frac{\partial}{\partial x_i}\left(\frac{p}{\rho_0} + gz\right) - \alpha g(\theta_0 - \theta)\delta_{i3} + \nu\Delta u_i, \tag{7.9}$$

where the Laplacian operator is given by $\Delta = \partial^2/\partial x_j^2$.

Next we must simplify the heat equation (7.4). Firstly, note that, if V is a representative velocity scale of the flow, d a length scale, and $\theta_0 - \theta_1$ a scale of temperature difference, then the ratio of the rate of production of heat by internal friction to the rate of transfer of heat is

$$\Phi \Big/ \rho\frac{D(c\theta)}{Dt} \approx \mu V^2 d^{-2}/\rho_0 c(\theta_0 - \theta_1)Vd^{-1} = \nu V/c(\theta_0 - \theta_1)d,$$

θ_1 being the temperature of the top of the layer of thickness d. Now, for a typical gas $\nu/c_v \approx 10^{-8}$ s K and for a typical liquid $\nu/c \approx 10^{-9}$ s K, which shows that the ratio is very small for both gases and liquids unless $V/(\theta_0 - \theta_1)d$ is very large. Therefore we shall neglect Φ. Secondly, note that the heating due to compression is

$$-p\frac{\partial u_j}{\partial x_j} = \frac{p}{\rho}\frac{\mathrm{D}\rho}{\mathrm{D}t} = \alpha p \frac{\mathrm{D}\theta}{\mathrm{D}t}.$$

For a perfect gas, $p = (c_p - c_v)\rho\theta$ and $\alpha = 1/\theta$. Therefore

$$\rho\frac{\mathrm{D}E}{\mathrm{D}t} + p\frac{\partial u_j}{\partial x_j} \cong c_p\rho\frac{\mathrm{D}\theta}{\mathrm{D}t},$$

and the heating due to compression is not negligible in comparison to the heat transfer, as approximation (7.7) might have led one to expect. For liquids, however, the heating is negligible at normal pressures. The reason for this difference between gases and liquids is chiefly because, although the heating due to compression is typically only an order of magnitude less for a liquid than for a gas, the heat transfer is proportional to the density of the fluid, and a typical liquid is 10^3 times more dense than a typical gas. With all of these approximations, the heat equation becomes

$$\frac{\mathrm{D}\theta}{\mathrm{D}t} = \kappa\,\Delta\theta, \tag{7.10}$$

where the thermal diffusivity $\kappa = k/\rho_0 c_p$ for a perfect gas and $k/\rho_0 c$ for a liquid. Equations (7.7), (7.9) and (7.10) are called the *Boussinesq equations* and describe the motion of a *Boussinesq fluid*.

8 The stability problem

8.1 *The linearized equations*

Rayleigh (1916a) modelled Bénard's experiments as the instability of a Boussinesq fluid at rest between two infinite horizontal planes at different temperatures. Let the planes have equations $z_* = 0$ and d, where the temperatures are θ_0 and θ_1 respectively. Here we denote a dimensional variable by a subscripted asterisk to prepare for our choice of dimensionless variables; e.g. we shall soon take $z = z_*/d$ to be the dimensionless variable of height. Then the equations of motion give the basic state with

$$\mathbf{U}_* = \mathbf{0}, \quad \Theta_* = \theta_0 - \beta z_*,$$
$$P_* = p_0 - g\rho_0(z_* + \tfrac{1}{2}\alpha\beta z_*^2) \quad \text{for } 0 \leqslant z_* \leqslant d, \tag{8.1}$$

where the basic temperature gradient $\beta = (\theta_0 - \theta_1)/d$. We anticipate that there can be instability only when $\theta_0 > \theta_1$, i.e. when there is an adverse temperature gradient and $\beta > 0$.

On putting

$$\mathbf{u}_* = \mathbf{u}'_*(\mathbf{x}_*, t_*),$$
$$\theta_* = \Theta_*(z_*) + \theta'_*(\mathbf{x}_*, t_*), \quad p_* = P_*(z_*) + p'_*(\mathbf{x}_*, t_*),$$

and linearizing the Boussinesq equations for small perturbations $\mathbf{u}'_*$, θ'_*, p'_*, it follows that

$$\boldsymbol{\nabla}_* \cdot \mathbf{u}'_* = 0, \tag{8.2}$$

$$\frac{\partial \mathbf{u}'_*}{\partial t_*} = -\frac{1}{\rho_0} \boldsymbol{\nabla}_* p'_* + \alpha g \theta'_* \mathbf{k} + \nu \Delta_* \mathbf{u}'_*, \tag{8.3}$$

$$\frac{\partial \theta'_*}{\partial t_*} - \beta w'_* = \kappa \Delta_* \theta'_*. \tag{8.4}$$

In the absence of any basic velocity, we seek convection driven by buoyancy and moderated by viscosity and thermal diffusivity, so it is convenient to use scales d of length, d^2/κ of time, and $\beta d = \theta_0 - \theta_1$ of temperature difference. (One may equivalently use d^2/ν as the time scale; this is somewhat simpler if $\nu \gg \kappa$.) Accordingly we define the dimensionless variables

$$\mathbf{x} = \mathbf{x}_*/d, \quad t = \kappa t_*/d^2, \quad \mathbf{u} = d\mathbf{u}'_*/\kappa,$$
$$\theta = \theta'_*/\beta d, \quad p = d^2 p'_*/\rho_0 \kappa^2. \tag{8.5}$$

Then the linearized stability equations (8.2)–(8.4) become

$$\boldsymbol{\nabla} \cdot \mathbf{u} = 0, \tag{8.6}$$

$$\frac{\partial \mathbf{u}}{\partial t} = -\boldsymbol{\nabla} p + R\, Pr \theta \mathbf{k} + Pr \Delta \mathbf{u}, \tag{8.7}$$

$$\frac{\partial \theta}{\partial t} - w = \Delta \theta, \tag{8.8}$$

respectively, where the dimensionless *Rayleigh number* is given by

$$R = g\alpha\beta d^4/\kappa\nu \tag{8.9}$$

and the *Prandtl number* by

$$Pr = \nu/\kappa. \tag{8.10}$$

Note that the Rayleigh number is positive when the lower boundary is the hotter one ($\theta_0 > \theta_1$) and is seen to be a characteristic ratio of the buoyancy to the viscous forces. Also note that the Prandtl number is an intrinsic property of the fluid, not of the flow; it measures the ratio of the rates of molecular diffusion of momentum and heat.

We can now easily eliminate all the dependent variables except w, to get a single stability equation. The curl of equation (8.7) gives

$$\frac{\partial \boldsymbol{\omega}}{\partial t} = R\, Pr(\nabla \theta \times \mathbf{k}) + Pr\Delta\boldsymbol{\omega}, \qquad (8.11)$$

where the vorticity $\boldsymbol{\omega} = \nabla \times \mathbf{u}$. The curl of equation (8.11) in turn gives, after use of equation (8.6),

$$\frac{\partial}{\partial t}\Delta\mathbf{u} = R\, Pr\left(\Delta\theta\mathbf{k} - \nabla\frac{\partial\theta}{\partial z}\right) + Pr\Delta^2\mathbf{u}. \qquad (8.12)$$

In particular,

$$\frac{\partial}{\partial t}\Delta w = R\, Pr\Delta_1\theta + Pr\Delta^2 w, \qquad (8.13)$$

where the horizontal Laplacian $\Delta_1 = \partial^2/\partial x^2 + \partial^2/\partial y^2$. Finally elimination of θ from equations (8.8), (8.13) gives

$$\left(\frac{\partial}{\partial t} - \Delta\right)\left(\frac{1}{Pr}\frac{\partial}{\partial t} - \Delta\right)\Delta w = R\Delta_1 w. \qquad (8.14)$$

Similarly it can be shown that θ satisfies the same equation.

It can be shown from the equation of continuity that

$$\Delta_1 u = -\frac{\partial^2 w}{\partial x\partial z} - \frac{\partial\omega_3}{\partial y}, \qquad (8.15)$$

$$\Delta_1 v = -\frac{\partial^2 w}{\partial y\partial z} + \frac{\partial\omega_3}{\partial x}, \qquad (8.16)$$

where $\omega_3 = \partial v/\partial x - \partial u/\partial y$ is the vertical component of vorticity. This is given by the vertical component of equation (8.11), namely

$$\frac{\partial\omega_3}{\partial t} = Pr\,\Delta\omega_3. \qquad (8.17)$$

So u, v can be found by solving the Poisson equations (8.15), (8.16) when w has been found by solving equation (8.14) and ω_3 by solving the diffusion equation (8.17).

8.2 *The boundary conditions*

Rayleigh considered both rigid and free surfaces. The no-slip condition on a rigid boundary $z = 0$ or 1 gives

$$u = v = w = 0. \tag{8.18}$$

The first two of these conditions imply that

$$\omega_3 = 0, \tag{8.19}$$

and, together with the continuity equation (8.6), that

$$\frac{\partial w}{\partial z} = 0. \tag{8.20}$$

The boundary conditions on a free surface are that the perturbations of the components of stress are zero. Let the upper surface be free, with elevation

$$z_* = d + \zeta'_*(x_*, y_*, t_*). \tag{8.21}$$

Then its outward normal

$$\mathbf{n} = (-\partial \zeta'_*/\partial x_*, -\partial \zeta'_*/\partial y_*, 1)\{1 + (\nabla_* \zeta'_*)^2\}^{-1/2}$$
$$= \mathbf{k} - \nabla_* \zeta'_* + O\{(\nabla_* \zeta'_*)^2\}. \tag{8.22}$$

Now the normal component of the stress on the deformed surface must equal its value, i.e. $-p_1 = -P_*(d)$, in the undisturbed state. Therefore

$$-p_1 = \sigma_{ij*} n_i n_j$$

$$= -P_*(z_*) - p'_* + \mu \left(\frac{\partial u'_{i*}}{\partial x_{j*}} + \frac{\partial u'_{j*}}{\partial x_{i*}} \right) n_i n_j$$

$$+ (\lambda - \tfrac{2}{3}\mu) \frac{\partial u'_{k*}}{\partial x_{k*}} \quad \text{at } z_* = d + \zeta'_*$$

$$= -P_*(d) - \frac{dP_*}{dz_*} \zeta'_* - p'_* + 2\mu \frac{\partial w'_*}{\partial z_*}$$

$$+ (\lambda - \tfrac{2}{3}\mu) \frac{\partial u'_{k*}}{\partial x_{k*}} \quad \text{at } z_* = d,$$

on linearization for small ζ'_*, u'_*, p'_*, because

$$P_*(d + \zeta'_*) = P_*(d) + \zeta'_* (dP_*/dz_*)_{z_* = d} + O(\zeta'^2_*),$$

etc. But

$$P_*(d) = p_1 \quad \text{and} \quad dP_*/dz_* = -g\rho_*.$$

Therefore

$$\zeta'_* = (g\rho_1)^{-1}\{p'_* - 2\mu\,\partial w'_*/\partial z_* - (\lambda - \tfrac{2}{3}\mu)\nabla_* \cdot \mathbf{u}'_*\} \quad \text{at } z_* = d. \tag{8.23}$$

In dimensionless variables, this becomes

$$\zeta = \zeta'_*/d = (\alpha\beta d\rho_0/R\rho_1)\{Pr^{-1}p - 2\partial w/\partial z + (\tfrac{2}{3} - \lambda/\mu)\,\nabla \cdot \mathbf{u}\}$$
$$\text{at } z = 1. \tag{8.24}$$

We have yet to make the Boussinesq approximation that $\alpha\beta d = (\rho_1 - \rho_0)/\rho_0 \ll 1$, with which equation (8.24) gives

$$\zeta = 0. \tag{8.25}$$

Now the exact nonlinear kinematic condition gives

$$w = D\zeta/Dt \quad \text{at } z = 1 + \zeta, \tag{8.26}$$

and so

$$w = 0 \quad \text{at } z = 1, \tag{8.27}$$

on linearization for a Boussinesq fluid. This result was assumed by Rayleigh (1916a), and supported by Davis & Segel (1968). It means that a free surface behaves as a rigid surface with tangential slip but without any tangential stress. The conditions that the tangential stresses vanish now reduce to

$$\frac{\partial u}{\partial z} + \frac{\partial w}{\partial x} = 0, \quad \frac{\partial w}{\partial y} + \frac{\partial v}{\partial z} = 0 \quad \text{at } z = 1. \tag{8.28}$$

In view of condition (8.27), these become

$$\frac{\partial u}{\partial z} = \frac{\partial v}{\partial z} = 0 \quad \text{at } z = 1. \tag{8.29}$$

This gives

$$\frac{\partial \omega_3}{\partial z} = 0 \quad \text{at } z = 1, \tag{8.30}$$

and, together with the equation of continuity (8.6),

$$\frac{\partial^2 w}{\partial z^2} = 0 \quad \text{at } z = 1. \tag{8.31}$$

Conditions similar to (8.27) and (8.29)–(8.31) hold on the lower boundary $z = 0$ when it is free. A free lower boundary is unrealistic, being susceptible to Rayleigh–Taylor instability, but it is sometimes assumed in order to simplify calculations in illustrative examples. Rayleigh (1916a) took boundaries at fixed temperatures and therefore with zero perturbations of temperature. Thus

$$\theta = 0 \quad \text{at } z = 0, 1. \tag{8.32}$$

In some experiments a boundary is better modelled as a perfect insulator, through which no heat flows; then $\partial\theta/\partial z = 0$ at the boundary (Jeffreys 1928, Sparrow, Goldstein & Jonsson 1964).

In summary, we shall take the boundary conditions that

$$\left.\begin{array}{l}
w = \dfrac{\partial w}{\partial z} = \theta = 0 \quad \text{(rigid boundary)} \\[2mm]
w = \dfrac{\partial^2 w}{\partial z^2} = \theta = 0 \quad \text{(free boundary)}
\end{array}\right\} \tag{8.33}$$

The various cases are called *rigid–rigid*, *rigid–free*, *free–rigid* or *free–free* according to whether the upper and lower boundaries respectively are rigid or free.

8.3 Normal modes

The linear equations (8.8), (8.13) and boundary conditions (8.33) are symmetric in x and y. So Rayleigh (1916a) was able to take normal modes of the form

$$w = W(z)f(x, y)\, e^{st}, \quad \theta = T(z)f(x, y)\, e^{st}, \tag{8.34}$$

where $s = \sigma + i\omega$. There will be a corresponding form for p. Then equation (8.8) becomes

$$(D^2 + \Delta_1 - s)fT = -fW, \tag{8.35}$$

where $D = d/dz$. It follows that

$$(D^2 - a^2 - s)T = -W, \tag{8.36}$$

and that

$$\Delta_1 f + a^2 f = 0, \tag{8.37}$$

where a^2 is an arbitrary constant arising out of separation of the variables. Equation (8.37) is well known, being called the *reduced wave equation*, the *Helmholtz equation* or the *membrane equation*. Thus a is interpreted as a real horizontal wavenumber. In general, a disturbance excites components for each real value of a. We shall

resolve f into sinusoidal components of x and y in § 11, but first we
will discuss the variation of the solution with z. Now equations
(8.13), (8.14) become

$$(D^2 - a^2)(D^2 - a^2 - s/Pr)W = a^2 RT, \qquad (8.38)$$

$$(D^2 - a^2)(D^2 - a^2 - s)(D^2 - a^2 - s/Pr)W = -a^2 RW. \qquad (8.39)$$

Boundary conditions (8.33) reduce to

$$\left.\begin{array}{l} W = DW = T = 0 \quad \text{(rigid)} \\ W = D^2 W = T = 0 \quad \text{(free)} \end{array}\right\} \quad \text{at } z = 0, 1, \qquad (8.40)$$

or, with equation (8.38),

$$\left.\begin{array}{l} W = DW = D^4 W - (2a^2 + s/Pr)D^2 W = 0 \quad \text{(rigid)} \\ W = D^2 W = D^4 W = 0 \qquad\qquad\qquad\qquad\quad \text{(free)} \end{array}\right\} \quad \text{at } z = 0, 1.$$

$$(8.41)$$

This gives three conditions at each end point for the sixth-order
linear equation (8.39) to determine the countable infinity of eigenvalues $s_j(a, R, Pr)$ and associated eigenfunctions W_j ($j = 1, 2, \ldots$).
Although this eigenvalue problem is self-adjoint, and the linear
equation has constant coefficients, its solution is not entirely trivial,
so we shall consider some general points first. For given values of
a, R and Pr, a complete set of solutions W_j satisfying the boundary
conditions is needed to represent an arbitrary initial disturbance,
which may vary quite generally with z. For given R and Pr the flow
is unstable if $\sigma > 0$ for any mode with any real value of a and stable
if $\sigma \leq 0$ for all modes. Hence the critical value of the Rayleigh
number is that value R_c for which $\sigma(a, R, Pr) > 0$ for some a
whenever $R > R_c$ and $\sigma(a, R, Pr) \leq 0$ for all a whenever $R \leq R_c$. It
will be shown in § 9 that the Prandtl number in fact does not affect
the conditions for marginal stability.

Equation (8.17) and the boundary conditions (8.19) or (8.30),
which determine the rotational component of the horizontal flow,
are uncoupled from those for the other variables. The properties of
the diffusion equation (8.17) are well known; in particular, $\omega_3 \to 0$ at
each point $\mathbf{x}$ as $t \to +\infty$, whatever the initial conditions. Therefore in
seeking a criterion for instability it is sufficient to take $\omega_3 \equiv 0$.

9 General stability characteristics

9.1 *Exchange of stabilities*

We shall now prove that the principle of exchange of stabilities is valid, namely that marginally stable modes with $\sigma = 0$ also have $\omega = 0$.

Multiply equation (8.36) by T^*, the complex conjugate of T, and integrate from $z = 0$ to 1. Then

$$\int_0^1 T^*(D^2 - a^2 - s)T \, dz = -\int_0^1 WT^* \, dz,$$

and on integrating by parts we obtain

$$-\int_0^1 \{|DT|^2 + (a^2 + s)|T|^2\} \, dz + [T^*DT]_0^1 = -\int_0^1 WT^* \, dz.$$

But T, and therefore T^*, vanishes on the boundaries. Therefore

$$\int_0^1 \{|DT|^2 + (a^2 + s)|T|^2\} \, dz = \int_0^1 WT^* \, dz. \tag{9.1}$$

We can write this as

$$sI_0 + I_1 = \int_0^1 WT^* dz, \tag{9.2}$$

where

$$I_0 = \int_0^1 |T|^2 \, dz, \; I_1 = \int_0^1 (|DT|^2 + a^2|T|^2) \, dz. \tag{9.3}$$

Similarly, multiply equation (8.38) by W^* and integrate from $z = 0$ to 1. Therefore

$$\int_0^1 W^*(D^2 - a^2)(D^2 - a^2 - s/Pr)W \, dz = a^2R \int_0^1 W^*T \, dz.$$

On repeated integration by parts, it follows that

$$\int_0^1 \{|D^2W|^2 + (2a^2 + s/Pr)|DW|^2 + a^2(a^2 + s/Pr)|W|^2\} \, dz$$

$$+ [W^*D^3W - DW^*D^2W - (2a^2 + s/Pr)W^*DW]_0^1$$

$$= a^2R \int_0^1 W^*T \, dz.$$

But W and either DW or D^2W vanish at each boundary, according to whether it is rigid or free respectively. Therefore

$$J_2 + sJ_1/Pr = a^2 R \int_0^1 W^*T \, dz, \qquad (9.4)$$

where

$$\left. \begin{aligned} J_1 &= \int_0^1 (|DW|^2 + a^2|W|^2) \, dz, \\ J_2 &= \int_0^1 (|D^2W|^2 + 2a^2|DW|^2 + a^4|W|^2) \, dz. \end{aligned} \right\} \qquad (9.5)$$

The integrals I_0, I_1, J_1, J_2 are positive. So, subtracting the complex conjugate of equation (9.4) from the product of a^2R and equation (9.2), we get

$$a^2 R I_1 - J_2 + sa^2 R I_0 - s^* J_1/Pr = 0. \qquad (9.6)$$

The real and imaginary parts of this equation are

$$\sigma(a^2 R I_0 - J_1/Pr) + a^2 R I_1 - J_2 = 0, \qquad (9.7)$$

and

$$\omega(a^2 R I_0 + J_1/Pr) = 0, \qquad (9.8)$$

respectively. If $R < 0$ then it follows from equation (9.7) that $\sigma < 0$. This confirms the physically obvious fact that there is stability when the upper boundary is hotter than the lower. If $R > 0$, equation (9.8) implies that $\omega = 0$, and in particular that there is exchange of stabilities. This argument is due to Pellew & Southwell (1940).

9.2 *A variational principle*

Accordingly, we now put $s = 0$ to seek the marginal states and the critical value of the Rayleigh number. Then equations (8.36) and (8.38) become

$$(D^2 - a^2)T = -W, \qquad (9.9)$$

and

$$(D^2 - a^2)^2 W = a^2 R T. \qquad (9.10)$$

This shows that conditions for marginal stability are independent of the Prandtl number. For what follows it is convenient to renormalize T by using instead

$$F = a^2 R T. \qquad (9.11)$$

Then equations (9.10) and (9.9) become

$$(D^2 - a^2)^2 W = F, \tag{9.12}$$

and

$$(D^2 - a^2)F = -a^2 R W. \tag{9.13}$$

We next prove that for marginally stable disturbances

$$R = \min \frac{\int_0^1 \{(DF)^2 + a^2 F^2\} \, dz}{a^2 \int_0^1 \{(D^2 - a^2) W\}^2 \, dz} \tag{9.14}$$

$$= \min (K_1 / a^2 K_2), \tag{9.15}$$

say, the minimum being taken over all real functions W for which

(a) the integrals K_1, K_2 exist,

(b) $F = (D^2 - a^2)^2 W$, and

(c) F, W, and either DW or $D^2 W$ vanish on each boundary according to whether it is rigid or free respectively.

(Note that some but not all such W are eigenfunctions. When W is an eigenfunction, $K_2 = J_2$, $K_1 = a^4 R^2 I_1$.) Further, for a given value of the wavenumber a, the minimizing function W is an eigenfunction for marginal stability. The proof will be in two parts, the first showing that the quotient is stationary, and the second that it is a minimum, for an eigenfunction.

To prove that the quotient is stationary, suppose that each of $W(z)$ and its small increment $\delta W(z)$ satisfy conditions (a), (b), (c), with $\delta F = (D^2 - a^2)^2 \delta W$. Then

$$\delta R = \frac{1}{a^2 K_2} \delta K_1 - \frac{K_1}{a^2 K_2^2} \delta K_2 = \frac{1}{a^2 K_2} (\delta K_1 - a^2 R \delta K_2). \tag{9.16}$$

But

$$\delta K_1 = 2 \int_0^1 (DF \delta DF + a^2 F \delta F) \, dz = -2 \int_0^1 \delta F (D^2 - a^2) F \, dz, \tag{9.17}$$

on integrating by parts. Similarly

$$\delta K_2 = 2 \int_0^1 \{(D^2 - a^2) W\}\{(D^2 - a^2) \delta W\} \, dz$$

$$= 2 \int_0^1 W(D^2 - a^2)^2 \delta W \, dz$$

$$= 2 \int_0^1 W \delta F \, dz. \tag{9.18}$$

Therefore

$$\delta R = -\frac{2}{a^2 K_2} \int_0^1 \delta F\{(D^2 - a^2)F + a^2 R W\} \, dz. \qquad (9.19)$$

If the quotient R is stationary with respect to small variations in W, then $\delta R = 0$ for arbitrary infinitesimal variations δF, and therefore

$$(D^2 - a^2)F = -a^2 R W.$$

Therefore W satisfies the stability equation as well as the boundary conditions (c), and is an eigenfunction. Conversely, if W is an eigenfunction, then R is stationary.

To show that the stationary value is a minimum, we shall assume that there is a countable infinity of marginal states for each real wavenumber a. Let these have real eigenfunctions F_j, W_j corresponding to Rayleigh number R_j $(j = 1, 2, \ldots)$ with $0 < R_1 < R_2 < \ldots$. We first show that F_j and W_k are orthogonal. Now equation (9.12) gives

$$(D^2 - a^2)^2 W_j = F_j. \qquad (9.20)$$

Multiply this by W_k, integrate from $z = 0$ to 1, and integrate by parts. Then

$$\int_0^1 F_j W_k \, dz = \int_0^1 W_k (D^2 - a^2)^2 W_j \, dz$$

$$= [W_k(D^3 W_j - a^2 D W_j) - D W_k D^2 W_j]_0^1$$

$$+ \int_0^1 (D^2 - a^2) W_j (D^2 - a^2) W_k \, dz.$$

But W and DW or $D^2 W$ vanish at each boundary. Therefore

$$\int_0^1 F_j W_k \, dz = \int_0^1 (D^2 - a^2) W_j (D^2 - a^2) W_k \, dz \qquad (9.21)$$

$$= \int_0^1 F_k W_j \, dz, \qquad (9.22)$$

by symmetry in j and k. Similarly,

$$(D^2 - a^2)F_j = -a^2 R_j W_j, \qquad (9.23)$$

and therefore

$$a^2 R_j \int_0^1 F_k W_j \, \mathrm{d}z = -\int_0^1 F_k (\mathrm{D}^2 - a^2) F_j \, \mathrm{d}z$$

$$= -[F_k \mathrm{D}F_j]_0^1 + \int_0^1 (\mathrm{D}F_j \mathrm{D}F_k + a^2 F_j F_k) \, \mathrm{d}z$$

$$= \int_0^1 (\mathrm{D}F_j \mathrm{D}F_k + a^2 F_j F_k) \, \mathrm{d}z. \qquad (9.24)$$

Interchange j and k in equation (9.24) and then subtract the new equation from (9.24) to get

$$a^2 (R_j - R_k) \int_0^1 F_j W_k \, \mathrm{d}z = 0, \qquad (9.25)$$

in view of the symmetry of equation (9.22). The desired ortho-gonality relation follows, i.e.

$$\int_0^1 F_j W_k \, \mathrm{d}z = 0 \quad (j \neq k). \qquad (9.26)$$

It is convenient to normalize the W_k so that

$$\int_0^1 F_j W_k \, \mathrm{d}z = \delta_{jk}. \qquad (9.27)$$

Next let W, F be any pair of real-valued functions satisfying conditions $(a), (b)$ and (c), but which are not necessarily eigen-functions. Then we can expand

$$W = \sum_{j=1}^\infty A_j W_j, \qquad (9.28)$$

for some constants A_j. In fact

$$A_j = \sum_{k=1}^\infty A_k \delta_{jk} = \sum_{k=1}^\infty A_k \int_0^1 W_k F_j \, \mathrm{d}z = \int_0^1 F_j \sum_{k=1}^\infty A_k W_k \, \mathrm{d}z$$

$$= \int_0^1 F_j W \, \mathrm{d}z. \qquad (9.29)$$

Also, on assuming uniform convergence to continuous functions of all series used,

$$F = (\mathrm{D}^2 - a^2)^2 W = \sum_{j=1}^\infty A_j (\mathrm{D}^2 - a^2)^2 W_j = \sum_{j=1}^\infty A_j F_j. \qquad (9.30)$$

Hence we obtain the expansion

$$K_1 = \int_0^1 \{(DF)^2 + a^2 F^2\} \, dz = -\int_0^1 F(D^2 - a^2) F \, dz \quad (9.31)$$

$$= -\int_0^1 \sum_{j=1}^{\infty} A_j F_j \cdot (D^2 - a^2) \sum_{k=1}^{\infty} A_k F_k \, dz$$

$$= -\sum_{j,k=1}^{\infty} A_j A_k \int_0^1 F_j (D^2 - a^2) F_k \, dz$$

$$= \sum_{j,k=1}^{\infty} a^2 A_j A_k R_k \int_0^1 F_j W_k \, dz$$

$$= \sum_{j=1}^{\infty} a^2 R_j A_j^2, \quad (9.32)$$

by orthonormality (9.27). Similarly

$$K_2 = \int_0^1 \{(D^2 - a^2) W\}^2 \, dz = \int_0^1 W(D^2 - a^2)^2 W \, dz$$

$$= \int_0^1 WF \, dz \quad (9.33)$$

$$= \sum_{j,k=1}^{\infty} A_j A_k \int_0^1 W_j F_k \, dz$$

$$= \sum_{j=1}^{\infty} A_j^2. \quad (9.34)$$

Therefore

$$\frac{K_1}{a^2 K_2} - R_1 = \frac{\sum_{j=1}^{\infty} (R_j - R_1) A_j^2}{K_2} \geq 0, \quad (9.35)$$

because $R_1 < R_2 < \cdots$, there being equality if and only if all the A_j except A_1 are zero. Therefore W_1, F_1 are the eigenfunctions that give the minimum R_1 of equation (9.14).

The stationary property was first proved by Pellew & Southwell (1940). They also stated the minimum property, which was subsequently proved by Chandrasekhar (1961).

10 Particular stability characteristics

10.1 *Free–free boundaries*

Rayleigh (1916a) solved the simplest, albeit unrealistic, case of two free boundaries. Here we have the eigenvalue problem

$$(D^2 - a^2)(D^2 - a^2 - s)(D^2 - a^2 - s/Pr)W = -a^2 RW, \quad (10.1)$$

$$W = D^2 W = D^4 W = 0 \quad \text{at } z = 0, 1. \quad (10.2)$$

By inspection, the solution is the complete set of eigenfunctions

$$W_j = \sin j\pi z \quad (j = 1, 2, \ldots), \quad (10.3)$$

where the eigenvalue relation is

$$(j^2\pi^2 + a^2)(j^2\pi^2 + a^2 + s)(j^2\pi^2 + a^2 + s/Pr) = a^2 R. \quad (10.4)$$

The solution of this quadratic gives the eigenvalues

$$s = -\tfrac{1}{2}(1 + Pr)(j^2\pi^2 + a^2)$$
$$\pm \{\tfrac{1}{4}(Pr - 1)^2(j^2\pi^2 + a^2)^2 + a^2 R \, Pr/(j^2\pi^2 + a^2)\}^{1/2}$$
$$(j = 1, 2, \ldots). \quad (10.5)$$

It can be seen that if $R < 0$ then $\sigma < 0$ and that if $R > 0$ then $\omega = 0$, in confirmation of the general result of § 9.1. For marginal stability $s = 0$, and equation (10.4) gives

$$R_j = (j^2\pi^2 + a^2)^3/a^2, \quad (10.6)$$

with $R_1 < R_2 < \ldots$. To find the minimum of $R_1(a^2)$, put

$$0 = \frac{dR}{da^2} = \frac{(2a^2 - \pi^2)(\pi^2 + a^2)^2}{a^4}.$$

This gives the minimum where $a = a_c = 2^{-1/2}\pi$, and

$$R_c = \min R_1(a^2) = (\tfrac{3}{2}\pi^2)^3/\tfrac{1}{2}\pi^2 = 27\pi^4/4 = 657.5. \quad (10.7)$$

If $R \leqslant R_c$ then all modes are stable, and if $R > R_c$ then at least one mode is unstable, as can be seen from the eigenvalue relation (10.5). When R is just supercritical, that is when R is just a little greater than R_c, thermal instability ensues with horizontal wavelength $2\pi d/a_c = 2^{3/2}d = 2.83d$. Note that the critical conditions, but not the rate of growth or damping, are independent of the Prandtl number.

Table 2.1. *The critical values of the Rayleigh number for the first two modes and various values of the wavenumber*

a	0	1.0	2.0	3.0	3.117	
R_1	∞	5854	2177	1711	1708	
R_2	∞	163 128	47 006	26 147	–	
a	4.0	5.0	5.365	6.0	7.0	∞
R_1	1879	2439	–	3418	4919	∞
R_2	19 685	17 731	17 610	17 933	19 576	∞

10.2 Rigid–rigid boundaries

When both boundaries are rigid, equation (10.1) has to be solved with the boundary conditions

$$W = DW = D^4 W - (2a^2 + s/Pr)D^2 W = 0 \quad \text{at } z = 0, 1. \quad (10.8)$$

Although equation (10.1) has exponential and sinusoidal solutions, the eigenvalue relation is transcendental and it is impossible to present a simple explicit solution for $s(a^2, R, Pr)$. A little computation is necessary to get particular results. The first reliable calculation of the critical Rayleigh number was made by Jeffreys (1928). Many good methods of computation have been used in the literature, but we shall not dwell on them here. The general structure of the solution can best be seen by analogy with Rayleigh's illustrative example for two free boundaries. In particular, the problem is symmetric with respect to the two boundaries, so that the eigenfunctions fall into two distinct classes, those with vertical velocity symmetric and those antisymmetric about the mid-plane $z = \frac{1}{2}$, just as the eigenfunctions given by equations (10.3) are even functions of $z - \frac{1}{2}$ for odd integers j and odd functions for even integers j.

There is exchange of stabilities, so one may put $s = 0$ to find R_c. Then

$$(D^2 - a^2)^3 W = -a^2 R W, \quad (10.9)$$

$$W = DW = (D^2 - a^2)^2 W = 0 \quad \text{at } z = 0, 1. \quad (10.10)$$

The values $R_1(a)$ for the first even mode $W_1(z)$ and $R_2(a)$ for the first odd mode $W_2(z)$ are given in Table 2.1, after Reid & Harris

(1958). The minimum of $R_1(a)$ is $R_c = 1708$ for $a_c = 3.117$. Thus instability occurs if and only if $R > 1708$ and the horizontal wavelength of the disturbance at the onset of instability is $2\pi d/a_c = 2.016d$. The graphs of $R_1(a)$, $R_2(a)$ and the eigenfunctions $W_1(z)$, $W_2(z)$ are shown in Fig. 2.2.

10.3 Free–rigid boundaries

Symmetry implies that the modes in this case are similar to the antisymmetric modes between two rigid planes $z_* = 0$, $2d$. To show this, note that those antisymmetric modes discussed in Section 10.2 satisfy the stability equation (10.1), the rigid conditions on $z = 0$, and the oddness conditions at the mid-plane, namely

$$W = D^2 W = D^4 W = 0 \quad \text{at } z = \tfrac{1}{2}. \tag{10.11}$$

These are just the properties which specify the modes of the present free–rigid case. So we can find a_c, R_c from Table 2.1 for the mode $R_2(a)$, remembering that distances must be scaled by a factor of two and that $a \propto d$ and $R \propto d^4$. Therefore

$$R_c = 17\,610/2^4 = 1101, \quad a_c = 5.365/2 = 2.682. \tag{10.12}$$

The stability equation (10.1) involves only operators even in $z - \tfrac{1}{2}$ so the free–rigid case is the same as the unrealistic rigid–free case with the free boundary below. Thus equations (10.12) give R_c, a_c for the rigid–free boundaries as well.

11 The cells

The classic model of thermal instability which we have described determines the horizontal wavenumber a_c at which instability arises. When $R > R_c$ the wavenumber of the fastest-growing mode is determined similarly, and according to linear theory this mode should be the dominant one. This specifies the horizontal wavelength of the observed disturbance, but the horizontal pattern of the disturbance should be determined by the initial horizontal pattern of the dominant mode. This pattern is associated with a solution of the reduced wave equation

$$\Delta_1 f + a^2 f = 0, \tag{11.1}$$

as it is for each mode. The wave equation is well known in the theories of light and sound, and also describes transverse oscillation

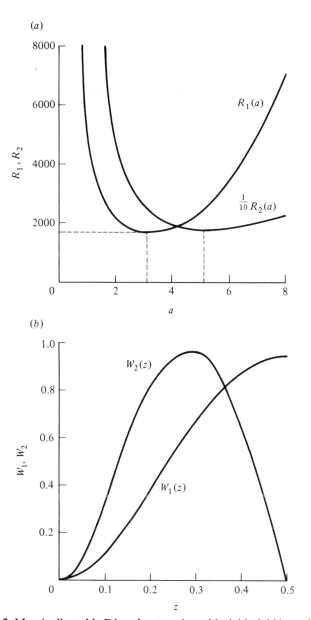

Fig. 2.2. Marginally stable Bénard convection with rigid–rigid boundaries. (a) The marginal curves of the first two modes (note the different scale of R_2). (b) The eigenfunctions W_1 and W_2, with an arbitrarily chosen scale, of the first two modes. (After Reid & Harris 1958.)

of a stretched membrane. In general there is no regular pattern in the initial disturbance. However, steady flow in polygonal cells is observed at slightly supercritical Rayleigh numbers. In fact the laminar flow in these cells is nonlinear, having equilibrated after a short time of linear instability with exponential growth. The magnitude and direction of flow is determined by the initial disturbance in linear theory, but nonetheless the nonlinear cellular flow has a regular horizontal pattern close to a certain solution of equation (11.1). This gives special interest to solutions of the wave equation corresponding to polygonal cells.

The general solution of equation (11.1) can be broken down into Fourier components of the form

$$f = \cos a_1 x \cos a_2 y, \tag{11.2}$$

where $a_1^2 + a_2^2 = a^2$ and where a cosine may be replaced by a sine for antisymmetric components. The displayed component has wavelength $2\pi/a_1$ in the x-direction and $2\pi/a_2$ in the y-direction. After a sufficiently long time the vertical vorticity ω_3 is negligible, so let us take $\omega_3 \equiv 0$ for all time. Then equations (8.15) and (8.16) give

$$u = -a_1 a^{-2} \sin a_1 x \cos a_2 y D W e^{st},$$
$$v = -a_2 a^{-2} \cos a_1 x \sin a_2 y D W e^{st}. \tag{11.3}$$

Horizontal areas of ascending and descending flow at any given height z are separated by lines on which $w = 0$, namely

$$x = (m + \tfrac{1}{2})\pi/a_1, \quad y = (n + \tfrac{1}{2})\pi/a_2 \quad (m, n = 0, \pm 1, \pm 2, \ldots). \tag{11.4}$$

For the lowest mode, W_1, w does not change sign with z between the boundaries for each given pair x, y, so that the vertical planes described by equation (11.4) separate regions of descending and ascending flow. When the flow is slightly supercritical it is effectively steady, so that the particle paths and streamlines coincide, having equations

$$\frac{dx}{u} = \frac{dy}{v} = \frac{dz}{w}. \tag{11.5}$$

These paths have orthogonal projections onto any horizontal plane with equation

$$\frac{dy}{dx} = \frac{v}{u} = \frac{\partial w}{\partial y} \Big/ \frac{\partial w}{\partial x}, \tag{11.6}$$

by equations (11.3). Therefore the derivative of w normal to the projections is zero:

$$\frac{\partial w}{\partial n} = 0. \tag{11.7}$$

On substituting from equation (11.3) for u and v, equation (11.6) becomes

$$\frac{\mathrm{d}y}{\mathrm{d}x} = \frac{a_2 \tan a_2 y}{a_1 \tan a_1 x}. \tag{11.8}$$

Therefore the paths are given by

$$(\sin a_2 y)^{a_2^2} = \text{constant} \times (\sin a_1 x)^{a_1^2}. \tag{11.9}$$

Definitions of a cell vary, but experiments suggest (Stuart 1964) that we seek cells with fixed vertical boundaries which no fluid particle crosses and, for the lowest mode, on which w does not change sign.

Bearing this in mind, look at the projected paths described by equation (11.9) in Fig. 2.3 for the case $a_1 = \frac{1}{2}\sqrt{3}a$, $a_2 = \frac{1}{2}a$ as an example. The rectangular grid defined by equation (11.4) on which $w = 0$ is denoted by broken lines. Some of the paths obtained using equation (11.9) are shown by the solid lines. Now $v = 0$ on FE ($y = \pi/a_2$) and GH ($y = -\pi/a_2$) and $u = 0$ on EH ($x = \pi/a_1$) and GF ($x = -\pi/a_1$). Therefore $\mathbf{u} \cdot \mathbf{n} = 0$ on the vertical column of fluid whose horizontal section is the rectangle $EFGH$, and no particle crosses this column. However, w changes sign eight times along $EFGH$ at each height z, so we must take the column with curved section $AJDMCLB$ as a cell, because $\mathbf{u} \cdot \mathbf{n} = 0$ but w does not change sign on it.

In spite of this, the solution (11.2) is sometimes said to give *rectangular cells*. *Rolls* and *square cells* are special cases. For rolls,

$$f = \cos ax, \tag{11.10}$$

$$u = -a^{-1}\mathrm{D}W \sin ax \, \mathrm{e}^{st}, \quad v = 0, \quad w = W \cos ax \, \mathrm{e}^{st}; \tag{11.11}$$

this gives a two-dimensional flow in long cells, with vertical boundaries at either $x = 2m\pi/a$ or $x = (2m+1)\pi/a$ ($m = 0, \pm 1, \pm 2, \ldots$). For squares, $a_1 = a_2 = a/\sqrt{2}$, and the typical cell boundary $AJDMCLBKA$ becomes the square with vertices $A(0, \sqrt{2}\pi/a)$,

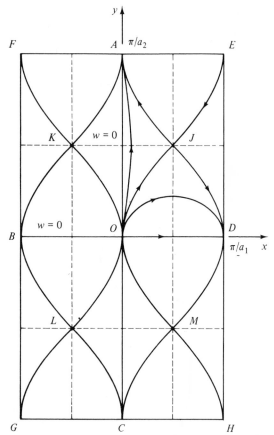

Fig. 2.3. Horizontal projection of some streamlines of a 'rectangular' cell with $a_1^2 = 3a_2^2$. Broken lines denote vertical planes on which $w = 0$. (After Stuart 1964.)

$B(-\sqrt{2}\pi/a, 0)$, $C(0, -\sqrt{2}\pi/a)$ and $D(\sqrt{2}\pi/a, 0)$ and sides of length $2\pi/a$.

Christopherson (1940) found the solution for *hexagonal cells*,

$$f = \cos \tfrac{1}{2}a(\sqrt{3}x + y) + \cos \tfrac{1}{2}a(\sqrt{3}x - y) + \cos ay. \quad (11.12)$$

It can be verified that

$$\Delta_1 f = -(\tfrac{1}{2}a)^2(3+1)\cos \tfrac{1}{2}a(\sqrt{3}x + y) - (\tfrac{1}{2}a)^2(3+1)\cos \tfrac{1}{2}a(\sqrt{3}x - y)$$
$$- a^2 \cos ay$$
$$= -a^2 f.$$

The solution is symmetric in $\pm x$, $\pm y$. There is periodicity in x, y with

$$f(x + 4m\pi/\sqrt{3}a, y + 4n\pi/a) \equiv f(x, y) \quad (m, n = 0, \pm 1, \pm 2, \ldots).$$

$$(11.13)$$

To verify that f is invariant under rotations of angle $60°$ about the vertical line through the origin, put $x = r \cos \theta$, $y = r \sin \theta$. Then $\frac{1}{2}(\sqrt{3}x \pm y) = r \sin (\frac{1}{3}\pi \pm \theta)$ and

$$f = \cos \{ar \sin (\theta + \tfrac{1}{3}\pi)\} + \cos \{ar \sin (\theta - \tfrac{1}{3}\pi)\} + \cos \{ar \sin \theta\}.$$

This gives $f(r, \theta + \frac{1}{3}\pi) \equiv f(r, \theta)$, the desired rotational symmetry. The vertical columns on which f, and therefore w, vanishes are in fact nearly circular, as shown with broken lines in Fig. 2.4. The typical column with centre O is surrounded by the regular hexagon $ABCDEF$ with sides of length $L = 4\pi/3a$. Now AF has equation $x = \sqrt{3}L/2 = 2\pi/\sqrt{3}a$, on which

$$u = a^{-2} \frac{\partial f}{\partial x} W e^{st}$$

$$= -a^{-1}[\tfrac{1}{2}\sqrt{3} \sin \{\tfrac{1}{2}a(\sqrt{3}x + y)\}$$

$$+ \tfrac{1}{2}\sqrt{3} \sin \{\tfrac{1}{2}a(\sqrt{3}x - y)\}]W e^{st}$$

$$= -a^{-1}\sqrt{3} \sin \pi \cos \tfrac{1}{2}ay W e^{st} = 0.$$

Therefore $\mathbf{u} \cdot \mathbf{n} = 0$ on AF. By rotational symmetry, $\mathbf{u} \cdot \mathbf{n} = 0$ on the other five sides of the hexagon $ABCDEF$. Therefore no particle crosses the surface of the vertical column with section $ABCDEF$. By the periodicity in x, y and by the rotational symmetry, interlocking hexagons of this kind cover the whole plane, as indicated in Fig. 2.4. The lowest mode W_1 does not change sign as z varies, and gives a flow which is either descending or ascending within the whole of each column; $w = 0$ at the middle of each hexagonal cell. The projections of some of the streamlines onto a horizontal plane for critical flow is shown in Fig. 2.4. An elevation of streamlines is shown in Fig. 2.5, where they are given in the vertical plane through AOD, after Reid & Harris (1959).

The pattern of the cells can also be affected by the side walls, though Rayleigh omitted them from his model. Pellew & Southwell

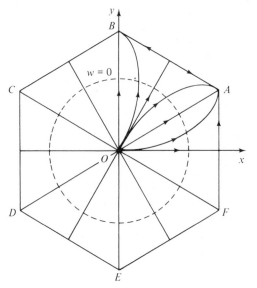

Fig. 2.4. Horizontal projection of some streamlines of a 'hexagonal' cell. Broken lines denote the surface on which $w = 0$. (After Stuart 1964.)

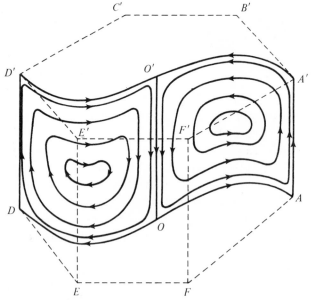

Fig. 2.5. An elevation of some streamlines of the 'hexagonal' cell. (After Reid & Harris 1959 and Stuart 1964.)

(1940) recognized the significance of side walls of experimental apparatus, but did not analyse the associated linear problem because then separation of variables is impossible. More recently S. H. Davis (1967) analysed linear instability in a rigid rectangular box at fixed temperatures, two faces being horizontal and cut by vertical side walls in a rectangle with sides l_1, l_2 ($>l_1$, say). There is a countable infinity of modes whose horizontal wavelengths are submultiples of l_1 and l_2, the growth rate varying from mode to mode. So the fastest-growing mode has a determinate horizontal as well as vertical behaviour. Davis's numerical results show that this preferred mode is some number of rolls of length l_1, with axes parallel to the shorter side of the rectangle, and there are only two non-zero velocity components varying with x, y and z. If $d < l_1 < l_2$ then the rolls have a cross-section which is nearly square in a vertical plane; otherwise the rolls are narrower. The critical value of the Rayleigh number is large when l_1, $l_2 \ll d$ but decreases to 1708 as d/l_1 and $d/l_2 \to 0$.

12 Experimental results

Of the many good experiments on Bénard convection, we shall select a few to relate to the theory, and in particular to describe observations of R_c, a_c and the cell shapes.

There are experimental difficulties in using a clean pure fluid, in fixing temperatures precisely, in observing flow between two rigid plates, and in accurately measuring any quantity within the fluid. One of the least difficult quantities to measure is the heat transfer. When the Boussinesq fluid is at rest, there is heat transfer at the rate $H = k(\theta_0 - \theta_1)/d$ across unit area of the layer of the fluid by thermal conduction. It is convenient to describe the heat transfer by the dimensionless *Nusselt number*,

$$Nu = Hd/k(\theta_0 - \theta_1).$$

This is the ratio of the actual heat transfer to what it would be if there were conduction but no convection. Thus $Nu = 1$ in stable conditions. In general, Nu varies with R, and weakly with Pr. The magnitude of Nu is determined in linear theory by the initial conditions, but it may be expected to increase suddenly in practice

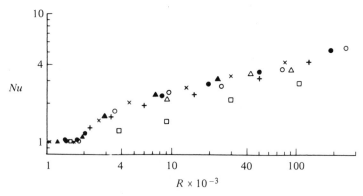

Fig. 2.6. Some experimental results on the heat transfer in various fluids in various containers. The Nusselt number is plotted against the Rayleigh number; ○ water; + heptane; × ethylene glycol; ● silicone oil AK 3; ▲ silicone oil AK 350; △ air; □ mercury. After Silveston 1958 and Rossby 1969.)

when convection ensues. The onset of instability may also be seen directly with visualization techniques.

Silveston's (1958) measurements of $Nu(R, Pr)$ for various liquids between two horizontal plates at distances d varying from 1.45 to 13 mm, together with Graaf & Held's (1953) measurements for air and Rossby's (1969) for mercury, for values of R up to 10^6 are shown in Fig. 2.6. Note the sudden increase of Nu near 1708 for a wide variety of fluids. In fact Silveston (1958) found the experimental value $R_c = 1700 \pm 51$. Some relevant physical quantities for these fluids are shown in Table 2.2.

Cells are made observable by various visualization techniques and photographed, but measurements of their wavelength are not very accurate. However, the wavelength is close to $2d$ at the onset of instability between two rigid plates (Schmidt & Saunders 1938, Silveston 1958). When the separation of the side walls is much greater than d, hexagons seem to predominate for supercritical R. As R increases they tend to join up, as if forming rolls. Disorder increases with R until the motion seems to be turbulent when $R \approx 5 \times 10^4$ (Schmidt & Saunders 1938), although more recently experimentalists have detected some cellular structure up to much higher values of R. Koschmieder (1966) has found that the side

Table 2.2. *Some physical constants of fluids at 20 °C and 10^5 Pa (i.e. 1000 mbar)*

	$\rho_0(\text{kg m}^{-3})$	$10^{-3}c$ $(\text{m}^2\,\text{s}^{-2}\,\text{K}^{-1})$	$10^7\nu$ $(\text{m}^2\,\text{s}^{-1})$	$10^7\kappa$ $(\text{m}^2\,\text{s}^{-1})$	$10^4\alpha$ (K^{-1})
Air	1.19	$1.01\ (10^{-3}c_\text{p})$	154	248	34.5
Heptane	684	2.22	6.16	0.875	12.4
Water	998	4.18	10.06	1.433	2.07
Silicone oil AK 3	912	1.61	32.0	0.779	10.6
Ethylene glycol	1113	2.38	191.5	0.942	6.4
Silicone oil AK 350	980	1.50	4670	1.061	9.2
Mercury	13550	0.139	1.15	44.0	1.82

walls affect the cell shapes in deep layers, and he has observed circular rolls within a circular side wall and linear rolls within rectangular side walls. (S. H. Davis's (1967) theory is consistent with these observations of linear rolls in so far as they are comparable, but better confirmed by the experiments of Stork & Müller (1972).) As R increases above R_c the wavelength of the cells increases.

On the basis of the linear theory just discussed, the cell pattern and the direction of flow is in principle uniquely determined by the initial conditions. In practice, however, observations of instability are made at values of the Rayleigh number slightly above the critical, and cell patterns and the direction of flow are largely independent of the unknown initial conditions. The facts that the motion has a preferred direction and is steady suggest that nonlinearity is significant.

It is also found, for example, that a liquid usually rises in the middle of a polygonal cell and a gas falls. Graham (1933) suggested that this is because the viscosity of a typical liquid decreases with temperature whereas that of a typical gas increases. This suggestion was subsequently confirmed by Tippelskirch's (1956) experiments on convection of liquid sulphur, for which the dynamic viscosity has a minimum near 153 °C. He found that the flow rises in the middle of the cells in convection of liquid sulphur below 153 °C and falls in the middle above 153 °C. It follows that the direction of flow is a non-Boussinesq as well as a nonlinear phenomenon.

In most recent experiments, carefully made apparatus has been used to elucidate nonlinear phenomena. As the Rayleigh number increases, a series of transitions from one complicated flow to the next more complicated one may be detected, even when to the eye the flow may look fully turbulent. Koschmieder (1974) has surveyed recent experiments, and we shall also discuss some of the nonlinear phenomena in Chapter 7.

13 Some applications

We have concentrated on Rayleigh's classic model of linear stability of a layer of Boussinesq fluid between two horizontal planes at fixed temperatures. We have briefly mentioned, either in the text or in problems, generalizations of this classic model with variable surface tension, non-Boussinesq fluids, insulated boundaries, heat generation within the fluid, variable density of a fluid due to a solute, side walls, rotating frames and nonlinearity. The basic ideas of the model have been applied to an even greater variety of phenomena. There are extensions to electric and magnetic fields, to gas containing a vapour that may condense and release latent heat, to radiation of heat, to water between planes at temperatures on either side of its temperature of minimum density (4 °C), to boundaries at fixed non-uniform temperatures, to a self-gravitating fluid in a spherical shell, etc. Some of these extensions and more are recorded by Chandrasekhar (1961), Saltzman (1962) and Gershuni & Zhukovitskii (1976), but lack of space allows us to discuss only nonlinearity (in Chapter 7). Many of these extensions have been inspired by engineering or scientific applications. For example, interest in the effect of the variation of surface tension with temperature was inspired by the uneven drying of paint; in magnetic fields and rotating fluids by convection in the sun; in convection in a gas containing water vapour by cloud formation; and in convection in a rotating spherical shell by the general circulation of the atmosphere and by the motion of the earth's core.

Problems for chapter 2

2.1. *Governing equations with internal heating.* Show that the heat equation takes the form

$$\frac{D\theta}{Dt} = \kappa \Delta\theta + \gamma$$

when there is uniform weak heating of the fluid at a constant rate. Show that there is then a basic state of rest with

$$\mathbf{U}_* = 0, \quad \Theta_* = \theta_0 - \beta z_* + \tfrac{1}{2}\gamma\kappa^{-1}z_*(d - z_*),$$
$$P_* = p_0 - g\rho_0 z_* - \tfrac{1}{2}\alpha g\rho_0 z_*^2 \{\beta - \tfrac{1}{2}\gamma\kappa^{-1}(d - \tfrac{2}{3}z_*)\}.$$

Deduce that the stability equation (8.36) becomes

$$(D^2 - a^2 - s)T = (D\Theta)W,$$

but that equation (8.38) is unchanged. [Sparrow, Goldstein & Jonsson (1964).]

2.2. *Governing equations with a solute.* Suppose a solute is dissolved in the fluid with concentration $c(\mathbf{x}, t)$ such that the density of the fluid is given by

$$\rho = \rho_0\{1 + \alpha(\theta_0 - \theta) + l(c - c_0)\}.$$

Neglect the diffusion of the solute, so that

$$\frac{Dc}{Dt} = 0.$$

Then show that the stability of a basic state of rest with concentration $C_*(z_*)$ and temperature $\Theta_* = \theta_0 - \beta z_*$ is governed by the equations

$$\left(\frac{\partial}{\partial t_*} - \kappa\Delta_*\right)\theta'_* = \beta w'_*$$

$$\frac{\partial}{\partial t_*}\left(\frac{\partial}{\partial t_*} - \nu\Delta_*\right)\Delta_* w'_* = g\Delta_{1*}\left(\alpha\frac{\partial\theta'_*}{\partial t_*} + l\frac{dC_*}{dz_*}w'_*\right).$$

Discuss the relation of the latter equation to Rayleigh–Taylor instability when $\beta = 0$, $\nu = 0$ and $\kappa = 0$. [Stern (1960), Rayleigh (1883b).]

2.3. *Governing equations with rotation.* If the whole system rotates with angular velocity $\Omega\mathbf{k}$, show that thermal instability is governed by the equations

$$\left(\frac{\partial}{\partial t_*} - \kappa\Delta_*\right)\theta'_* = \beta w'_*,$$

$$\left(\frac{\partial}{\partial t_*} - \nu\Delta_*\right)\omega'_{3*} = 2\Omega\frac{\partial w'_*}{\partial z_*},$$

$$\left(\frac{\partial}{\partial t_*} - \nu\Delta_*\right)\Delta_* w'_* = \alpha g\Delta_{1*}\theta'_* - 2\Omega\frac{\partial\omega'_{3*}}{\partial z_*}.$$

Hence show that

$$(D^2 - a^2 - s)\{(D^2 - a^2 - s/Pr)^2(D^2 - a^2) + \mathcal{T}D^2\}W$$
$$= -a^2 R(D^2 - a^2 - s/Pr)W$$

for a normal mode, where the *Taylor number* for this problem is given by $\mathcal{T} = 4\Omega^2 d^4/\nu^2$. [Chandrasekhar (1961), § 26; Problem 2.10.]

2.4. *Conservation of energy at marginal stability.* Consider the energetics of Bénard convection at marginal stability. Show that the average rate at which energy is released by buoyancy in a vertical column of fluid with horizontal section of unit area is given by

$$\varepsilon_g = g\alpha\beta\kappa\rho_0 d \int_0^1 WT \, dz \int_A f^2 \, dx \, dy/A,$$

where A is a large horizontal section of a column of fluid. Taking

$$f = \cos a_1 x \cos a_2 y,$$

deduce that

$$\varepsilon_g = \tfrac{1}{4}g\alpha\beta\kappa\rho_0 dI_1.$$

Similarly, show that the average rate at which energy is dissipated by viscosity is given by

$$\varepsilon_\nu = \kappa^2\nu\rho_0 J_2/4a^2 d^3,$$

when the linearized flow is steady. Hence show that at marginal stability equation (9.6) expresses the conservation of energy. [Note that I_1 is defined in equations (9.3) and J_2 in equations (9.5). Chandrasekhar (1961), § 14.]

2.5. *A simple experiment on thermal instability.* Pour corn oil in a clean frying pan (i.e. skillet), so that there is a layer of oil about 2 mm deep. Heat the bottom of the pan gently and uniformly. To visualize the instability, drop in a little powder (cocoa serves well). Sprinkling powder on the surface reveals the polygonal pattern of the steady cells. The movement of individual particles of powder may be seen, with rising near the centre of a cell and falling near the sides. Tilt the pan a little to find the critical depth. Verify that the size of the cells and the critical Rayleigh number are of the order of magnitude predicted by theory.

2.6. *A simple experiment on thermohaline convection.* Half fill a glass beaker with hot dyed brine (a few grams of salt and a little ink or methylene blue dissolved in a half litre of water in a litre beaker serve well). Insert cold fresh water under the brine. To restrict mixing of the water and brine, pour the water slowly through a glass tube close to the bottom of the beaker, for example as shown in Fig. 2.7. Rest the beaker on white paper (to see clearly), keep it steady (to avoid setting up internal gravity waves) and

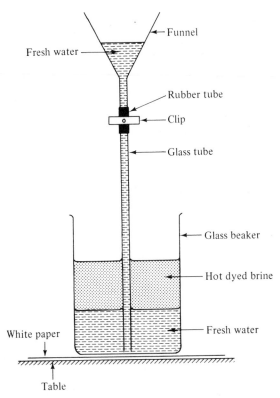

Fig. 2.7. Sketch of the apparatus for a simple experiment on thermohaline convection (Problem 2.6).

ensure that there is no swirl (to avoid twisting the instability). The density of the brine should be less than that of the water, so that the fluids are in hydrostatically stable equilibrium. However, *thermohaline instability* will soon arise owing to the different rates of diffusion of temperature and salt, the diffusion of salt being very slow. Observe the finger-like cells grow at a rate of order of magnitude 10^{-1} mm s^{-1}, with a horizontal scale of about 3 mm. Discuss this as an example of *double-diffusive convection*, the more general phenomenon when two quantities, heat and one solute or two solutes, are diffused differently. [Problem 2.2; Stern (1960), Baines & Gill (1969).]

2.7. *Analytic solution for rigid–rigid boundaries.* Show that the general solution of equation (10.9), which is an even function of $z - \frac{1}{2}$, has the form

$$W = A_0 \cos \{q_0(z - \tfrac{1}{2})\} + A \cosh \{q(z - \tfrac{1}{2})\} + A^* \cosh \{q^*(z - \tfrac{1}{2})\},$$

where A_0 is any real and A any complex constant, and where

$$q_0 = a(\tau - 1)^{1/2}, \quad q = q_+ + iq_-,$$
$$q_\pm = a\{\tfrac{1}{2}(1 + \tau + \tau^2)^{1/2} \pm (1 + \tfrac{1}{2}\tau)\}^{1/2}, \quad \text{and} \quad \tau = (R/a^4)^{1/3}.$$

Hence use the rigid–rigid conditions given by equation (10.10) to show that the eigenvalue relation for the even modes is

$$\text{Im}\{(\sqrt{3} + i)q \tanh \tfrac{1}{2}q\} + q_0 \tan \tfrac{1}{2}q_0 = 0.$$

[Low (1929), Pellew & Southwell (1940).]

2.8. *Computation of the critical Rayleigh number for rigid–rigid boundaries by use of a variational principle.* Taking $F = \cos \pi(z - \tfrac{1}{2})$ as a trial function for the variational principle given by equation (9.14), show that

$$W = (\pi^2 + a^2)^{-2}\{\cos \pi(z - \tfrac{1}{2}) + A \cosh a(z - \tfrac{1}{2}) + B(z - \tfrac{1}{2}) \sinh a(z - \tfrac{1}{2})\}$$

for rigid–rigid boundaries, where

$$A = -\pi \frac{\sinh \tfrac{1}{2}a}{\sinh a + a}, \quad B = 2\pi \frac{\cosh \tfrac{1}{2}a}{\sinh a + a}.$$

Hence deduce that

$$K_1 = \tfrac{1}{2}(\pi^2 + a^2),$$
$$K_2 = \tfrac{1}{2}(\pi^2 + a^2)^{-2}[1 - 16\pi^2 a \cosh^2 \tfrac{1}{2}a/\{(\pi^2 + a^2)^2(\sinh a + a)\}].$$

With a little computation find that $K_1/a^2K_2 = 2185$, 1718, 1888 for $a = 2, 3, 4$ respectively and thence that $R_c < 1716$. [Pellew & Southwell (1940), Reid & Harris (1958), Chandrasekhar (1961), § 17.]

2.9. *An initial-value problem.* It is given that $\theta(\mathbf{x}, 0) = \delta(x)z(1 - z)$, $\mathbf{u}(\mathbf{x}, 0) = 0$ initially for a linearized disturbance in thermal convection between two free boundaries. Use Fourier analysis to show that

$$\theta(\mathbf{x}, t) = \int_{-\infty}^{\infty} \sum_j \theta_{aj}\, da \quad \text{for } t > 0,$$

where the summation is over the odd positive integers,

$$\theta_{aj} = 4j^{-3}\pi^{-4}\{B_{+j} \exp(s_{+j}t) + B_{-j} \exp(s_{-j}t)\} \cos ax \sin j\pi z,$$

$B_{+j} + B_{-j} = 1$, and $s_{\pm j}$ are the roots of equation (10.5). Similarly show that

$$w(\mathbf{x}, t) = \int_{-\infty}^{\infty} \sum_j w_{aj}\, da,$$

where

$$w_{aj} = 4j^{-3}\pi^{-4}\{(j^2\pi^2 + a^2 + s_{+j})B_{+j} \exp(s_{+j}t)$$
$$+ (j^2\pi^2 + a^2 + s_{-j})B_{-j} \exp(s_{-j}t)\} \cos ax \sin \pi j z.$$

Now find

$$B_{\pm j} = \mp (j^2\pi^2 + a^2 + s_{\mp j})/(s_{+j} - s_{-j}),$$

and hence determine $\theta(\mathbf{x}, t)$, $w(\mathbf{x}, t)$ for all $t > 0$. Show that $\omega_3 \equiv 0$ and use equations (8.15), (8.16) to find u, v for $t > 0$.

2.10. *Instability with rotation.* Show that the solution of Problem 2.3 with both surfaces free is given by $W = \sin j\pi z$ and

$$R = \frac{(j^2\pi^2 + a^2 + s)\{(j^2\pi^2 + a^2 + s/Pr)^2(j^2\pi^2 + a^2) + j^2\pi^2\mathscr{T}\}}{a^2(j^2\pi^2 + a^2 + s/Pr)}$$

for $j = 1, 2, \ldots$. Hence prove that the principle of exchange of stabilities is *not* valid when the Taylor number

$$\mathscr{T} > (1 + Pr)(j^2\pi^2 + a^2)^3/(1 - Pr)j^3\pi^3.$$

[Problem 2.3; Chandrasekhar (1961), § 29.]

2.11. *Annular cells.* Show that a solution of the reduced wave equation (11.1) is given by $f = J_0(ar)$, where $r = (x^2 + y^2)^{1/2}$, and J_0 is the Bessel function of order zero. Hence show that concentric annular cells may occur, with ascension or descension alternating between consecutive radii $r = j_{0,n}/a$ for $n = 1, 2, \ldots$, where $j_{0,n}$ is the nth positive zero of J_0. [Pellew & Southwell (1940), Zierep (1961).]

2.12. *Annular cells with a side wall.* For marginal stability of the first mode of free–free convection, show that equation (10.6) becomes

$$(a^2 + \pi^2)^3 = a^2R,$$

and has three roots $a^2 = a_1^2(R)$, $a_2^2(R)$ and $a_3^2(R)$ say, where $a_1^2 > a_2^2 > 0 > a_3^2$ if $R > R_c = 27\pi^4/4$.

Hence show that, if there is axisymmetric convection in a circular cylinder with a perfectly insulating rigid side wall at $r = l$, then the eigenfunctions are of the form

$$w = \left\{ \frac{A_1 J_0(a_1 r)}{J_0(a_1 l)} + \frac{A_2 J_0(a_2 r)}{J_0(a_2 l)} + \frac{A_3 I_0(|a_3|r)}{I_0(|a_3|l)} \right\} \sin \pi z,$$

where

$$A_1 \Big/ \left(\frac{T_3}{|a_3|} - \frac{T_2}{a_2} \right) = A_2 \Big/ \left(\frac{T_1}{a_1} - \frac{T_3}{|a_3|} \right) = A_3 \Big/ \left(\frac{T_2}{a_2} - \frac{T_1}{a_1} \right),$$

and the discrete values of R for marginal stability satisfy the relation

$$\frac{a_1 T_1}{a_1^2 + \pi^2} \left(\frac{T_3}{|a_3|} - \frac{T_2}{a_2} \right) + \frac{a_2 T_2}{a_2^2 + \pi^2} \left(\frac{T_1}{a_1} - \frac{T_3}{|a_3|} \right)$$

$$+ \frac{|a_3| T_3}{|a_3|^2 - \pi^2} \left(\frac{T_2}{a_2} - \frac{T_1}{a_1} \right) = 0.$$

Here we have used the definitions

$$T_1 = J_1(a_1 l)/J_0(a_1 l), \quad T_2 = J_1(a_2 l)/J_0(a_2 l),$$
$$T_3 = I_1(|a_3|l)/I_0(|a_3|l).$$

This problem is a mathematical illustration of S. H. Davis's results on the quantization in the number of cells and of the values of R, and on the associated stabilizing effects of side walls. [Hales (1937), Joseph (1971), § 5.]

CHAPTER 3

CENTRIFUGAL INSTABILITY

It seems doubtful whether we can expect to understand fully the instability of fluid flow without obtaining a mathematical representation of the motion of a fluid in some particular case in which instability can actually be observed, so that a detailed comparison can be made between the results of analysis and those of experiment.

– G. I. Taylor (1923)

14 Introduction

Instability also occurs in a homogeneous fluid owing to the dynamical effects of rotation or of streamline curvature. Three important examples of flows which exhibit this type of centrifugal instability are shown in Fig. 3.1. They are Couette flow, in which the fluid is contained between two rotating coaxial cylinders; flow in a curved channel due to a pressure gradient acting around the channel; and the flow in a boundary layer on a concave wall. In the absence of curvature, the last two of these examples may exhibit the type of instability associated with parallel shear flows, which will be discussed in Chapter 4.

The instability of rotating flows was first considered by Rayleigh (1880, 1916b). He considered a basic swirling flow of an inviscid fluid which moves with angular velocity $\Omega(r)$, an arbitrary function of the distance r from the axis of rotation. By a simple physical argument Rayleigh then derived his celebrated criterion for stability. *Rayleigh's circulation criterion* states that a necessary and sufficient condition for stability to axisymmetric disturbances is that the square of the circulation does not decrease anywhere, i.e. that $\Phi \geqslant 0$ everywhere in the field of flow, where Φ is the *Rayleigh discriminant* defined by

$$\Phi(r) = \frac{1}{r^3} \frac{\mathrm{d}}{\mathrm{d}r} (r^2 \Omega)^2. \tag{14.1}$$

He further observed that there is an analogy between the stability of rotating flows and the stability of a stratified fluid at rest in a gravitational field, the analogue of Φ in fact being $- g \, \mathrm{d}\bar{\rho}/\bar{\rho} \, \mathrm{d}z$, where $\bar{\rho}(z)$ is the basic density of the stratified fluid.

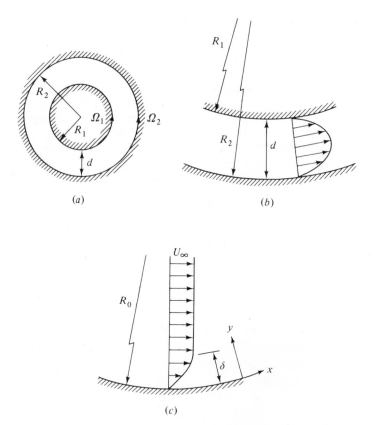

Fig. 3.1. (a) Couette flow. (b) Flow in a curved channel. (c) Flow in a boundary layer on a concave wall.

The stability of viscous flow between rotating cylinders has been widely studied, both theoretically and experimentally. In some early experiments by Couette (1890) to measure viscosity, the inner cylinder was fixed and the outer one was rotated at angular velocity Ω_2, say. Measurements of the torque exerted by the fluid on the inner cylinder showed that it was proportional to Ω_2 provided Ω_2 did not exceed a certain value. For larger values of Ω_2, however, the torque increased more rapidly, and this change was attributed to a transition from laminar to turbulent flow. In subsequent experiments Mallock (1896) fixed the outer cylinder and rotated the inner cylinder at angular velocity Ω_1, say, and found instability for each value of Ω_1. These circumstances were re-examined by Taylor. Now Rayleigh's criterion for an inviscid fluid predicts instability if $\Omega_1 R_1^2 > \Omega_2 R_2^2$, where R_1 and R_2 are the radii of the inner and outer cylinders, respectively, and so gives instability for Mallock's but not Couette's method of rotation. However, Mallock's instability seemed inconsistent with the stability of slow flow of a viscous fluid (Rayleigh 1913). In 1921 Taylor used an apparatus in which the cylinders could rotate independently, and in preliminary experiments he showed that the motion which appears at the onset of instability is axisymmetric, as assumed by Rayleigh, but that Rayleigh's criterion does not hold when the cylinders rotate in opposite directions. This culminated in the outstanding theoretical and experimental paper of Taylor (1923) on the instability of Couette flow of a viscous fluid, which forms the substance of § 17. But, before describing this in detail, we shall build up the theory for an inviscid fluid.

15 Instability of an inviscid fluid

We begin our discussion of the stability of rotating flows by neglecting the effects of viscosity. The equations of motion in cylindrical polar coordinates (r, θ, z) are then given by

$$\frac{\mathrm{D}u_r}{\mathrm{D}t} - \frac{u_\theta^2}{r} = -\frac{1}{\rho}\frac{\partial p}{\partial r}, \tag{15.1}$$

$$\frac{\mathrm{D}u_\theta}{\mathrm{D}t} + \frac{u_r u_\theta}{r} = -\frac{1}{\rho}\frac{1}{r}\frac{\partial p}{\partial \theta}, \tag{15.2}$$

and

$$\frac{\mathrm{D}u_z}{\mathrm{D}t} = -\frac{1}{\rho}\frac{\partial p}{\partial z}, \tag{15.3}$$

where

$$\frac{\mathrm{D}}{\mathrm{D}t} = \frac{\partial}{\partial t} + u_r\frac{\partial}{\partial r} + \frac{u_\theta}{r}\frac{\partial}{\partial \theta} + u_z\frac{\partial}{\partial z}. \tag{15.4}$$

Also the equation of mass conservation is given by

$$\frac{\partial u_r}{\partial r} + \frac{u_r}{r} + \frac{1}{r}\frac{\partial u_\theta}{\partial \theta} + \frac{\partial u_z}{\partial z} = 0. \tag{15.5}$$

These equations admit a steady basic solution of the form

$$u_r = u_z = 0, \quad u_\theta = V(r) \quad \text{and} \quad p = P(r), \qquad (15.6)$$

where $V(r) = r\Omega(r)$ is an arbitrary function of r, and the pressure distribution is determined by

$$P(r) = \rho \int \frac{V^2}{r} \, dr. \qquad (15.7)$$

In studying the stability of the steady flow given by equation (15.6), we seek to distinguish between stable and unstable distributions of basic angular velocity.

Consider first the physical argument given by Rayleigh (1916b). He considered only axisymmetric disturbances, for which equations (15.1)–(15.3) reduce to

$$\frac{\mathrm{D}u_r}{\mathrm{D}t} - \frac{u_\theta^2}{r} = -\frac{1}{\rho} \frac{\partial p}{\partial r}, \qquad (15.8)$$

$$\frac{\mathrm{D}u_\theta}{\mathrm{D}t} + \frac{u_r u_\theta}{r} = 0, \qquad (15.9)$$

and

$$\frac{\mathrm{D}u_z}{\mathrm{D}t} = -\frac{1}{\rho} \frac{\partial p}{\partial z}, \qquad (15.10)$$

where now

$$\frac{\mathrm{D}}{\mathrm{D}t} = \frac{\partial}{\partial t} + u_r \frac{\partial}{\partial r} + u_z \frac{\partial}{\partial z} \qquad (15.11)$$

and the equation of mass conservation is

$$\frac{\partial u_r}{\partial r} + \frac{u_r}{r} + \frac{\partial u_z}{\partial z} = 0. \qquad (15.12)$$

He then observed that equation (15.9) can be written in the form

$$\frac{\mathrm{D}(ru_\theta)}{\mathrm{D}t} = 0 \qquad (15.13)$$

and hence, by Kelvin's circulation theorem, that the angular momentum per unit mass of a fluid element $H (= ru_\theta)$ about the axis remains constant. (Note that $2\pi H$ is the circulation round the circle with equations $r = \text{constant}$ and $z = \text{constant}$.) Therefore the motion in planes through the axis is as if u_θ were absent, but there is instead a centrifugal force density $\rho u_\theta^2/r = \rho H^2/r^3$ acting in the radial direction. This centrifugal force density can be associated with a potential energy density $\frac{1}{2}\rho H^2/r^2$, which is simply the kinetic energy of the azimuthal motion $\frac{1}{2}\rho u_\theta^2$. Now consider the change in this kinetic energy which results when two fluid elements are interchanged. For this purpose take two elemental rings with equations $r = r_1$, $z = z_1$ and $r = r_2$, $z = z_2$ and with equal volumes dV of fluid. Therefore the kinetic energy of the azimuthal motion of the fluid contained in these two elements is $\frac{1}{2}\rho(H_1^2 r_1^{-2} + H_2^2 r_2^{-2})dV$. If we now interchange the fluid in these two elements, then the resulting kinetic energy is $\frac{1}{2}\rho(H_1^2 r_2^{-2} + H_2^2 r_1^{-2})dV$, because the angular momentum of an element of the inviscid fluid is constant. The change in the kinetic energy is therefore proportional to

$$(H_2^2 - H_1^2)(r_1^{-2} - r_2^{-2}). \tag{15.14}$$

If we suppose that $r_2 > r_1$, then this change in the kinetic energy is positive or negative according as $H_2^2 \gtrless H_1^2$. Thus, if H^2 decreases with r anywhere this could liberate kinetic energy and this would imply instability. However, it must be emphasized that if H^2 is non-decreasing then the inference of stability applies only to axisymmetric disturbances and that the flow may be unstable to other types of disturbances. Since H^2 is proportional to the square of the circulation around a ring, this is simply Rayleigh's criterion. Kármán (1934) has also discussed this result in terms of the centrifugal force and the pressure gradient.

Although physical arguments of the type given by Rayleigh and Kármán are helpful in understanding the mechanism of the instability, it is also of interest to follow Synge's mathematical derivation of Rayleigh's criterion and thus to deduce the important role played by the Rayleigh discriminant $\Phi(r)$. Therefore we shall begin by briefly discussing three-dimensional disturbances and then lead, by further simplifying assumptions, to the important special cases of axisymmetric and two-dimensional disturbances.

15.1 Three-dimensional disturbances

Consider an infinitesimal perturbation of the basic flow given by equations (15.6) and describe the perturbed flow by

$$\mathbf{u} = (u_r', V + u_\theta', u_z') \quad \text{and} \quad p = P + p'. \tag{15.15}$$

The linearized equations of motion then follow from equations (15.1)–(15.5) in the form

$$\frac{\partial u_r'}{\partial t} + \Omega \frac{\partial u_r'}{\partial \theta} - 2\Omega u_\theta' = -\frac{1}{\rho} \frac{\partial p'}{\partial r}, \qquad (15.16)$$

$$\frac{\partial u_\theta'}{\partial t} + \Omega \frac{\partial u_\theta'}{\partial \theta} + \left(\frac{dV}{dr} + \frac{V}{r}\right) u_r' = -\frac{1}{\rho} \frac{1}{r} \frac{\partial p'}{\partial \theta}, \qquad (15.17)$$

$$\frac{\partial u_z'}{\partial t} + \Omega \frac{\partial u_z'}{\partial \theta} = -\frac{1}{\rho} \frac{\partial p'}{\partial z}, \qquad (15.18)$$

and

$$\frac{\partial u_r'}{\partial r} + \frac{u_r'}{r} + \frac{1}{r} \frac{\partial u_\theta'}{\partial \theta} + \frac{\partial u_z'}{\partial z} = 0. \qquad (15.19)$$

These are the equations which describe the linearized problem for an inviscid fluid. In spite of the enormous simplification which results from linearization, a general analysis of these equations is still lacking. They do, however, provide the starting point for a number of important investigations based on various additional simplifying assumptions.

A substantial simplification can be achieved by making a normal-mode analysis in which all of the disturbances are assumed to depend on t, θ and z through a factor of the form $\exp\{st + i(n\theta + kz)\}$, where n is an integer (positive, negative or zero) and k is the wavenumber in the axial direction. Thus, if we let

$$(u_r', u_\theta', u_z', p'/\rho) = (u, v, w, \varpi) \exp\{st + i(n\theta + kz)\} \qquad (15.20)$$

then u, v, w and ϖ are functions of r only; they represent the amplitudes of the normal modes and satisfy the equations

$$(s + in\Omega)u - 2\Omega v = -D\varpi, \qquad (15.21)$$

$$(s + in\Omega)v + (D_* V)u = -inr^{-1}\varpi, \qquad (15.22)$$

$$(s + in\Omega)w = -ik\varpi, \qquad (15.23)$$

and

$$D_* u + inr^{-1}v + ikw = 0, \qquad (15.24)$$

where

$$D = d/dr \quad \text{and} \quad D_* = d/dr + 1/r. \qquad (15.25)$$

It is then possible to eliminate v, w and ϖ from these equations and obtain the single stability equation

$$\gamma^2 D\left(\frac{r^2 D_* u}{n^2 + k^2 r^2}\right)$$

$$-\left\{\gamma^2 + \frac{k^2 r^2 \Phi}{n^2 + k^2 r^2} + in\gamma r D\left(\frac{Z}{n^2 + k^2 r^2}\right)\right\} u = 0, \quad (15.26)$$

where

$$\gamma = s + in\Omega, \qquad (15.27)$$

$\Phi(r)$ is the Rayleigh discriminant defined by equation (14.1) and $Z = D_* V$ is the basic vorticity. If the fluid is contained between two rigid coaxial cylinders of radii R_1 and R_2, then equation (15.26) must be considered together with the boundary conditions

$$u = 0 \quad \text{at } r = R_1 \quad \text{and} \quad R_2. \qquad (15.28)$$

The perturbation equations (15.21)–(15.24) can also be expressed in terms of the *Lagrangian displacement* $\boldsymbol{\xi}(\mathbf{x}, t)$ which describes the displacement of a fluid element in the perturbed flow relative to its location $\mathbf{x}$ at time t in the unperturbed flow.† (In this description the basic flow need not be steady.) If we let q be any quantity that can be associated with a fluid element, then the *Lagrangian change* in q due to the perturbation is defined by

$$\Delta q = q(\mathbf{x} + \boldsymbol{\xi}(\mathbf{x}, t), t) - Q(\mathbf{x}, t), \qquad (15.29)$$

where $Q(\mathbf{x}, t)$ is the distribution of q in the unperturbed basic flow, and the corresponding *Eulerian change* is defined by

$$q' = q(\mathbf{x}, t) - Q(\mathbf{x}, t). \qquad (15.30)$$

To first order in $\boldsymbol{\xi}$, these changes are related by

$$\Delta q = q' + \boldsymbol{\xi} \cdot \boldsymbol{\nabla} q, \qquad (15.31)$$

and this relation is valid for any quantity q.

Consider then the Lagrangian change in the velocity, which is defined as the velocity of a fluid element at the position $\mathbf{x} + \boldsymbol{\xi}$ in the

† For a more general discussion of Lagrangian displacements see Chandrasekhar (1969, § 13).

perturbed flow relative to its velocity at **x** in the unperturbed flow. Thus we have

$$\Delta \mathbf{u} = \frac{\partial \boldsymbol{\xi}}{\partial t} + \mathbf{U} \cdot \nabla \boldsymbol{\xi}, \tag{15.32}$$

where **U** is the velocity field of the unperturbed flow, and

$$\mathbf{u}' = \frac{\partial \boldsymbol{\xi}}{\partial t} + \mathbf{U} \cdot \nabla \boldsymbol{\xi} - \boldsymbol{\xi} \cdot \nabla \mathbf{U} \tag{15.33}$$

for the Eulerian change from equation (15.31). We also note that the incompressibility condition, $\nabla \cdot \mathbf{u} = 0$, implies that $\nabla \cdot \boldsymbol{\xi} = 0$.

We can now proceed directly to the equations of the normal modes by letting $\boldsymbol{\xi} = (\xi, \eta, \zeta) \exp \{st + i(n\theta + kz)\}$. Then from equation (15.33) we have

$$u = \gamma\xi, \quad v = \gamma\eta - r(D\Omega)\xi \quad \text{and} \quad w = \gamma\zeta. \tag{15.34}$$

On substituting these expressions for u, v and w into equations (15.21)–(15.24) we obtain

$$(\gamma^2 + 2r\Omega D\Omega)\xi - 2\Omega\gamma\eta = -D\varpi, \tag{15.35}$$

$$\gamma^2\eta + 2\Omega\gamma\xi = -inr^{-1}\varpi, \tag{15.36}$$

$$\gamma^2\zeta = -ik\varpi, \tag{15.37}$$

and

$$D_*\xi + inr^{-1}\eta + ik\zeta = 0. \tag{15.38}$$

These equations can be somewhat simplified by eliminating η and ζ to obtain

$$D_*\xi - \frac{2in\Omega}{\gamma r}\xi = -\frac{1}{\gamma^2}\left(k^2 + \frac{n^2}{r^2}\right)\varpi \tag{15.39}$$

and

$$D\varpi + \frac{2in\Omega}{\gamma r}\varpi = -(\gamma^2 + \Phi)\xi. \tag{15.40}$$

These equations must be considered together with the boundary conditions

$$\xi = 0 \quad \text{at } r = R_1 \quad \text{and} \quad R_2. \tag{15.41}$$

In the special case when Ω = constant, equations (15.39) and (15.40) can easily be combined to give

$$\left(D_*D - \frac{n^2}{r^2}\right)\varpi = k^2\left(1 + \frac{4\Omega^2}{\gamma^2}\right)\varpi \qquad (15.42)$$

and conditions (15.41) can be put in the form

$$D\varpi + \frac{2in\Omega}{\gamma r}\varpi = 0 \qquad \text{at } r = R_1 \text{ and } R_2, \qquad (15.43)$$

and this is the eigenvalue problem which determines the frequencies of oscillation of a rotating column of liquid. This classical problem of inertial oscillations in rotating fluid is due to Kelvin (1880a), and is illustrated by a film loop of Fultz (FL 1964); the theory is introduced by Problem 3.1.

For a general basic angular velocity Ω, Rayleigh's criterion is known to be invalid for non-axisymmetric disturbances. Howard (1962) has given a simple counter-example of a basic flow which is unstable to two-dimensional disturbances, even though the circulation is non-decreasing outwards (see also Problem 3.5). The non-axisymmetric case has also been studied by Howard & Gupta (1962), who concluded that no general stability criterion was known when neither n nor k vanishes. However, more can be said about axisymmetric and two-dimensional disturbances, which we shall now discuss in turn.

15.2 Axisymmetric disturbances

To restrict consideration to axisymmetric disturbances, we set $\partial/\partial\theta = 0$ in equations (15.16)–(15.19) and obtain the equation

$$\frac{\partial^2}{\partial t^2}\left(\frac{\partial^2}{\partial r^2} + \frac{1}{r}\frac{\partial}{\partial r} - \frac{1}{r^2} + \frac{\partial^2}{\partial z^2}\right)u'_r + \Phi\frac{\partial^2 u'_r}{\partial z^2} = 0, \qquad (15.44)$$

which governs the initial-value problem. A general discussion of this initial-value problem for axisymmetric disturbances has not yet been given, and we proceed therefore to analyse the normal modes. By letting $u'_r = u \exp(st + ikz)$ in equation (15.44) or by setting $n = 0$ in equation (15.26), we obtain

$$(DD_* - k^2)u - \frac{k^2}{s^2}\Phi u = 0. \qquad (15.45)$$

This equation can also be derived from the θ-component of the linearized vorticity equation,

$$\frac{\partial(r\omega_\theta')}{\partial t} - 2V\frac{\partial u_\theta'}{\partial z} = 0, \quad \text{where } \omega_\theta' = \frac{\partial u_r'}{\partial z} - \frac{\partial u_z'}{\partial r}, \quad (15.46)$$

together with the θ-component of the linearized momentum equation and the equation of conservation of mass. It therefore represents a balance between the rate of increase of ω_θ' and the stretching of the vortex lines of the basic flow by the disturbance.

Synge (1933) noted that equation (15.45), together with the boundary conditions (15.28), is an eigenvalue problem of the classical Sturm–Liouville type and, from the well-known theorems of that subject (cf. Ince 1927, Chapter X), concluded that *the eigenvalues, k^2/s^2, are all negative if $\Phi > 0$ throughout the interval $R_1 < r < R_2$ and they are all positive if $\Phi < 0$. However, if Φ changes sign then we have both positive and negative eigenvalues with limit points at $\pm\infty$.* Moreover, the corresponding eigenfunctions are complete in the sense that any square-integrable function is equal almost everywhere to some infinite sum of the products of these eigenfunctions and appropriate coefficients.

Thus, if Φ is negative somewhere for a given basic flow, then we may immediately conclude that the flow is unstable in accordance with Rayleigh's criterion. However, if Φ is positive everywhere, then we cannot conclude that the flow is stable without also considering its stability with respect to non-axisymmetric disturbances. Nevertheless, when $\Phi > 0$ we may sometimes describe the basic flow as stable, omitting an explicit restriction to axisymmetric disturbances for brevity.

The foregoing results have been derived under the assumption that, in the absence of viscosity, Ω is an arbitrary function of r. When viscosity is present, however, the allowed forms for Ω are restricted. For example, for flow between rotating cylinders – Couette flow – the basic angular velocity distribution must be of the form

$$\Omega(r) = A + B/r^2, \quad (15.47)$$

where we have

$$A = \Omega_1 \frac{\mu - \eta^2}{1 - \eta^2} \quad \text{and} \quad B = \Omega_1 R_1^2 \frac{1 - \mu}{1 - \eta^2} \quad (15.48)$$

in terms of the parameters

$$\mu = \Omega_2/\Omega_1 \quad \text{and} \quad \eta = R_1/R_2 \qquad (15.49)$$

(see Fig. 3.1(a)).

Consider now the application of Rayleigh's criterion to the basic flow given by equation (15.47). If, for simplicity, we measure r in units of the radius R_2 of the outer cylinder then we have

$$\Phi = 4A\Omega = 4\Omega_1^2 \frac{(1-\mu)(\mu-\eta^2)}{(1-\eta^2)^2} \left(\frac{\eta^2}{r^2} + \frac{\mu-\eta^2}{1-\mu} \right), \quad (15.50)$$

where r is now a dimensionless variable and lies in the interval $\eta \leqslant r \leqslant 1$. When the cylinders rotate in the same direction it is easily shown that for all r in $\eta \leqslant r \leqslant 1$

and
$$\begin{array}{ll} \Phi > 0 & \text{if} \quad \mu > \eta^2 \\ \Phi < 0 & \text{if} \quad 0 < \mu < \eta^2. \end{array} \qquad (15.51)$$

Therefore the stability boundary for Couette flow of inviscid fluid is given by the *Rayleigh line* $\mu = \eta^2$ or $\Omega_1 R_1^2 = \Omega_2 R_2^2$. However, when the cylinders rotate in opposite directions we have

and
$$\left. \begin{array}{ll} \Phi > 0 & \text{for all } r \text{ in } \eta_0 < r \leqslant 1 \\ \Phi < 0 & \text{for all } r \text{ in } \eta \leqslant r < \eta_0, \end{array} \right\} \qquad (15.52)$$

where the *nodal surface* $r = \eta_0$ on which Ω and Φ vanish is given by

$$\eta_0 = \eta \left(\frac{1+|\mu|}{\mu^2+|\mu|} \right)^{1/2}. \qquad (15.53)$$

Thus when there is counter-rotation (i.e. $\mu < 0$) the flow is unstable, but it is only the layer of fluid between the inner cylinder and the nodal surface which is unstable by Rayleigh's criterion. Such a flow is said to be *locally unstable* at each point of the region $\eta < r < \eta_0$ in the sense that the disturbance is much larger in the unstable layer than it is in the stable layer. In the extreme case when $\mu \to -\infty$, $\eta_0 \to \eta$ and the unstable layer becomes vanishingly thin.

When Ω has the form described by equation (15.47) for Couette flow, the solution of equation (15.45) can be expressed in terms of Bessel functions of imaginary order and argument (Chandrasekhar 1961, pp. 290–2) but, despite the importance of this problem, the eigenvalue relation has not yet been analysed generally. The problem can be further simplified, however, by making the *narrow-gap approximation*, and the solution in that approximation will be given in § 16.

15.3 *Two-dimensional disturbances*

To illustrate the importance of non-axisymmetric disturbances, we shall now consider two-dimensional disturbances which depend only on r, θ and t. For these disturbances it is convenient to introduce a stream function ψ' of the perturbation such that

$$u_r' = \frac{1}{r} \frac{\partial \psi'}{\partial \theta} \quad \text{and} \quad u_\theta' = -\frac{\partial \psi'}{\partial r}. \tag{15.54}$$

On setting $\partial/\partial z = 0$ in equations (15.16)–(15.19), expressing u_r' and u_θ' in terms of ψ', and then eliminating the pressure, we obtain

$$\left(\frac{\partial}{\partial t} + \Omega \frac{\partial}{\partial \theta} \right) \Delta \psi' - \frac{DZ}{r} \frac{\partial \psi'}{\partial \theta} = 0 \quad \text{and} \quad \left(\frac{\partial}{\partial t} + \Omega \frac{\partial}{\partial \theta} \right) u_z' = 0, \tag{15.55}$$

where $Z (= rD\Omega + 2\Omega)$ is the basic vorticity and the Laplacian is given by

$$\Delta = \frac{\partial^2}{\partial r^2} + \frac{1}{r} \frac{\partial}{\partial r} + \frac{1}{r^2} \frac{\partial^2}{\partial \theta^2}. \tag{15.56}$$

The two equations (15.55) govern the initial-value problem for two-dimensional disturbances. The second equation shows that if $u_z'(r, \theta, 0) = F(r, \theta)$ then $u_z'(r, \theta, t) = F(r, \theta - \Omega t)$. Therefore we can set $u_z' \equiv 0$ without loss of generality in seeking conditions for instability. A general discussion of the initial-value problem for ψ' has not yet been given. We may note, however, that when Ω is of the form $A + B/r^2$ it follows immediately that $DZ = rD^2\Omega + 3D\Omega \equiv 0$ and hence that $\Delta \psi' = G(r, \theta - \Omega t)$ for an arbitrary function G which is differentiable with respect to θ. Apart from technical difficulties associated with the cylindrical geometry, the solution for ψ' could then be found by the method developed by Orr (1907, pp.26–27) for studying the stability of plane Couette flow. For more general distributions of angular velocity, however, the initial-value problem could be treated by the use of Fourier–Laplace transforms along the lines developed, for example, by Case (1960a) and Dikii (1960b) for parallel basic flows.

We proceed now to consider the normal-mode analysis for two-dimensional disturbances. Here, on letting $\psi'(r, \theta, t) =$

$\phi(r) \exp(st + in\theta)$, we have $u = inr^{-1}\phi$ and $v = -D\phi$; and from the first of equations (15.55) or by setting $k = 0$ in equation (15.26), we obtain

$$(s + in\Omega)(D_*D - n^2/r^2)\phi - inr^{-1}(DZ)\phi = 0. \qquad (15.57)$$

This equation can also be derived from the z-component of the linearized vorticity equation,

$$\left(\frac{\partial}{\partial t} + \Omega\frac{\partial}{\partial \theta}\right)\omega_z' + (DZ)u_r' = 0, \qquad (15.58)$$

where

$$\omega_z' = \frac{1}{r}\frac{\partial}{\partial r}(ru_\theta') - \frac{1}{r}\frac{\partial u_r'}{\partial \theta} = -\Delta\psi', \qquad (15.59)$$

and it therefore represents a balance between the rate of increase of ω_z' and the stretching of the vortex lines of the basic flow by the disturbance.

Equation (15.57) has a quite different structure from equation (15.45) for axisymmetric disturbances and is more closely related to the Rayleigh stability equation for parallel flow (see Chapter 4), to which it reduces in the narrow-gap approximation. When the flow is confined between two concentric cylinders so that $\phi = 0$ at $r = R_1$ and R_2, we have the following result due to Rayleigh (1880): *A necessary condition for instability with respect to two-dimensional disturbances is that the gradient of the basic vorticity,* $DZ = rD^2\Omega + 3D\Omega$, *must change sign at least once in the interval* $R_1 < r < R_2$. This result is the analogue for curved and rotating flows of Rayleigh's well-known inflexion-point theorem, which will be proved in Chapter 4. The proof of the present result is similar, so we shall omit it here (but set it as Problem 3.2).

In the case of Couette flow, for which $DZ \equiv 0$, we see that if $s + in\Omega \neq 0$ then equation (15.57) reduces to $(D_*D - n^2r^{-2})\phi = 0$, the general solution of which is $\phi = C_1r^n + C_2r^{-n}$. Application of the boundary conditions then shows that ϕ vanishes identically. Therefore there are no discrete modes in the spectrum for this problem; there is, however, a continuous spectrum of stable singular modes for which $s + in\Omega$ vanishes somewhere in the interval $R_1 < r < R_2$.

16 Instability of Couette flow of an inviscid fluid

We now consider the stability of Couette flow with respect to axisymmetric disturbances in the *narrow-gap approximation*. This approximation is based on the geometrical assumption that the difference in radii of the two cylinders is small compared with their mean, i.e. that

$$d = R_2 - R_1 \ll \tfrac{1}{2}(R_1 + R_2). \tag{16.1}$$

This condition is equivalent to the assumption that $1 - \eta \ll 1$ and thus the angular velocity distribution (15.47) can be approximated by the linear distribution

$$\Omega(r) \sim \Omega_1 \{1 - (1 - \mu)\zeta\} \quad \text{as } \eta \to 1, \tag{16.2}$$

where $\zeta = (r - R_1)/d$ and $0 \leqslant \zeta \leqslant 1$. One of the dynamical consequences of this approximation is that centrifugal effects are entirely neglected in the description of the basic flow.

Consider first the case of solid-body rotation, for which $\mu = 1$, $A = \Omega_1$ and $B = 0$. Then $\Phi(r) = 4\Omega_1^2$ and equation (15.45) becomes

$$(DD_* - k^2)u - 4\Omega_1^2 \frac{k^2}{s^2} u = 0. \tag{16.3}$$

To first order in $1 - \eta$ we can approximate D_* by D, and equation (16.3) can then be written in the dimensionless form

$$(D^2 - a^2)u - \frac{4\Omega_1^2}{s^2} a^2 u = 0, \tag{16.4}$$

where D now stands for $d/d\zeta$ and $a = kd$, and the boundary conditions are $u = 0$ at $\zeta = 0$ and 1. The solutions of equation (16.4) which satisfy these boundary conditions are

$$u_m = \text{constant} \times \sin m\pi\zeta, \tag{16.5}$$

where m is an integer; the corresponding eigenvalues are given by

$$s_m = \pm \frac{2i\Omega_1 a^2}{a^2 + m^2\pi^2}. \tag{16.6}$$

For $\mu \neq 1$, the Rayleigh discriminant is

$$\Phi(r) \sim -2\Omega_1^2 \frac{1 - \mu}{1 - \eta} \{1 - (1 - \mu)\zeta\}, \tag{16.7}$$

and the flow is therefore stable or unstable according to whether $\mu - 1$ is positive or negative. When the cylinders rotate in the same direction, i.e. $0 < \mu < 1$, the flow is locally unstable for each ζ in $0 \leq \zeta \leq 1$; but for counter-rotating cylinders ($\mu < 0$) the flow is locally unstable only in the layer $0 \leq \zeta < \zeta_0$, where $\zeta_0 = 1/(1 + |\mu|)$. The narrow-gap form of equation (15.45) now becomes

$$(D^2 - a^2)u = -\frac{a^2}{\sigma^2}\{1 - (1 - \mu)\zeta\}u, \tag{16.8}$$

where

$$\sigma = \frac{s}{(-4A\Omega_1)^{1/2}} \sim \frac{s}{2\Omega_1}\left\{\frac{2(1 - \eta)}{1 - \mu}\right\}^{1/2}. \tag{16.9}$$

The factor $(1 - \eta)^{1/2}$ or $(d/R_1)^{1/2}$ in this expression shows that in the narrow-gap approximation the appropriate scaling for s is not simply Ω_1 but rather $\Omega_1(R_1/d)^{1/2}$.

From the form of equation (16.8) it is clear that the solution can be expressed in terms of Airy functions (Reid 1960). For this purpose we use the new variable

$$x = \left(\frac{a\sigma^2}{1 - \mu}\right)^{2/3}\left[1 - \frac{1}{\sigma^2}\{1 - (1 - \mu)\zeta\}\right] \tag{16.10}$$

to transform equation (16.8) into the standard form of Airy's equation,

$$\left(\frac{d^2}{dx^2} - x\right)u = 0, \tag{16.11}$$

and the boundary conditions into the form

$$u = 0 \quad \text{at} \quad x = x_0 \quad \text{and} \quad x_1, \tag{16.12}$$

where

$$x_0 = \left(\frac{a\sigma^2}{1 - \mu}\right)^{2/3}\left(1 - \frac{1}{\sigma^2}\right) \quad \text{and} \quad x_1 = \left(\frac{a\sigma^2}{1 - \mu}\right)^{2/3}\left(1 - \frac{\mu}{\sigma^2}\right). \tag{16.13}$$

The general solution of equation (16.11) can be expressed as a linear combination of the Airy functions $\text{Ai}(x)$ and $\text{Bi}(x)$, and the boundary conditions then lead to the eigenvalue relation,

$$\frac{\text{Ai}(x_0)}{\text{Bi}(x_0)} = \frac{\text{Ai}(x_1)}{\text{Bi}(x_1)}, \tag{16.14}$$

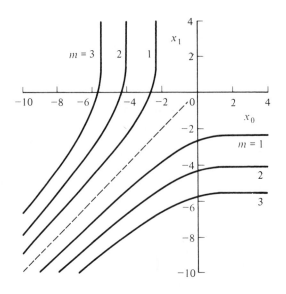

Fig. 3.2. The solutions of equation (16.14) for $m = 1, 2$ and 3. (From Reid 1960.)

the form of which is entirely independent of the parameters of the stability problem. The solutions of equation (16.14) are shown in Fig. 3.2 for the first three modes. It may be noted that $x_1 \to a_m$ as $x_0 \to \infty$ for some positive integer m and, conversely, where a_m (for $m = 1, 2, \ldots$) are the zeros of $\text{Ai}(x)$. With the relationship between x_0 and x_1 determined in this manner, the corresponding values of a^2 and σ^2 are then given parametrically by

$$a^2 = (x_1 - x_0)^2 \frac{x_1 - \mu x_0}{1 - \mu} \quad \text{and} \quad \sigma^2 = \frac{x_1 - \mu x_0}{x_1 - x_0}. \quad (16.15)$$

Since a is real, a^2 must be non-negative, and this restricts the admissible values of x_0 and x_1. When $\mu > 1$ we must require that $x_1 - \mu x_0 \leqslant 0$ and we then find that σ^2 is positive for all modes and that $\sigma^2 \to \mu$ as $a \to \infty$; for these modes s^2 is negative and there is therefore stability, as anticipated. When $\mu < 1$, however, we must require that $x_1 - \mu x_0 \geqslant 0$ and there is then a set of unstable modes for which σ^2 and s^2 are positive with $\sigma^2 \to 1$ as $a \to \infty$; when $\mu < 0$, there is also a second set of stable modes for which σ^2 and s^2 are negative with $\sigma^2 \to -|\mu|$ as $a \to \infty$. These results are shown in Fig. 3.3 for three values of μ.

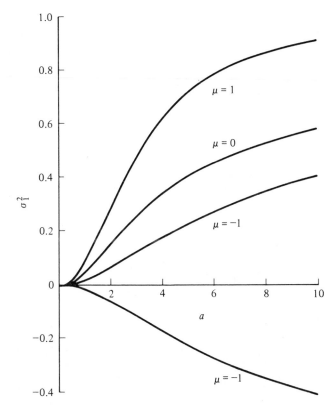

Fig. 3.3. The eigenvalues σ^2 for $m = 1$ and for $\mu = 1$, 0 and -1. (From Reid 1960.)

When the cylinders rotate in opposite directions, it is only the layer of fluid between the inner cylinder and the nodal surface that is locally unstable, and this suggests that if $-\mu$ is quite large then the relevant length scale for the unstable modes is no longer the distance between the cylinders but rather the thickness of the unstable layer. To study these modes in the limit as $\mu \to -\infty$, we therefore introduce the scaling

$$\bar{a} = a/(1-\mu) \quad \text{and} \quad \bar{\zeta} = (1-\mu)\zeta. \qquad (16.16)$$

With the radial coordinate stretched in this manner, the nodal surface is fixed at $\bar{\zeta} = 1$, and equation (16.8) becomes

$$(\bar{D}^2 - \bar{a}^2)u = -\frac{\bar{a}^2}{\sigma^2}(1 - \bar{\zeta})u, \qquad (16.17)$$

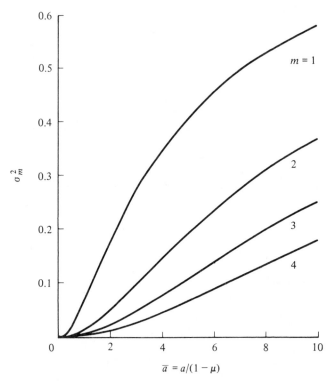

Fig. 3.4. The eigenvalues σ^2 for $m = 1, 2, 3$ and 4 for $\mu = -\infty$. (From Reid 1960.)

with the boundary conditions $u = 0$ at $\bar{\zeta} = 0$ and $1 - \mu$. By means of the transformation

$$\bar{x} = (\bar{a}\sigma^2)^{2/3}\{1 - \sigma^{-2}(1 - \bar{\zeta})\}, \qquad (16.18)$$

equation (16.17) can be reduced to Airy's equation (16.11) with x replaced by $\bar{x}$. In the limit as $\mu \to -\infty$, the solution must remain bounded as $\bar{x} \to \infty$ and it must therefore be of the form

$$u(\bar{\zeta}) = \text{constant} \times \text{Ai}(\bar{x}). \qquad (16.19)$$

The boundary condition at $\bar{\zeta} = 0$ then leads to the eigenvalue relation

$$(\bar{a}\sigma^2)^{2/3}(1 - \sigma^{-2}) = a_m \quad \text{for} \quad m = 1, 2, \ldots, \qquad (16.20)$$

where a_m are the zeros of $\text{Ai}(\bar{x})$. The dependence of σ^2 on $\bar{a}$ obtained from this equation is shown in Fig. 3.4. The eigenfunction

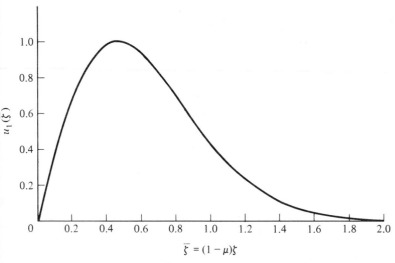

Fig. 3.5. The eigenfunction $u(\bar{\zeta})$ for $m = 1$, $\mu = -\infty$ and $\bar{a} = 2.03$. (From Reid 1960.)

for the first mode, normalized to unity, is shown in Fig. 3.5 for $\bar{a} = 2.03$, this value of $\bar{a}$ having been chosen because it is the wavenumber at which instability occurs in the corresponding viscous problem (see § 17). Fig. 3.5 also illustrates the concept of local stability, because the disturbance is significantly larger in the unstable part of the flow than it is in the stable part, and, for values of $-\mu$ greater than about two, the solution is largely independent of μ.

The motion in planes through the axis can conveniently be described by a Stokes stream function ψ such that

$$u_r = -\frac{1}{r}\frac{\partial\psi}{\partial z} \quad \text{and} \quad u_z = \frac{1}{r}\frac{\partial\psi}{\partial r}, \qquad (16.21)$$

and we then have

$$\psi = -(ik)^{-1}ru(r)\,e^{st+ikz}. \qquad (16.22)$$

Since the phase of the linearized disturbance in the axial direction is arbitrary, it is customary to define the spatial pattern of the motion by an equation of the form $ru(r)\cos kz = $ constant. In the narrow-gap approximation this becomes simply $u(\zeta)\cos(az/d) = $ constant;

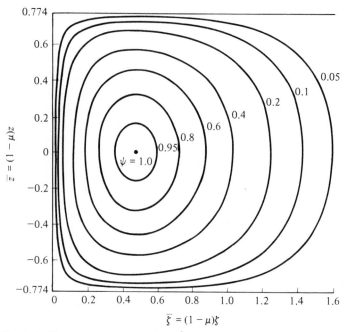

Fig. 3.6. The cell pattern in the narrow-gap approximation for $m = 1$, $\mu = -\infty$ and $\bar{a} = 2.03$: eigenfunction $\psi = u_1(\zeta) \cos(\bar{a}\bar{z})$. (From Reid 1960.)

and in Fig. 3.6 the cell pattern is shown in the limit as $\mu \rightarrow -\infty$ with $\bar{a} = 2.03$.

The stability of Couette flow of an inviscid fluid with respect to non-axisymmetric disturbances has also been studied by Krueger & DiPrima (1962) and by Bisshopp (1963) in the narrow-gap approximation. Their results taken together cover the entire range of rotation $-\infty < \mu \leqslant 1$ and show that the non-axisymmetric modes always have a smaller growth rate than the corresponding axisymmetric modes. When viscous effects are included, however, the most unstable mode need not be an axisymmetric one. This aspect of the Taylor problem will be discussed further in § 17.2.

17 The Taylor problem

We now consider the effects of viscosity on the instability of rotating and curved flows. Viscous effects in a rotating fluid are usually characterized by a dimensionless *Taylor number T*, the precise definition of which may vary. If $\Omega = $ constant, for example, then one usually takes $T = 4\Omega^2 L^4/\nu^2$, where L is a convenient length scale, and, in general, the Taylor number is proportional to the square of a Reynolds number. A basic flow which is stable according to Rayleigh's circulation criterion would be expected to remain stable when viscous effects are considered. However, if the flow is unstable according to Rayleigh's criterion, then we would still expect viscosity to have a stabilizing effect, in the sense that the flow is stable provided the Taylor number is less than a certain critical value T_c (say). We are interested therefore in determining T_c and, more generally, its dependence on the other parameters of the problem.

In cylindrical polar coordinates $\mathbf{x} = (r, \theta, z)$ and $\mathbf{u} = (u_r, u_\theta, u_z)$, the Navier–Stokes equations for a viscous incompressible fluid are

$$\frac{Du_r}{Dt} - \frac{u_\theta^2}{r} = -\frac{1}{\rho}\frac{\partial p}{\partial r} + \nu\left(\Delta u_r - \frac{u_r}{r^2} - \frac{2}{r^2}\frac{\partial u_\theta}{\partial \theta}\right), \qquad (17.1)$$

$$\frac{Du_\theta}{Dt} + \frac{u_r u_\theta}{r} = -\frac{1}{\rho}\frac{1}{r}\frac{\partial p}{\partial \theta} + \nu\left(\Delta u_\theta - \frac{u_\theta}{r^2} + \frac{2}{r^2}\frac{\partial u_r}{\partial \theta}\right), \qquad (17.2)$$

$$\frac{Du_z}{Dt} = -\frac{1}{\rho}\frac{\partial p}{\partial z} + \nu\Delta u_z, \qquad (17.3)$$

where

$$\frac{D}{Dt} = \frac{\partial}{\partial t} + u_r\frac{\partial}{\partial r} + \frac{u_\theta}{r}\frac{\partial}{\partial \theta} + u_z\frac{\partial}{\partial z} \qquad (17.4)$$

and

$$\Delta = \frac{\partial^2}{\partial r^2} + \frac{1}{r}\frac{\partial}{\partial r} + \frac{1}{r^2}\frac{\partial^2}{\partial \theta^2} + \frac{\partial^2}{\partial z^2}. \qquad (17.5)$$

Also the equation of mass conservation is given by

$$\frac{\partial u_r}{\partial r} + \frac{u_r}{r} + \frac{1}{r}\frac{\partial u_\theta}{\partial \theta} + \frac{\partial u_z}{\partial z} = 0, \qquad (17.6)$$

as for an inviscid incompressible fluid. If we now assume that the basic flow is of the form

$$u_r = u_z = 0, \quad u_\theta = V(r) = r\Omega(r) \quad \text{and} \quad p = P(r), \qquad (17.7)$$

then equations (17.1) and (17.2) give $\quad D_* = \frac{d}{dr} + \frac{1}{r}$

$$\frac{1}{\rho}\frac{dP}{dr} = \frac{V^2}{r} \quad \text{and} \quad \nu DD_* V = 0. \tag{17.8}$$

Thus in a viscous fluid V can no longer be an arbitrary function of r, but must be of the form $Ar + B/r$, which describes Couette flow.

If we again suppose that the perturbed flow is of the form

$$\mathbf{u} = (u'_r, \, V + u'_\theta, \, u'_z) \quad \text{and} \quad p = P + p', \tag{17.9}$$

then the linearized equations of motion are given by

$\rho(V + u'_\theta)^2/r = 2V u'_\theta/r$

$$\frac{\partial u'_r}{\partial t} + \Omega\frac{\partial u'_r}{\partial \theta} - 2\Omega u'_\theta = -\frac{1}{\rho}\frac{\partial p'}{\partial r} + \nu\left(\Delta u'_r - \frac{u'_r}{r^2} - \frac{2}{r^2}\frac{\partial u'_\theta}{\partial \theta}\right), \tag{17.10}$$

$$\frac{\partial u'_\theta}{\partial t} + \Omega\frac{\partial u'_\theta}{\partial \theta} + (D_* V)u'_r = -\frac{1}{\rho}\frac{1}{r}\frac{\partial p'}{\partial \theta} + \nu\left(\Delta u'_\theta - \frac{u'_\theta}{r^2} + \frac{2}{r^2}\frac{\partial u'_r}{\partial \theta}\right), \tag{17.11}$$

$$\frac{\partial u'_z}{\partial t} + \Omega\frac{\partial u'_z}{\partial \theta} = -\frac{1}{\rho}\frac{\partial p'}{\partial z} + \nu\Delta u'_z, \tag{17.12}$$

and

$$\frac{\partial u'_r}{\partial r} + \frac{u'_r}{r} + \frac{1}{r}\frac{\partial u'_\theta}{\partial \theta} + \frac{\partial u'_z}{\partial z} = 0. \tag{17.13}$$

These equations have not yet been analysed in general, but they have been studied on the basis of various simplifying assumptions, some of which will be considered next.

17.1 *Axisymmetric disturbances*

In Taylor's (1921) early experiments he first observed that instability led to a steady secondary flow in the form of toroidal vortices which were regularly spaced along the axis of the cylinders (see Fig. 3.7). These observations suggested some important simplifications in his theoretical treatment (Taylor 1923) of the problem and have been widely used since. Thus consider first the case of axisymmetric disturbances, for which equations (17.10)–(17.13) can be reduced to the pair of coupled equations,

$$\left(\frac{\partial}{\partial t} - \nu\Delta_*\right)\Delta_* u'_r = 2\Omega\frac{\partial^2 u'_\theta}{\partial z^2} \tag{17.14}$$

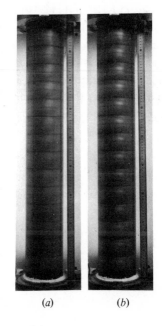

(a)   (b)

Fig. 3.7. Taylor vortices for values of T near T_c. Outer cylinder at rest; free surface at top, non-rotating bottom end plate; fluid column length $= 86.30$ cm; $\nu = 1.420$ cm^2 s^{-1}; $\eta = 0.505$. (a) $T/T_c = 1.000$, (b) $T/T_c = 2.050$. (Photograph: J. E. Burkhalter & E. L. Koschmieder.)

and

$$\left(\frac{\partial}{\partial t} - \nu \Delta_*\right) u'_\theta = -(\mathbf{D}_* V)u'_r, \tag{17.15}$$

where

$$\Delta_* = \Delta - \frac{1}{r^2} \quad \text{and} \quad \Delta = \frac{\partial^2}{\partial r^2} + \frac{1}{r}\frac{\partial}{\partial r} + \frac{\partial^2}{\partial z^2}. \tag{17.16}$$

Since $\mathbf{D}_* V (= 2A)$ is constant for Couette flow, equations (17.14) and (17.15) could, if desired, be combined into a single equation for u'_θ. Also these equations can be further simplified by analysing the disturbance into normal modes. For this purpose we let

$$(u'_r, u'_\theta) = (u, v)\, \mathrm{e}^{st+ikz}, \tag{17.17}$$

where k is the axial wavenumber. Equations (17.14) and (17.15) then become

$$\{\nu(\mathbf{DD}_* - k^2) - s\}(\mathbf{DD}_* - k^2)u = 2k^2 \Omega v \tag{17.18}$$

and

$$\{\nu(\mathbf{DD}_* - k^2) - s\}v = (\mathbf{D}_* V)u, \tag{17.19}$$

which formally reduce to equation (15.45) on setting $\nu = 0$.

It is convenient here to rewrite equations (17.18) and (17.19) in an appropriate dimensionless form. This can be done in a number of ways depending primarily on the choice of L. When the cylinders rotate in the same direction, the natural length scale is the gap width $d = R_2 - R_1$ and, following Walowit, Tsao & DiPrima (1964), we shall therefore choose $L = d$. This choice of L also has the advantage that it remains a relevant length scale in the narrow-gap approximation. Thus we let

$$x = (r - R_0)/d, \tag{17.20}$$

where $R_0 = \frac{1}{2}(R_1 + R_2)$ is the mean radius of the cylinders and $-\frac{1}{2} \leq x \leq \frac{1}{2}$ is the domain of the flow. The basic angular velocity distribution can then be written in the form

$$\Omega(r) = \Omega_1 g(x), \tag{17.21}$$

where

$$g(x) = A_1 + \frac{B_1}{\xi^2}, \quad A_1 = \frac{\mu - \eta^2}{1 - \eta^2}, \quad B_1 = \eta^2 \frac{1 - \mu}{1 - \eta^2} \tag{17.22}$$

and

$$\xi = r/R_2 = \eta + (1 - \eta)(x + \tfrac{1}{2}). \tag{17.23}$$

If we further let

$$a = kd \quad \text{and} \quad \sigma = sd^2/\nu, \tag{17.24}$$

then equations (17.18) and (17.19) become

$$(DD_* - a^2 - \sigma)(DD_* - a^2)u = a^2 \frac{2\Omega_1 d^2}{\nu} g(x)v \tag{17.25}$$

and

$$(DD_* - a^2 - \sigma)v = \frac{2Ad^2}{\nu} u, \tag{17.26}$$

where D and D_* now stand for the dimensionless derivatives

$$D = \frac{d}{dx} \quad \text{and} \quad D_* = \frac{d}{dx} + \frac{1 - \eta}{\xi}. \tag{17.27}$$

On replacing $(2Ad^2/\nu)u$ by u, these equations reduce to the standard dimensionless forms

$$(DD_* - a^2 - \sigma)(DD_* - a^2)u = -a^2 T g(x)v \tag{17.28}$$

and

$$(DD_* - a^2 - \sigma)v = u, \tag{17.29}$$

where the Taylor number T is defined by

$$T = -\frac{4A\Omega_1 d^4}{\nu^2} = \frac{4\Omega_1^2 R_1^4}{\nu^2} \frac{\eta^2 - \mu}{1 - \eta^2}\left(\frac{1-\eta}{\eta}\right)^4. \tag{17.30}$$

These equations must be considered together with the boundary conditions,

$$u = Du = v = 0 \quad \text{at} \quad x = \pm\tfrac{1}{2}. \tag{17.31}$$

Some authors, including Chandrasekhar (1961), have chosen $L = R_2$, and this leads to a somewhat different definition of the Taylor number:

$$T_1 = -\frac{4ABR_2^2}{\nu^2} = \frac{4\Omega_1^2 R_1^4}{\nu^2} \frac{(1-\mu)(1-\mu/\eta^2)}{(1-\eta^2)^2}. \tag{17.32}$$

These two Taylor numbers are related by

$$\frac{T_1}{T} = \frac{(1-\mu)\eta^2}{(1-\eta^2)(1-\eta)^4}. \tag{17.33}$$

It should also be noted that we have made s dimensionless with respect to the viscous time scale d^2/ν. For some purposes, however, it is preferable to use the inertial time scale $1/\Omega_1$ so that the coefficient of kinematic viscosity then appears only in the Taylor number. This is equivalent (cf. equation (16.9)) to letting

$$\frac{sd^2}{\nu} = \sigma_1 T^{1/2} \quad \text{or} \quad \sigma_1 = \frac{s}{(-4A\Omega_1)^{1/2}} = \frac{s}{2\Omega_1}\left(\frac{1-\eta^2}{\eta^2 - \mu}\right)^{1/2}. \tag{17.34}$$

If we now replace u by $-(-A/\Omega_1)^{1/2}u$ in equations (17.25) and (17.26) or, equivalently, by $T^{1/2}u$ in equations (17.28) and (17.29), then we obtain

$$(DD_* - a^2 - \sigma_1 T^{1/2})(DD_* - a^2)u = a^2 T^{1/2}g(x)v \tag{17.35}$$

and

$$(DD_* - a^2 - \sigma_1 T^{1/2})v = T^{1/2}u. \tag{17.36}$$

On formally letting $T \to \infty$ in these two equations we recover the dimensionless form of the inviscid equation for Couette flow.

Synge (1938) has shown that Re (σ) can be expressed as the ratio of certain positive definite integrals and that Re $(\sigma) < 0$ when $\mu > \eta^2$. Thus the viscous flow is stable when Rayleigh's circulation criterion $\mu > \eta^2$ for inviscid flow is satisfied. Also Wood (1964) has proved this result by use of a modified energy integral for the perturbed velocity field. However, neither of these proofs leads to any conclusion about the behaviour of Im (σ).

Consider now equations (17.28) and (17.29) in the narrow-gap approximation. To first order in $1 - \eta$ we have

$$(D^2 - a^2 - \sigma)^2 (D^2 - a^2)v = -a^2 T\{1 - (1 - \mu)(x + \tfrac{1}{2})\}v,$$
$$(17.37)$$

which must be considered together with the boundary conditions,

$$v = (D^2 - a^2 - \sigma)v = D(D^2 - a^2 - \sigma)v = 0 \quad \text{at } x = \pm \tfrac{1}{2}.$$
$$(17.38)$$

We can then get an eigenvalue relation of the form $\mathcal{F}(a, T, \sigma, \mu) = 0$.

On the basis of experimental evidence, Taylor and many subsequent workers assumed that the principle of exchange of stabilities is valid for this problem (at least if the cylinders rotate in the same direction). The analysis of the problem for an inviscid fluid given in § 16 is consistent with the principle being valid when $\mu > 0$ but casts doubt on its validity when $\mu < 0$. For a viscous fluid, S. H. Davis (1969) has developed a perturbation procedure which shows that the principle is valid when the gap is small and μ lies in the range $0.68 < \mu < 1$, and Yih (1972) later showed that the principle is valid provided $\mu > 0$ but with no restriction on η. Yih's proof is rather complicated and a simpler proof would certainly be desirable.

Thus, on setting $\sigma = 0$ in equations (17.37) and (17.38), we have

$$(D^2 - a^2)^3 v = -a^2 T\{1 - (1 - \mu)(x + \tfrac{1}{2})\}v \qquad (17.39)$$

and

$$v = (D^2 - a^2)v = D(D^2 - a^2)v = 0 \quad \text{at} \quad x = \pm \tfrac{1}{2}. \quad (17.40)$$

Because the independent variable x appears linearly in equation (17.39), it is clear that the general solution can be expressed in terms of Laplace integrals,[†] but the application of the boundary conditions (17.40) and subsequent analysis of the eigenvalue relation is difficult and has not been carried through in detail.

A simple but powerful Galerkin-type method to solve this problem was developed by Chandrasekhar (1954a). It is somewhat similar to a method applied by Malurkar & Srivastava (1937) to Bénard convection, and since 1954 it has been used for a wide variety of problems in hydrodynamic and magnetohydrodynamic stability. We shall therefore describe the method in the context of the somewhat more general problem,

$$(D^2 - a^2)^2 u = f(x)v \qquad (17.41)$$

and

$$(D^2 - a^2)v = -a^2 Tg(x)u, \qquad (17.42)$$

where u and v satisfy the boundary conditions (17.31). The basic idea in this method is to expand v in a complete set of functions which vanish at $x = \pm\frac{1}{2}$. Consider, for example, the Fourier series

$$v = \sum_{m=1}^{\infty} (A_m \cos p_m x + B_m \sin q_m x), \qquad (17.43)$$

where $p_m = (2m-1)\pi$ and $q_m = 2m\pi$. When this expression for v is substituted into equation (17.41) we obtain the equation

$$(D^2 - a^2)^2 u = f(x) \sum_{m=1}^{\infty} (A_m \cos p_m x + B_m \sin q_m x), \qquad (17.44)$$

which must be *solved* subject to the boundary conditions $u = Du = 0$ at $x = \pm\frac{1}{2}$. The required solution can be written in the form

$$u = \sum_{m=1}^{\infty} \{A_m f_m(x) + B_m g_m(x)\}. \qquad (17.45)$$

To complete the solution we now require that the error in equation (17.42) be orthogonal to $\cos p_n x$ and $\sin q_n x$ for each positive integer n. Thus the expressions for u and v above are substituted into equation (17.42) and the result multiplied by $\cos p_n x$ and $\sin q_n x$ in turn and then integrated between $x = -\frac{1}{2}$ and $x = \frac{1}{2}$. In this way we

[†] This approach has been considered by Granoff & Bleistein (1972) and Granoff (1972), but their results are not directly applicable to the stability problem.

obtain a doubly infinite system of linear homogeneous equations for the constants A_m and B_m. In order that these constants do not vanish identically, the determinant of the system must vanish, and this condition yields the eigenvalue relation

$$\begin{vmatrix} \frac{1}{2}(p_m^2 + a^2)(1/a^2 T)\,\delta_{mn} - A_{mn} & -D_{mn} \\ -C_{mn} & \frac{1}{2}(q_m^2 + a^2)\delta_{mn} - B_{mn} \end{vmatrix} = 0,$$

(17.46)

where

$$A_{mn} = \int_{-1/2}^{1/2} f_m(x)\cos p_n x\,\mathrm{d}x, \qquad B_{mn} = \int_{-1/2}^{1/2} g_m(x)\sin q_n x\,\mathrm{d}x,$$

$$C_{mn} = \int_{-1/2}^{1/2} f_m(x)\sin q_n x\,\mathrm{d}x, \qquad D_{mn} = \int_{-1/2}^{1/2} g_m(x)\cos p_n x\,\mathrm{d}x.$$

This eigenvalue relation is in the form of a determinant of infinite order. In practice, of course, one considers only a finite number of terms (M say) in the expansions of u and v, and with $M = 1, 2, \ldots$ this leads to a sequence of approximations, the convergence of which has been examined by DiPrima & Sani (1965) and Sani (1968).

The success of this method can be traced to the fact that equation (17.41) and all of the boundary conditions are satisfied exactly. The solution of equation (17.44), however, may be difficult (or at least tedious) if $f(x)$ is at all complicated and this is perhaps the major limitation of the method. Other methods have also been developed to deal with equations like (17.41) and (17.42) and some of these will be discussed in §§ 18 and 19. Also the development and spread of electronic computers has changed the criteria by which we judge a numerical method desirable, so that different methods might be applied if the Taylor problem were new today.

We now apply this method to the narrow-gap Taylor problem, for which $f(x) = 1 - (1 - \mu)(x + \frac{1}{2})$ and $g = 1$. A first approximation, obtained by retaining only the term $\cos \pi x$ in the expansion (17.43), leads to the result

$$\frac{1}{2}(1 + \mu)T = \frac{(\pi^2 + a^2)^3}{a^2[1 - 16\pi^2 a \cosh^2 \frac{1}{2}a/\{(\pi^2 + a^2)^2 (\sinh a + a)\}]}.$$

(17.47)

Table 3.1. *Taylor numbers for* $0 \leq \mu \leq 1$

		$\bar{T}_c$	
μ	a_c	'Exact'	Equation (17.50)
1.0	3.117	1707.76	1707.76
0.5	3.12	1706.5	1706.3
0.25	3.12	1703.3	1703.1
0.0	3.127	1694.95	1694.76

This expression for $\frac{1}{2}(1+\mu)T$ is identical with the result obtained in Problem 2.8 for the Bénard problem with two rigid boundaries. Thus in this approximation we have

$$T_c = 1715/\tfrac{1}{2}(1+\mu) \quad \text{and} \quad a_c = 3.12, \qquad (17.48)$$

which is formally equivalent to replacing $f(x)$ by its average value $\langle f(x) \rangle = \frac{1}{2}(1+\mu)$. This *averaging approximation* leads to a substantial simplification by reducing the narrow-gap Taylor problem to the Bénard problem with T replaced by $\frac{1}{2}(1+\mu)T = \bar{T}$ (say). The analogy between the two problems is exact in the limit as $\mu \to 1$, as first shown by Jeffreys (1928). More generally, however, there is now considerable evidence to suggest that the averaging approximation is a good one provided that the cylinders rotate in the same direction. This approximation further suggests that when $0 \leq \mu \leq 1$ the relevant Taylor number is not T but rather $\bar{T}$ and that $\bar{T}_c$ and a_c are nearly constant with the values

$$\bar{T}_c = 1708 \quad \text{and} \quad a_c = 3.12 \quad \text{for} \quad 0 \leq \mu \leq 1. \qquad (17.49)$$

A comparison of (17.49) with the results of more accurate calculations is given in Table 3.1 and shows that the error in this approximation is less than one per cent. It is also interesting that a result of this type was first derived by Taylor (1923) himself, who obtained the value $\bar{T}_c = \pi^4/0.0571 = 1706$; but it was not until five years later that the critical value of the Rayleigh number for the Bénard problem with two rigid boundaries was first obtained accurately by Jeffreys (1928).

Some further insight into the averaging approximation can be obtained by considering the limit as $\mu \to 1$. The Taylor problem can then be treated as a small perturbation of the Bénard problem. A systematic calculation of this type has been made by Chandrasekhar (1961), § 71(d), who showed that

$$\bar{T}_c \sim 1707.76\{1 - 7.61 \times 10^{-3}\left(\frac{1-\mu}{1+\mu}\right)^2 + \cdots\} \quad \text{as } \mu \to 1.$$

$$(17.50)$$

Thus the change in $\bar{T}_c$ is of the order of $(1-\mu)^2$ with a coefficient which is very small. Although this result was derived on the assumption that $1 - \mu$ is small, the predictions of equation (17.50) are in very good agreement with the 'exact' results over the entire range (see Table 3.1). This agreement, together with the smallness of the coefficient of $\{(1-\mu)/(1+\mu)\}^2$, seems to suggest that the result has an asymptotic character. Meksyn (1946a, 1961) tried to solve the problem by asymptotic methods, treating the wavenumber as a large parameter and thereby obtaining $\bar{T}_c = \pi^4/0.0569$, in close agreement with Taylor's result. More generally, however, what is required here is an algorithm by which we can approximate the solution of the narrow-gap Taylor problem asymptotically in terms of the solution of the Bénard problem with two rigid boundaries.

When the cylinders rotate in opposite directions, the variable coefficient on the right-hand side of equation (17.39) changes sign at the nodal surface $x_0 = (1-\mu)^{-1} - \frac{1}{2}$, which now lies in the interval $-\frac{1}{2} < x_0 < \frac{1}{2}$. As in the inviscid problem, the relevant length scale is no longer the gap width but rather the distance from the inner cylinder to the nodal surface. For $\mu < 0$, it is therefore convenient to renormalize the problem by letting (cf. equation (16.16))

$$\bar{a} = a/(1-\mu), \quad \tau = T/(1-\mu)^4 \quad \text{and} \quad \bar{\zeta} = (1-\mu)(x + \tfrac{1}{2}).$$

$$(17.51)$$

Equation (17.39) then becomes

$$(\mathrm{D}^2 - \bar{a}^2)^3 v = -\bar{a}^2 \tau (1 - \bar{\zeta}) v, \qquad (17.52)$$

where D now stands for $d/d\bar{\zeta}$, and the boundary conditions are

$$v = (\mathrm{D}^2 - \bar{a}^2)v = \mathrm{D}(\mathrm{D}^2 - \bar{a}^2)v = 0 \quad \text{at} \quad \bar{\zeta} = 0 \quad \text{and} \quad 1 - \mu.$$

$$(17.53)$$

Table 3.2. *Critical values of the wavenumber and the Taylor*
number for $\mu \leqslant 0$. (Harris & Reid 1964.)

μ	$\bar{a}_c$	τ_c	a_c	T_c
0	3.127	3389.9	3.127	3 389.9
−0.5	2.133	1266.9	3.199	6 413.7
−1.0	2.000	1166.4	3.999	18 663
−1.5	2.040	1180.4	5.099	46 111
−2.0	2.034	1178.6	6.101	95 470

The limiting values of $\bar{a}_c$ and τ_c as $\mu \to -\infty$

$\bar{a}_c$	τ_c	Reference	Method
2	1132	Meksyn (1946b, 1961)	WKBJ
2.03	1180	Chandrasekhar (1954b, 1961)	Fourier expansion
2.125	1075	DiPrima (1955)	Galerkin
2.002	1186.4	Duty & Reid (1964)	Comparison equation
2.034	1178.6	Harris & Reid (1964)	Numerical

The eigenvalue problem posed by equations (17.52) and (17.53) has been studied by various methods. For example, the Fourier expansion method described above has been used successfully by Chandrasekhar (1954b; 1961, § 71(b)) for values of μ in the range $-3 \leqslant \mu \leqslant 0$. When $-\mu$ is large, however, the eigenfunctions are very asymmetrical and a large number of terms must be used in the expansions to obtain accurate results. Fortunately, as shown in Figs. 3.8 and 3.9 the values of a_c and τ_c become nearly independent of μ as soon as $-\mu$ becomes greater than about 2. Some numerical results are also given in Table 3.2, including the limiting values of $\bar{a}_c$ and τ_c as $\mu \to -\infty$, which have been obtained by various methods. The radial eigenfunction for the first mode is shown in Fig. 3.10 and the cell pattern at the onset of instability is shown in Fig. 3.11. These results for the viscous problem should be compared with the corresponding inviscid results shown in Figs. 3.5 and 3.6.

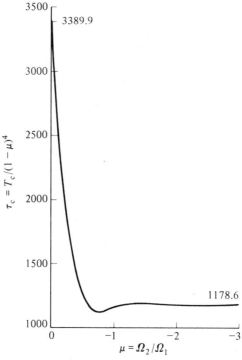

Fig. 3.8. The critical Taylor number as a function of μ. (From Harris & Reid 1964.)

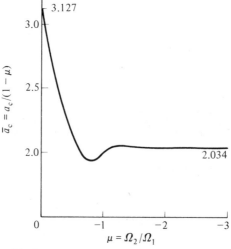

Fig. 3.9. The critical wavenumber as a function of μ. (From Harris & Reid 1964.)

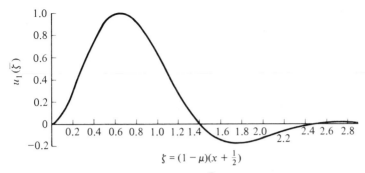

Fig. 3.10. The radial eigenfunction $u_1(\bar{\zeta})$ at the onset of instability for $m = 1$, $\mu = -\infty$, $\bar{a}_c = 2.034$ and $\tau = 1178.6$. (From Duty & Reid 1964.)

Although the numerical results for this problem are quite complete there are some important mathematical aspects of the problem which require further study. For example, if τ is treated as a large parameter, then the problem becomes one of obtaining asymptotic expansions to the solutions of equation (17.52) which are uniformly valid in the interval $0 \leqslant \bar{\zeta} \leqslant 1 - \mu$. The major difficulty in this approach arises from the fact that equation (17.52) has *two* nearly coincident turning points, at $\bar{\zeta} = 1$ and $1 - \bar{a}^4/\tau$, which coalesce as $\bar{a} \to 0$ and both lie in the interval, $0 \leqslant \bar{\zeta} \leqslant 1 - \mu$. The problem has not been solved completely, but some interesting partial results have been obtained by several different methods. Meksyn (1946b, c, 1961) used the WKBJ method and obtained the required connexion formula for the intervals $-\infty < \bar{\zeta} < 1 - \bar{a}^4/\tau$ and $1 < \bar{\zeta} < \infty$. The comparison equation method was later used by Duty & Reid (1964), who expressed the solutions of equation (17.52) in terms of the solutions of the comparison equation $d^6y/dx^6 = xy$, where now $x = (\bar{a}^2\tau)^{1/7}(\bar{\zeta} - 1)$. More recently Granoff & Bleistein (1972) and Granoff (1972) used the method of steepest descents to obtain uniformly valid asymptotic expansions of the solutions of equation (17.52), but they did not consider any boundary conditions.

In spite of the mathematical interest in this problem, it would now appear that it is not the relevant problem physically when $-\mu$ becomes sufficiently large. There is growing evidence, both theoretical and experimental (see § 17.3 below), that when $\mu < -0.78$ (in the narrow-gap approximation) instability sets in at a

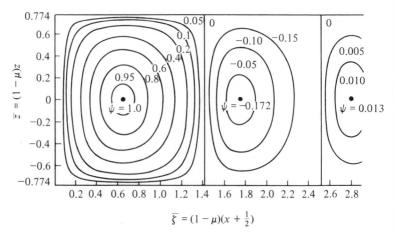

$$\bar{\zeta} = (1 - \mu)(x + \tfrac{1}{2})$$

Fig. 3.11. The cell pattern in the narrow-gap approximation for $\mu = -\infty$, $\bar{a}_c = 2.034$ and $\tau = 1178.6$: streamfunction $\psi = u_1(\bar{\zeta}) \cos(\bar{a}\bar{z})$. (From Duty & Reid 1964.)

smaller value of the Taylor number with respect to non-axisymmetric disturbances. Thus, for large values of $-\mu$, the axisymmetric problem is irrelevant physically† and one must consider the full three-dimensional problem.

Although the narrow-gap approximation is entirely satisfactory for the study of many aspects of the Taylor problem, experimental observations are necessarily made for fixed values of η and it is important therefore to determine the dependence of T_c not only on μ but also on η. On the assumption that $1 - \eta$ is small, Taylor (1923) obtained the simple formula,

$$\bar{T}_c \cong 1708\{1 + 0.652\frac{1 - \mu}{1 + \mu}(1 - \eta)\} \qquad (17.54)$$

and

$$a_c \cong 3.12 \quad \text{for} \quad 0 \leqslant \mu \leqslant 1 \quad \text{and} \quad 1 - \eta \ll 1.$$

A similar result was also obtained more recently by Witting (1958), who developed a systematic expansion procedure in powers of $1 - \eta$. When $1 - \eta$ is not small, however, analytical results are difficult to obtain. Lin (1955), § 2.2 (ii) has suggested that when the cylinders rotate in the same direction some form of the averaging

† For other reasons the two-dimensional problem is also irrelevant (see § 17.2).

approximation should be applicable. Walowit, Tsao & DiPrima (1964) argued that the averaging should be applied to the velocity rather than the angular velocity and they obtained the formula

$$T_c \cong 1708 \frac{1-\eta^2}{\mu - \eta^2 + 2(1-\mu)\eta^2 \ln(1/\eta)/(1-\eta^2)}. \quad (17.55)$$

Even a formal justification of this result has not yet been given but it is interesting to note that as $\eta \to 1$, equation (17.55) reduces to equation (17.54) with 0.652 replaced by $\frac{2}{3}$. More detailed numerical results for various values of η have been obtained by Chandrasekhar (1961), § 73, Kirchgässner (1961) and Walowit, Tsao & DiPrima (1964).

17.2 Two-dimensional disturbances

The governing equation for two-dimensional disturbances is of some interest because of its connexion, in the narrow-gap approximation, with the governing equation for the stability of plane parallel flow, namely the Orr–Sommerfeld equation (see Chapter 4). Thus we now suppose, without further loss of generality, that $u_z = 0$ and that u_r and u_θ depend only on r, θ and t. In the notation of § 15.3 it then follows from equations (17.10)–(17.13) that

$$\nu\left(D_* D - \frac{n^2}{r^2}\right)^2 \phi = (s + in\Omega)\left(D_* D - \frac{n^2}{r^2}\right)\phi - \frac{in}{r}(DZ)\phi.$$

$$(17.56)$$

Alternatively, this equation can be derived from the vorticity equation satisfied by the stream function in plane polar coordinates. This equation has not been extensively studied and we shall therefore consider only some limiting forms of it. For this purpose it is unnecessary to rewrite the equation in dimensionless form. In the inviscid limit the equation reduces immediately to equation (15.57) and in the narrow-gap approximation it reduces to (a dimensional form of) the Orr–Sommerfeld equation. For Couette flow, $DZ = rD^2\Omega + 3D\Omega = 0$ and the structure of equation (17.56) is then very similar to that of the equation which governs the stability of plane Couette flow. Since plane Couette flow is known to be stable it seems reasonable to conjecture that circular Couette flow is also stable with respect to two-dimensional disturbances.

17.3 *Three-dimensional disturbances*

Although Taylor (1923) assumed in all his theoretical work that the disturbances were axisymmetric, he did observe three-dimensional disturbances in some of his experiments. He noticed that the axisymmetric vortices became wavy and then broke up as the angular speeds of the cylinders were increased beyond their critical values for fixed values of μ less than about -1. It seems that this was in fact a secondary nonlinear instability. The investigation of this phenomenon was taken up by later experimentalists, and led DiPrima (1961) to treat three-dimensional disturbances mathematically. He posed the appropriate linear eigenvalue problem and solved it on the basis of the narrow-gap approximation and an averaging approximation for $\mu = 0$ and 0.5. He found that the critical value of the Taylor number did increase with the azimuthal wavenumber n, but that three-dimensional disturbances are only slightly less unstable than axisymmetric ones. This conclusion was extended by Roberts (1965), who took $\mu = 0$ but did not make the narrow-gap approximation. Krueger, Gross & DiPrima (1966) went on to consider the linear problem for cylinders rotating in opposite directions, and found that in the narrow-gap approximation, when μ is less than about -0.78, the most unstable disturbance is no longer axisymmetric. As μ decreases from this value the most unstable mode at first has azimuthal wavenumber $n = 1$ but then takes higher values in rapid succession. At the least value of μ considered, namely -1.25, it appears that the most unstable mode has $n = 5$. Krueger, Gross & DiPrima (1966) confirmed these results qualitatively for the particular case of a finite gap with $\eta = 0.95$ and $\mu = 0$ and -1.

17.4 *Some experimental results*

Taylor's (1923) paper was remarkably complete experimentally as well as theoretically. In his experiments he used a single outer cylinder, its internal radius being $R_2 = 4.035$ cm. It was made of glass so the flow of water, rendered visible by injected dye, between the cylinders could be seen. The glass was carefully turned, bored, polished and mounted so that R_2 was accurate to within 0.1 mm along its length. He used various inner cylinders in different runs of his experiments, the gap width d varying between 0.235 cm and 1.105 cm. To avoid end effects (which Taylor suspected were

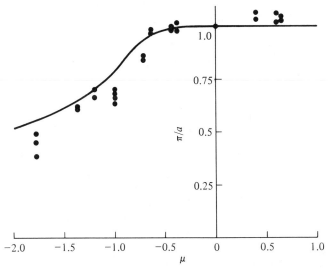

Fig. 3.12. Observations and narrow-gap calculations of the spacing of Taylor vortices at the onset of instability in water for $R_1 = 3.80$ cm and $R_2 = 4.035$ cm. (Experimental data are from Taylor (1923), Table 8.)

responsible for the discrepancy between theoretical predictions and Mallock's (1896) experimental results), the cylinders were as long as 90 cm. The ratio μ of the angular velocities of the inner and outer cylinders was fixed during each run by setting a variable-speed gear, and then the angular velocities were together gradually increased in magnitude until the onset of instability. To detect instability Taylor injected the dye into the water through six small holes near the middle of the inner cylinder. The instability could then be viewed through the glass wall of the outer cylinder. The angular velocity at the onset of instability seemed quite definite, the measurement being repeated on different runs with an accuracy of about one per cent.

Taylor's chief aims were to measure the wavelength of the axisymmetric mode of instability, now called *Taylor vortices* (see Fig. 3.7), and the angular velocity at the onset of instability, which would now be expressed in terms of a critical value of the Taylor number. The closeness of the agreement between his theoretical and experimental results was without precedent in the history of fluid mechanics.

In Fig. 3.12 we have compared Taylor's observations of the spacing of the vortices (i.e. their half wavelength) with the results of

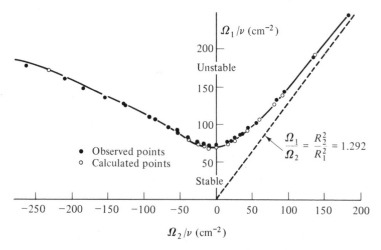

Fig. 3.13. Observations and narrow-gap calculations of the curve of marginal stability for water when $R_1 = 3.55$ cm and $R_2 = 4.035$ cm. (From Taylor 1923.)

later narrow-gap calculations. These results are for a pair of cylinders with $R_1 = 3.80$ cm and $R_2 = 4.035$ cm, so that $\eta = 0.9418$. He also obtained similar results for an inner cylinder with $R_1 = 2.93$ cm. (The discrepancies between observations and calculation were on the whole less than five per cent and often much less.)

Taylor also compared the observed and theoretical critical values of the angular velocities of the cylinders at the onset of instability for various values of μ and three values of η. To illustrate this for a given pair of cylinders, in our Fig. 3.13 we reproduce Taylor's figure for $R_1 = 3.55$ cm and $R_2 = 4.035$ cm. Thus the value of $\eta = 0.8798$ is fixed. He then plotted experimental points and the theoretical marginal curve in the plane of Ω_1/ν and Ω_2/ν. This way of plotting results is better suited to the design of Taylor's experiments than to his theory, but leaves no doubt that 'the accuracy with which these points fall on the curves appears remarkable when it is remembered how complicated was the analysis employed in obtaining them'. It may be noted that the region of instability lies well within the Rayleigh line $\mu = \eta^2$.

There have been many good later experiments, including some by Taylor (1936) himself. The chief development of technique after

Taylor's first work was the measurement of the torque between the cylinders. This seems to have been first done by Wendt (1933). One may suspend a section of the outer cylinder and compare the measurement of the torque G on it necessary to maintain its angular velocity with the theoretical prediction for laminar Couette flow,

$$G = \frac{4\pi\rho\nu H R_1^2 R_2^2 \Omega_1}{R_2^2 - R_1^2}, \tag{17.57}$$

where H is the length of the suspended section. This allows a sensitive measurement of the onset of instability, but is not practical when Ω_2 has to be varied in an experiment. When $\Omega_2 = 0$, the torque can be expressed in the dimensionless form

$$\frac{G}{\rho H \Omega_1^2 R_1^4} = f(Re, \eta), \tag{17.58}$$

where $Re = \Omega_1 R_1 (R_2 - R_1)/\nu$. Experimental measurements of the torque, such as those shown in Fig. 3.14, are in good agreement with the theoretical values of T_c (or Re_c).

There have also been many refinements of technique and extension of results for wider ranges of μ and η. Of the many experiments, we cite those of Donnelly (1958) and Donnelly & Fultz (1960), because they are both reliable and closely linked with theoretical results of Chandrasekhar. Snyder (1968b) has given more recent results, over wider ranges of μ and η. Snyder (1968a) also found three-dimensional spiral modes at the onset of instability when μ was negative and sufficiently large in magnitude. These experimental results agree with the theoretical results of Krueger, Gross & DiPrima (1966), but go further, Snyder having used an apparatus with $\eta = 0.2, 0.5, 0.8$ and 0.959.

Most recent experiments, however, have been made to investigate nonlinear phenomena. Donnelly & Simon (1960) analysed measurements of the mean torque over a wide range of supercritical as well as subcritical values of the Taylor number. Coles (1965) made a careful study of how the Taylor vortices themselves become unstable, tracing an intricate succession of steady cellular motions as the Taylor number increases above its critical value. Finally turbulence develops, when the Taylor number may be more than two orders of magnitude greater than its critical value. Like Taylor,

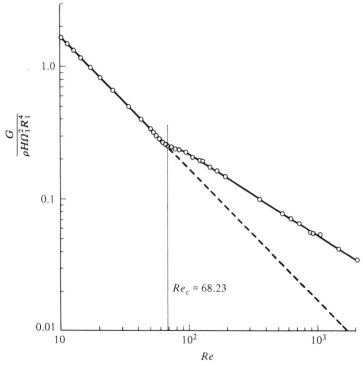

Fig. 3.14. Variation of the torque G with the Reynolds number $Re = \Omega_1 R_1 (R_2 - R_1)/\nu$ for $\Omega_2 = 0$, $R_1 = 1.0$ cm, $R_2 = 2.0$ cm, $H = 5.0$ cm, $\nu = 0.1226$ cm^2 s^{-1}, $\rho = 0.8404$ g cm^{-3}. G is measured in dyn cm ($= 10^{-7}$ N m). (Experimental data are from Donnelly & Simon (1960), Table 2.)

Coles (1965) found three-dimensional motion when the Taylor vortices broke up. Snyder & Lambert (1966) have examined the higher harmonics of the supercritical instabilities. Burkhalter & Koschmieder (1973) also conducted experiments at large values of the Taylor number, examining end effects and initial effects. Some of these phenomena can be seen in the film loops by Coles (FL 1963a,b).

18 The Dean problem

18.1 *The Dean problem*

In the last section we considered the instability of the flow of viscous fluid between rotating cylinders. A similar type of instability can also occur when a viscous fluid flows in a curved channel owing to a pressure gradient acting round the channel, as shown in Fig. 3.1(b). This problem was first studied by Dean (1928) for a channel formed by two coaxial cylinders in the narrow-gap approximation and it has been considered again by Hämmerlin (1958) and Reid (1958). The finite-gap problem also has been studied by Walowit, Tsao & DiPrima (1964).

If we assume that the basic flow for this problem is of the form

$$u_r = u_z = 0 \quad \text{and} \quad u_\theta = V(r), \tag{18.1}$$

then equations (17.1) and (17.2) give

$$\frac{1}{\rho}\frac{\partial p}{\partial r} = \frac{V^2}{r} \quad \text{and} \quad \nu \mathrm{DD}_* V = \frac{1}{\rho r}\left(\frac{\partial p}{\partial \theta}\right)_0, \tag{18.2}$$

where $(\partial p/\partial \theta)_0$ is the constant pressure gradient acting in the azimuthal direction. Thus V must be of the general form

$$V(r) = \frac{1}{2\rho\nu}\left(\frac{\partial p}{\partial \theta}\right)_0\left(r \ln r + Cr + \frac{E}{r}\right), \tag{18.3}$$

where the constants of integration C and E are determined by the boundary conditions. If we suppose that the cylinders are at rest, then V must vanish at $r = R_1$ and R_2, and we obtain

$$C = -\frac{R_2^2 \ln R_2 - R_1^2 \ln R_1}{R_2^2 - R_1^2} \quad \text{and} \quad E = \frac{R_1^2 R_2^2}{R_2^2 - R_1^2}\ln\frac{R_2}{R_1}. \tag{18.4}$$

In the narrow-gap approximation, the velocity distribution given by equation (18.3) reduces to the familiar parabolic form for plane Poiseuille flow. Thus, with x defined by equation (17.20), we have

$$V(r) \cong \tfrac{3}{2}V_\mathrm{m}(1 - 4x^2), \tag{18.5}$$

where

$$V_\mathrm{m} = -\frac{d^2}{12\rho\nu R_1}\left(\frac{\partial p}{\partial \theta}\right)_0 \tag{18.6}$$

is the mean velocity across the channel. In this approximation, therefore, centrifugal effects are neglected in the description of the basic flow but they are partially retained in the disturbance equations. On applying Rayleigh's criterion to the velocity distribution given by equation (18.5), we see that the flow is locally stable at each point in the interval $-\frac{1}{2} < x < 0$ but unstable in the interval $0 < x < \frac{1}{2}$. The linearized equations (17.18) and (17.19), for axisymmetric disturbances, are applicable to this problem provided V is of the form given by equation (18.3). In the narrow-gap approximation, with u replaced by $(3V_m d^2/R_1 \nu)a^2 u$, the standard form of the governing equations for axisymmetric disturbances becomes

$$(D^2 - a^2 - \sigma)(D^2 - a^2)u = (1 - 4x^2)v, \qquad (18.7)$$

and

$$(D^2 - a^2 - \sigma)v = -a^2 \Lambda x u, \qquad (18.8)$$

where†

$$\Lambda = 36R^2 d/R_1 \quad \text{and} \quad R = V_m d/\nu. \qquad (18.9)$$

These equations must be considered together with the boundary conditions

$$u = Du = v = 0 \quad \text{at } x = \pm \tfrac{1}{2}. \qquad (18.10)$$

The parameter Λ thus plays the role of a Taylor number in the present problem; for historical reasons, however, numerical results are usually given in terms of the equivalent parameter $R(d/R_1)^{1/2}$, sometimes called the *Dean number*.

Although it has not been proved that the principle of exchange of stabilities governs this problem, the available experimental evidence (Brewster, Grosberg & Nissan 1959) suggests that the onset of instability is marked by the appearance of a steady secondary motion in the form of toroidal vortices, as in the Taylor problem. We shall assume, therefore, that $\sigma = 0$. Then equations (18.7) and (18.8) become

$$(D^2 - a^2)^2 u = (1 - 4x^2)v, \qquad (18.11)$$

and

$$(D^2 - a^2)v = -a^2 \Lambda x u. \qquad (18.12)$$

A crude approximation to the solution of this problem can be obtained by replacing the factor $1 - 4x^2$ which appears in equation (18.11) by its average value (namely $\tfrac{2}{3}$). The resulting equations then have the same form as the adjoint Taylor problem for $\mu = -1$, and from Table 3.2 we obtain $a_c \cong 4.00$, $\Lambda_c \cong 3T_c \cong 56\,000$ and $R(d/R_1)^{1/2} \cong 39.4$. More accurate results have also been obtained

---

† This definition of Λ differs by a factor of two from the one adopted by Chandrasekhar (1961).

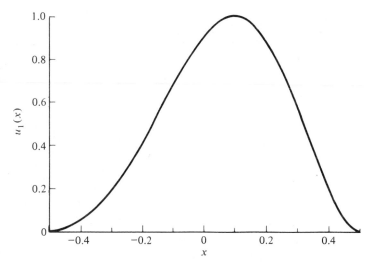

Fig. 3.15. The radial eigenfunction $u_1(x)$ at the onset of instability for $m = 1$, $a_c = 3.950$ and $\Lambda_c = 46\,456$. (From Hughes & Reid 1964.)

by quite different methods by Hämmerlin (1958) and Reid (1958). The method used by Hämmerlin is similar to the one first used by Görtler in his study of the stability of laminar boundary layers on concave walls. In this method the governing equations are first transformed into an equivalent pair of coupled integral equations, and an approximate solution is then obtained by iteration. One of the methods used by Reid (1958) was the Fourier-expansion method which was described in § 17.1. In applying that method to equations (18.11) and (18.12), however, it should be noted that at least one even and one odd function must be included in the expansion (17.43).

The curve of marginal stability for this problem has the same general shape as the corresponding curves for the Bénard and Taylor problems. Accurate values of a_c and Λ_c have been computed by Gibson & Cook (1974), using a Chebyshev collocation method, and they found that

$$a_c = 3.95, \quad \Lambda_c = 46\,458 \quad \text{and} \quad R(d/R_1)^{1/2} = 35.92.$$
$$(18.13)$$

The radial component of the velocity perturbation and the cell pattern at the onset of instability are shown in Figs. 3.15 and 3.16.

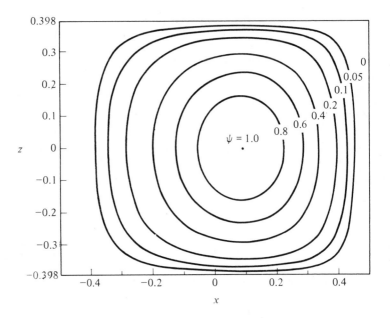

Fig. 3.16. The cell pattern in the narrow-gap approximation for $a_c = 3.98$ and $\Lambda_c = 47\,494$: streamfunction $\psi = u_1(x) \cos(az)$. (From Reid 1958.)

It is also of some interest to consider the stability of viscous flow in a curved channel with respect to two-dimensional disturbances. In the narrow-gap approximation it is easy to see that the problem is governed by the equation for stability of parallel flow, namely the Orr–Sommerfeld equation, and it thus becomes identical with the stability problem for plane Poiseuille flow. From the results, given in Chapter 4, for that problem, it follows that instability with respect to two-dimensional disturbances may occur when

$$(\tfrac{3}{2}V_m)(\tfrac{1}{2}d)/\nu > 5772, \quad \text{i.e. when } R > R_c = 7696. \qquad (18.14)$$

On comparing this result with equation (18.13) we see that this two-dimensional instability would arise only if $R_1/d > 4.59 \times 10^4$, and this requires a very straight channel indeed! This aspect of the problem has been considered in greater detail by Gibson & Cook (1974), who also included the effects of a finite gap.

18.2 The Taylor–Dean problem

When rotation and an azimuthal pressure gradient are both present, the problem has some distinctive features which are absent from either limiting case. This problem was first studied experimentally by Brewster & Nissan (1958) and by Brewster, Grosberg & Nissan (1959), and it has since been studied theoretically by DiPrima (1959), Meister (1962), Hughes & Reid (1964), and Raney & Chang (1971). In the narrow-gap approximation the velocity distribution of the basic flow can be written as the sum of two terms: a linear term due to the rotation and a quadratic term due to the azimuthal pressure gradient. As a measure of the relative importance of these two terms, it is convenient to introduce the parameter

$$\lambda = 6V_m/\Omega_1 R_1. \tag{18.15}$$

The basic velocity distribution is then given (cf. equations (16.2) and (18.5)) by

$$V(r) \cong \Omega_1 R_1\{1 - (1-\mu)(x + \tfrac{1}{2}) + \tfrac{1}{4}\lambda(1 - 4x^2)\}, \tag{18.16}$$

the average value of which is

$$\langle V(r) \rangle = \Omega_1 R_1\{\tfrac{1}{2}(1+\mu) + \tfrac{1}{6}\lambda\}. \tag{18.17}$$

There is then no mean flow when $\lambda = -3(1+\mu)$.

The application of Rayleigh's circulation criterion to the basic velocity distribution given by equation (18.16) shows that the flow is locally unstable where

$$\{1 - (1-\mu)(x + \tfrac{1}{2}) + \tfrac{1}{4}\lambda(1 - 4x^2)\}(1 - \mu + 2\lambda x) > 0. \tag{18.18}$$

The zeros of this expression, therefore, define the boundaries between locally stable and unstable layers of fluid. If the first factor vanishes we have $\lambda = -\{1 - (1-\mu)(x + \tfrac{1}{2})\}/(\tfrac{1}{4} - x^2)$ and if the second factor vanishes we have $\lambda = -(1-\mu)/2x$. These results are illustrated in Fig. 3.17 for $\mu = 0$ and -1.

In the narrow-gap approximation, with u replaced by $(2\Omega_1 d^2/\nu)a^2 u$, the governing equations for axisymmetric disturbances are

$$(D^2 - a^2 - \sigma)(D^2 - a^2)u = \{1 - (1-\mu)(x + \tfrac{1}{2}) + \tfrac{1}{4}\lambda(1 - 4x^2)\}v \tag{18.19}$$

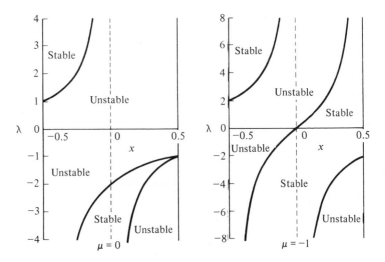

Fig. 3.17. The Rayleigh diagrams for $\mu = 0$ and -1. (From Hughes & Reid 1964.)

and

$$(D^2 - a^2 - \sigma)v = -a^2 T \left(1 + \frac{2\lambda}{1-\mu} x \right) u, \qquad (18.20)$$

which must be considered together with the boundary conditions (18.10). The parameters a, σ and T have the same meanings as in equations (17.24) and (17.30). When $\lambda = 0$ we recover the Taylor problem and in the limit as $\lambda \to \pm\infty$ (after a suitable renormalization) we recover the Dean problem. In this latter limit we may also note that

$$\lim_{\lambda \to \pm\infty} \tfrac{1}{2} T \frac{\lambda^2}{1-\mu} = \Lambda. \qquad (18.21)$$

Most of the calculations made for this problem have been for the case when $\mu = 0$, and it has usually been assumed that the principle of exchange of stabilities holds. It is then found that for values of λ near $\lambda_* = -3.667$ the curve of marginal stability for stationary modes has two branches as shown in Fig. 3.18. For $\lambda < \lambda_*$ the critical value of the Taylor number is determined by the left-hand branch and for $\lambda > \lambda_*$ by the right-hand branch, and this leads to the results for a_c and T_c shown in Fig. 3.19.

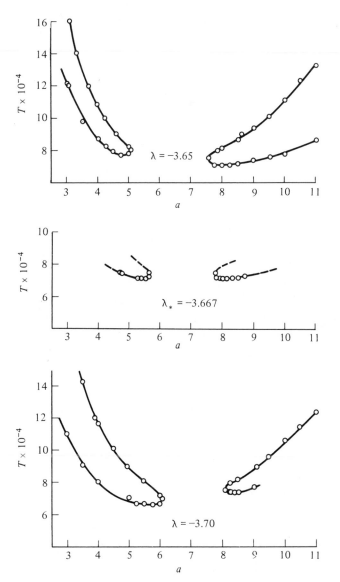

Fig. 3.18. The curves of marginal stability for $\mu = 0$ and λ in the neighbourhood of λ_*. (From Hughes & Reid 1964.)

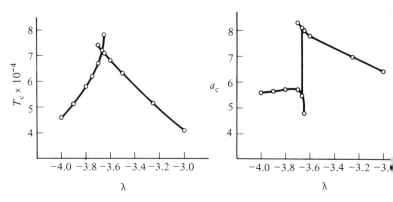

Fig. 3.19. The behaviour of the critical Taylor number T_c and the critical wavenumber a_c as a function of λ for $\mu = 0$. (From Hughes & Reid 1964.)

A more detailed analysis of the problem by Raney & Chang (1971), however, shows that there also exist oscillatory marginal modes lying between the two branches for the stationary modes shown in Fig. 3.18. They have further shown that when $-3.850 < \lambda < -3.635$ the most critical mode is a stationary three-dimensional one. Thus, although the resulting reduction in the values of T_c is not great, their results do emphasize the need for caution when only axisymmetric modes are considered or when the principle of exchange of stabilities is assumed to hold.

19 The Görtler problem

A third example of centrifugal instability was discovered by Görtler (1940), who showed that this type of instability can also occur in boundary layers along a concave wall, as illustrated in Figs. 3.1(c) and 3.20. This problem has also been considered by Meksyn (1950), Hämmerlin (1955) and Witting (1958).

The theory is based on three approximations. Firstly, we assume that the boundary-layer thickness δ is much smaller than the radius of curvature of the wall R_0, this being equivalent to the narrow-gap approximation. Secondly, we assume that the basic flow is nearly parallel to the wall; this means that centrifugal effects are neglected in the description of the basic flow, although they are partially retained in the disturbance equations. Thirdly, we consider a local

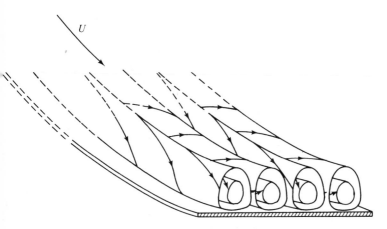

Fig. 3.20. The form of the secondary flow which occurs at the onset of instability in boundary layers along a concave wall. (From Görtler 1940.)

stability analysis for which the basic flow is assumed to be independent of x and the y-component of the basic flow is neglected.

To write the governing equations in dimensionless form we identify the length scale L and velocity scale V with δ and U_∞, respectively. The basic velocity then has the form $\{U(\eta), 0, 0\}$, where $\eta = y/\delta$ and $U(\eta) \to 1$ as $\eta \to \infty$. A simple calculation shows that the Rayleigh discriminant given by equation (14.1) is proportional to $-U \, dU/d\eta$ and instability is to be expected therefore locally, near the wall, in a thin layer, the thickness of which is of order δ. Consider now the perturbed flow with velocity $(U + u', v', w')$ and suppose that it can be resolved into normal modes of the form

$$(u', v', w', p') = (u, v, w, p)\, e^{st + ikz}. \tag{19.1}$$

On substituting equation (19.1) into the equations of motion and then linearizing, we obtain a set of four coupled equations for u, v, w and p. On eliminating w and p from these four equations, we obtain, to first order in δ/R_0, the governing equations in the standard form

$$(D^2 - a^2)(D^2 - a^2 - \sigma)v = -a^2 \mu U u \tag{19.2}$$

and

$$(D^2 - a^2 - \sigma)u = U'v, \tag{19.3}$$

where

$$D = d/d\eta, \quad a = k\delta, \quad \sigma = s\delta^2/\nu,$$

$$\mu = 2(U_\infty \delta/\nu)^2(\delta/R_0) \quad \text{and} \quad U' = DU. \tag{19.4}$$

We also have the boundary conditions

$$u = v = Dv = 0 \quad \text{at} \quad \eta = 0 \quad \text{and as} \quad \eta \to \infty. \tag{19.5}$$

The parameter μ is essentially the narrow-gap Taylor number appropriate to the present problem. It is common practice, however, to express the results in terms of the *Görtler number*

$$G = (U_\infty \vartheta/\nu)(\vartheta/R_0)^{1/2}, \tag{19.6}$$

where ϑ is the momentum thickness of the boundary layer. The two parameters are related by $G^2 = \frac{1}{2}(\vartheta/\delta)^3\mu$. Alternatively the displacement thickness δ_1 is sometimes used in the definition of the Görtler number.

It is usually assumed that the principle of the exchange of stabilities holds for this problem. Although this has not been proved, the existing experimental evidence (see, for example, Gregory & Walker (1950) and Bippes & Görtler (1972)) lends some support to it. Thus we set $\sigma = 0$, and find

$$(D^2 - a^2)^2 v = -a^2\mu Uu \tag{19.7}$$

and

$$(D^2 - a^2)u = U'v. \tag{19.8}$$

These equations are very similar to those which govern the Taylor and Dean problems. The present problem, however, is defined on a semi-infinite interval, and this leads to some important differences, particularly in the behaviour of the curve of marginal stability for small values of a.

Görtler (1940) transformed equations (19.7) and (19.8) into an equivalent pair of coupled integral equations, from which an approximate solution could be obtained by iteration. Also Meksyn (1950) treated the equations by the same asymptotic methods he had previously developed for the Taylor problem. Both Görtler and Meksyn obtained results which suggested that the curve of marginal stability for this problem was qualitatively similar to the marginal curves for the Bénard and Taylor problems, but their numerical

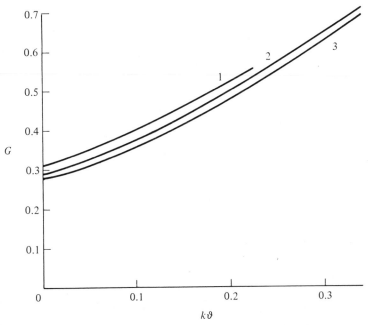

Fig. 3.21. The critical values of the Görtler number for (1) the Blasius profile, (2) the asymptotic suction profile and (3) the straight-line profile. (After Hämmerlin 1955.)

results disagreed substantially. To clarify this situation, Hämmerlin (1955) studied the problem in greater detail, and showed that $\mu(a)$ is a monotone increasing function of a. Some of his results are shown in Fig. 3.21 for the three representative velocity profiles:

(a) straight-line profile

$$U(\eta) = \begin{cases} \eta & \text{for } 0 \leq \eta \leq 1 \\ 1 & \text{for } 1 \leq \eta < \infty; \end{cases} \tag{19.9}$$

(b) Blasius profile;

(c) asymptotic suction profile

$$U(\eta) = 1 - e^{-\eta}. \tag{19.10}$$

Later Witting (1958) showed that the limiting value of μ as $a \downarrow 0$ is given by the simple formula

$$\mu(0) = 8\left\{ \int_0^\infty \eta^3 U'(\eta) \, d\eta \right\}^{-1}. \tag{19.11}$$

Table 3.3. *Stability characteristics of three representative velocity profiles*

	ϑ/δ	$\mu(0)$	$G(0)$
Straight-line profile	$\frac{1}{6}$	32	0.2772
Blasius profile	0.4696	1.8689	0.3111
Asymptotic suction profile	$\frac{1}{2}$	$\frac{4}{3}$	0.2887

The values of $\mu(0)$ and the corresponding values of $G(0)$ are given in Table 3.3 for the three profiles. In the case of the Blasius profile, a simple estimate for $\mu(0)$ can be obtained by using Weyl's second approximation to $U'(\eta)$ in equation (19.11) (see, for example, Lagerstrom (1964), p. 123). This gives $\mu(0) \cong 6\sqrt{2}\{\Gamma(\frac{1}{3})\}^{-3/2} = 1.9352$ and hence $G(0) \cong 0.3166$. The values given in Table 3.3 were obtained by Hämmerlin, but not from equation (19.11). From Fig. 3.21 and Table 3.3 it is evident that when the results are expressed in terms of G they are not sensitive to the detailed form of the basic velocity profile.

Since the minimum value of $G(a)$ occurs at $a = 0$, this corresponds to an infinite wavelength and is rather unsatisfactory physically. Some attempts, however, have been made to remedy this situation. Witting (1958), for example, has included additional curvature terms in the disturbance equations and thereby introduced δ/R_0 as an additional parameter. He then found that the curve of marginal stability has a single minimum and that the values of a and G associated with this minimum depend directly on δ/R_0. Smith (1955) has also derived a much more complicated set of disturbance equations which include not only additional curvature terms but also certain normal-flow terms associated with the streamwise growth of the boundary layers. Kobayashi (1972) has reconsidered the stability of the asymptotic suction profile including the effect of the suction velocity (see Problem 3.9 for the relevant disturbance equations). As might have been expected, the suction has a stabilizing effect for small wavenumbers and it also leads to a curve of marginal stability with a single minimum at which $G_c = 1.17$ and $k\vartheta = 0.22$.

Problems for chapter 3

3.1. *The oscillations of a rotating column of liquid.* Consider a cylindrical column of liquid of radius R_0 rotating about its axis with a constant angular velocity Ω_0. From equations (15.42) and (15.43), show that the frequencies of oscillation are given by

$$\frac{s}{\Omega_0} = \pm \frac{2i}{(1+\alpha^2/a^2)^{1/2}} - in,$$

where $a = kR_0$ and α is any root of the equation

$$\alpha J_n'(\alpha) \pm n(1+\alpha^2/a^2)^{1/2} J_n(\alpha) = 0.$$

When $n = 0$ this simplifies to $s = \pm 2i\Omega_0/(1+j_{1,m}^2/a^2)$, where $j_{1,m}$ is the mth positive zero of $J_1(\alpha)$. [Chandrasekhar (1961), § 68; Fultz (FL 1964).]

3.2. *The analogue of Rayleigh's inflexion-point theorem for two-dimensional disturbances of rotating and curved flows.* Consider two-dimensional disturbances bounded by rigid cylinders at $r = R_1$ and R_2. By multiplying equation (15.57) by $r\phi^*/(s+in\Omega)$ and integrating from $r = R_1$ to R_2, where ϕ^* is the complex conjugate of ϕ, deduce that

$$\int_{R_1}^{R_2} r\left\{ |D\phi|^2 + \frac{n^2}{r^2}|\phi|^2 + \frac{in(DZ)|\phi|^2}{s+in\Omega} \right\} dr = [r\phi^*D\phi]_{R_1}^{R_2}.$$

Hence show that if there is instability (i.e. Re $s > 0$) then

$$n \int_{R_1}^{R_2} \frac{r(DZ)|\phi|^2}{|s+in\Omega|^2} dr = 0,$$

and that Rayleigh's theorem (§ 15.3) follows. [Rayleigh (1880), p. 70.]

3.3. *The 'jump' conditions at a discontinuity of Ω or $D\Omega$.* Consider the inviscid problem for two-dimensional disturbances and suppose that Ω or $D\Omega$ is discontinuous at R_0 (say). By requiring continuity of the normal velocity and the pressure at the deformed interface, show that ϕ must satisfy the 'jump' conditions

$$\Delta[\phi/(s+in\Omega)] = 0$$

and

$$\Delta[r(s+in\Omega)D\phi - inZ\phi - n^2\Omega^2\phi/(s+in\Omega)] = 0,$$

where

$$\Delta f = f(R_0+0) - f(R_0-0) \quad \text{and} \quad Z = rD\Omega + 2\Omega.$$

3.4. *The oscillations of a columnar vortex.* Consider the inviscid problem for the basic flow

$$\Omega = \begin{cases} \Omega_0 & \text{for } 0 < r < R_0 \\ \Omega_0(R_0/r)^2 & \text{for } R_0 < r < \infty, \end{cases}$$

which represents a flow with uniform vorticity $Z = 2\Omega_0$ for $r < R_0$ and an irrotational vortex for $r > R_0$. Show that this flow is stable with respect to both axisymmetric and two-dimensional disturbances, and that the frequencies of the latter are given by Im $s = -\Omega_0(n-1)$. By considering the motion of the perturbed boundary of the vortex, which may be taken to have the equation $r = R_0\{1 + \varepsilon \exp(st + in\theta)\}$ for small ε, show that such a disturbance represents a system of waves travelling round the boundary of the vortex with an angular velocity $\Omega_0(1 - n^{-1})$. [Kelvin (1880a).]

3.5. *The instability of a cylindrical vortex sheet.* Consider the inviscid problem for the basic flow

$$\Omega = \begin{cases} 0 & \text{for } 0 \leqslant r < R_0 \\ \Omega_0(R_0/r)^2 & \text{for } R_0 < r < \infty \end{cases}$$

which represents a cylindrical vortex sheet of radius R_0. Show that this flow is stable with respect to axisymmetric disturbances but unstable to two-dimensional disturbances. For the latter, show that

$$\frac{s}{\Omega_0} = \pm\tfrac{1}{2}(n^2 - 2n)^{1/2} - \tfrac{1}{2}in \quad (n = 1, 2, \ldots).$$

[Rotunno (1978).]

*3.6. *The instability of a cylindrical shear layer.* Consider the inviscid problem for the basic flow

$$\Omega = \begin{cases} 0 & \text{for } 0 \leqslant r \leqslant R_1 \\ A + B/r^2 & \text{for } R_1 \leqslant r \leqslant R_2 \\ \Omega_0(R_2/r)^2 & \text{for } R_2 \leqslant r < \infty, \end{cases}$$

where A and B are determined so that Ω is continuous, i.e.

$$A = \Omega_0 R_2^2/(R_2^2 - R_1^2) \quad \text{and} \quad B = -\Omega_0 R_1^2 R_2^2/(R_2^2 - R_1^2).$$

For axisymmetric disturbances, use Rayleigh's circulation criterion to show that the flow is stable. For two-dimensional disturbances, show that the eigenvalue relation is

$$(s + \tfrac{1}{2}in\Omega_0)^2 = -\left\{\tfrac{1}{4}(n-2)^2 + \frac{\eta^2}{1-\eta^2}\left[\frac{1-\eta^{2n-2}}{1-\eta^2} - (n-1)\right]\right\}\Omega_0^2,$$

where $\eta = R_1/R_2$. Show that the modes with $n = 1$ and 2 are stable for all values of η. Otherwise, show that the normal modes are stable if $0 \leqslant \eta \leqslant \tfrac{1}{2}$ but unstable if $\tfrac{1}{2} < \eta < 1$. As $\eta \uparrow 1$, show that $s = \{\pm\tfrac{1}{2}(n^2 - 2n)^{1/2} - \tfrac{1}{2}in\}\Omega_0$, which is simply the eigenvalue relation for a cylindrical vortex sheet. [Michalke & Timme (1967), Busse (1968).]

*3.7. *The effect of a radial flow on the stability of Couette flow.* Assume that the cylinders are porous and that the basic velocity has a radial component $U(r)$. Show that $U = M/r$, which represents the flow from a line source of strength $2\pi M$ per unit length. If the fluid is inviscid and $M \neq 0$, show that $V(r)$ is no longer an arbitrary function but must be of the form $V = K/r$, which represents the flow due to a line vortex of circulation $2\pi K$. If the fluid is viscous, however, show that $V = Ar^{1+\lambda} + B/r$, where $\lambda \equiv M/\nu$ ($\neq -2$) is a radial Reynolds number, $A = -\Omega_1 R_2^{-\lambda} (\eta^2 - \mu)/(1 - \eta^{2+\lambda})$, and $B = \Omega_1 R_1^2 (1 - \mu \eta^\lambda)/(1 - \eta^{2+\lambda})$. Suppose now that the radial Reynolds number is large, i.e. $\lambda \gg 1$; then $V \cong \Omega_1 R_1^2/r$ except in a thin layer adjacent to the outer cylinder, in which $V \cong \Omega_1 R_2 \eta^2 \{1 - (1 - \mu/\eta^2) e^{-\xi}\}$, where $\xi = \lambda(1 - r/R_2)$. A first-order composite approximation† to V, formed by additive composition, is therefore $V \cong \Omega_1 R_1 \{(R_1/r) - \eta(1 - \mu/\eta^2) e^{-\xi}\}$ for $R_1 \leqslant r \leqslant R_2$ and $\lambda \gg 1$. Hence show that the Rayleigh discriminant is sensibly different from zero only in the thin layer adjacent to the outer cylinder in which $1 - r/R_2 = O(\lambda^{-1})$ as $\lambda \to \infty$, and that the flow in this layer is stable if $\mu > \eta^2$ but unstable if $\mu < \eta^2$. Consider also the case when $-\lambda \gg 1$. [Chang & Sartory (1967), Bahl (1970).]

3.8. *Inviscid modes of instability for boundary-layer flows along a concave wall.* In the inviscid limit, equations (19.2) and (19.3) reduce to

$$(D^2 - a^2)v = -a^2 \lambda UU'v,$$

where $\lambda = 2(U_\infty/s\delta)^2(\delta/R_0)$, and the boundary conditions are that $v = 0$ at $\eta = 0$ and v is bounded as $\eta \to \infty$. Derive the eigenvalue relation for the straight-line profile given by equation (19.9). [Take v and Dv continuous at $\eta = 1$. D'Arcy (1951), Reid (1961).]

*3.9. *The Görtler problem with suction.* Consider the stability of the asymptotic suction boundary-layer profile with (dimensionless) velocity components $U(\eta) = 1 - e^{-\eta}$ and $V = -(U_\infty \delta/\nu)^{-1}$. Show that the disturbance equations (with $\sigma = 0$) are then

$$(D^2 - a^2)(D^2 + D - a^2)v = -a^2 \mu Uu,$$

and

$$(D^2 + D - a^2)u = U'v.$$

Discuss the relationship between this problem and the Taylor problem with a radial flow and $\mu = 0$. [Chang & Sartory (1967), Kobayashi (1972).]

---

† For a discussion of composite approximations see, for example, the book by Van Dyke (1975). These results can also be derived by singular perturbation methods directly from the differential equation satisfied by $V(r)$.

CHAPTER 4

PARALLEL SHEAR FLOWS

The transition of laminar flow, with its clean layers of flow tubes, to strongly mixed, irregular turbulent flow is one of the principal problems of modern hydrodynamics. It is certain that this fundamental change in type of motion of the fluid is traceable to an instability in the laminar flow, for laminar flows of themselves would always be possible solutions of the hydrodynamic equations.   – W. Tollmien (1935)

20 Introduction

In this chapter we wish to consider the stability of steady two-dimensional or axisymmetric flows with parallel streamlines. Flows of this type were first studied experimentally by Reynolds (1883), who observed that instability could occur in quite different ways depending on the form of the basic velocity distribution. Thus, when the velocity profile is of the form shown in Fig. 4.1(a) he observed that 'eddies showed themselves reluctantly and irregularly' whereas when the profile is as shown in Fig. 4.1(b) the 'eddies appeared in the middle regularly and readily'. From these observations he was led to consider the role of viscosity in flows of this type. By comparing the flow of a viscous fluid with that of an inviscid fluid, both flows being assumed to have the same basic velocity distribution, he was led to formulate two fundamental hypotheses which can be stated as follows:

> *First Hypothesis.* The inviscid fluid may be unstable and the viscous fluid stable. The effect of viscosity is then purely stabilizing.
> *Second Hypothesis.* The inviscid fluid may be stable and the viscous fluid unstable. In this case viscosity would be the cause of the instability.

Although Reynolds was unable to suggest a physical mechanism by which viscosity could cause instability he refused to exclude such a possibility. In this chapter, therefore, we will examine the circumstances under which these hypotheses are valid and give, so far as possible, a physical description of the mechanism of instability.

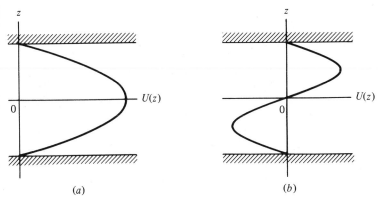

Fig. 4.1. The velocity profiles considered by Reynolds (1883).

The analytical study of inviscid flows of this type had been initiated somewhat earlier by Helmholtz (1868), Kelvin (1871) and Rayleigh (1880), who considered the purely inertial instability of an incompressible fluid of constant density. Perhaps the most important result of this period, of which Reynolds was apparently unaware, was Rayleigh's famous theorem on the role of inflexion points in the velocity profile. According to this theorem an inviscid flow of the form shown in Fig. 4.1(a) would be stable but the one shown in Fig. 4.1(b) would be unstable. This part of the theory would now appear to be reasonably complete, both physically and mathematically, and provides quite general criteria under which the First Hypothesis is valid. A general review of the inviscid theory has been given by Drazin & Howard (1966) who also discussed the effects of rotation and stratification.

Progress has been much slower, however, in our understanding of the stability of viscous flows. One of the main steps in this direction was not taken until much later by Orr (1907) and Sommerfeld (1908), who derived the celebrated equation which now bears their names. The importance of this equation can hardly be exaggerated, and much of the subsequent work on the stability of viscous flows has concentrated on it. At the time the Orr–Sommerfeld equation was derived the existing methods of asymptotic analysis were not sufficiently well developed to deal with it effectively, and various heuristic methods of approximation were later suggested by

Heisenberg (1924), Tollmien (1929, 1947), and Lin (1945, 1955). These methods have been widely used for many computational purposes and they will therefore be discussed in this chapter together with an examination of their defects and limitations. The mathematical challenge provided by the Orr–Sommerfeld equation has also stimulated substantial further developments in the asymptotic solution of ordinary differential equations and some of these more recent developments will be discussed in the following chapter. The existing viscous theory is thus not nearly as complete or general as the inviscid theory, and it provides only a partial understanding of the role of viscosity in those circumstances when it is the cause of instability.

THE INVISCID THEORY

21 The governing equations

For a parallel two-dimensional flow, the basic steady flow is of the form

$$\mathbf{U}_* = U_*(z_*)\mathbf{i} \quad (z_{1*} \leqslant z_* \leqslant z_{2*}), \tag{21.1}$$

where $\mathbf{i}$ denotes a unit vector in the x_*-direction and the asterisks denote dimensional quantities. For an inviscid fluid $U_*(z_*)$ can be an arbitrary function of z_*. The flow is assumed to be bounded by the two planes $z_* = z_{1*}$ and z_{2*} which may be either rigid or free. On a rigid boundary the normal component of the velocity must vanish and on a free boundary the pressure must be constant. More generally, one of the boundaries may be at infinity as in the case of boundary layers or they may both be at infinity as in the case of shear layers, jets and wakes.

It is convenient, as usual, to write the governing equations in terms of dimensionless quantities, and for this purpose we introduce a characteristic length L and a characteristic velocity V associated with the basic flow. The choice of L and V is, of course, not unique; considerable variation in their definition exists in the literature and some care is required therefore in comparing the results of different writers. In the present discussion, however, we will usually take

$$V = \max_{z_{1*} \leqslant z \leqslant z_{2*}} |U_*(z_*)|$$

and, for flows in a channel, $L = \frac{1}{2}(z_{2*} - z_{1*})$. If we now let

and
$$t = t_* V/L, \quad \mathbf{x} = \mathbf{x}_*/L, \quad \mathbf{u} = \mathbf{u}_*/V, \quad p = p_*/\rho V^2,$$
$$\mathbf{U} = \mathbf{U}_*/V \equiv U(z)\mathbf{i}, \tag{21.2}$$

where ρ is the (constant) density of the fluid, then the Euler equations of motion and the equation of continuity become

$$\frac{\partial \mathbf{u}}{\partial t} + \mathbf{u} \cdot \nabla \mathbf{u} = -\nabla p \quad \text{and} \quad \nabla \cdot \mathbf{u} = 0. \tag{21.3}$$

The basic flow $\mathbf{U} = U(z)\mathbf{i}$ with $U(z)$ an arbitrary function of z automatically satisfies both of the boundary conditions and equations (21.3) are also satisfied provided $\nabla p = 0$, i.e. the pressure is constant. To study the stability of this flow we let

and
$$\mathbf{u}(\mathbf{x}, t) = U(z)\mathbf{i} + \mathbf{u}'(\mathbf{x}, t)$$
$$p(\mathbf{x}, t) = \text{constant} + p'(\mathbf{x}, t), \tag{21.4}$$

where $\mathbf{u}'$ is the disturbance velocity and p' is the disturbance pressure. On substituting these expressions into equations (21.3) and neglecting the term $\mathbf{u}' \cdot \nabla \mathbf{u}'$, which is quadratic in the disturbance velocity, we then obtain the *linearized equations of motion*

$$\left(\frac{\partial}{\partial t} + U\frac{\partial}{\partial x}\right) \mathbf{u}' + w'\frac{dU}{dz}\mathbf{i} = -\nabla p' \quad \text{and} \quad \nabla \cdot \mathbf{u}' = 0. \tag{21.5}$$

In studying the linearized stability problem for an inviscid fluid, one way to proceed would be to consider a suitably posed *initial-value problem*. Although the formulation of such an initial-value problem is not difficult, the subsequent analysis, even for basic flows of simple form, rapidly becomes complicated. Since one of the major aims of the inviscid theory is to provide general criteria by which one can decide whether a given basic flow is stable or not, it would be desirable if we could avoid having to solve the initial-value problem in detail. This can be partially achieved by considering a *normal-mode analysis* of equations (21.5). These two approaches are equivalent if it can be shown that the normal modes are complete and that an arbitrary initial disturbance can be expanded in terms of them. As we will see, however, the normal modes of the linearized inviscid stability problem are, in general, not complete, but they do play an important role in the initial-value approach, further discussion of which will be given in § 24. There it will be shown that instability, if it exists, is always associated with the discrete part of the spectrum; thus, in seeking general criteria for instability, it is sufficient to consider only the normal modes.

Since the coefficients in equations (21.5) depend only on z, the equations admit solutions which depend on x, y, and t exponentially. We consider therefore solutions of the form

and
$$\mathbf{u}'(\mathbf{x}, t) = \hat{\mathbf{u}}(z) \exp\left[i(\alpha x + \beta y - \alpha ct)\right]$$
$$p'(\mathbf{x}, t) = \hat{p}(z) \exp\left[i(\alpha x + \beta y - \alpha ct)\right], \tag{21.6}$$

in which it is understood that the real parts of these expressions must be taken to obtain physical quantities. The requirement that the solutions remain bounded as x, $y \to \pm\infty$ implies that the wavenumbers α and β must be real. The wave speed c may be complex, i.e. $c = c_r + ic_i$, and the expressions (21.6) thus represent waves which travel in the direction $(\alpha, \beta, 0)$ with phase speed $\alpha c_r/(\alpha^2 + \beta^2)^{1/2}$ and which grow or decay in time like $\exp(\alpha c_i t)$. Such a wave is said to be stable if $\alpha c_i \leqslant 0$, unstable if $\alpha c_i > 0$, and neutrally stable if $\alpha c_i = 0$.

If we now let $D = d/dz$, then on substituting the expressions (21.6) into equations (21.5) we obtain the system of ordinary differential equations:

$$
\begin{aligned}
i\alpha(U - c)\hat{u} + U'\hat{w} &= -i\alpha\hat{p}, \\
i\alpha(U - c)\hat{v} &= -i\beta\hat{p}, \\
i\alpha(U - c)\hat{w} &= -D\hat{p}, \\
i(\alpha\hat{u} + \beta\hat{v}) + D\hat{w} &= 0.
\end{aligned}
\tag{21.7}
$$

and

For rigid boundaries, this being the usual case, we impose the boundary conditions

$$\hat{w} = 0 \quad \text{at } z = z_1 \quad \text{and} \quad z_2. \tag{21.8}$$

Thus it is clear from equations (21.7) and the boundary conditions (21.8) that we have an eigenvalue problem which will lead to an eigenvalue relation of the form

$$\mathcal{F}(\alpha, \beta, c) = 0. \tag{21.9}$$

For a given basic flow $U(z)$ and a given vector wavenumber $(\alpha, \beta, 0)$ this eigenvalue relation determines the allowable values of c. Since the allowed values of c are, in general, finite in number the normal modes are obviously not complete and, in addition to this discrete part of the spectrum, there is also a continuous part which arises from the singularity in the equations where $U - c = 0$. The continuous part of the spectrum must be included in order to represent an arbitrary initial disturbance and will be discussed further in § 24 in connexion with the initial-value problem.

In deriving equations (21.7) we have considered general three-dimensional disturbances and we now wish to show how the three-dimensional problem defined by equations (21.7) and (21.8) can be reduced to an *equivalent two-dimensional problem*. For this purpose we use the transformation first introduced by Squire (1933) for the more general viscous problem. Thus if we let

$$\tilde{\alpha} = (\alpha^2 + \beta^2)^{1/2}, \quad \tilde{\alpha}\tilde{u} = \alpha\hat{u} + \beta\hat{v},$$
$$\tilde{p}/\tilde{\alpha} = \hat{p}/\alpha, \quad \tilde{w} = \hat{w}, \text{ and } \tilde{c} = c, \tag{21.10}$$

then equations (21.7) can be combined to give

$$\begin{aligned}
\mathrm{i}\tilde{\alpha}(U - \tilde{c})\tilde{u} + U'\tilde{w} &= -\mathrm{i}\tilde{\alpha}\tilde{p}, \\
\mathrm{i}\tilde{\alpha}(U - \tilde{c})\tilde{w} &= -\mathrm{D}\tilde{p}, \\
\mathrm{i}\tilde{\alpha}\tilde{u} + \mathrm{D}\tilde{w} &= 0,
\end{aligned} \right\} \tag{21.11}$$

and

and the boundary conditions are

$$\tilde{w} = 0 \quad \text{at } z = z_1 \text{ and } z_2. \tag{21.12}$$

These equations have exactly the same mathematical form as the original equations with $\beta = \hat{v} = 0$ and they thus define the equivalent two-dimensional problem. It is sufficient, therefore, to consider only two-dimensional disturbances; for, once the solution of equations (21.7) and (21.8) with $\beta = \hat{v} = 0$ has been obtained, we can immediately obtain the corresponding solution of the equivalent two-dimensional problem by a trivial change in notation and from this, by means of Squire's transformation, we can then obtain the solution of the original three-dimensional problem. From these results we can now easily prove

Squire's theorem for an inviscid fluid. To each unstable three-dimensional disturbance there corresponds a more unstable two-dimensional one.

To prove the theorem observe first that if $c = f(\alpha)$ is the solution of the two-dimensional problem, i.e. the solution of equations (21.7) and (21.8) with $\beta = \hat{v} = 0$, then $\tilde{c} = f(\tilde{\alpha})$ is the solution of the equivalent two-dimensional problem and, by Squire's transformation, $c = f((\alpha^2 + \beta^2)^{1/2})$ is the solution of the three-dimensional problem. Thus, to each unstable three-dimensional disturbance with growth rate αc_i there corresponds a two-dimensional disturbance with growth rate $\tilde{\alpha} c_i$ which is more unstable since $\tilde{\alpha} > \alpha$ if $\beta \neq 0$.

An important consequence of this theorem is that in seeking sufficient criteria for instability we need consider only two-dimensional disturbances, and it is then convenient to introduce a stream function $\psi'(x, z, t)$ such that the two components of the disturbance velocity are given by

$$u' = \partial\psi'/\partial z \quad \text{and} \quad w' = -\partial\psi'/\partial x. \quad (21.13)$$

If we next let

$$\psi'(x, z, t) = \phi(z) \, e^{i\alpha(x - ct)} \quad (21.14)$$

then

$$\hat{u} = \phi' \quad \text{and} \quad \hat{w} = -i\alpha\phi, \quad (21.15)$$

and the first of equations (21.7) gives

$$\hat{p} = U'\phi - (U - c)\phi'. \quad (21.16)$$

On substituting this result for $\hat{p}$ into the third of equations (21.7) we obtain *Rayleigh's stability equation*

$$(U - c)(\phi'' - \alpha^2\phi) - U''\phi = 0 \quad (21.17)$$

which, together with the boundary conditions

$$\alpha\phi = 0 \quad \text{at } z = z_1 \quad \text{and} \quad z_2, \quad (21.18)$$

defines the basic eigenvalue problem for inviscid parallel shear flows. In fact Rayleigh's stability equation is the vorticity equation of the disturbance (see Problem 4.1).

Note that Rayleigh's equation and the boundary conditions are unchanged when α is replaced by $-\alpha$. Thus, without loss of generality, we can take $\alpha \geq 0$, and the criterion for instability then becomes that there exists a solution with $c_i > 0$ for some $\alpha > 0$. Furthermore, if ϕ is an eigenfunction with eigenvalue c for some α, then so too is ϕ^* with eigenvalue c^* for the same α. Thus, to each unstable mode there is a corresponding stable mode, and it is convenient therefore to adopt the convention of taking $c_i > 0$ as the criterion of instability and to ignore the complex conjugate eigenvalue with $c_i < 0$.

Rayleigh's equation is not self-adjoint but its adjoint is easily found to be

$$(D^2 - \alpha^2)(U - c)\phi^\dagger - U''\phi^\dagger = 0, \qquad (21.19)$$

where $\phi^\dagger$ must satisfy the same boundary conditions. On comparing this equation with equation (21.17) it immediately follows that $\phi^\dagger = \text{constant} \times \phi/(U - c)$. Equation (21.19) can also be written in the self-adjoint form

$$D[(U - c)^2 D\phi^\dagger] - \alpha^2(U - c)^2\phi^\dagger = 0, \qquad (21.20)$$

and this equation will be used later in the proof of Howard's semicircle theorem.

22 General criteria for instability

A distinctive feature of the velocity profiles shown in Fig. 4.1 is that one has an inflexion point but the other does not. The importance of this fact and its bearing on the stability or instability of the flow was first recognized by Rayleigh (1880), who proved

Rayleigh's inflexion-point theorem. A necessary condition for instability is that the basic velocity profile should have an inflexion point.

To prove this theorem first rewrite Rayleigh's equation in the form

$$\phi'' - \alpha^2\phi - \frac{U''}{U - c}\phi = 0, \qquad (22.1)$$

and suppose that $c_i > 0$ so that the equation is non-singular. On multiplying this equation by ϕ^*, integrating from z_1 to z_2, and then integrating the first term by parts, we obtain

$$\int_{z_1}^{z_2} (|D\phi|^2 + \alpha^2|\phi|^2)\, dz + \int_{z_1}^{z_2} \frac{U''}{U - c}|\phi|^2\, dz = 0. \qquad (22.2)$$

The imaginary part of this equation is

$$c_i \int_{z_1}^{z_2} \frac{U''}{|U-c|^2} |\phi|^2 \, dz = 0, \tag{22.3}$$

from which it follows that U'' must change sign at least once in the open interval (z_1, z_2).

A stronger form of this condition was obtained later by Fjørtoft (1950) who proved

Fjørtoft's theorem. A *necessary* condition for instability is that $U''(U-U_s) < 0$ somewhere in the field of flow, where z_s is a point at which $U'' = 0$ and $U_s = U(z_s)$.

To prove this theorem consider the real part of equation (22.2):

$$\int_{z_1}^{z_2} \frac{U''(U-c_r)}{|U-c|^2} |\phi|^2 \, dz = -\int_{z_1}^{z_2} (|D\phi|^2 + \alpha^2 |\phi|^2) \, dz. \tag{22.4}$$

If we now add

$$(c_r - U_s) \int_{z_1}^{z_2} \frac{U''}{|U-c|^2} |\phi|^2 \, dz = 0$$

to the left-hand side of equation (22.4) we obtain

$$\int_{z_1}^{z_2} \frac{U''(U-U_s)}{|U-c|^2} |\phi|^2 \, dz = -\int_{z_1}^{z_2} (|D\phi|^2 + \alpha^2 |\phi|^2) \, dz < 0, \tag{22.5}$$

from which the result follows. Thus, if $U(z)$ is a monotone function with only one inflexion point then a necessary condition for instability is that $U''(U-U_s) \le 0$ for $z_1 \le z \le z_2$ with equality only at $z = z_s$, and this condition is illustrated in Fig. 4.2.

Unfortunately, neither of these conditions for instability is sufficient. This is perhaps most clearly seen from Tollmien's (1935) simple counter-example with $U = \sin z$ which will be discussed later in this section. Tollmien also gave a heuristic argument, however, which suggests that the conditions are sufficient for symmetric profiles in a channel and for monotone profiles of the boundary-layer type. His argument was based on first showing the existence of a neutrally stable eigensolution

$$\phi = \phi_s, \quad \alpha = \alpha_s > 0, \quad c = c_s \tag{22.6}$$

and then, by perturbing this solution, to construct neighbouring unstable modes for α close to α_s with $\alpha < \alpha_s$.

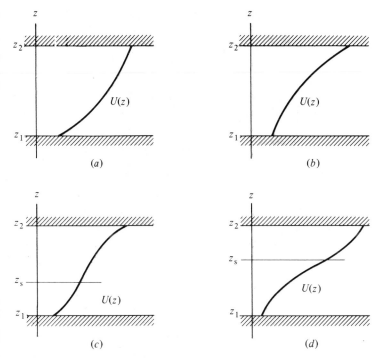

Fig. 4.2. (a) Stable: $U'' < 0$; (b) stable: $U'' > 0$; (c) stable: $U'' = 0$ at z_s but $U''(U - U_s) \geq 0$; (d) possibly unstable: $U'' = 0$ at z_s but $U''(U - U_s) \leq 0$. (From Drazin & Howard 1966.)

It is worth noting that if $\alpha = 0$ then $\phi = U - c$ is a neutral mode provided c can be chosen so as to satisfy the boundary conditions. This apparently trivial solution is merely a perturbed form of the basic flow but it does play a minor role in the viscous theory in determining the asymptotic behaviour of the curve of marginal stability. To demonstrate the existence of a neutrally stable eigensolution with $\alpha_s > 0$ we suppose, following Friedrichs (Mises & Friedrichs 1942), that $K(z) = -U''/(U - U_s)$ is regular at z_s, i.e. $U_s'' = 0$, and let $\lambda = -\alpha^2$. If $c = c_s = U_s$ then Rayleigh's stability equation can be rewritten in the form

$$\phi'' + \{\lambda + K(z)\}\phi = 0, \tag{22.7}$$

which, together with the boundary conditions (21.18), is a standard Sturm–Liouville problem for which there exists an infinite sequence of eigenvalues with limit point at $+\infty$. The least eigenvalue of this problem is given by the variational principle

$$\lambda_s = \min \left\{ \int_{z_1}^{z_2} (f'^2 - Kf^2)\, dz \Big/ \int_{z_1}^{z_2} f^2\, dz \right\}, \qquad (22.8)$$

where the minimum is to be taken for functions f that satisfy the boundary conditions and have square-integrable derivatives. From the well-known inequality

$$(z_2 - z_1)^2 \int_{z_1}^{z_2} f'^2\, dz \geqslant \pi^2 \int_{z_1}^{z_2} f^2\, dz \qquad (22.9)$$

we see that if $K(z) > \pi^2/(z_2 - z_1)^2$ everywhere then $\lambda_s < 0$ and hence $\alpha_s > 0$. Drazin & Howard (1966) have also proved that there is instability only for $\alpha < \alpha_s$. To show this suppose that $K(z) > 0$ throughout the flow and that $c_i \neq 0$. Then the real part of equation (22.2) plus $(U_s - c_r)/c_i$ times equation (22.3) gives

$$\int_{z_1}^{z_2} (|D\phi|^2 + \alpha^2 |\phi|^2)\, dz = \int_{z_1}^{z_2} \frac{(U - c_r)^2 - (U_s - c_r)^2}{(U - c_r)^2 + c_i^2} K |\phi|^2\, dz$$

$$< \int_{z_1}^{z_2} K |\phi|^2\, dz, \qquad (22.10)$$

from which it immediately follows that

$$\alpha^2 < -\lambda_s = \alpha_s^2. \qquad (22.11)$$

Hence instability is possible only when $0 < \alpha < \alpha_s$ and we have stability $(c_i = 0)$ for $\alpha \geqslant \alpha_s$. But this argument does *not* show that if $0 < \alpha < \alpha_s$ then $c_i \neq 0$, for we have not excluded the possibility of the eigensolution defined by equation (22.6) being an isolated neutral mode.

If, however, we *assume* the existence of unstable modes for α close to α_s with $\alpha < \alpha_s$, whose limit as $c_i \downarrow 0$ is the neutrally stable eigensolution defined by equation (22.6), then, as Tollmien (1935) and Lin (1945, 1955) have shown, they can be found by a simple perturbation procedure. The requirement that c_i tend to zero through positive values is a consequence of considering either the inviscid initial-value problem or the inviscid limit of the viscous problem. We follow here a method suggested by Hughes & Reid (1965b) which, with minor modifications, lends itself to certain applications in the viscous theory. For this purpose it is necessary to consider a second solution, ψ_s (say), of Rayleigh's equation with $\alpha = \alpha_s$ and $c = U_s = c_s$. (This solution does not, of course, satisfy the boundary conditions.) A standard form of this solution can conveniently be defined by

$$\psi_s(z) = \phi_s(z) \int_{z_s}^{z} \{\phi_s(z)\}^{-2} \, dz \qquad (22.12)$$

provided $\phi_s(z_s) \neq 0$ and a few of its properties may be briefly noted. The Wronskian of the two solutions is given by $\mathcal{W}(\phi_s, \psi_s) = 1$. We also have

$$\psi_s(z_1) = -1/\phi'_s(z_1), \quad \psi_s(z_2) = -1/\phi'_s(z_2),$$
$$\psi_s(z_s) = 0 \quad \text{and} \quad \psi'_s(z_s) = 1/\phi_s(z_s). \qquad (22.13)$$

For (α, c) near (α_s, c_s) we now assume that $\phi(z; \alpha, c)$ can be expanded in powers of both $\alpha - \alpha_s$ and $c - c_s$ in the form

$$\phi(z) = \phi_s(z) + \Phi_1(z)(\alpha - \alpha_s) + \Phi_2(z)(c - c_s) + \cdots, \qquad (22.14)$$

where Φ_1 and Φ_2 must satisfy the equations

$$\text{and} \quad \begin{aligned} (U - c_s)(\Phi''_1 - \alpha_s^2 \, \Phi_1) - U'' \Phi_1 &= 2\alpha_s (U - c_s)\phi_s \\ (U - c_s)(\Phi''_2 - \alpha_s^2 \, \Phi_2) - U'' \Phi_2 &= U''(U - c_s)^{-1}\phi_s. \end{aligned} \Bigg\} \quad (22.15)$$

The solutions of these equations that vanish at $z = z_1$ (say) are

$$\Phi_1 = 2\alpha_s \left(\psi_s \int_{z_1}^{z} \phi_s^2 \, dz - \phi_s \int_{z_1}^{z} \phi_s \psi_s \, dz \right) \qquad (22.16)$$

and

$$\Phi_2 = \psi_s \int_{z_1}^{z} \frac{U''}{(U - c_s)^2} \phi_s^2 \, dz - \phi_s \int_{z_1}^{z} \frac{U''}{(U - c_s)^2} \phi_s \psi_s \, dz. \qquad (22.17)$$

In accordance with the requirement that c_i tend to zero through positive values, the path of integration for the first integral in equation (22.17) must lie below z_s if $U'_s > 0$ and above z_s if $U'_s < 0$; the integrand of the second integral, however, is regular at z_s. At $z = z_2$, Φ_1 and Φ_2 have the values

$$\text{and} \quad \begin{aligned} \Phi_1(z_2) &= -\frac{2\alpha_s}{\phi'_s(z_2)} \int_{z_1}^{z_2} \phi_s^2 \, dz \\ \Phi_2(z_2) &= -\frac{1}{\phi'_s(z_2)} \int_{z_1}^{z_2} \frac{U''}{(U - c_s)^2} \phi_s^2 \, dz, \end{aligned} \Bigg\} \quad (22.18)$$

and are thus independent of ψ_s. It may be noticed that $\Phi_1(z_2)$ is real but that $\Phi_2(z_2)$ is complex with real and imaginary parts given by

$$\Phi_{2r}(z_2) = -\frac{1}{\phi'_s(z_2)} \mathcal{P} \int_{z_1}^{z_2} \frac{U''}{(U-c_s)^2} \phi_s^2 \, dz$$

and
$$(22.19)$$

$$\Phi_{2i}(z_2) = -\pi \frac{U'''_s}{U'^2_s} \frac{\phi_s^2(z_s)}{\phi'_s(z_2)} \, \text{sgn} \, U'_s,$$

where $\mathcal{P}$ denotes the Cauchy principal value of the integral. With Φ_1 and Φ_2 determined in this manner, ϕ in equation (22.14) automatically vanishes at $z = z_1$; the requirement that it also vanish at $z = z_2$ shows that

$$c - c_s \sim \frac{\Phi_1(z_2)\Phi_2^*(z_2)}{|\Phi_2(z_2)|^2} (\alpha_s - \alpha) \text{ as } \alpha \uparrow \alpha_s, \qquad (22.20)$$

and this result is equivalent to Lin's formula (1955), p.123 for $(\partial c/\partial \alpha^2)_{\alpha = \alpha_s}$. In particular, the imaginary part of equation (22.20) is

$$c_i \sim -\frac{\Phi_1(z_2)\Phi_{2i}(z_2)}{|\Phi_2(z_2)|^2} (\alpha_s - \alpha) \text{ as } \alpha \uparrow \alpha_s. \qquad (22.21)$$

The sign of the coefficient in this expression is determined by the sign of $U'''_s \text{sgn} \, U'_s$; alternatively if $K(z_s) = -U'''_s/U'_s > 0$ then c_i is positive for α just less than α_s and we have instability. Perturbation formulae of this type can also be derived without difficulty for semi-bounded flows of the boundary-layer type (Hughes & Reid 1965b) and unbounded flows of the jet and shear-layer type.

To illustrate some of the consequences of these results consider a *sinusoidal basic flow* with $U = \sin z$ $(z_1 \leq z \leq z_2)$. The points of inflexion for this flow are where $z = z_s = n\pi$ $(n = 0, \pm 1, \pm 2, \ldots)$. If there are no values of z_s in the interval (z_1, z_2) then the flow is stable by Rayleigh's theorem. Next suppose that there is at least one value of z_s in the interval which, without loss of generality, we can take to be $z_s = 0$ so that $z_1 < 0 < z_2$. Thus, with $c = c_s = 0$ Rayleigh's equation becomes

$$\sin z\{\phi'' + (1-\alpha^2)\phi\} = 0 \quad \text{with } \phi = 0 \quad \text{at } z = z_1 \quad \text{and} \quad z_2.$$
$$(22.22)$$

If we now simply drop the factor $\sin z$, thereby ignoring the continuous spectrum, then we have

and
$$\left. \begin{array}{l} \phi_s = \sin\{n\pi(z - z_1)/(z_2 - z_1)\} \\ \alpha_s = \{1 - n^2\pi^2/(z_2 - z_1)^2\}^{1/2} \end{array} \right\} \qquad (22.23)$$

for each integer $n < (z_2 - z_1)/\pi$. Thus, if $z_2 - z_1 < \pi$ the flow is stable even though it has an inflexion point and this is Tollmien's counter-example to the sufficiency of Rayleigh's theorem. If $z_2 - z_1 > \pi$, however, then the flow is unstable; this condition also follows from the inequality (22.9) on noting that $K(z) \equiv 1$ for this velocity profile. Suppose now that $z_1 = -\pi$ and $z_2 = \pi$ so that the flow is like the one shown in Fig. 4.1(b). We then have the neutral mode

$$\phi_s = \cos\tfrac{1}{2}z, \quad \alpha_s = \tfrac{1}{2}\sqrt{3}, \quad c_s = 0, \qquad (22.24)$$

and from equation (22.21) we have $c_i \sim \sqrt{3}(\alpha_s - \alpha)$ as $\alpha \uparrow \alpha_s$. In addition there is also the trivial neutral mode $\phi_s = \sin z$, $\alpha_s = 0$, $c_s = 0$ and for this mode $\phi_s(z_s) = 0$.

Tollmien's inviscid solutions. Consider now the solutions of Rayleigh's equation when c is not necessarily equal to c_s. A point $z = z_c$ where $U - c = 0$ and $U'_c \neq 0$ is a regular singular point of equation (21.17) with exponents 0 and 1. Thus, in a neighbourhood of z_c there exists one solution which is analytic at $z = z_c$. It is convenient to write this solution in the form

$$\phi_1(z) = (z - z_c)P_1(z), \qquad (22.25)$$

where $P_1(z)$ is analytic at z_c and $P_1(z_c) \neq 0$. For convenience, we shall choose $P_1(z_c) = 1$. The second linearly independent solution of equation (21.17), however, has a logarithmic branch point at $z = z_c$ and is of the form

$$\phi_2(z) = P_2(z) + (U''_c/U'_c)\phi_1(z) \ln(z - z_c), \qquad (22.26)$$

where $P_2(z)$ is also analytic at z_c with $P_2(z_c) = 1$. To make this second solution definite, it is convenient to suppose that $\phi_2(z)$ contains no multiple of $\phi_1(z)$, i.e. that the coefficient of $z - z_c$ in the power series expansion of $P_2(z)$ is zero. The solutions of Rayleigh's equation were first given in this form by Tollmien (1929) in connexion with his discussion of the Orr–Sommerfeld equation and they are often referred to as Tollmien's inviscid solutions. A more descriptive terminology will be helpful in our later discussion of the viscous problem and we will therefore call $\phi_1(z)$ the 'regular inviscid solution', $\phi_2(z)$ the 'singular inviscid solution', and $P_2(z)$ the 'regular part' of the singular inviscid solution.

In the case of neutral stability c, and hence z_c, is real and it is then necessary to specify the correct branch of the multivalued solution given by equation (22.26). By again letting c_i tend to zero through positive values we see that if $U_c' > 0$ and we let $\ln(z - z_c) = \ln|z - z_c|$, for $z > z_c$ then we have $\ln(z - z_c) = \ln|z - z_c| - \pi i$ for $z < z_c$. Later, in our discussion of the viscous problem, we will be concerned with the circumstances under which these solutions of Rayleigh's equation provide approximations to the solutions of the Orr–Sommerfeld equation.

The first few terms in the power series expansion of P_1 and P_2 are

$$
\left.
\begin{aligned}
P_1(z) &= 1 + \frac{U_c''}{2U_c'}(z - z_c) + \frac{1}{6}\left(\frac{U_c'''}{U_c'} + \alpha^2\right)(z - z_c)^2 + \cdots \\
\text{and} \\
P_2(z) &= 1 + \left(\frac{U_c'''}{2U_c'} - \frac{U_c''^2}{U_c'^2} + \frac{1}{2}\alpha^2\right)(z - z_c)^2 + \cdots .
\end{aligned}
\right\} \quad (22.27)
$$

For velocity profiles with a sufficiently simple analytical form, the summation of these series is often feasible; more generally, however, ϕ_1 and the regular part of ϕ_2 can be obtained by direct numerical integration. When this latter method is used, ϕ_1 can conveniently be defined as the solution of Rayleigh's equation that satisfies the initial conditions $\phi_1(z_c) = 0$ and $\phi_1'(z_c) = 1$. Similarly, as suggested by Conte & Miles (1959), P_2 can be obtained as the solution of the inhomogeneous equation

$$
(U - c)(P_2'' - \alpha^2 P_2) - U'' P_2 = -\frac{U_c''}{U_c'} \cdot \frac{U - c}{z - z_c}\{2(z - z_c)P_1' + P_1\} \tag{22.28}
$$

that satisfies the initial conditions $P_2(z_c) = 1$ and $P_2'(z_c) = 0$. The Wronskian of the solutions ϕ_1 and ϕ_2 is a constant with the value $\mathcal{W}(\phi_1, \phi_2) = -1$, and this relation often provides a useful check on numerical work.

The corresponding solutions of the adjoint equation can conveniently be taken in the form

$$
\phi_j^\dagger = \frac{U_c'}{U - c}\phi_j \quad (j = 1, 2) \tag{22.29}
$$

from which we have immediately

$$\phi_1^\dagger(z) = P_1^\dagger(z)$$

and

$$\phi_2^\dagger(z) = (z - z_c)^{-1} P_2^\dagger(z) + (U_c''/U_c') \phi_1^\dagger(z) \ln(z - z_c), \qquad (22.30)$$

where $P_1^\dagger$ and $P_2^\dagger$ are again analytic at z_c and $P_1^\dagger(z_c) = P_2^\dagger(z_c) = 1$. Since the power series expansion of $P_1^\dagger$ has no linear term, $\phi_2^\dagger$ contains no multiple of $\phi_1^\dagger$. The solution $\phi_1^\dagger$, like ϕ_1, is regular at z_c, but $\phi_2^\dagger$ is more singular than ϕ_2 and this has important consequences in the asymptotic theory of the adjoint Orr–Sommerfeld equation.

The modified Heisenberg expansions. The solutions of Rayleigh's equation had been obtained even earlier by Heisenberg (1924) as power series in α^2 of the form

$$\phi_j(z; \alpha^2, c) = \frac{U - c}{U_c'} \sum_{n=0}^{\infty} \alpha^{2n} q_{jn}(z; c) \quad (j = 1, 2), \quad (22.31)$$

where

$$q_{10} = 1, \quad q_{20} = \int (U - c)^{-2} \, dz$$

and

$$(22.32)$$

$$q_{j,n+1} = \int (U - c)^{-2} \, dz \int (U - c)^2 q_{jn} \, dz \ (n \geq 0).$$

The lower limits of integration in these expressions are, of course, arbitrary but they are usually taken as z_1. When the limits of integration are fixed in this manner the relationship between the expansions (22.31) and the Tollmien solutions is neither simple nor direct. This difficulty can be avoided, however, by a simple redefinition of the coefficients q_{jn} as discussed by Nield (1972). Thus we let

$$q_{10} = 1, q_{20} = \frac{1}{z - z_c} + \frac{U_c''}{U_c'} \ln(z - z_c) - \frac{U_c''}{2U_c'}$$

$$- \int_{z_c}^{z} \left\{ \left(\frac{U_c'}{U - c} \right)^2 - \frac{1}{(z - z_c)^2} + \frac{U_c''}{U_c'(z - z_c)} \right\} dz \quad (22.33)$$

and

$$q_{j,n+1} = \int_{z_c}^{z} (U - c)^{-2} \, dz \int_{z_c}^{z} (U - c)^2 q_{jn}(z) \, dz.$$

It is then an easy matter to show that these modified Heisenberg expansions are simply different representations of the Tollmien solutions. The series (22.31) are uniformly convergent for bounded values of α and fixed values of $z \neq z_c$. They are particularly useful in the viscous theory in connexion with the determination of the asymptotic behaviour of the curves of marginal stability. For unbounded or semi-bounded flows, however, different forms of approximations are needed and these will be discussed later in this section.

Suppose now that $U(z)$ is monotone with a single inflexion point at z_s ($z_1 < z_s < z_2$). Then both ϕ_1 and ϕ_2 are regular at z_s, and ϕ_s must be a linear combination of ϕ_1 and P_2. Since $\phi_s(z_s) \neq 0$ (except when $\alpha = 0$), it is convenient to let

$$\phi_s(z) = A\phi_1(z) + P_2(z), \qquad (22.34)$$

so that $\phi_s(z_s) = 1$ and the two boundary conditions then determine A and α_s. This method of determining ϕ_s remains applicable to monotone profiles with more than one inflexion point and to non-monotone profiles for which $U'' = 0$ whenever $U - U_s = 0$ (in particular, this includes symmetric flows). For non-monotone profiles which do not satisfy this condition† the determination of the neutrally stable eigensolution is more difficult for then ϕ_s must necessarily be singular at points where $U - c_s = 0$, and the following argument, due to Foote & Lin (1950), is helpful in such cases.

The average of the (dimensionless) Reynolds stress over one wavelength is

$$\tau = -\langle u'w' \rangle = -\frac{\alpha}{2\pi} \int_0^{2\pi/\alpha} u'w' \, dx$$
$$= \tfrac{1}{4} i\alpha (\phi D\phi^* - \phi^* D\phi) \exp(2\alpha c_i t) \qquad (22.35)$$

and on using Rayleigh's equation we have

$$\frac{d\tau}{dz} = \frac{1}{2} \alpha c_i \frac{U''}{|U - c|^2} |\phi|^2 \exp(2\alpha c_i t). \qquad (22.36)$$

The integral of $d\tau/dz$ over (z_1, z_2) must vanish since $\tau = 0$ at the boundaries. Suppose now, however, that we have an unstable mode for some value of α, and that, as $\alpha \uparrow \alpha_s$ (say), the corresponding value of c becomes real, i.e. $c_i(\alpha) \downarrow 0$. For this neutral mode $d\tau/dz = 0$ everywhere, i.e. $\tau = $ constant, except possibly at $z = z_c$ where

---

† A simple example is the free-convection boundary layer with suction for which $U = z \, e^{-z}$. This profile has an inflexion point at $z_s = 2$ but $c_s \neq U_s = 2 \, e^{-2}$.

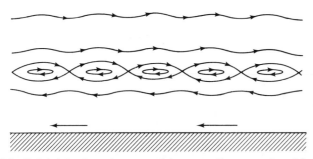

Fig. 4.3. Kelvin's 'cat's eye' pattern of the streamlines near the critical level as viewed by an observer moving with the wave.

$U - c = 0$. On integrating equation (22.36) across the critical layer $z = z_c$ and then letting $c_i \downarrow 0$ we find that the 'jump' in τ is given by

$$\Delta \tau = \tfrac{1}{2} \alpha \pi (U_c''/U_c') |\phi_c|^2, \tag{22.37}$$

where $\Delta \tau = \tau(z_c + 0) - \tau(z_c - 0)$. Thus, if the profile is monotone, then there can be only one jump and this implies that $U_c'' = 0$ since $\phi_c \neq 0$ except possibly when $\alpha = 0$. More generally, the algebraic sum of all such jumps must be zero. In the case of non-monotone profiles for which $c_s \neq U_s$ this condition must be used in the determination of α_s and c_s and the corresponding eigenfunction ϕ_s will be (mildly) singular whenever $U - c_s = 0$.

The form of the streamlines in the neighbourhood of the critical level where $U - c = 0$ has been given by Kelvin (1880b). If we impose a velocity equal to the phase velocity on the whole system then the motion becomes steady and the streamlines are identical with the particle paths. The physical stream function for this steady flow is

$$\Psi(z) + A \operatorname{Re} \{\phi(z) \, e^{i\alpha x}\},$$

where

$$\Psi(z) = \int_{z_c}^{z} (U - c) \, dz$$

and A is proportional to the amplitude of the wave. Near the critical level $z = z_c$ the equation for the streamlines is

$$\tfrac{1}{2} U_c' (z - z_c)^2 + A\phi(z_c) \cos \alpha x = \text{constant}, \tag{22.38}$$

where we have taken $\phi(z_c)$ to be real. The streamlines therefore have the famous 'cat's eye' pattern shown in Fig. 4.3.

Rayleigh had proved that if $c_i \neq 0$ then c_r must lie in the range $U_{min} < c_r < U_{max}$, and this result has been generalized by Howard (1961) who proved

Howard's semicircle theorem. For unstable waves c must lie in the semicircle

$$\{c_r - \tfrac{1}{2}(U_{max} + U_{min})\}^2 + c_i^2 \leq \{\tfrac{1}{2}(U_{max} - U_{min})\}^2 \quad (c_i > 0). \quad (22.39)$$

To prove this result we multiply equation (21.20) by $\phi^{\dagger *}$ and integrate from z_1 to z_2 to obtain

$$\int_{z_1}^{z_2} (U - c)^2 Q \, dz = 0 \quad \text{where } Q \equiv |D\phi^\dagger|^2 + \alpha^2 |\phi^\dagger|^2 > 0, \quad (22.40)$$

and we have assumed that $c_i \neq 0$ so that $\phi^\dagger$ is non-singular. The real and imaginary parts of this integral are

$$\int_{z_1}^{z_2} \{(U - c_r)^2 - c_i^2\} Q \, dz = 0 \quad \text{and} \quad 2c_i \int_{z_1}^{z_2} (U - c_r) Q \, dz = 0.$$

$$(22.41)$$

The second integral here gives Rayleigh's result that c_r must lie in the range of U. Observe further that

$$0 \geq \int_{z_1}^{z_2} (U - U_{min})(U - U_{max}) Q \, dz$$

$$= \int_{z_1}^{z_2} \{(c_r^2 + c_i^2) - (U_{max} + U_{min})c_r + U_{max} U_{min}\} Q \, dz$$

and hence that

$$c_r^2 + c_i^2 - (U_{max} + U_{min})c_r + U_{max} U_{min} \leq 0,$$

from which the theorem follows.

Unbounded flows. For unbounded flows the Heisenberg expansions of the solutions of Rayleigh's equation are not uniformly convergent as $z \to \pm\infty$ and, following Drazin & Howard (1962), we now wish to consider the long-wave approximation for such flows. We assume, firstly, that U approaches constant values $U(\pm\infty)$ as $z \to \pm\infty$. If $U(+\infty) = U(-\infty)$ then, by a simple re-definition of c, we can normalize U so that $U(\pm\infty) = 0$ and this will be referred to as the *jet case*. Similarly, if $U(+\infty) \neq U(-\infty)$, then we can normalize U so that $U(\pm\infty) = \pm 1$ and this will be referred to as the *shear-layer case*. Secondly, we assume that $U \to$ constant as $z \to \pm\infty$ sufficiently rapidly so that (at least for $c \neq 0$) the solutions of Rayleigh's equation are asymptotic to $e^{\mp\alpha z}$ as $z \to \pm\infty$. Now let $W = U - c$ so that Rayleigh's equation becomes

$$W(\phi'' - \alpha^2\phi) - W''\phi = 0 \qquad (22.42)$$

and consider two solutions of the form

$$\phi_\pm(z) = e^{\mp\alpha z}\chi_\pm(z). \qquad (22.43)$$

It is convenient to normalize these solutions so that $\chi_\pm(\pm\infty) = W(\pm\infty) \equiv W_{\pm\infty}$. This normalization does not require that $\phi_+(0) = \phi_-(0)$. But if $\phi = \phi_+$ for $z > 0$ then its continuation to $z < 0$ must be a multiple of ϕ_-, i.e.

$$\phi_+(0) = K\phi_-(0) \quad \text{and} \quad \phi'_+(0) = K\phi'_-(0),$$

where $K = K(\alpha, c)$. Eliminating K between these equations then gives the eigenvalue relation

$$\phi_+(0)\phi'_-(0) - \phi'_+(0)\phi_-(0) = 0, \qquad (22.44)$$

which is simply the Wronskian of $\phi_\pm(z)$. For small values of α we now let

$$\chi_\pm(z) = \sum_{n=0}^{\infty} (\pm\alpha)^n \chi_{\pm n}(z). \qquad (22.45)$$

The coefficients in this expansion can be expressed explicitly in terms of W and repeated integrals of W, and on expanding the eigenvalue relation (22.44) for small values of α we obtain

$$\alpha(W_\infty^2 + W_{-\infty}^2) + \alpha^2 \int_{-\infty}^{\infty} (W^2 - W_\infty^2)(W^2 - W_{-\infty}^2)W^{-2}\,dz$$
$$+ O(\alpha^3) = 0. \qquad (22.46)$$

In a first approximation we have $(U_\infty - c)^2 + (U_{-\infty} - c)^2 = 0$ and, on taking the root with $c_i > 0$, this gives

$$c \to \tfrac{1}{2}(U_\infty + U_{-\infty}) + \tfrac{1}{2}i(U_\infty - U_{-\infty}) \text{ as } \alpha \downarrow 0. \qquad (22.47)$$

Thus, in the shear-layer case we get $c = i$. For the jet case it is necessary to go to a second approximation to get

$$c \sim i\alpha^{1/2}\left(\tfrac{1}{2}\int_{-\infty}^{\infty} U^2\,dz\right)^{1/2} \text{ as } \alpha \downarrow 0. \qquad (22.48)$$

23 Flows with piecewise-linear velocity profiles

The eigenvalue problem defined by equations (21.17) and (21.18) is difficult to solve explicitly when $U(z)$ is a smoothly varying function. When $U(z)$ is piecewise linear, however, the solutions of Rayleigh's equation are simple exponential or hyperbolic functions which must satisfy certain matching conditions at a discontinuity of $U(z)$ or $U'(z)$. The use of piecewise-linear profiles thus provides a simple method of modelling some features of smoothly varying profiles.

Suppose then that U or U' are discontinuous at $z = z_0$ (say) and let $\Delta f = f(z_0 + 0) - f(z_0 - 0)$ denote the 'jump' in $f(z)$ at z_0. To derive the first matching condition rewrite Rayleigh's equation in the form

$$[(U-c)\phi' - U'\phi]' - \alpha^2(U-c)\phi = 0. \tag{23.1}$$

On integrating this equation across the discontinuity from $z_0 - \varepsilon$ to $z_0 + \varepsilon$ and then letting $\varepsilon \downarrow 0$ we obtain

$$\Delta[(U-c)\phi' - U'\phi] = 0 \quad \text{at } z = z_0. \tag{23.2}$$

This condition also follows immediately from equation (21.16) by requiring that the pressure be continuous across the material interface, i.e. $\Delta \hat{p} = 0$ at $z = z_0 + \zeta(x, t)$, where $\zeta(x, t) = \zeta_0 \exp\{i\alpha(x - ct)\}$. To first order in the small perturbation ζ_0 this gives equation (23.2).

To derive the second matching condition divide the pressure equation (21.16) by $(U-c)^2$ to get

$$\left(\frac{\phi}{U-c}\right)' = -\frac{\hat{p}}{(U-c)^2}. \tag{23.3}$$

On integrating across the discontinuity, we have

$$\Delta\left(\frac{\phi}{U-c}\right) = -\lim_{\varepsilon \downarrow 0} \int_{z_0-\varepsilon}^{z_0+\varepsilon} \frac{\hat{p}}{(U-c)^2} \, dz. \tag{23.4}$$

The right-hand side of this equation vanishes provided either $c_i \neq 0$ as $\varepsilon \downarrow 0$ or, if $c_i = 0$, then $U - c \neq 0$ for z in the interval $[z_0 - \varepsilon, z_0 + \varepsilon]$. Thus the second matching condition becomes

$$\Delta[\phi/(U-c)] = 0 \quad \text{at } z = z_0. \tag{23.5}$$

This condition can also be derived from the requirement of streamline tangency, i.e. by requiring that the normal velocity of the fluid be continuous at the material interface.

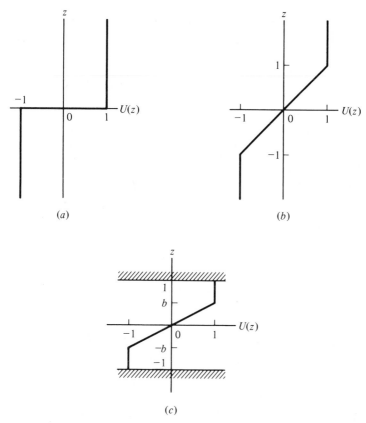

Fig. 4.4. (a) Unbounded vortex sheet; (b) unbounded shear layer; (c) bounded shear layer.

Since $\phi^\dagger = \phi/(U-c)$, the corresponding matching conditions for the adjoint problem are

$$\Delta\phi^\dagger = \Delta[(U-c)^2 D\phi^\dagger] = 0 \quad \text{at } z = z_0. \tag{23.6}$$

To illustrate some of the general stability characteristics discussed in § 22, consider the three flows shown in Fig. 4.4.

23.1 Unbounded vortex sheet

The Kelvin–Helmholtz instability of a vortex sheet has already been discussed in some detail in § 4 on the assumption that the disturbed flow was irrotational. When $U'' \equiv 0$ Rayleigh's equation reduces to

$(U-c)(\phi''-\alpha^2\phi)=0$ and if we ignore the continuous part of the spectrum then we have simply $\phi''-\alpha^2\phi=0$, which is equivalent to the vanishing of the y-component of the perturbation vorticity. Thus, with $U(z)=\text{sgn }z$ we take $\phi=Ae^{-\alpha z}$ for $z>0$ and $\phi=Be^{\alpha z}$ for $z<0$. The matching conditions then lead to the eigenvalue relation $c^2+1=0$ from which we have $c_r=0$ and, in accordance with our convention, $c_i=1$. The instability is therefore in the form of a standing wave which grows like $e^{\alpha t}$. The eigenvalue relation for this problem is independent of α and is in agreement with the long-wave approximation (22.47). We also have $A-iB=0$ so that if we normalize the solution by letting $\phi=e^{-\alpha z}$ for $z>0$ then $\phi=-ie^{\alpha z}$ for $z<0$.

23.2 Unbounded shear layer

Consider next the shear layer shown in Fig 4.4(b) for which $U(z)=z$ for $|z|<1$ and $U(z)=z/|z|$ for $|z|>1$. For this problem it is convenient to take the solution in the form

$$\phi=\begin{cases} Ae^{-\alpha(z-1)} & (z>1) \\ Be^{-\alpha(z-1)}+Ce^{\alpha(z+1)} & (|z|<1) \\ De^{\alpha(z+1)} & (z<-1). \end{cases}$$

On applying the matching conditions at $z=\pm1$ we obtain the eigenvalue relation (Rayleigh 1894, vol. II, p. 393)

$$c^2=(4\alpha^2)^{-1}\{(1-2\alpha)^2-e^{-4\alpha}\}. \tag{23.7}$$

If we let $\alpha_s\cong0.64$ be the root of $1-2\alpha+e^{-2\alpha}=0$ then c is purely imaginary for $0<\alpha<\alpha_s$ and we have instability. The growth rate αc_i is greatest when $\alpha\cong0.40$, and in the long-wave limit we again recover equation (22.47). The eigenfunction ϕ_s associated with the neutral mode $\alpha=\alpha_s$ and $c=0$ is also of some interest. If we fix the normalization of ϕ_s by letting $A=1$ then we have

$$\phi=\begin{cases} \exp\{-\alpha_s(z-1)\} & (z>1) \\ (\cosh\alpha_s)^{-1}\cosh\alpha_s z & (|z|<1) \\ \exp\{\alpha_s(z+1)\} & (z<-1) \end{cases}$$

so that ϕ_s is an even function of z. In the long-wave limit $\phi\to$ constant $\times U(z)$ as $\alpha\downarrow0$.

23.3 *Bounded shear layer*

To further illustrate the lack of sufficiency in Rayleigh's inflexion-point theorem consider the shear layer shown in Fig. 4.4(*c*) for which

$$U(z) = \begin{cases} 1 & (b < z \leqslant 1) \\ z/b & (|z| < b) \\ -1 & (-1 \leqslant z < -b) \end{cases}$$

with $0 < b \leqslant 1$. In the limit as $b \to 0$ we have a vortex sheet which is unstable for all values of α. When $b = 1$, however, we have plane Couette flow which has no discrete eigenvalues and for which, as discussed in § 24, the continuous spectrum is stable. If we take the solution in the form

$$\phi = \begin{cases} A \sinh \alpha(1-z) & (b < z \leqslant 1) \\ B \sinh \alpha z + C \cosh \alpha z & (|z| < b) \\ D \sinh \alpha(1+z) & (-1 \leqslant z < -b) \end{cases}$$

which automatically satisfies the boundary conditions at $z = \pm 1$, then on applying the matching conditions at $z = \pm b$ we obtain the eigenvalue relation (Rayleigh 1894, vol. II, p. 388)

$$c^2 = 1 - \frac{\alpha b(1 + X^2) Y^2 + 2\alpha b XY - XY^2}{\alpha^2 b^2 \{(1 + X^2) Y + X(1 + Y^2)\}}, \tag{23.8}$$

where $X = \tanh \alpha b$ and $Y = \tanh \alpha(1-b)$. A little analysis of this result shows that $c^2 \to 2b - 1$ as $\alpha \to 0$ and that $c^2 \to 1$ as $\alpha \to \infty$. Since c^2 is a monotone increasing function of α it then follows that the flow is unstable provided $b < \frac{1}{2}$.

24 The initial-value problem

In this section we shall digress a little from the topic of normal modes, which forms the basis of the development of this chapter. In Chapter 1 we introduced the idea of seeking the development in time of an arbitrary small disturbance of some basic flow to test whether the flow was stable. This led to the linearization of the equations of motion and to the resolution of the arbitrary disturbance into independent wave components, each a normal mode.

Even if one accepts the approximation of linearization without reservation, however, one would expect to see in practice not only one normal mode but some superposition of many normal modes which is determined by the nature of the initial disturbance. Here we shall elaborate these ideas in the context of the linear stability of steady parallel flows.

We model this situation first by supposing that at some instant, say $t = 0$, there are given arbitrary smooth distributions of $\mathbf{u}'$ and p' in space subject to the constraint of incompressibility. We then seek to trace the subsequent development of the solutions $\mathbf{u}'(\mathbf{x}, t)$ and $p'(\mathbf{x}, t)$ of the linearized equations for $t > 0$. This is the classic *initial-value problem*. If $\mathbf{u}'$ and p' remain bounded for all time we deem the disturbance stable. Of course the initial-value problem is closely related to the spectrum of normal modes, because if the spectrum is complete then any solution of the initial-value problem may be represented as sums and integrals of the normal modes.

We noted in § 21 that for a given flow there is in fact only a finite number of non-singular normal modes at each value of the wavenumber. Therefore it is impossible in general to represent any smooth initial disturbance as a superposition of those normal modes, and we may say that they are incomplete. We found in § 22 the finiteness of the number of non-singular normal modes at marginal stability, and showed how to find their eigenfunctions ϕ_s. For many flows there is only one non-singular normal mode, and for any stable flow there is none. However, the spectrum of normal modes is in fact made complete by the inclusion of a continuous spectrum of singular normal modes associated with the singularity of the Rayleigh stability equation where $U(z) = c$. Because this singularity arises only when c is real (and lies within the range of U over the domain of flow), this continuous spectrum is composed of only stable modes and may be ignored when seeking only a criterion for stability.

These ideas are well exemplified by the simple case of plane Couette flow, even though plane Couette flow is stable and so has no discrete spectrum of non-singular normal modes for any value of the wavenumber. Thus we take

$$U(z) = z \quad \text{for } -1 \le z \le 1. \tag{24.1}$$

Orr (1907), pp. 26–29 was the first to treat this problem. He noted that the equation governing two-dimensional disturbances was simply of the form

$$\left(\frac{\partial}{\partial t}+z\frac{\partial}{\partial x}\right)\left(\frac{\partial^2\psi'}{\partial x^2}+\frac{\partial^2\psi'}{\partial z^2}\right)=0. \tag{24.2}$$

(This is the linearized vorticity equation, which takes this simple form because the basic flow has uniform vorticity. On taking the normal mode described by equation (21.14), it reduces to the Rayleigh stability equation.) This equation gives

$$\frac{\partial^2\psi'}{\partial x^2}+\frac{\partial^2\psi'}{\partial z^2}=F(x-zt,z), \tag{24.3}$$

where F is an arbitrary function of integration which is differentiable with respect to x. Any initial disturbance specified by $\psi'(x,z,0)$ may be used to determine F. Orr went on to solve the Poisson equation (24.3) for $\psi'(x,z,t)$ by the use of Fourier series in z.

An alternative, and more generally applicable, approach to the problem was developed by Eliassen, Høiland & Riis (1953) and Case (1960a,b). They took the Laplace transform of equation (24.2) with respect to t, and the Fourier transform with respect to x. This leads to an inhomogeneous linear ordinary differential equation in z for the transform of ψ', the inhomogeneity being due to the initial disturbance. The solution of this inhomogeneous equation and its boundary conditions can be found in terms of a Green's function which has a discontinuous derivative. Finally the transforms may be inverted in the usual way to give $\psi'(x,z,t)$. In fact it is found that in general $\psi'=O(t^{-2})$ and $u'=\partial\psi'/\partial z=O(t^{-1})$ as $t\to\infty$ for fixed x and z, and that the flow is stable.

This method of Fourier–Laplace transforms has been applied to a general basic flow by Case (1960a) and Dikii (1960a). Their argument gives a solution of the form

$$\psi'(x,z,t)=\frac{1}{2\pi}\int_{-\infty}^{\infty}d\alpha\,\frac{1}{2\pi i}\int_{\varepsilon-i\infty}^{\varepsilon+i\infty}ds\,e^{i\alpha x+st}\phi_{s\alpha}(z) \tag{24.4}$$

for $t>0$, $-\infty<x<\infty$, $z_1<z<z_2$, and ε is chosen so that the Bromwich contour of integration in the complex s-plane is to the

right of all singularities of the integrand. The Fourier–Laplace transform is then determined by

$$\phi_{s\alpha}(z) = \int_{z_1}^{z_2} G(z, z_0; s) \frac{\Psi(z_0; 0)}{s + i\alpha U(z_0)} \, dz_0, \qquad (24.5)$$

where G is the appropriate Green's function, and Ψ is determined from the initial distribution of the stream function $\psi'(x, z, 0)$. In the inversion of the Laplace transform the discrete spectrum emerges from the residues at the poles of G, if any, the poles occurring wherever the normal-mode problem gives an eigenvalue for a non-singular eigenfunction. Also the continuous spectrum is associated with the zeros of $s + i\alpha U(z_0)$.

Yet another method is to use generalized functions, directly or indirectly. Eliassen, Høiland & Riis (1953) solved the initial-value problem for plane Couette flow in this way, attributing their solution to unpublished work by Fjørtoft and Høiland. We shall follow the later treatment of Case (1960a), who invoked the method of Fourier–Laplace transforms to justify his use of generalized functions. For plane Couette flow defined by equation (24.1) the normal-mode eigenvalue problem becomes

$$(z - c)(D^2 - \alpha^2)\phi = 0, \qquad (24.6)$$

and

$$\phi = 0 \quad \text{at } z = \pm 1. \qquad (24.7)$$

Now if c does not lie in the range $(-1, 1)$ of the basic velocity given by equation (24.1) then we may divide both sides of equation (24.6) by $z - c$ to get

$$(D^2 - \alpha^2)\phi = 0. \qquad (24.8)$$

It follows at once that there is no non-trivial solution of the system described by equations (24.8) and (24.7); thus, there is no non-singular normal mode and the discrete spectrum is empty. If $-1 < c < 1$, however, then division of equation (24.6) by $z - c$ gives

$$(D^2 - \alpha^2)\phi = \delta(z - c), \qquad (24.9)$$

where an arbitrary normalization has been chosen and δ is the Dirac delta function. The solution of the system described by equations

(24.9) and (24.7) is readily seen to be

$$\phi = G(z, c) \equiv \begin{cases} \dfrac{\sinh \alpha(c-1) \sinh \alpha(z+1)}{\alpha \sinh 2\alpha} & \text{for } -1 \leqslant z \leqslant c \\[2ex] \dfrac{\sinh \alpha(c+1) \sinh \alpha(z-1)}{\alpha \sinh 2\alpha} & \text{for } c \leqslant z \leqslant 1. \end{cases} \tag{24.10}$$

This function is related to the Green's function in the Fourier–Laplace transform method for this problem, and can be seen to have a discontinuous derivative at $z = c$. It gives the continuous spectrum of singular eigenfunctions corresponding to values of c in the interval $(-1, 1)$ for each value of α. The physical quantity $\psi'(x, z, t)$, however, which is composed of a double integral of the eigenfunctions given by equations (24.10) is non-singular and may be used to represent the development in time of any smooth initial disturbance.

Although the formal treatment of the inviscid initial-value problem is straightforward in principle, detailed analysis of the solution, even for simple basic flows and specially chosen initial conditions, is technically complicated (Case 1960a). For the viscous initial-value problem, Gaster (1975) has made a theoretical study of the development of a model wavepacket in a laminar boundary layer and good agreement was obtained with the experimental results of Gaster & Grant (1975) during the initial stages of growth. In physical terms, the linear instability may be described qualitatively as follows. If the initial disturbance is localized in space, then the disturbance propagates much like a wavepacket of the most unstable modes, travelling with their group velocity and growing exponentially like them as it travels. (One may note that the group velocity may be zero, but is typically in the direction of the basic flow, and that interference of the most unstable modes leads to algebraic moderation of the exponential growth of a wavepacket.) The disturbance decays rapidly up-stream and down-stream of its centre as it travels. Thus, relative to an observer moving with the group velocity, the disturbance grows exponentially in time. But the disturbance measured at any fixed station x eventually decays rapidly with time because the localized disturbance travels away; indeed, at a distant station the disturbance may appear first to grow and then decay rapidly as its centre passes by. This instability is called *convected instability* by plasma physicists; if, however, the group velocity is zero, the instability remains stationary as it grows exponentially and then is called *absolute* or *non-convected*.

So far we have treated the classic initial-value problem. But this is a poor model of many experiments for which some device, such as a small loudspeaker or an oscillating ribbon, is introduced into a basic flow to produce a controlled wave disturbance. At some instant the device is turned on, and thereafter it is maintained as a wave source of fixed amplitude form and fixed angular frequency, ω (say). If the group velocity for that frequency has the same direction as the basic flow, the resultant forced oscillation of the flow is observed to have the real frequency $\omega = -\alpha c$ downstream of the source after the transients have propagated away. This situation and its mathematical model will be discussed more fully for the viscous problem in § 32, but it suffices to state now that what little work has been done on the model suggests that the forced disturbance of frequency ω develops at any fixed station x as $t \to \infty$ and is composed of modes each of which varies downstream like $e^{i\alpha x}$, where $\alpha = \alpha(\omega)$ is real if the mode is stable and has negative imaginary part if it is unstable. At each instant the disturbance is bounded as $x \to \pm\infty$ because the transients travel with their group velocities, but the disturbance may seem unbounded if one examines it by letting $t \to \infty$ before one lets $x \to +\infty$.

Closely related to this kind of initial-value problem is the use of *spatially growing modes*, sometimes called simply *spatial modes*. They were first applied to hydrodynamic stability by Watson (1962) and Gaster (1962), but seem to have been first used by Landau (1946) for a problem of plasma physics, a subject for which they have since been used extensively (see, for example, Clemmow & Dougherty (1969), Chap. 6).

The essential theoretical ideas behind the use of spatial modes are as follows. The dispersion relation can be written in the form

$$\mathscr{F}(\alpha, \omega) = 0, \tag{24.11}$$

where $\omega = -\alpha c$, and may be regarded as determining either the complex value or values of ω for any real value of α or of α for any real value of ω, or, indeed, as determining the complete functional relationship between the complex variables ω and α. Hitherto we have treated only *temporally growing modes*, or *temporal modes*, for which we take a real value of the wavenunber α and seek complex c or ω to determine the temporal growth of the mode through the factor $e^{-i\alpha ct} = e^{i\omega t}$. For spatial modes we take a real value of the frequency ω and seek the complex eigenvalues $\alpha = \alpha_r + i\alpha_i$. Consequently the spatial modes behave like $\exp\{i(\alpha x + \omega t)\} = \exp\{-\alpha_i x + i(\alpha_r x + \omega t)\}$; and so they grow or decay exponentially with x unless $\alpha_i = 0$ and also oscillate in x with wavelength $2\pi/\alpha_r$. These spatial modes resemble the forced oscillation due to a source of frequency ω, so one might think that the criterion of instability of the basic flow is simply that $\alpha_i \neq 0$ for some real value of ω, but unfortunately the criterion is not always as simple as that.

It must be remembered that spatial modes (and, indeed, temporal modes) are significant because of their role in describing solutions of initial-value problems. A spatial mode is itself inadmissible as a solution when it is unbounded at infinity, but may nevertheless describe the solution a long time after a localized source of given frequency ω has been turned on. In this way the physical properties of spatial modes are closer than those of temporal modes to the instability observed in most experiments on parallel flows, and for that reason spatial modes are often used when making comparisons with experimental results as discussed in § 32.

To interpret spatial modes without ambiguity, however, one must go to the initial-value problem, and this is not easy. One can see that to evaluate Fourier–Laplace integrals such as in equation (24.4) one may seek to integrate first with respect to either α or s ($= i\omega$). Also one can change the contours of integration in the complex planes of α and ω if one knows the nature of the singularities, as is often desirable in evaluating the integral asymptotically as $t \to \infty$. It is for these reasons that the relationship (24.11) between the complex variables α and ω is important.

These ideas are discussed further by Clemmow & Dougherty (1969), Chap. 6 and applied to viscous fluid by Gaster (1965); see also § 47.

THE VISCOUS THEORY

25 The governing equations

In discussing the governing equations for a viscous fluid we shall again suppose that the basic steady flow is of the form given by equation (21.1). For a viscous fluid, however, $U_*(z_*)$ can no longer be an arbitrary function of z_* but must satisfy the equation of motion

$$\nu \frac{d^2 U_*}{dz_*^2} = \frac{1}{\rho} \frac{dP_*}{dx_*}, \qquad (25.1)$$

where $dP_*/dx_* = $ constant. The class of strictly parallel flows is thus somewhat limited since U_* can at most be quadratic in z_*. This includes, however, two important special cases which in dimensionless form are:

Plane Couette flow:

$$U(z) = z, \quad P = \text{constant} \, (-1 \leq z \leq 1), \qquad (25.2)$$

where $V = $ the velocity of the upper plate and $L = $ half the channel width; and

Plane Poiseuille flow:

$$U(z) = 1 - z^2, \quad dP/dx = \text{constant} \, (-1 \leq z \leq 1), \qquad (25.3)$$

where $V = $ the maximum velocity at the centre of the channel $= -(L^2/2\rho\nu)dP_*/dx_*$ and $L = $ half the channel width. We have therefore a one-parameter family of strictly parallel flows which can be thought of as a linear combination of plane Couette flow and plane Poiseuille flow. The governing equations will be derived in this section on the assumption that the basic flow is strictly parallel. More generally, however, the resulting equations can often be used to discuss the stability of so-called *nearly parallel flows*. A two-dimensional basic flow $\mathbf{U} = \{U(x, z), 0, W(x, z)\}$ is said to be nearly parallel if

$$W(x, z) \ll U(x, z) \text{ and } \partial U/\partial x \ll 1. \qquad (25.4)$$

The class of nearly parallel flows thus includes many important flows, for example boundary layers, jets and shear layers.

The linearized equations of motion are then easily found to be

$$\left(\frac{\partial}{\partial t} + U\frac{\partial}{\partial x}\right)\mathbf{u}' + w'\frac{dU}{dz}\mathbf{i} = -\boldsymbol{\nabla}p' + R^{-1}\Delta\mathbf{u}' \text{ and } \boldsymbol{\nabla}\cdot\mathbf{u}' = 0,$$

$$(25.5)$$

where $R = VL/\nu$ is the Reynolds number. These equations differ from equations (21.5) only in the additional viscous terms $R^{-1}\Delta\mathbf{u}'$.

In studying the solutions of equations (25.5) we shall proceed directly with the usual normal-mode analysis and later we will consider briefly the more general initial-value problem. Thus, when $\mathbf{u}'$ and p' have the forms (21.6) we immediately obtain

$$\left.\begin{aligned}
\{D^2 - (\alpha^2 + \beta^2) - i\alpha R(U - c)\}\hat{u} &= RU'\hat{w} + i\alpha R\hat{p}, \\
\{D^2 - (\alpha^2 + \beta^2) - i\alpha R(U - c)\}\hat{v} &= i\beta R\hat{p}, \\
\{D^2 - (\alpha^2 + \beta^2) - i\alpha R(U - c)\}\hat{w} &= RD\hat{p},
\end{aligned}\right\} \quad (25.6)$$

and
$$i(\alpha\hat{u} + \beta\hat{v}) + D\hat{w} = 0.$$

For rigid boundaries we have the usual viscous boundary conditions

$$\hat{u} = \hat{v} = \hat{w} = 0 \quad \text{at} \quad z = z_1 \quad \text{and} \quad z_2. \quad (25.7)$$

In particular, the eigenvalue relation for this problem must therefore be of the form

$$\mathcal{F}(\alpha, \beta, c, R) = 0. \quad (25.8)$$

The three-dimensional problem defined by equations (25.6) and (25.7) can, however, be reduced to an equivalent two-dimensional problem by the use of Squire's (1933) transformation. Thus if we use the relations (21.10) and, in addition, let $\tilde{\alpha}\tilde{R} = \alpha R$ then we obtain

$$\left.\begin{aligned}
\{D^2 - \tilde{\alpha}^2 - i\tilde{\alpha}\tilde{R}(U - \tilde{c})\}\tilde{u} &= \tilde{R}U'\tilde{w} + i\tilde{\alpha}\tilde{R}\tilde{p}, \\
\{D^2 - \tilde{\alpha}^2 - i\tilde{\alpha}\tilde{R}(U - \tilde{c})\}\tilde{w} &= \tilde{R}D\tilde{p}, \\
i\tilde{\alpha}\tilde{u} + D\tilde{w} &= 0,
\end{aligned}\right\} \quad (25.9)$$

and

$$\tilde{u} = \tilde{w} = 0 \quad \text{at } z = z_1 \quad \text{and} \quad z_2. \quad (25.10)$$

These equations have the same mathematical structure as equations (25.6) and (25.7) with $\beta = \hat{v} = 0$ and they thus define the equivalent two-dimensional problem. In particular, since $\tilde{\alpha} \geq \alpha$ it immediately follows that $\tilde{R} \leq R$ and hence we have

Squire's theorem. To obtain the minimum critical Reynolds number it is sufficient to consider only two-dimensional disturbances.

In the subsequent discussion we shall therefore restrict our attention to two-dimensional disturbances and it is then convenient to express the governing equations in terms of the stream function $\psi'(x, y, z, t)$ or its amplitude $\phi(z)$ (cf. equations (21.13) and (21.14)). The linearized equation of vorticity for a two-dimensional disturbance can then be written in the form

$$\frac{\partial}{\partial t}\Delta\psi' + U\frac{\partial}{\partial x}\Delta\psi' - U''\frac{\partial\psi'}{\partial x} = R^{-1}\Delta^2\psi', \qquad (25.11)$$

where $\Delta\psi'$ is the y-component of the vorticity. From this equation, or from equations (25.6) with $\beta = \hat{v} = 0$, it follows that $\phi(z)$ satisfies the celebrated *Orr–Sommerfeld equation*

$$(i\alpha R)^{-1}(D^2 - \alpha^2)^2\phi = (U - c)(D^2 - \alpha^2)\phi - U''\phi \quad (25.12)$$

with boundary conditions

$$\alpha\phi = D\phi = 0 \quad \text{at } z = z_1 \quad \text{and} \quad z_2. \qquad (25.13)$$

The eigenvalue relation for the two-dimensional problem is then simply of the form

$$\mathcal{F}(\alpha, c, R) = 0. \qquad (25.14)$$

Although the Orr–Sommerfeld equation was derived on the assumption that the basic flow is strictly parallel, in discussing certain properties of the equation it is convenient to suppose more generally that $U(z)$ is an analytic function of the complex variable z. Equation (25.12) is then regular not only in the independent variable z but also in the parameters α, c, R considered as complex variables. The solutions of equation (25.12) are therefore integral functions of z, α, c, R and if the boundary conditions (25.13) are imposed for finite values of z then the eigenvalue relation (25.14) is also an integral function of α, c, R. Thus if there were a continuous set of eigenvalues for c with given values of α and R then, by the theory of analytic functions, equation (25.14) would reduce to an identity. The eigenvalue spectrum for the viscous problem must therefore be purely *discrete*. This argument is due to Lin (1961). For unbounded flows of, for example, the boundary-layer type, this argument is not applicable. Recent work by Jordinson (1971), Mack (1976), Murdock & Stewartson (1977), Antar & Benek (1978), and Grosch & Salwen (1978) has shown that the spectrum for the Blasius boundary layer consists of a finite number of discrete eigenvalues, the number of which increases with R, and a continuous spectrum for which the eigenfunctions vary sinusoidally as $z \to \infty$.

In the usual temporal stability problem, in which the growth of a disturbance in time is considered, we take α and R to be real and fixed. The eigenvalue relation (25.14) then defines a discrete set of eigenvalues c_j ($j = 1, 2, \cdots$) which may, for convenience, be arranged in order of increasing damping, i.e. $c_{1i} \geqslant c_{2i} \geqslant \cdots$. The dependence of the eigenvalues c_j on the parameters α and R is exceedingly intricate, however, even for such simple flows as plane Couette or plane Poiseuille flow and it has been found that this ordering may change as α and R are varied.

For some purposes it is convenient to suppose that the eigenvalue relation (25.14) has been solved for c and to write it in the form

$$c = f(\alpha, R) + i g(\alpha, R). \tag{25.15}$$

If one or more neutrally stable modes exist for which $c_i = 0$ then $g(\alpha, R) = 0$ and the curves of neutral stability are given by $R = h(\alpha)$ (say). To establish that these neutral curves represent stability boundaries it is necessary to show that c_i changes sign on crossing the neutral curves. This requires a more detailed analysis of the eigenvalue relation as discussed, for example, by Lin (1945).

When the eigenvalue spectrum is discrete it is natural to attempt to treat the initial-value problem by an expansion in terms of the eigenfunctions $\phi_j(z)$. On considering a single wavenumber in the x-direction, the expansion would take the form

$$\psi'(x, z, t) = \sum_{j=1}^{\infty} a_j(\alpha, R) \phi_j(z, \alpha, R) \exp\{i\alpha[x - c_j(\alpha, R)t]\},$$
$$\tag{25.16}$$

where the dependence of a_j, c_j and ϕ_j on α and R have been indicated explicitly. The behaviour of this series is extraordinarily complicated and difficult to analyse for large values of R because the eigenfunctions $\phi_j(z, \alpha, R)$ do not in general have limits as $R \to \infty$. For fixed values of α and R, however, an important question concerns the conditions under which a function $f(z)$ can be expanded in the form

$$f(z) = \sum_{j=1}^{\infty} a_j \phi_j(z). \tag{25.17}$$

This problem was first considered by Schensted (1960), using classical methods of analysis, and later by DiPrima & Habetler (1969), who proved that if f has a continuous first derivative and vanishes at $z = z_1$ and z_2 then the series given by equation (25.17) converges uniformly with pointwise limit f. Assuming that the eigenvalues c_j are all simple, the expansion coefficients can then be obtained formally in the following manner. Let $\phi_j^\dagger(z)$ be the eigenfunctions of the adjoint problem, which consists of the equation

$$(i\alpha R)^{-1}(D^2 - \alpha^2)^2 \phi^\dagger = (D^2 - \alpha^2)(U - c)\phi^\dagger - U''\phi^\dagger$$
(25.18)

together with the boundary conditions

$$\alpha\phi^\dagger = D\phi^\dagger = 0 \quad \text{at } z = z_1 \quad \text{and} \quad z_2. \tag{25.19}$$

It then follows that

$$\int_{z_1}^{z_2} \phi_j(D^2 - \alpha^2)\phi_k^\dagger \, dz = 0 \text{ for } j \neq k. \tag{25.20}$$

If we suppose that the eigenfunctions have been normalized so that

$$\int_{z_1}^{z_2} \phi_j(D^2 - \alpha^2)\phi_j^\dagger \, dz = 1 \tag{25.21}$$

then the expansion coefficients are given by

$$a_j = \int_{z_1}^{z_2} f(z)(D^2 - \alpha^2)\phi_j^\dagger \, dz. \tag{25.22}$$

26 The eigenvalue spectrum for small Reynolds numbers

For sufficiently small values of the Reynolds number we expect viscosity to have a purely stabilizing effect. Since c has been made dimensionless with respect to V we would expect the damping rate αc_i to become large (and negative) like $1/R$ as $R \to 0$ so that the product αRc is independent of V in this limit. Thus, as $R \to 0$ it is clear that we must recover Rayleigh's (1892a) results for the small oscillations of a fluid at rest, and this is confirmed by the present discussion.

The Orr–Sommerfeld equation has been studied in detail for small values of the Reynolds number by Southwell & Chitty (1930) for the particular case of plane Couette flow. A more general discussion was given later by Pekeris (1936), who also gave detailed results for both plane Couette and plane Poiseuille flow. The problem has been considered again independently by Birikh, Gershuni & Zhukhovitskii (1965). These results all show that there are important qualitative differences in the spectrum depending upon whether the basic flow is odd or not.

26.1 *A perturbation expansion*

Consider then, following Pekeris, an expansion of the solution in powers of $i\alpha R$ of the form

$$\phi(z) = \sum_{s=0}^{\infty} (i\alpha R)^s \phi^{(s)}(z) \quad \text{and} \quad i\alpha Rc = \sum_{s=0}^{\infty} (i\alpha R)^s c^{(s)}.$$

$$(26.1)$$

For simplicity we will restrict the present discussion to the determination of $c^{(0)}$ and $c^{(1)}$ which give the limiting values of the damping rate and wave speed respectively.

To the first order, therefore, we have the equation

$$(D^2 - \alpha^2 + c^{(0)})(D^2 - \alpha^2)\phi^{(0)} = 0 \qquad (26.2)$$

which is independent of the basic flow. It is convenient to consider the even and odd solutions of this equation separately and for this purpose we shall suppose that the boundary conditions are imposed at $z = \pm 1$. The even solutions are then given by

$$c^{(0)} = \alpha^2 + p_n^2$$

and

$$\phi_n^{(0)}(z) = \frac{\cosh \alpha z}{\cosh \alpha} - \frac{\cos p_n z}{\cos p_n} \quad (n = 0, 2, 4, \ldots), \quad (26.3)$$

where the $p_n(\alpha)$ are the positive roots of

$$\alpha \tanh \alpha + p \tan p = 0. \qquad (26.4)$$

For $\alpha = 0$, $p_n(0) = \frac{1}{2}(n+2)\pi$ and, for example, $\phi_0^{(0)}(z) = 1 + \cos \pi z$. Similarly, the odd solutions are given by

$$c^{(0)} = \alpha^2 + q_n^2$$

and

$$\phi_n^{(0)}(z) = \frac{\sinh \alpha z}{\sinh \alpha} - \frac{\sin q_n z}{\sin q_n} \quad (n = 1, 3, 5, \ldots), \quad (26.5)$$

where the $q_n(\alpha)$ are the positive roots of

$$\alpha \coth \alpha - q \cot q = 0. \tag{26.6}$$

For $\alpha = 0$, $q_n(0)/\pi \cong 1.430, 2.459, 3.471, \ldots$ with $q_n(0) \sim \frac{1}{2}(n+2)\pi$ for large values of n and, for example, $\phi_1^{(0)}(z) = z - (\sin q_1)^{-1} \sin q_1 z$. These results thus give the damping rates for a fluid at rest in a channel, as first found by Rayleigh (1892a).

To determine $c^{(1)}$ we must now consider the equation satisfied by $\phi^{(1)}(z)$:

$$(D^2 - \alpha^2 + c^{(0)})(D^2 - \alpha^2)\phi^{(1)} = \{(U - c^{(1)})(D^2 - \alpha^2) - U''\}\phi^{(0)}. \tag{26.7}$$

The solvability condition for this equation is

$$\int_{-1}^{1} \phi^{(0)}\{(U - c^{(1)})(D^2 - \alpha^2) - U''\}\phi^{(0)} \, dz = 0 \tag{26.8}$$

and this determines $c^{(1)}$. For odd profiles, therefore, $c^{(1)} = 0$ and, more generally, it can be shown that $c^{(n)} = 0$ for $n = 1, 3, 5, \ldots$ Modes of this type, for which $c_r = 0$, are often called 'standing' waves. For plane Couette flow, Southwell & Chitty (1930) and many others have shown† that c_r remains zero for $0 \leqslant \alpha R \leqslant (\alpha R)_*$ (say); for $\alpha R > (\alpha R)_*$, $c_r \neq 0$ in general and $c_r \to \pm 1$ as $\alpha R \to \infty$. A more complete discussion of the modal structure for plane Couette flow is given in § 31. From the present results, however, it seems likely, as Pekeris first suggested, that the series in equation (26.1) converge only up to $(\alpha R)_*$.

For profiles which are not odd, $c^{(1)} \neq 0$ in general and the modes are then of the 'travelling' wave type. Even for profiles of simple form, however, the calculation of $c^{(1)}$ is rather long and not very illuminating. For plane Poiseuille flow with $\alpha = 0$ we may note that

$$c^{(1)} = \begin{cases} \dfrac{2}{3} - \dfrac{5}{2p_n^2} & (n = 0, 2, 4, \ldots) \\[2mm] \dfrac{2}{3} + \dfrac{5}{6q_n^2} & (n = 1, 3, 5, \ldots). \end{cases} \tag{26.9}$$

†These statements have been expressed in terms of αR rather than R so as to include the case when $\alpha = 0$ but αR is finite. For example, for plane Couette flow $(\alpha R)_* \cong 37.6$ when $\alpha = 0$.

Hence, for the higher modes (i.e. n large), $c^{(1)} \to \frac{2}{3}$ which is simply the mean velocity of the basic flow.

26.2 Sufficient conditions for stability

A simple method for obtaining sufficient conditions for stability has been given by Synge (1938) and his results have recently been extended and improved by Joseph (1968, 1969). In this method we first multiply the Orr–Sommerfeld equation by ϕ^* and integrate over the interval $(-1, 1)$ to obtain

$$I_2^2 + 2\alpha^2 I_1^2 + \alpha^4 I_0^2 = -i\alpha RQ + i\alpha Rc(I_1^2 + \alpha^2 I_0^2), \quad (26.10)$$

where we take $z_1 = -1$, $z_2 = 1$,

$$I_n^2 = \int_{-1}^{1} |D^n\phi|^2 \, dz \quad (n = 0, 1, 2)$$

and

$$Q = \int_{-1}^{1} \{U|D\phi|^2 + (\alpha^2 U + U'')|\phi|^2\}dz + \int_{-1}^{1} U'(D\phi)\phi^* \, dz.$$

From this equation we immediately have

$$c_r = \text{Re}\,(Q)/(I_1^2 + \alpha^2 I_0^2) \qquad (26.11)$$

and

$$c_i = \{\text{Im}\,(Q) - (\alpha R)^{-1}(I_2^2 + 2\alpha^2 I_1^2 + \alpha^4 I_0^2)\}/(I_1^2 + \alpha^2 I_0^2), \qquad (26.12)$$

where

$$\text{Re}\,(Q) = \int_{-1}^{1} \{U|D\phi|^2 + (\alpha^2 U + \tfrac{1}{2}U'')|\phi|^2\} \, dz$$

and

$$\text{Im}\,(Q) = \tfrac{1}{2}i \int_{-1}^{1} U'\{\phi(D\phi^*) - (D\phi)\phi^*\} \, dz.$$

Equation (26.12) is simply the energy equation for two-dimensional disturbances propagating in the direction of the basic flow. Observe now that

$$|\text{Im}\,(Q)| \leqslant \int_{-1}^{1} |U'| \cdot |\phi| \cdot |D\phi| \, dz$$

and hence, by Schwarz's inequality,

$$|\text{Im}\,(Q)| \leqslant qI_0I_1, \quad \text{where } q = \max_{-1 \leqslant z \leqslant 1} |U'|.$$

This gives an upper bound for c_i:

$$c_i \leqslant \{qI_0I_1 - (\alpha R)^{-1}(I_2^2 + 2\alpha^2 I_1^2 + \alpha^4 I_0^2)\}/(I_1^2 + \alpha^2 I_0^2), \quad (26.13)$$

from which it follows that a sufficient condition for stability is

$$\alpha Rq < (I_2^2 + 2\alpha^2 I_1^2 + \alpha^4 I_0^2)/I_0I_1. \quad (26.14)$$

A more convenient and explicit form of these conditions has been obtained by Joseph (1969) who proved[†]

Theorem 1. Let $c(\alpha, R)$ be any eigenvalue of the Orr–Sommerfeld equation with $\phi(\pm 1) = \phi'(\pm 1) = 0$. Then

$$c_i \leqslant \frac{q}{2\alpha} - \left\{ \frac{\pi^2(\pi^2 + \alpha^2)}{\pi^2 + 4\alpha^2} + \alpha^2 \right\} \frac{1}{\alpha R}. \quad (26.15)$$

Furthermore, no amplified disturbances exist if

$$\alpha Rq < f(\alpha) \equiv \max \{f_1(\alpha), f_2(\alpha)\}, \quad (26.16)$$

where

$$f_1(\alpha) = \lambda_3^2 \pi + 2^{3/2}\alpha^3 \quad \text{and} \quad f_2(\alpha) = (\lambda_3^2 + \alpha^2)\pi.$$

The proof of this theorem will not be given here, but it may be noted that the proof does depend in an essential way on the use of the inequalities

$$I_1^2 \geqslant \tfrac{1}{4}\pi^2 I_0^2, \quad I_2^2 \geqslant \tfrac{1}{4}\pi^2 I_1^2, \quad \text{and} \quad I_2^2 \geqslant \lambda_3^4 I_0^2, \quad (26.17)$$

where λ_3^4 is the least eigenvalue of a vibrating rod with clamped ends at $z = \pm 1$, i.e. $\lambda_3 \cong 2.3650$. For $\alpha = 0$ we have stability if $\alpha Rq < 17.6$; otherwise, $f(\alpha)$ has a single minimum at $\alpha \cong 1.46$ which then gives stability if $Rq < 18.1$.

Some improvement in the bound given by equation (26.16) has been obtained by Georgescu (1970) but the best estimates are obtained by considering the variational problem associated with equation (26.12) when $c_i = 0$. This shows that the flow is stable if

[†]Joseph's results were derived for the interval $(0, 1)$ and they have been restated here for the interval $(-1, 1)$

$R < R_1(\alpha)$, where R_1 is the least eigenvalue of R for which ϕ satisfies

$$(D^2 - \alpha^2)^2 \phi + i\alpha R(U'D\phi + \tfrac{1}{2}U''\phi) = 0, \qquad (26.18)$$

with $\phi(\pm 1) = \phi'(\pm 1) = 0$. This equation was first derived by Orr (1907) in a different manner which will be discussed in § 53.1. Equation (26.18) might appear to be somewhat simpler than the Orr–Sommerfeld equation, but its structure is quite different, and it is not easy to solve even in the case of plane Couette flow. Orr's calculations show, however, that we have stability if

$$\left. \begin{array}{l} R < 44.3 \text{ for plane Couette flow (with } U = z) \\ R < 87.7 \text{ for plane Poiseuille flow (with } U = 1 - z^2). \end{array} \right\} \quad (26.19)$$

and

This result for plane Poiseuille flow has also been confirmed by MacCreadie (1931).

For an inviscid fluid we recall that the wave speed c_r must lie in the range of U. The generalization of this result for a viscous fluid has also been considered by Joseph (1968) who proved

Theorem 2. Let $c(\alpha, R)$ be any eigenvalue of the Orr–Sommerfeld equation with $\phi(\pm 1) = \phi'(\pm 1) = 0$. Then

(a) if $U''_{min} \geqslant 0$, $U_{min} < c_r < U_{max} + \dfrac{2U''_{max}}{\pi^2 + 4\alpha^2}$,

(b) if $U''_{min} \leqslant 0 \leqslant U''_{max}$,

$$U_{min} + \frac{2U''_{min}}{\pi^2 + 4\alpha^2} < c_r < \frac{2U''_{max}}{\pi^2 + 4\alpha^2},$$

(c) if $U''_{max} \leqslant 0$, $U_{min} + \dfrac{2U''_{min}}{\pi^2 + 4\alpha^2} < c_r < U_{max}$.

Here $U_{min} = \min U$ and $U''_{min} = \min U''$, etc. These results follow from equation (26.11) on noting that

$$c_r = U(z_1) + \frac{U''(z_2)}{2\{(I_1/I_0)^2 + \alpha^2\}},$$

where z_1 and z_2 are mean values in the interval $(-1, 1)$, and using the first of the inequalities (26.17). For a viscous fluid, therefore, the wave speed c_r is restricted to an interval which is slightly larger than the range of U. In situations covered by (c), e.g. plane Poiseuille flow, negative wave speeds (or, rather, $c_r < U_{min}$) have not been found. For Jeffery–Hamel flow in a diverging channel with back flow, however, Eagles (1966) found negative wave speeds lying outside the range of U. Although these results are not fully understood, they are consistent with (b).

Similar results have also been obtained by Joseph (1969) for flows in boundary layers and in circular pipes. For jets, however, a different approach is required and for this case reference may be made to Howard (1959).

27 Heuristic methods of approximation

In studying the behaviour of the solutions of the Orr–Sommerfeld equation for large values of αR, our ultimate goal is to obtain asymptotic approximations which are uniformly valid in a bounded domain of the (complex) z-plane containing not only the turning point z_c but also the boundary points z_1 and z_2 and which can then be used to derive approximations to the eigenvalue relation. In this section, however, we will consider the heuristic methods of approximation which were developed mainly by Heisenberg (1924), Tollmien (1929, 1947) and Lin (1945, 1955). Although the resulting approximations are not uniformly valid, they are adequate for many computational purposes and they do provide valuable insight into the general structure of the problem. In addition, these approximations continue to play an important role in some of the more recent developments, as discussed in Chapter 5.

The heuristic theory can perhaps best be understood from the point of view of the *method of matched asymptotic expansions*. The essential elements of the theory include the derivation of first approximations of inner and outer type, the matching of these approximations, and finally the combining of them to form composite approximations. To simplify the present discussion we shall now restrict attention to the case of marginal stability for which c and hence U'_c is real. It is then convenient to suppose that $U(z)$ is monotone increasing so that U'_c is positive. To insure that we are dealing with a simple turning point we must also require that U'_c be bounded away from zero. This condition is not satisfied, for example, by a boundary layer at separation for which $U'_c \downarrow 0$ as $c \downarrow 0$ (Hughes & Reid 1965b).

27.1 *The reduced equation and the inviscid approximations*

From the point of view of singular perturbation theory, a natural starting point would be to consider the inviscid or *reduced equation*

$$(U-c)(D^2-\alpha^2)\phi^{(0)} - U''\phi^{(0)} = 0, \qquad (27.1)$$

which is obtained by formally setting $(i\alpha R)^{-1} = 0$ in equation (25.12). Alternatively, $\phi^{(0)}$ can be regarded as the leading term in a formal expansion of ϕ in inverse powers of $i\alpha R$. Equation (27.1) is, of course, simply Rayleigh's stability equation and, since it is of only second order, it can provide approximations to at most two solutions of the Orr–Sommerfeld equation. Furthermore, it is singular at the critical point z_c (unless $U_c'' = 0$) whereas the full equation is regular at z_c. This singular character of the reduced equation is one of the chief sources of difficulty in constructing uniformly valid asymptotic approximations to the solutions of the Orr–Sommerfeld equation.

As discussed in § 22, the solutions of Rayleigh's equation were expressed by Heisenberg (1924) as power series in α^2 and later by Tollmien (1929) in a form more suitable for the present purposes. Thus, from equations (22.25) and (22.26) we have

$$\phi_1^{(0)}(z) = (z - z_c)P_1(z) \qquad (27.2)$$

and

$$\phi_2^{(0)}(z) = P_2(z) + (U_c''/U_c')\phi_1^{(0)}(z) \ln (z - z_c), \qquad (27.3)$$

where $P_1(z)$ and $P_2(z)$ are analytic at z_c with $P_1(z_c) = P_2(z_c) = 1$ and $P_2'(z_c) = 0$. Since $\phi_1^{(0)}(z)$ is analytic at $z = z_c$, we would expect it to provide an approximation to some solution, $\phi_1(z)$ (say), of the Orr–Sommerfeld equation with an error of the order of $(\alpha R)^{-1}$. But $\phi_2^{(0)}(z)$ has a logarithmic branch point at $z = z_c$ and it cannot, therefore, provide an approximation to *any* solution of the Orr–Sommerfeld equation in a full neighbourhood of z_c. It does, however, provide an approximation of outer type to some solution, $\phi_2(z)$ (say), in a certain restricted domain which must necessarily exclude z_c. An important problem, therefore, is the determination of the domain in which $\phi_2(z) \sim \phi_2^{(0)}(z)$, and this will be considered later in this section and again, in greater detail, in Chapter 5. In the case of marginal stability with z_c real this problem is often posed in the following form: if we take $\ln (z - z_c) = \ln |z - z_c|$ for $z - z_c > 0$,

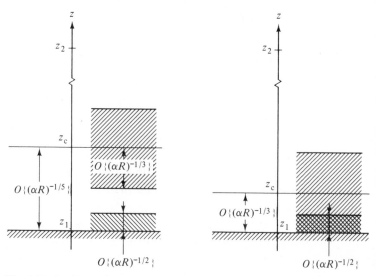

Fig. 4.5. A schematic representation of the inner and outer viscous layers along the upper branch (left) and the lower branch (right) of the curve of marginal stability for plane Poiseuille flow. For the Blasius boundary-layer profile $U''(z_1) = 0$ and then $z_c - z_1 = O\{(\alpha R)^{-1/9}\}$ along the upper branch.

what branch of $\ln(z - z_c)$ must we take for $z - z_c < 0$? This question was first answered in a remarkable piece of analysis by Tollmien (1929) who showed that if $U'_c > 0$ we must take $\ln(z - z_c) = \ln|z - z_c| - \pi i$ for $z - z_c < 0$.

For large values of αR, viscous effects may be expected to be important in at least two regions: near the critical point z_c where $\phi_2^{(0)}(z)$ is singular and near a rigid boundary where the boundary conditions must be applied. These viscous regions are often called the inner and outer viscous layers respectively. In these viscous regions there must be a balance between the term $(i\alpha R)^{-1}\phi^{iv}$ and the remaining terms in the equation, and it is this balance which determines the thicknesses of the layers. A more subtle question concerns the circumstances in which these two viscous layers are well separated or not. This question will be considered in § 28 where it will be shown that the layers remain well separated along the upper branch of the curve of marginal stability but that they merge into a single viscous layer along the lower branch. The relationship between these viscous layers is shown schematically in Fig. 4.5.

In discussing the viscous approximations to the solutions of the Orr–Sommerfeld equation there are several interrelated approaches which will now be considered.

27.2 The boundary-layer approximation near a rigid wall

If we assume that the inner and outer viscous layers are well separated then we would expect a boundary layer of simple exponential type to occur near a rigid wall. Let $z = z_1$ denote the position of the wall with $z \geqslant z_1$ and suppose that $U(z_1) = 0$ which is the usual case. Then sufficiently close to z_1 the boundary-layer approximation to the Orr–Sommerfeld equation becomes

$$(i\alpha R)^{-1}\phi^{iv} = -c\phi'', \tag{27.4}$$

and the boundary-layer solution of this equation, i.e. the solution which decays exponentially for $z > z_1$, is simply

$$\phi_{\mathrm{BL}}(z) = \text{constant} \times \exp\{-(z - z_1)(\alpha Rc)^{1/2}\,e^{-\pi i/4}\}. \tag{27.5}$$

When this approximation is used in the eigenvalue relation (see § 28) only one branch of the curve of marginal stability is obtained. More precisely, the asymptote to the upper branch is obtained correctly but no lower branch is found. This approximation does show, however, that for values of c bounded away from zero the thickness of the viscous layer near a rigid wall is of order $(\alpha R)^{-1/2}$.

27.3 The WKBJ approximations

The inviscid approximations derived earlier in this section provide approximations of outer type to two solutions of the Orr–Sommerfeld equation. To obtain approximations of outer type to two other solutions we consider, following Heisenberg (1924), the approximations of Liouville–Green or WKBJ type. These approximations are valid in certain domains of the (complex) z-plane excluding the immediate neighbourhood of the critical point. For this purpose let

$$\phi(z) = \exp\{\textstyle\int g(z)\,dz\}, \tag{27.6}$$

where $g(z)$ is a slowly varying function of αR and satisfies the third-order nonlinear equation

$$g^4 + 6g^2 g' + 4gg'' + 3g'^2 + g''' - 2\alpha^2(g^2 + g') + \alpha^4$$
$$= i\alpha R\{(U - c)(g^2 + g' - \alpha^2) - U''\}. \tag{27.7}$$

Approximations to the solutions of this equation can then be obtained in the usual way by assuming a formal expansion of the form

$$g(z) = (i\alpha R)^{1/2} g_0(z) + g_1(z) + (i\alpha R)^{-1/2} g_2(z) + \cdots. \quad (27.8)$$

On substituting this expansion into the Orr–Sommerfeld equation and comparing corresponding powers of $(i\alpha R)^{1/2}$ we obtain a sequence of equations from which all of the coefficients $g_n(z)$ ($n = 0, 1, 2, \ldots$) can be determined algebraically. In a first approximation of this type only g_0 and g_1 are required and they satisfy the equations

$$g_0^2\{g_0^2 - (U - c)\} = 0 \quad (27.9)$$

and

$$2g_0\{2g_0^2 - (U - c)\}g_1 = -g_0'\{6g_0^2 - (U - c)\}. \quad (27.10)$$

From the first of these equations we see that *either* $g_0^2 = 0$ *or* $g_0^2 = U - c$. If $g_0 = 0$ then equation (27.10) vanishes identically and it is necessary to consider the equation for g_2 which is found to reduce to

$$(U - c)(g_1' + g_1^2 - \alpha^2) - U'' = 0. \quad (27.11)$$

This result also follows immediately on substituting the expansion (27.8) with $g_0 = 0$ into equation (27.7). Equation (27.11) is simply a first-order nonlinear equation which is equivalent to the reduced equation and we thus recover the inviscid approximations. If $g_0 \neq 0$, however, we have

$$g_0 = \mp(U - c)^{1/2} \quad \text{and} \quad g_1 = -\frac{5}{4}\frac{U'}{U - c}, \quad (27.12)$$

and on substituting these results into equation (27.6) we obtain the WKBJ approximations

$$\phi_{3,4}(z) \sim C_{3,4}(U - c)^{-5/4} \exp\{\mp(\alpha R)^{1/2}Q(z)\}, \quad (27.13)$$

where

$$Q(z) = e^{\pi i/4} \int_{z_c}^{z} (U - c)^{1/2} \, dz. \quad (27.14)$$

The constants C_3 and C_4 are arbitrary; they will, however, be fixed later, after we have discussed the approximations of inner type, in a

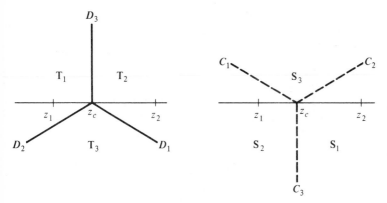

Fig. 4.6. The local behaviour of the Stokes lines (left) and the anti-Stokes lines (right) in the z-plane for U'_c real and positive.

way that would appear to be particularly convenient. The lower limit of integration in equation (27.14) has been chosen, for convenience, as z_c; any other choice for this limit would merely change the approximations by constant multiplicative factors.

In discussing the domains of validity of these approximations the *Stokes lines* and *anti-Stokes lines* (also sometimes called principal curves), which are associated with the exponential factors in approximation (27.13), play a particularly important role. These curves are defined by the conditions $\text{Im}\,\{Q(z)\} = 0$ and $\text{Re}\,\{Q(z)\} = 0$ respectively. Now, near the critical point $U - c = U'_c(z - z_c) + O\{(z - z_c)^2\}$ and hence locally the Stokes lines are given by

$$\text{Im}\,\{e^{\pi i/4}(U'_c)^{1/2}(z - z_c)^{3/2}\} = 0. \qquad (27.15)$$

Thus there are three distinct Stokes lines radiating outwards from the critical point with equal angular spacings of $2\pi/3$ and they are located in the directions given by

$$\text{ph}(z - z_c) = (\tfrac{1}{2}\pi, \tfrac{7}{6}\pi, \tfrac{11}{6}\pi) - \tfrac{1}{3}\,\text{ph}\,U'_c. \qquad (27.16)$$

In the case of marginal stability U'_c is real and if we continue to suppose that $U(z)$ is monotone increasing with $\text{ph}\,U'_c = 0$ then the Stokes and anti-Stokes lines have the arrangement shown in Fig. 4.6.

We can now describe the behaviour of the approximations (27.13) in the various sectors of the complex plane. The approximation to $\phi_3(z)$ is *recessive*, i.e. exponentially small, in the sector S_1. Phase-integral theory (Heading 1962) then suggests that the approximation can be traced across the anti-Stokes lines C_2 and C_3 into the sectors S_2 and S_3 in which it is *dominant*, i.e. exponentially large, up to but excluding the anti-Stokes line C_1. Similarly, the approximation to $\phi_4(z)$ is recessive in S_2 and it can therefore be traced across C_1 and C_3 into the sectors S_1 and S_3 in which it is dominant up to but excluding C_2. The approximations (27.13) which are said to be of dominant-recessive type are valid therefore in the usual Poincaré sense in *open* sectors of angle 2π.

In differentiating these approximations, only the term which arises from differentiation of the rapidly varying exponential part of the approximation can consistently be retained. Thus, for example, we have

$$\phi'_{3,4}(z) \sim \mp C_{3,4}(i\alpha R)^{1/2}(U-c)^{-3/4}\exp\{\mp(\alpha R)^{1/2}Q(z)\}. \quad (27.17)$$

This is, however, not inconsistent with retaining two terms in the expansion (27.8) for $g = \phi'/\phi$.

Near the lower boundary z_1 we have

$$Q(z) = Q(z_1) + e^{-\pi i/4}c^{1/2}(z-z_1) + O\{(z-z_1)^2\}, \quad (27.18)$$

where $\mathrm{Re}\{Q(z_1)\} < 0$. It is evident, therefore, that $\phi_3(z)$ is related to the boundary-layer approximation (27.5). More precisely, if the WKBJ approximation to $\phi_3(z)$ is expanded to first order for z near z_1 then we recover the boundary-layer approximation (27.5). When the WKBJ approximation to ϕ_3 is used in the eigenvalue relation, however, two branches of the curve of marginal stability are found. The asymptote to the upper branch is again obtained correctly but the asymptote to the lower branch is not and, furthermore, the minimum critical Reynolds number is unrealistically low. The failure of the WKBJ approximations to give the boundary values of the solutions with sufficient accuracy along the lower branch of the marginal curve is, of course, largely due to the fact that they are not valid in a domain of the z-plane that contains the critical point. To overcome these limitations we must now consider other forms of approximation that are valid in domains containing the critical point.

27.4 *The local turning-point approximations*

To obtain approximations to ϕ_3 and ϕ_4 that are valid in a small neighbourhood of z_c, only the dominant terms in the Orr–Sommerfeld equation for small values of $|z - z_c|$ need be retained. Thus, with $U - c$ approximated by $U'_c(z - z_c)$, we have the equation

$$(i\alpha R)^{-1}\phi^{iv} = U'_c(z - z_c)\phi'' \qquad (27.19)$$

which provides first approximations of inner type. In discussing the solutions of this equation it is convenient to let

$$\phi(z) = \chi^{(0)}(\xi), \qquad (27.20)$$

where

$$\xi = (z - z_c)/\varepsilon \quad \text{and} \quad \varepsilon = (i\alpha R U'_c)^{-1/3}. \qquad (27.21)$$

We will suppose, of course, that $0 < |\varepsilon| \ll 1$ and that

$$\text{ph } \varepsilon = -\tfrac{1}{6}\pi - \tfrac{1}{3}\text{ph } U'_c, \quad \text{where } 0 \leqslant \text{ph } U'_c < 2\pi. \qquad (27.22)$$

Equation (27.19) can then be written in the parameter-free form

$$\left(\frac{d^2}{d\xi^2} - \xi\right)\frac{d^2}{d\xi^2}\chi^{(0)} = 0 \qquad (27.23)$$

which is simply Airy's equation for $\chi^{(0)\prime\prime}$. The stretching of the independent variable which we have introduced here shows that the thickness of the inner viscous layer is of order $(\alpha R)^{-1/3}$ and hence thicker than the outer viscous layer.

Two of the solutions of equation (27.23) are extremely simple, being merely solutions of $\chi^{(0)\prime\prime} = 0$. If they are taken in the form

$$\chi^{(0)}_1(\xi) = \varepsilon\xi \quad \text{and} \quad \chi^{(0)}_2(\xi) = 1 \qquad (27.24)$$

then they can be immediately identified with the leading terms in the inviscid approximations $\phi^{(0)}_1(z)$ and $\phi^{(0)}_2(z)$ respectively. The other two solutions have a rapidly varying character and can be expressed in terms of double integrals of Airy functions. In the older literature of hydrodynamic stability it was customary to express these solutions in terms of the modified Hankel functions of order one-third. As Jeffreys (1942) has remarked, however, 'Bessel functions of order $\tfrac{1}{3}$ seem to have no application except to provide an inconvenient way of expressing [the function $Ai(z)$].' For comparison purposes, however, we note the relations

$$Ai(z \ e^{\pi i/6}) = \tfrac{1}{2}i \times 12^{-1/6} h_1(iz),$$
$$Ai(z \ e^{5\pi i/6}) = e^{-2\pi i/3} A_2(z \ e^{\pi i/6}) = -\tfrac{1}{2}i \times 12^{-1/6} h_2(iz)$$

and

$$h_1(z) = x^{1/3} H_{1/3}^{(1)}(x), \quad h_2(z) = x^{1/3} H_{1/3}^{(2)}(x),$$

where $x = \tfrac{2}{3} z^{3/2}$. For tables of h_1 and h_2 see: Harvard University Computation Laboratory 1945 *Tables of the modified Hankel functions of order one-third and of their derivatives*. Cambridge, Mass.: Harvard University Press.

The present discussion can be substantially simplified, however, if we use the generalized Airy functions discussed in the Appendix. One of the required solutions can be taken in the form

$$\chi_3^{(0)}(\xi) = A_1(\xi, 2). \tag{27.25}$$

For $|\mathrm{ph} \ \xi| < \pi/3$ this solution is recessive; it represents, therefore, the first term in the inner expansion of $\phi_3(z)$. The Stokes and anti-Stokes lines associated with the Airy and generalized Airy functions can be obtained from Fig. 4.6 by a positive rotation through an angle of $\pi/6$. If we now let ξ_1 and ξ_2 denote the values of ξ corresponding to z_1 and z_2 then, at least in the case of marginal stability, ξ_1 and ξ_2 lie in the sectors S_2 and S_1 respectively. Thus, $\chi_3^{(0)}(\xi)$ is exponentially small at ξ_2.

In choosing a second viscous solution of equation (27.23), considerations of symmetry suggest that it should be exponentially small at ξ_1 and exponentially large at ξ_2. These requirements can be satisfied if we let

$$\chi_4^{(0)}(\xi) = A_2(\xi, 2). \tag{27.26}$$

With this choice, the set of solutions $\{\chi_1^{(0)}, \chi_2^{(0)}, \chi_3^{(0)}, \chi_4^{(0)}\}$ is *numerically satisfactory* in the sense of J. C. P. Miller; cf. Olver (1974), pp. 154–155. The solution (27.26) is not used in the derivation of the eigenvalue relation for reasons explained in § 28 and it need not therefore be discussed further here.

The solution (27.25) has been tabulated by Holstein (1950) and his results have been reproduced by Stuart (1963). The function actually tabulated by Holstein, which he denotes by $\phi_{31}(\eta)$, is given in the present notation by

$$\phi_{31}(\eta) = -2\sqrt{3}\, e^{-\pi i/12} A_1(\eta\, e^{\pi i/6}, 2) \qquad (27.27)$$

with η real. Holstein tabulates the real and imaginary parts of $\phi_{31}(\eta)$ and its first two derivatives for $\pm\eta = 0(0.5)8$. The tables by Singh, Lumley & Betchov (1963) can also be used for the computation of such functions.

When the inner approximation (27.25) is used in the eigenvalue relation (see § 28) the asymptotes to both branches of the neutral curve are obtained correctly and, for example, for plane Poiseuille flow, the error in the minimum critical Reynolds number is about 6.5 per cent. This approximation is therefore adequate for many purposes and because of its simplicity it has been widely used for computational purposes. Nevertheless, from a theoretical point of view the situation is less satisfactory. The inner approximations of viscous type are valid only in a small neighbourhood of z_c that shrinks to zero as $|\varepsilon| \to 0$. Thus, strictly speaking, they are not valid at the points z_1 and z_2 where the boundary conditions must be imposed. In Chapter 5, therefore, we shall be interested, on the one hand, in examining the reasons for the undoubted success of these approximations and, on the other, in deriving uniform approximations that are valid in a bounded domain containing not only z_c but also z_1 and z_2.

By using the idea of a composite approximation which is commonly used in connexion with the method of matched asymptotic expansions we can, however, combine the foregoing approximations to form composite approximations which are valid in domains that do not shrink to zero as $|\varepsilon| \to 0$. Since the WKBJ approximations are valid only in an open sector of angle 2π, the resulting composite approximations are not valid in a full neighbourhood of z_c. By considering the second term in the approximations of WKBJ type (see Problem 4.7) it is found that the WKBJ approximations are valid provided $|z - z_c| \gg |\varepsilon|$. Thus, if $|z - z_c|$ is small or, more precisely, if $|\varepsilon| \ll |z - z_c| \ll 1$ and $-7\pi/6 < \mathrm{ph}(z - z_c) < 5\pi/6$, then from equation (27.13) we have

$$\phi_3(z) \sim C_3\{U'_c(z - z_c)\}^{-5/4} \exp\{-\tfrac{2}{3}(z - z_c)^{3/2}(i\alpha R U'_c)^{1/2}\}.$$
$$(27.28)$$

Similarly, by considering the second term in the approximations of inner type (see Problem 4.8) it is found that the inner approximations are valid provided $|z - z_c| \ll |\varepsilon|^{3/5}$. For unbounded values of $|\xi|$ or, more precisely, for $1 \ll |\xi| \ll |\varepsilon|^{-2/5}$ and $|\text{ph } \xi| < \pi$, equation (27.25) can be expanded to give

$$\chi_3^{(0)}(\xi) \sim \tfrac{1}{2}\pi^{-1/2}\xi^{-5/4} \exp\left(-\tfrac{2}{3}\xi^{3/2}\right). \qquad (27.29)$$

These limiting forms of $\phi_3(z)$ and $\chi_3^{(0)}(\xi)$ are both valid in the *overlap domain*: $|\varepsilon| \ll |z - z_c| \ll |\varepsilon|^{3/5}$ with $|\text{ph } \xi| < \pi$. In that domain they can be made identical, i.e. matched, by choosing

$$C_3 = \tfrac{1}{2}\pi^{-1/2}(\varepsilon U_c')^{5/4}. \qquad (27.30)$$

The expressions (27.28) and (27.29) are then called the *common part* (=cp) of the inner and outer approximations. Consider now the formation of composite approximations according to the usual rule for multiplicative composition (Van Dyke 1975) which we will denote by

$$C^{\times}\{f(z)\} = (\text{inner } f)(\text{outer } f)/(\text{cp } f). \qquad (27.31)$$

Application of this rule to $\phi_3(z)$ then gives

$$C^{\times}\{\phi_3(z)\} = \left\{\frac{U-c}{U_c'(z-z_c)}\right\}^{-5/4} \exp\left\{\tfrac{2}{3}\xi^{3/2} - (\alpha R)^{1/2}Q(z)\right\}$$
$$\times A_1(\xi, 2), \qquad (27.32)$$

which reduces to the WKBJ approximation (27.13) when $|z - z_c| \gg |\varepsilon|$ and to the inner approximation (27.25) when $|z - z_c| \ll |\varepsilon|^{3/5}$. To obtain a composite approximation to $\phi_3'(z)$ we cannot simply differentiate equation (27.32). Instead, by a direct calculation using the derivatives of the inner and outer approximations, we have

$$C^{\times}\{\phi_3'(z)\} = \varepsilon^{-1}\left\{\frac{U-c}{U_c'(z-z_c)}\right\}^{-3/4} \exp\left\{\tfrac{2}{3}\xi^{3/2} - (\alpha R)^{1/2}Q(z)\right\}$$
$$\times A_1(\xi, 1). \qquad (27.33)$$

When these composite approximations are used in the eigenvalue relation only their ratio is needed. The exponential factors in the approximations then cancel, and it was largely for this reason that we used the rule for multiplicative rather than additive composition.

27.5 *The truncated equation and Tollmien's improved viscous approximations*

In all of the approximations to the solutions of the Orr–Sommerfeld equation of viscous type which have been considered thus far, the only terms in the equation that contributed to the approximations were $(i\alpha R)^{-1}\phi^{iv}$ and $(U-c)\phi''$, and this suggests (Shen 1964, Reid 1965) a unified approach in which the Orr–Sommerfeld equation is approximated at the outset by the *truncated equation*

$$(i\alpha R)^{-1}\phi^{iv} = (U-c)\phi''. \tag{27.34}$$

In spite of the obvious limitations of such an approach, one of its main advantages is that equation (27.34) is a second-order equation for ϕ'' to which all of the standard asymptotic theory for second-order equations is directly applicable (see, for example, Olver (1974)). It also provides a somewhat more natural derivation of Tollmien's (1947) improved viscous approximations which yield, in appropriate limits, both the WKBJ and the local turning-point approximations.

Consider then a preliminary transformation of both dependent and independent variables of the form

$$\phi''(z) = \dot{z}^{1/2}\Phi(\eta), \tag{27.35}$$

where the dot denotes differentiation with respect to η and the exponent of $\dot{z}(\eta)$ has been chosen so that the transformed equation is in normal form, i.e. it does not contain Φ'. Thus, without further approximation, equation (27.34) becomes

$$d^2\Phi/d\eta^2 = \{i\alpha R\dot{z}^2(U-c)+\psi(\eta)\}\Phi, \tag{27.36}$$

where

$$\psi(\eta) = \dot{z}^{1/2}\frac{d^2}{d\eta^2}(\dot{z}^{-1/2}).$$

In specifying the relationship between z and η we must require that they be analytic functions of each other at the turning point z_c. This requirement can be satisfied if we let

$$\dot{z}^2(U-c)/U_c' = \eta, \tag{27.37}$$

which, on integration, gives the *Langer variable*

$$\eta(z) = \left[\frac{3}{2}\int_{z_c}^z \left(\frac{U-c}{U_c'}\right)^{1/2}dz\right]^{2/3}. \tag{27.38}$$

Near $z = z_c$, we have

$$\eta(z) = z - z_c + \tfrac{1}{10}(U_c''/U_c')(z - z_c)^2 + O\{(z - z_c)^3\},$$
$$(27.39)$$

so that $\eta(z)$ is analytic at $z = z_c$.

Because of the approximations that have already been made in starting with the truncated equation, we will not attempt to exploit the full generality of the existing asymptotic theory for second-order differential equations with a simple turning point. In a first approximation, however, that theory shows that we can simply neglect $\psi(\eta)$ in equation (27.36) and we then have

$$d^2\Phi/d\zeta^2 = \zeta\Phi, \qquad (27.40)$$

where $\zeta = \eta/\varepsilon$ is a stretched Langer variable. Thus, Φ satisfies (a parameter-free form of) Airy's equation which then plays the role of a comparison equation in this theory. The solutions of equation (27.40) can conveniently be taken in the form

$$\Phi_{3,4}(\eta) = \varepsilon^{-2} A_{1,2}(\zeta). \qquad (27.41)$$

Equation (27.35) must now be integrated twice to obtain approximations to $\phi_{3,4}(z)$. In carrying out these integrations, the slowly varying coefficient $\dot{z}^{1/2}$ must be treated as a constant, and in this way we obtain Tollmien's improved viscous approximations:

$$\phi_{3,4}(z) = \left(\frac{U-c}{U_c'\eta}\right)^{-5/4} A_{1,2}(\zeta, 2). \qquad (27.42)$$

For large values of $|\zeta|$ these approximations reduce, on expansion of the Airy functions and noting that $\tfrac{2}{3}\zeta^{3/2} \equiv (\alpha R)^{1/2} Q(z)$, to the WKBJ approximations (27.13) with C_3 again given by equation (27.30) and $C_4 = iC_3$. Similarly, for small values of $|z - z_c|$ we have $U - c = U_c'(z - z_c) + O\{(z - z_c)^2\}$, $\eta = z - z_c + O\{(z - z_c)^2\}$, $\zeta = \xi\{1 + O(z - z_c)\}$, and they reduce therefore to the local turning-point approximations (27.25) and (27.26).

When this approximation to $\phi_3(z)$ is used in the eigenvalue relation the asymptotes to both branches of the marginal curve are again obtained correctly and for plane Poiseuille flow the error in the minimum critical Reynolds number is reduced to about 1.3 per cent. This approximation is therefore entirely adequate for most computational purposes. From a theoretical point of view, however, the situation is still not entirely satisfactory. The use of the truncated equation (27.34) to derive the approximations (27.42) does not suggest an algorithm for obtaining higher-order approximations nor does it permit a discussion of the circumstances in which $\phi_2(z)$ exhibits viscous behaviour.

27.6 *The viscous correction to the singular inviscid solution*

In discussing the approximations of inviscid type it was found that one of the approximations, which we denoted by $\phi_2^{(0)}(z)$, has a logarithmic branch point at z_c and hence that viscous effects must play an important role in determining the behaviour of the solution $\phi_2(z)$, at least in the neighbourhood of z_c. This is perhaps the most difficult part of the heuristic theory and was first treated by Tollmien (1929); it has been discussed further by Tollmien (1947), Holstein (1950) and DiPrima (1954). The present discussion has been adapted from the account given by Reid (1965).

To obtain the viscous correction to $\phi_2^{(0)}(z)$ in the neighbourhood of z_c it is necessary to consider the inner approximations to second order. This can be done in a systematic way by letting $\phi(z) = \chi(\xi)$, where $\chi(\xi)$ satisfies the equation

$$\chi^{\text{iv}} - 2\alpha^2\varepsilon^2\chi'' + \alpha^4\varepsilon^4\chi = i\alpha R\varepsilon^2\{(U-c)(\chi'' - \alpha^2\varepsilon^2\chi) - \varepsilon^2 U''\chi\}.$$
(27.43)

We then consider an expansion of the solutions in powers of ε of the form

$$\chi(\xi, \varepsilon) = \chi^{(0)}(\xi) + \varepsilon\chi^{(1)}(\xi) + \cdots,$$
(27.44)

where $\chi^{(0)}$ satisfies equation (27.23) and $\chi^{(1)}$ satisfies

$$\left(\frac{d^2}{d\xi^2} - \xi\right)\frac{d^2}{d\xi^2}\chi^{(1)} = \frac{U_c''}{U_c'}\left(\frac{1}{2}\xi^2\frac{d^2}{d\xi^2} - 1\right)\chi^{(0)}.$$
(27.45)

The solution $\chi_1^{(1)}(\xi)$ must clearly be $\chi_1^{(1)}(\xi) = \frac{1}{2}\varepsilon(U_c''/U_c')\xi^2$ in order to match with the second term in the power series expansion of $\phi_1^{(0)}(z)$, and the solution $\chi_2^{(1)}(\xi)$ must be a suitably chosen particular integral of the equation

$$\left(\frac{d^2}{d\xi^2} - \xi\right)\frac{d^2}{d\xi^2}\chi_2^{(1)} = -\frac{U_c''}{U_c'}.$$
(27.46)

The solution of this equation must be chosen so that in sectors of the z-plane containing the boundary points and $|\xi| \gg 1$, $\chi_2^{(0)}(\xi) + \varepsilon \chi_2^{(1)}(\xi)$ is asymptotic to $\phi_2^{(0)}(z)$ for $|z - z_c| \ll 1$. Alternatively, this condition can be expressed in terms of the usual *asymptotic matching principle* (Van Dyke 1975)

$$\text{inner } \{\phi_2^{(0)}(z)\} = \text{outer } \{\chi_2^{(0)}(\xi) + \varepsilon \chi_2^{(1)}(\xi)\}. \qquad (27.47)$$

From equation (27.3) we have

$$\begin{aligned}
\text{inner}\{\phi_2^{(0)}(z)\} &= 1 + (U_c''/U_c')(z - z_c) \ln (z - z_c) \\
&\quad + O\{(z - z_c)^2 \ln (z - z_c)\} \\
&\equiv 1 + \varepsilon (U_c''/U_c')\xi(\ln \varepsilon + \ln \xi) + O(\varepsilon^2 \ln \varepsilon). \qquad (27.48)
\end{aligned}$$

From the results given in the Appendix we see immediately that the required solution of equation (27.46) must be of the form

$$\chi_2^{(1)}(\xi) = -(U_c''/U_c')\{B_3(\xi, 2, 1) + A\xi + B\}, \qquad (27.49)$$

where the constants A and B must be chosen to satisfy the matching condition (27.47). On expanding (27.49) for $|\xi| \gg 1$ and $-\pi < \text{ph } \xi < \pi/3$ we have

$$\chi_2^{(1)}(\xi) \sim -(U_c''/U_c')\{-\xi \ln \xi + [\psi(2) + 2\pi i]\xi + A\xi + B + O(\xi^{-3})\} \qquad (27.50)$$

and on comparing this result with equation (27.48) it immediately follows that

$$A = -[\ln \varepsilon + \psi(2) + 2\pi i] \quad \text{and} \quad B = 0. \qquad (27.51)$$

Thus we have

$$\chi_2^{(1)}(\xi) = -(U_c''/U_c')\{B_3(\xi, 2, 1) - [\ln \varepsilon + \psi(2) + 2\pi i]\xi\}. \qquad (27.52)$$

This result shows that $\chi_2^{(1)}(\xi)$ is exponentially large in the sector $\mathbf{S}_3$ and hence that $\phi_2(z)$ exhibits viscous behaviour not only in the immediate neighbourhood of z_c but also throughout the whole of the sector $\mathbf{S}_3$ as indicated in Fig. 4.7. The present approximations, however, are not adequate to describe the behaviour of $\phi_2(z)$ in $\mathbf{S}_3$ for finite values of $|z - z_c|$.

The viscous correction $\chi_2^{(1)}(\xi)$ has also been tabulated by Holstein (1950) and his results have been reproduced by Stuart (1963). In

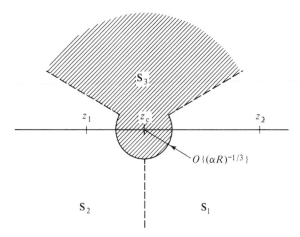

Fig. 4.7. The domain in which $\phi_2(z)$ exhibits viscous behaviour. For finite values of $|z - z_c|$ in S_3, $\phi_2(z)$ is exponentially large.

Stuart's notation the function actually tabulated is denoted by $N(\eta)$ which, in the present notation, is given by

$$N(\eta) = -e^{-\pi i/6}B_3(\eta \, e^{\pi i/6}, 2, 1) + [\psi(2) + \tfrac{11}{6}\pi i]\eta \quad (27.53)$$

with η real. Holstein tabulates the real and imaginary parts of $N(\eta)$ and its first two derivatives for $\pm\eta = 0(0.5)8$; 4D. Tables of $N''(\eta)$, $e^{-\pi i/6}N''(\eta)$, $N'''(\eta)$ and $e^{-\pi i/3}N'''(\eta)$ have also been given by Reid (1965) for $-\eta = 0(0.02)10$; 4S. We may also note the initial values

$$\begin{aligned} N(0) &= 3^{-1/3}\Gamma(\tfrac{2}{3})i, \\ N'(0) &= 1 - \tfrac{1}{3}(2\gamma + \ln 3) - \tfrac{1}{2}\pi i, \\ N''(0) &= 3^{-2/3}\Gamma(\tfrac{1}{3})i. \end{aligned} \right\} \quad (27.54)$$

and

With $\chi_2^{(1)}(\xi)$ given by equation (27.52), Tollmien (1947) has suggested that a 'corrected' approximation for $\phi_2(z)$ can be obtained by simply replacing the logarithmic term $(U_c''/U_c')(z - z_c)$ $\times\ln(z - z_c)$ in equation (27.3) by the 'viscous correction' $\varepsilon\chi_2^{(1)}(\xi)$, and this procedure then gives

$$C^T\{\phi_2(z)\} = P_2(z) + \varepsilon P_1(z)\chi_2^{(1)}(\xi), \quad (27.55)$$

where $P_1(z)$ and $P_2(z)$ are still given by equations (22.27). For $|\xi| \gg 1$ and $-\pi < \text{ph } \xi < \pi/3$ this reduces to the outer approximation

$\phi_2^{(0)}(z)$ and for $|z - z_c| \ll 1$ it reduces to the inner approximation $\chi_2^{(0)}(\xi) + \varepsilon\chi_2^{(1)}(\xi)$. Nevertheless, it is not a composite approximation of either additive or multiplicative type but is closely related to the uniform approximations which will be discussed in Chapter 5.

From the point of view of the method of matched asymptotic expansions, however, a somewhat more natural procedure would be to form composite approximations according to the usual rules. Although multiplicative composition leads to particularly simple first-order composite approximations for $\phi_{3,4}(z)$, additive composition seems to be preferable for $\phi_2(z)$. According to the usual rule for additive composition (Van Dyke 1975) we have

$$C^+\{f(z)\} = (\text{inner} + \text{outer} - \text{cp})f. \qquad (27.56)$$

Applying this rule to $\phi_2(z)$ then gives

$$C^+\{\phi_2(z)\} = P_2(z) + \varepsilon\chi_2^{(1)}(\xi) + (U_c''/U_c')\{P_1(z) - 1\}$$
$$\times (z - z_c) \ln (z - z_c) \qquad (27.57)$$

in which the remaining singularity is as weak as $(z - z_c)^2 \ln (z - z_c)$. The corresponding approximations for $\phi_2'(z)$ are

$$C^T\{\phi_2'(z)\} = P_2'(z) - (U_c''/U_c')(z - z_c)P_1'(z)$$
$$+ \{(z - z_c)P_1(z)\}'\chi_2^{(1)\prime}(\xi) \qquad (27.58)$$

and

$$C^+\{\phi_2'(z)\} = P_2'(z) + \chi_2^{(1)\prime}(\xi) + (U_c''/U_c')\{[P_1(z) - 1][\ln (z - z_c) + 1]$$
$$+ P_1'(z)(z - z_c) \ln (z - z_c)\}. \qquad (27.59)$$

The approximation (27.59) is singular like $(z - z_c) \ln (z - z_c)$, i.e. it has the same singular behaviour as $\phi_2^{(0)}(z)$.

28 Approximations to the eigenvalue relation

Having now obtained approximations to four solutions of the Orr–Sommerfeld equation, we can now use the properties of these approximations to derive simplified forms of the eigenvalue relation for large values of αR. The precise form of the eigenvalue relation depends not only on the class of flows considered but also on the viscous approximations that are used, and there are therefore a number of cases to be considered.

28.1 *Symmetrical flows in a channel*

The most important flow of this type is, of course, plane Poiseuille flow for which $U(z) = 1 - z^2$ over $-1 \le z \le 1$ and this flow will be used for illustrative purposes throughout this section. More generally, however, it is convenient to suppose that L has been chosen so that $z_1 = -1$ and $z_2 = 0$, and that $U(z)$ is monotone increasing on this interval with $U(z_1) = 0$. Because of the symmetry of the basic flow we can consider the even† and odd solutions separately. It is found that the odd solutions are stable and we will therefore concentrate our attention on the even solutions which must then satisfy the boundary conditions

$$\phi = D\phi = 0 \text{ at } z = z_1 \quad \text{and} \quad D\phi = D^3\phi = 0 \text{ at } z = z_2. \quad (28.1)$$

The solutions ϕ_3 and ϕ_4 are dominant at z_1 and z_2 respectively. But if z_c is significantly closer to z_1 than to z_2, as it happens to be in most problems, then $|\phi_4(z_2)| \gg |\phi_3(z_1)|$ and, as shown in § 40, we can then reject ϕ_4 with an error of the order of $(\alpha R)^{-2}$. This approximation merely reflects the fact that, except near z_1 and z_c, the solution has a largely inviscid character. Thus we let

$$\phi = A\phi_1 + \phi_2 + C\phi_3, \quad (28.2)$$

where the coefficient of ϕ_2 has been chosen to be unity to fix the normalization.

In applying the boundary conditions at $z = z_2$ we can, consistent with the rejection of ϕ_4, also neglect ϕ_3 and its derivatives since they are exponentially small compared with all other terms retained in the subsequent analysis. Thus, near z_2, we have

$$\phi = A\phi_1^{(0)} + \phi_2^{(0)} + O\{(\alpha R)^{-1}\} \text{ as } R \to \infty. \quad (28.3)$$

It is then convenient to let $\Phi = A\phi_1^{(0)} + \phi_2^{(0)}$, where $A = -\phi_2^{(0)\prime}(z_2)/\phi_1^{(0)\prime}(z_2)$. On differentiating Rayleigh's equation we see that if $\Phi'(z_2) = 0$ and $U(z)$ is an even function of z then $\Phi'''(z_2)$ vanishes automatically. Near z_1, therefore, we have

$$\phi = \Phi + C\phi_3 \quad (28.4)$$

† An even solution for ϕ corresponds, by equations (21.15), to a two-dimensional disturbance for which $\hat{u}$ is odd and $\hat{w}$ is even.

and the remaining two boundary conditions then lead to the eigen-value relation

$$\frac{\Phi'(z_1)}{\Phi(z_1)} = \frac{\phi_3'(z_1)}{\phi_3(z_1)}. \tag{28.5}$$

In this approximation to the eigenvalue relation we have considered only the inviscid approximations to ϕ_1 and ϕ_2. As a result, equation (28.5) is of 'separable' form in the sense that the left-hand side of the equation is independent of R and depends only on α and c.

In discussing the various approximations to equation (28.5) it is convenient to isolate the inviscid part of the equation by letting[†]

$$W(\alpha, c) = -\frac{U_1'}{c} \frac{\Phi(z_1)}{\Phi'(z_1)}, \tag{28.6}$$

where $U_1' = U'(z_1)$, and the eigenvalue relation then becomes

$$W(\alpha, c) = -\frac{U_1'}{c} \frac{\phi_3(z_1)}{\phi_3'(z_1)}. \tag{28.7}$$

To determine the asymptotes to the upper and lower branches of the marginal curve for flows without an inflexion point,[‡] we must first obtain the limiting behaviour of W for small values of α and c. By using the modified Heisenberg expansions (22.31) to evaluate $\phi_1^{(0)\prime}$ and $\phi_2^{(0)\prime}$ at z_2 it can be shown that (see, for example, Nield (1972))

where
$$\begin{aligned} A(\alpha, c) &= (U_1'^2/I_2)\alpha^{-2}\{1 + O(\alpha^2, c)\}, \\ I_2 &= \int_{z_1}^{z_2} U^2 \, dz. \end{aligned} \right\} \tag{28.8}$$

For small values of c, an easy calculation shows that

$$z_1 - z_c = -\frac{c}{U_1'} \left\{ 1 - \frac{U_1''}{2U_1'^2} c + O(c^2) \right\} \tag{28.9}$$

[†] For flows with $U_1 = U(z_1) \neq 0$, the factor $-U_1'/c$ in this equation must be replaced by $U_1'/(U_1 - c)$.

[‡] For flows with an inflexion point, see Problem 4.10.

and by using the Tollmien expansions (27.2) and (27.3) we have

$$\left.\begin{aligned}
\phi_1^{(0)}(z_1) &= -(c/U_1')\{1 + O(c)\}, \\
\phi_2^{(0)}(z_1) &= 1 + O(c \ln c) + i\pi(U_1''/U_1'^2)c\{1 + O(c)\}, \\
\phi_1^{(0)\prime}(z_1) &= 1 + O(c), \\
\text{and} \\
\phi_2^{(0)\prime}(z_1) &= (U_1''/U_1') \ln c + O(1) \\
&\quad - i\pi(U_1''/U_1')\{1 + O(c)\}.
\end{aligned}\right\} \tag{28.10}$$

As $\alpha R \to \infty$ along the upper and lower branches of the marginal curve, α and c both tend to zero but α^2/c tends to a finite limit. Thus, if we let $K_1 = (I_2/U_1')(\alpha^2/c)$ then we have

$$W(\alpha, c) = 1 - K_1 + o(1) - i\pi(U_1''I_2/U_1'^3)\alpha^2 K_1\{1 + o(1)\} \tag{28.11}$$

as α and c tend to zero.

28.2 Flows of the boundary-layer type

For flows of the boundary-layer type we take $z_1 = 0$ and $z_2 = \infty$, and it is also convenient to suppose that V has been chosen to be the free-stream speed so that $U(\infty) = 1$. As $z \to \infty$ for fixed values of α, c and R, the Orr–Sommerfeld equation becomes

$$(D^2 - \alpha^2)(D^2 - \beta^2)\phi = 0, \tag{28.12}$$

where $\beta^2 = i\alpha R(1 - c) + \alpha^2$. If we choose $\mathrm{Re}(\beta) > 0$ then ϕ must be asymptotic to a linear combination of the inviscid solution $e^{-\alpha z}$ and the viscous solution $e^{-\beta z}$ as $z \to \infty$. Since $\mathrm{Re}(\beta)$ is large like $(\alpha R)^{1/2}$, the viscous solution decays very much more rapidly than the inviscid one and it can therefore be neglected provided $|z - z_c| \gg (\alpha R)^{-1/2}$. The inviscid solution $e^{-\alpha z}$ is, of course, simply the first term of the long-wave approximation (22.43).

Thus, if we now let Φ denote the solution of the inviscid equation that is bounded as $z \to \infty$, i.e.

$$\left.\begin{aligned}
\Phi &= A\phi_1^{(0)} + \phi_2^{(0)} \\
&\sim B\,e^{-\alpha z} \quad \text{as } z \to \infty,
\end{aligned}\right\} \tag{28.13}$$

then near z_1 the required approximation is again of the form given in equation (28.4) and the boundary conditions at z_1 lead to the eigenvalue relation (28.5). Although the eigenvalue relation has the same form for both classes of flows it should be emphasized that the definition of Φ is different in the two cases.

To determine the asymptotes to the upper and lower branches of the marginal curve for boundary-layer flows without an inflexion point, we must again obtain the limiting behaviour of W for small values of α and c. By using the long-wave approximation (22.43) together with the modified Heisenberg solutions of the inviscid equation it can be shown that (Nield 1972)

and
$$A(\alpha, c) = U_1'^2 \alpha^{-1}\{1 + O(\alpha, c)\} \\ B(\alpha, c) = U_1' \alpha^{-1}\{1 + O(\alpha, c)\}. \quad (28.14)$$

As $\alpha R \to \infty$ along the upper and lower branches of the marginal curve, α and c again tend to zero but now it is the ratio α/c that tends to a finite limit. Thus, if we let $K_2 = (1/U_1')(\alpha/c)$ then we have

$$W(\alpha, c) = 1 - K_2 + o(1) - i\pi(U_1''/U_1'^3)\alpha K_2\{1 + o(1)\} \quad (28.15)$$

as α and c tend to zero.

28.3 *The boundary-layer approximation to $\phi_3(z)$*

The simplest approximation to $\phi_3(z)$ is the boundary-layer approximation (27.5). When this approximation is used in the eigenvalue relation (28.7) we have

$$W(\alpha, c) = X^{-3/2} e^{\pi i/4}, \quad (28.16)$$

where

$$X = c(\alpha R/U_1'^2)^{1/3}. \quad (28.17)$$

Calculations based on this approximation to the eigenvalue relation lead to only one branch of the marginal curve as shown in Fig. 4.8.

In spite of the obvious limitations of this approximation, it does give the asymptote to the upper branch correctly. Thus, for symmetrical flows in a channel we have

$$1 - K_1 - i\pi(U_1''I_2/U_1'^3)\alpha^2 K_1 \sim X^{-3/2} e^{\pi i/4} \quad (28.18)$$

as α and c tend to zero. It can be verified *a posteriori* that c is of order $(\alpha R)^{-1/5}$ as $R \to \infty$ along the upper branch, and hence $X \to \infty$ as $\alpha R \to \infty$. As $X \to \infty$, therefore, the real part of equation (28.18) shows that $K_1 \to 1$, i.e.

$$c \sim (I_2/U_1')\alpha^2, \quad (28.19)$$

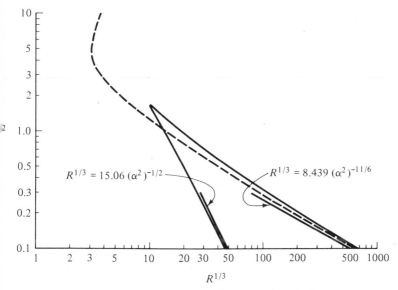

Fig. 4.8. The curves of marginal stability for plane Poiseuille flow based on equations (28.16) (dashed curve) and (28.23) (solid curve).

and the imaginary part then gives

$$R \sim \tfrac{1}{2}\pi^{-2}(U_1'^{11}/I_2^5 U_1''^2)\alpha^{-11}. \tag{28.20}$$

A similar calculation for flows of the boundary-layer type gives†

$$c \sim (1/U_1')\alpha \quad \text{and} \quad R \sim \tfrac{1}{2}\pi^{-2}(U_1'^{11}/U_1''^2)\alpha^{-6} \quad \text{as } \alpha \downarrow 0 \tag{28.21}$$

along the upper branch, if $U_1'' > 0$. For the Blasius boundary layer $U_1'' = 0$ and this last result must then be modified (see Problem 4.11).

28.4 *The WKBJ approximation to* $\phi_3(z)$

Consider now the WKBJ approximation to $\phi_3(z)$. Since, by equation (27.6), $g = \phi'/\phi$ we have immediately from equations (27.8) and (27.12)

$$\frac{\phi_3'(z)}{\phi_3(z)} = -\{i\alpha R(U-c)\}^{1/2} - \frac{5}{4}\frac{U'}{U-c} + \cdots. \tag{28.22}$$

From the manner in which this result was derived, the second term on the right-hand side of this equation may be expected to be

---

† It might appear that these results are not dimensionally correct, but this is merely a consequence of having set $U(\infty) = 1$.

unreliable. More precisely the two terms become comparable when $|z - z_c| = O\{(\alpha R)^{-1/3}\}$ and the WKBJ approximation to $\phi_3(z)$ cannot therefore give the correct asymptotic behaviour of the lower branch of the marginal curve on which $|z_1 - z_c| = O\{(\alpha R)^{-1/3}\}$. Because of the simplicity of this approximation, however, it is worth while exploring its limitations a little further. In evaluating equation (28.22) at the boundary z_1 we must use the fact that the WKBJ approximation to $\phi_3(z)$ is valid only in the sector $-7\pi/6 <$ ph $(z - z_c) < 5\pi/6$, and the eigenvalue relation can then be written in the form

$$W(\alpha, c) = G(X), \tag{28.23}$$

where

$$G(X) = (X^{3/2} e^{-\pi i/4} - \tfrac{5}{4})^{-1}. \tag{28.24}$$

This approximation to the eigenvalue relation differs from equation (28.16) only in the constant term $\tfrac{5}{4}$ and they agree therefore as $X \to \infty$.

In many of the older stability calculations, an equation such as equation (28.23) was often solved by means of a graphical construction in which the real and imaginary parts of both sides of the equation were plotted on the same graph. Although this procedure is no longer used for computational purposes it does provide a convenient way of comparing various approximations to the viscous part of the eigenvalue relation. Consider then the behaviour of $G(X)$ in a (G_r, G_i)-plane. On solving equation (28.24) for $X^{3/2}$ we have

$$X^{3/2} = \left[\frac{1}{G(X)} + \frac{5}{4} \right] e^{\pi i/4}. \tag{28.25}$$

Since X is real (and positive) in the case of marginal stability, the imaginary part of the right-hand side of this equation must vanish, and this condition gives

$$(G_r + \tfrac{2}{5})^2 + (G_i - \tfrac{2}{5})^2 = (\tfrac{2}{5}\sqrt{2})^2, \tag{28.26}$$

which is simply a circle of radius $\tfrac{2}{5}\sqrt{2}$ centred at the point $(-\tfrac{2}{5}, \tfrac{2}{5})$. Since $X^{3/2}$ must be positive, the real part of equation (28.25) then shows that it is only that part of the circle for which $G_i \geq 0$ that is relevant, as shown in Fig. 4.9.

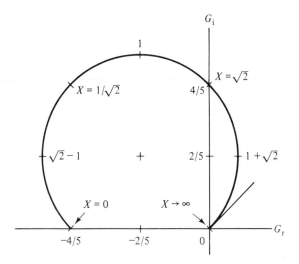

Fig. 4.9. The viscous function $G(X)$. (From Reid 1965.)

Calculations based on the approximation (28.23) lead to the marginal curve shown in Fig. 4.8. Two branches of the marginal curve are found but the asymptotic behaviour of the lower branch as $R \to \infty$ is incorrect and the minimum critical Reynolds number is much too low. It is worth noting, however, how the asymptotes are obtained from equation (28.23). As before, for small values of α and c, we have

$$1 - K_1 - i\pi(U_1'' I_2 / U_1'^3)\alpha^2 K_1 \sim G(X). \qquad (28.27)$$

As $X \to \infty$ we recover the previous results for the upper asymptote given by equations (28.19) and (28.20). As $X \to 0$, however, a second solution is possible and corresponds to the lower asymptote. On noting that

$$G(X) = -\tfrac{4}{5}\{1 + \tfrac{4}{5}X^{3/2}\,e^{-\pi i/4} + O(X^3)\} \quad \text{as } X \to 0, \qquad (28.28)$$

we see that $K_1 = \tfrac{9}{5}$ and hence that

$$c \sim \tfrac{5}{9}(I_2 / U_1')\alpha^2 \quad \text{and} \quad R \sim \tfrac{1}{5}2^{-7}3^{10}\pi^2(U_1''^2 / I_2 U_1')\alpha^{-3}. \qquad (28.29)$$

This shows that c, and hence $|z_1 - z_c|$, are of order $(\alpha R)^{-1}$ as $X \to 0$ whereas the WKBJ approximations from which these results were derived are only valid if $|z - z_c| \gg (\alpha R)^{-1/3}$.

28.5 *The local turning-point approximation to $\phi_3(z)$*

Consider next the local turning-point approximation to $\phi_3(z)$ which is given by equation (27.25). At the lower boundary this approximation gives

$$\frac{\phi_3(z_1)}{\phi_3'(z_1)} = (z_1 - z_c)\frac{A_1(\xi_1, 2)}{\xi_1 A_1(\xi_1, 1)}, \qquad (28.30)$$

where ξ_1 is the value of ξ at $z = z_1$. It is then convenient to let

$$\xi_1 = Y e^{-5\pi i/6}, \quad \text{where } Y = (z_c - z_1)(\alpha R U_c')^{1/3} \quad (28.31)$$

and Y is real (and positive) in the marginal case. If we further let

$$1 + \lambda_1(c) = (U_1'/c)(z_c - z_1), \qquad (28.32)$$

where $\lambda_1(c) = -(U_1''/2U_1'^2)c + O(c^2)$ as $c \downarrow 0$, then the eigenvalue relation can be written in the form

$$W(\alpha, c) = \{1 + \lambda_1(c)\}F(Y), \qquad (28.33)$$

where

$$F(Y) = \frac{A_1(\xi_1, 2)}{\xi_1 A_1(\xi_1, 1)}. \qquad (28.34)$$

The function $F(Y)$, called the Tietjens function, was first tabulated by Tietjens (1925) and has since been recomputed many times. Some authors have preferred to use the modified Tietjens function

$$\mathscr{F}(Y) = \{1 - F(Y)\}^{-1}, \qquad (28.35)$$

which has some advantages in estimating the minimum critical Reynolds number. The most extensive existing tables are those computed by Miles (1960), who tabulated the real and imaginary parts of F, $\mathscr{F}$ and $\mathscr{F}'$ for $Y = -6(0.1)10$; 4S. The general behaviour of $F(Y)$ is shown in Fig. 4.10. A comparison of Figs. 4.9 and 4.10 shows that $F(Y)$ and $G(X)$ agree only near the origin, where both X and Y are large.

The behaviour of $F(Y)$ in the neighbourhood of Y_0 ($\cong 2.297$) where $F_i(Y_0) = 0$ is also of some interest since it determines the asymptotic behaviour of the lower branch of the marginal curve. Since $F(Y)$ is analytic at Y_0 we have

$$F(Y) = F(Y_0) + F'(Y_0)(Y - Y_0) + O\{(Y - Y_0)^2\}, \quad (28.36)$$

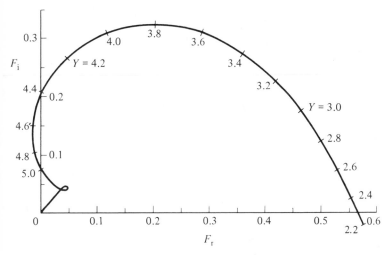

Fig. 4.10. The Tietjens function $F(Y)$. (From Reid 1965.)

where

$$F(Y_0) \cong 0.5645 \text{ and } F'(Y_0) \cong -0.1197 + 0.2307i. \qquad (28.37)$$

The asymptotic expansion of $F(Y)$ as $|Y| \to \infty$ is

$$F(Y) \sim e^{\pi i/4} Y^{-3/2} + \tfrac{5}{4} e^{\pi i/2} Y^{-3} + \tfrac{151}{32} e^{3\pi i/4} Y^{-9/2} + \cdots \qquad (28.38)$$

and is valid in the sector $-\pi/6 < \text{ph } Y < 11\pi/6$. In the sector $-5\pi/6 < \text{ph } Y < \pi/2$ which contains the ray ph $Y = -\pi/6$, the asymptotic expansion of $F(Y)$ has a more complicated form (see, for example, Lakin & Reid (1970), equation (8.23)).

In solving the eigenvalue relation (28.33) it is often more efficient to recompute the Tietjens function than to use the existing tables. For real values of Y, the Airy functions which appear in the definition of $F(Y)$ are exponentially large for Y positive and exponentially small for Y negative. To avoid the numerical difficulties caused by this exponential behaviour of the Airy functions, Hughes & Reid (1968) suggested the possibility of direct numerical integration of the differential equation satisfied by $F(Y)$ itself. More precisely, since $F(Y)$ has a simple pole at $Y = 0$, it is convenient to let

$$H(Y) = YF(Y), \qquad (28.39)$$

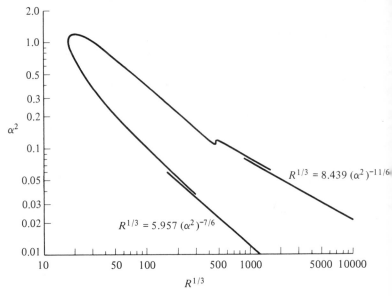

Fig. 4.11. The curve of marginal stability for plane Poiseuille flow based on equation (28.33). (From Reid 1965.)

and it then follows from equation (A5) that $H(Y)$ satisfies the second-order nonlinear equation

$$HH'' - 2H'^2 + 3H' - 1 + i(H^3 - YH^2) = 0 \qquad (28.40)$$

with the initial conditions

$$H(0) = 3^{2/3}\{\Gamma(\tfrac{1}{3})\}^{-1}\,e^{-\pi i/6} \quad \text{and} \quad H'(0) = 1 - \tfrac{1}{2}3^{3/2}\pi^{-1}. \tag{28.41}$$

Equation (28.40) can also be used to obtain the higher-order terms in the expansion (28.38).

Calculations based on the approximation (28.33) lead to the results shown in Figs. 4.11 and 4.12. The 'kink' along the upper branch of the marginal curve can be traced to the loop in the Tietjens function when plotted in the (F_r, F_i)-plane (see Fig. 4.10). The existence of such a kink has also been confirmed by direct numerical integration of the Orr–Sommerfeld equation (Hughes 1972), thus showing that it is not merely a consequence of the asymptotic approximations which have been used. Nevertheless, it seems unlikely that this kink has any physical importance. The

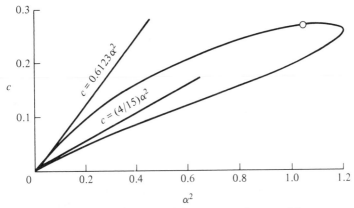

Fig. 4.12. The relationship between the wavenumber α and the wavespeed c along the curve of marginal stability for plane Poiseuille flow based on equation (28.33). The circled point corresponds to the minimum critical Reynolds number. There is also a very small loop (not shown) near $\alpha^2 = 0.112$ and $c = 0.029$.

values of the parameters associated with the minimum critical Reynolds number are given in Table 4.1 together with the 'exact' values obtained by Orszag (1971) by a direct numerical solution of the Orr–Sommerfeld equation.

The correct asymptotic behaviour of the lower branch of the marginal curve can now be obtained from equation (28.27) on replacing $G(X)$ by $F(Y)$. For symmetrical flows in a channel we have $K_1 \to 1 - F_r(Y_0)$ as $Y \to Y_0$ and hence

$$c \sim 2.296(I_2/U_1')\alpha^2 \quad \text{and} \quad R \sim 1.002(U_1'^5/I_2^3)\alpha^{-7};$$
$$(28.42)$$

similarly, for flows of the boundary-layer type we have

$$c \sim 2.296(1/U_1')\alpha \quad \text{and} \quad R \sim 1.002U_1'^5\alpha^{-4}. \quad (28.43)$$

These results show that $|z_1 - z_c|$ is of order $(\alpha R)^{-1/3}$ as $\alpha R \to \infty$ along the lower branch of the marginal curve and hence that the inner and outer viscous layers are no longer distinct in this limit.

28.6 Tollmien's improved approximation to $\phi_3(z)$

When the approximation (27.42) is used we have

$$\frac{\phi_3(z_1)}{\phi_3'(z_1)} = \frac{\eta(z_1)}{\eta'(z_1)} \frac{A_1(\zeta_1, 2)}{\zeta_1 A_1(\zeta_1, 1)}, \quad (28.44)$$

Table 4.1. *The values of the parameters associated with the minimum critical Reynolds number for plane Poiseuille flow*

	α	c	R	Y
Equation (28.33)	1.022	0.2672	5397.1	3.043
Equation (28.47)	1.010	0.2607	5697.3	3.009
'Exact' values	1.021	0.2640	5772.2	3.072

where ζ_1 is the value of ζ at $z = z_1$. For small values of c, it can be seen immediately that this result reduces to equation (28.30). If we now let

$$\zeta_1 = Z e^{-5\pi i/6},$$

where

$$Z = \left[\frac{3}{2} \int_{z_1}^{z_c} (c - U)^{1/2} \, dz \right]^{2/3} (\alpha R)^{1/3} \qquad (28.45)$$

and

$$1 + \lambda_2(c) = -\frac{U_1'}{c} \frac{\eta(z_1)}{\eta'(z_1)}$$

$$= \frac{3}{2} (U_1'/c^{3/2}) \int_{z_1}^{z_c} (c - U)^{1/2} \, dz, \qquad (28.46)$$

then the eigenvalue relation can be written in the form

$$W(\alpha, c) = \{1 + \lambda_2(c)\} F(Z). \qquad (28.47)$$

Calculations based on this approximation to the eigenvalue relation lead to a substantially more accurate value of R_c with very little additional complexity in the calculations. As indicated in Table 4.1, the error in the value of R_c has been reduced to less than two per cent. It is of some interest, therefore, to examine more closely the precise differences between the approximations (28.33) and (28.47).

Equation (28.47) differs from equation (28.33) in two respects: in the term $\lambda_2(c)$ and in the use of Z rather than Y. In evaluating $\lambda_2(c)$ it is convenient to let

$$\mu(c) = \int_{z_1}^{z_c} (c - U)^{1/2} \, dz \qquad (28.48)$$

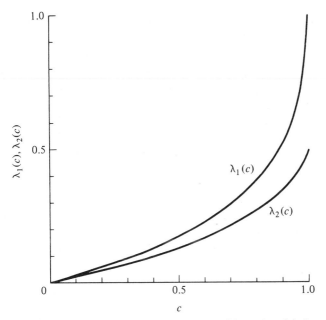

Fig. 4.13. A comparison of the functions $\lambda_1(c)$ and $\lambda_2(c)$ for plane Poiseuille flow.

so that

$$1 + \lambda_2(c) = \tfrac{3}{2}(U_1'/c^{3/2})\mu(c). \qquad (28.49)$$

For plane Poiseuille flow an easy calculation gives

$$\mu(c) = \tfrac{1}{2}\{\sqrt{c} - (1-c)\tanh^{-1}\sqrt{c}\}, \qquad (28.50)$$

and this result for $\lambda_2(c)$ is compared with $\lambda_1(c)$ in Fig. 4.13. Since c does not exceed about 0.3 on the marginal curve for plane Poiseuille flow, the differences between $\lambda_1(c)$ and $\lambda_2(c)$, though small, would appear to be significant.

To compare the variables X, Y and Z let

$$X = f_0(c)(\alpha R)^{1/3}, \quad Y = f_1(c)(\alpha R)^{1/3} \quad \text{and} \quad Z = f_2(c)(\alpha R)^{1/3}, \qquad (28.51)$$

where

and

$$\left.\begin{array}{l} f_0(c) = c(U_1')^{-2/3}, \\ f_1(c) = (z_c - z_1)(U_c')^{1/3}, \\ f_2(c) = \{\tfrac{3}{2}\mu(c)\}^{2/3}. \end{array}\right\} \qquad (28.52)$$

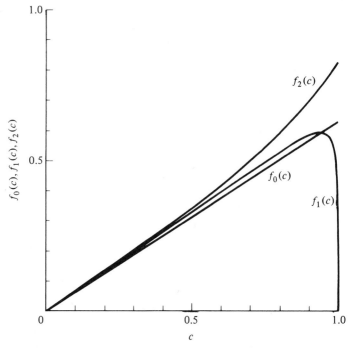

Fig. 4.14. A comparison of the functions $f_0(c)$, $f_1(c)$ and $f_2(c)$ for plane Poiseuille flow.

The behaviour of these functions for plane Poiseuille flow are shown in Fig. 4.14. For values of c less than about 0.3 the differences between them are extremely small. Thus, although there are strong theoretical reasons for using the Langer variable η and hence Z rather than $z - z_c$ and Y, the major differences occur indirectly through the λs.

Based on the values of R_c given in Table 4.1, the approximation (28.47) is clearly superior to approximation (28.33) and it has been generally accepted that the former should be used in preference. A more detailed comparison of the curves of marginal stability based on these two approximations is shown in Fig. 4.15. The results for equations (28.33) and (28.47) are due to B. S. Ng; the circled points are from Reynolds & Potter (1967); and the dashed curve has been completed using additional unpublished results from Hughes. Thus,

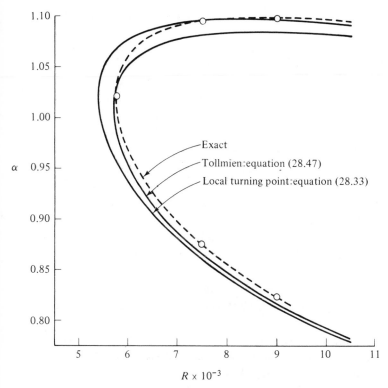

Fig. 4.15. A comparison of the curves of marginal stability for plane Poiseuille flow based on equations (28.33) and (28.47). (The circled points are from Reynolds & Potter (1967) and the dashed curve is based on additional 'exact' numerical values obtained by T. H. Hughes (unpublished).)

although equation (28.47) leads to a good approximation for R_c, such good agreement must be regarded as partly fortuitous and similar remarks apply to equation (28.33) near α_{max}. These results further emphasize the importance of uniformity in approximating not only the solutions of the Orr–Sommerfeld equation but also the eigenvalue relation. In Chapter 5 it will be shown that the use of uniform approximations to ϕ_2 and ϕ_3 leads to an approximate form of the eigenvalue relation which differs substantially from both equations (28.33) and (28.47).

29 The long-wave approximation for unbounded flows

The stability characteristics of unbounded flows are substantially different from those of bounded or semi-bounded flows. These flows are of two basic types. For the shear-layer type, $U(-\infty) \neq U(\infty)$, and without loss of generality we can choose the origin of velocity so that $U(-\infty) = -U(\infty)$. This may be effected by a Galilean transformation which will change c_r, but not c_i, and hence not the growth rate of the disturbance. We also choose the characteristic velocity V so that $U(\infty) = 1$. For flows of the jet type, $U(-\infty) = U(\infty)$, and in this case we choose the origin of velocity so that $U(\infty) = 0$. The characteristic velocity V is then usually chosen so that $\max\{U(z)\} = 1$.

In studying the stability of unbounded flows an important question that immediately arises is whether or not the basic flow can be treated as nearly parallel in the sense of equations (25.4). A general and precise answer to this question is not presently available but it is a matter of continuing concern and study, especially in connexion with flows of the boundary-layer type. The nature of the difficulties, however, can easily be illustrated by considering Bickley's (1937) solution of the boundary-layer equations for a two-dimensional jet. If we let M denote the (constant) momentum flux across any plane normal to the x-axis then Bickley's solution can be written in the dimensionless form

and
$$\left.\begin{array}{l} U = \mathrm{sech}^2 z \\ W = (2/R)(2z\,\mathrm{sech}^2 z - \tanh z), \end{array}\right\} \tag{29.1}$$

where
$$\left.\begin{array}{l} L = 2(6\nu^2 \rho x_*^2/M)^{1/3}, \\ V = \tfrac{1}{4}(6M^2/\nu\rho^2 x_*)^{1/3}, \end{array}\right\} \tag{29.2}$$

and
$$R = VL/\nu = \tfrac{1}{2}(36Mx_*/\nu^2\rho)^{1/3}. \tag{29.3}$$

The first of the conditions (25.4) is thus seen to be satisfied provided R is large, for then W/U is of order R^{-1}. Since this flow does not have a geometrical length scale, both L and V depend on x_*, and the second of the conditions (25.4) must therefore be stated in dimensional form. This gives $(1/U_*)(\partial U_*/\partial x_*) = O(1/x_*)$ which shows that U_* is a slowly varying function of x_* for large values of x_* which, in turn, are generally associated with large values of R^3. In treating this flow as nearly parallel we then neglect W, consider U as a function of z only, and continue to use the Orr–Sommerfeld equation in the subsequent analysis.

Unfortunately, the minimum critical Reynolds number is found (see § 31) to be about 4 and for such small values of the Reynolds number the parallel-flow assumption is of questionable validity. A similar question could also be raised concerning the basic flow itself since Bickley's solution given by equations (29.1) was obtained from the usual boundary-layer equations and is not an exact solution of the Navier–Stokes equations. Thus, in dealing with the stability of non-parallel flows, there are two major problems: one is concerned with obtaining a suitably accurate description of the basic flow and the other is concerned with obtaining approximations to the solutions of the partial differential equations which govern the disturbance flow. These two problems will be discussed some-what further in § 32 in connexion with the Blasius boundary-layer flow. For the unbounded flows with which we are concerned in this section, however, the parallel-flow approximation provides a basis for obtaining some simple qualitative results for such flows.

A further difficulty in dealing with unbounded flows is to obtain approximations to the solutions of the Orr–Sommerfeld equation for small values of α which remain uniformly valid as $z \to \pm\infty$. For flows of the jet type, however, methods for overcoming this difficulty have been developed independently by Tatsumi & Kakutani (1958) and by Howard (1959). In both methods the Orr–Sommerfeld equation is first rewritten in the form

$$(\mathrm{D}^2 - \alpha^2)(\mathrm{D}^2 - \beta^2)\phi = i\alpha R\{U(\mathrm{D}^2 - \alpha^2) - U''\}\phi, \qquad (29.4)$$

where

$$\beta = (\alpha^2 - i\alpha Rc)^{1/2} \quad \text{with} \quad \mathrm{Re}(\beta) > 0, \qquad (29.5)$$

and this equation must be solved subject to the boundary conditions

$$\alpha\phi = \mathrm{D}\phi = 0 \text{ at } z = \pm\infty. \qquad (29.6)$$

More particularly, we wish to determine the behaviour of c and R as $\alpha \downarrow 0$. For large values of $|z|$ the right-hand side of equation (29.4) becomes negligibly small and this feature of the equation plays an essential role in obtaining an approximate solution for small values of α. Throughout this discussion of jets we shall assume that $U(z)$ is even, for simplicity. Further, we shall assume that the mode is sinuous, with even eigenfunction ϕ, because the most unstable mode is sinuous.

In the method developed by Tatsumi & Kakutani (1958) we consider an expansion of the form

$$\phi(z) = \sum_{n=0}^{\infty} (i\alpha R)^n \phi^{(n)}(z; \alpha, \beta), \qquad (29.7)$$

where

$$(D^2 - \alpha^2)(D^2 - \beta^2)\phi^{(0)} = 0 \qquad (29.8)$$

and

$$(D^2 - \alpha^2)(D^2 - \beta^2)\phi^{(n)} = \{U(D^2 - \alpha^2) - U''\}\phi^{(n-1)} \quad (n \geqslant 1). \quad (29.9)$$

It is found that on the marginal curve R is of order $\alpha^{-1/2}$ as $\alpha \downarrow 0$ and the expansion (29.7) is therefore effectively one in powers of $\alpha^{1/2}$. For an even mode it is convenient to consider the problem on the interval $0 \leqslant z < \infty$ with $D\phi = D^3\phi = 0$ at $z = 0$. The relevant solutions of equation (29.8) are simply $e^{-\alpha z}$ and $e^{-\beta z}$, and from equation (29.9) the higher approximations can all be obtained by quadrature. To obtain the limiting behaviour of c and R as $\alpha \downarrow 0$, however, it is necessary to retain four terms in the expansion (29.7). The term of order $(\alpha R)^3$ was not considered by Tatsumi & Kakutani, but they found later (cf. Drazin 1961) that it must be included and thereby obtained agreement with Howard's (1959) results, which were obtained from an equivalent integral equation.

The determination of the higher-order terms in the expansion (29.7) or, alternatively, in Howard's approximate solution of the equivalent integral equation is straightforward in principle but complicated in practice. In the case of the Bickley jet $U = \operatorname{sech}^2 z$, however, Howard found that as $\alpha \downarrow 0$ for fixed values of $X \equiv \alpha R^2$ we have

$$c = \alpha^2\{2(2X^{-1} - 1) - 2iX^{-3/2}[(\tfrac{1}{6}\pi^2 - 1)X^2 - 9X + 8]\alpha^{1/2} + O(\alpha)\}.$$
$$(29.10)$$

In this limit therefore $c_i(\alpha, R) = 0$ has asymptotes

$$(\tfrac{1}{6}\pi^2 - 1)X^2 - 9X + 8 = 0. \tag{29.11}$$

Thus, there are two branches of the curve of marginal stability on which $\alpha R^2 \to 0.954$ or 13.0 and $c/\alpha^2 \to 2.19$ or -1.69 respectively. But the significance of the negative values of c on the second branch is not fully understood. Equation (29.10) also shows that we have instability if $0.954 < \alpha R^2 < 13.0$ and stability if αR^2 lies outside that interval. For *general* (even) profiles of the jet type Tatsumi & Kakutani have shown† that as $\alpha \downarrow 0$ the two branches of the curve of marginal stability are given by

$$AX^2 - 18BX + 8 = 0 \quad \text{and} \quad c/\alpha^2 = (4/I_1 X)(1 - BX), \tag{29.12}$$

where

$$I_1 = \int_0^\infty U \, dz,$$

$$A = 4I_1\left[\int_0^\infty z^2 U \int_z^\infty U \int_z^\infty U \, dz \, dz \, dz - I_1 \int_0^\infty z^2 U \int_z^\infty U \, dz \, dz \right]$$
$$+ 10\left[\int_0^\infty U \int_z^\infty zU \, dz \, dz \right]^2$$
$$- 10I_1 \int_0^\infty U \int_z^\infty zU \int_z^\infty zU \, dz \, dz \, dz,$$

and

$$B = \int_0^\infty U \int_z^\infty zU \, dz \, dz.$$

When $U = \text{sech}^2 z$, we find that $I_1 = 1$, $A = \tfrac{1}{6}\pi^2 - 1$, and $B = \tfrac{1}{2}$ in agreement with equation (29.11).

For flows of the shear-layer type it was first shown by Esch (1957) in a particular example that $R \to 0$ as $\alpha \downarrow 0$ and hence that there is no lower branch of the curve of marginal stability. The long-wave approximation for general flows of the shear-layer type was subsequently studied by Tatsumi & Gotoh (1960) using a modification of the method of Tatsumi & Kakutani (1958)

---

† In an unpublished letter to P. G. Drazin in 1960.

described above. In a second approximation they found that as $R \to 0$

$$c_r = 0,$$

$$\alpha = \frac{(1 - \sqrt{3}c_i)^2}{4(\sqrt{3} - c_i)}$$

$$\times R \left\{ 1 - \frac{(1 - 3c_i^2)(7 - 2\sqrt{3}c_i + c_i^2)}{4\sqrt{3}(\sqrt{3} - c_i)^3} (W_+ - W_-)R^2 + O(R^3) \right\}$$

$$(29.13)$$

where

$$W_\pm = \int_0^{\pm\infty} z\{U(z) - U(\pm\infty)\} \, \mathrm{d}z.$$

The limiting behaviour of the curve of marginal stability is therefore given by

$$\alpha = \frac{R}{4\sqrt{3}} \{1 - \tfrac{7}{36}(W_+ - W_-)R^2 + O(R^3)\}. \qquad (29.14)$$

For shear layers it is customary to choose the characteristic velocity V so that $U(\pm\infty) = \pm 1$ but the choice of the characteristic length L is still at our disposal. It could, for example, be chosen so that $W_+ - W_- = 1$ and the results given by equations (29.13) and (29.14) then become universal up to order R^3. This would be a natural choice for small values of R; for large values of R, however, it has the effect of exaggerating the differences in the curves of marginal stability for different velocity profiles. This effect can be illustrated by considering the two profiles:

(a)
$$U(z) = \begin{cases} kz & (|kz| < 1) \\ \pm 1 & (\pm kz > 1) \end{cases} \qquad (29.15)$$

(b)
$$U(z) = \tanh kz. \qquad (29.16)$$

If we fix k and hence L by requiring that $W_+ - W_- = 1$ then $k = 1/\sqrt{3}$ and $\pi/2\sqrt{3}$ respectively. As $R \to \infty$ it is then found that $\alpha \uparrow \alpha_s \cong 0.37$ and 0.91 respectively. Alternatively, if we choose $k = 1$ so that $U'(0) = 1$ for both profiles then $\alpha \uparrow \alpha_s \cong 0.64$ and 1 respectively as $R \to \infty$.

An important feature of these results for both jets and shear layers is that they depend only on integrals of the basic velocity

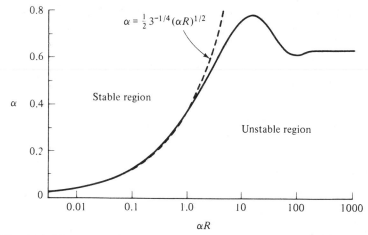

Fig. 4.16. The curve of marginal stability for the shear layer defined by equations (29.15) with $k = 1$. (From Esch 1957.)

distribution. This suggests the possibility of modelling the stability characteristics of smoothly varying velocity profiles for small values of α by the use of piecewise linear profiles along the lines discussed in § 23 for an inviscid fluid. Suppose then that U and/or U' are discontinuous at $z = z_0$ (say). By repeated integration of the Orr–Sommerfeld equation across the discontinuity, Drazin (1961) has shown that for bounded values of α and αR the required matching conditions are

$$\left. \begin{array}{c} \Delta\phi = 0, \quad \Delta[D\phi] = 0, \\[4pt] \Delta[D^2\phi + i\alpha R(U - c)\phi] = 0, \\[4pt] \Delta[D^3\phi - i\alpha R\{(U - c)\,D\phi - U'\phi\}] = 0. \end{array} \right\} \qquad (29.17)$$

When U is constant, the solutions of the Orr–Sommerfeld equation are simple exponentials; when U is linear, however, the solutions involve both exponential functions and integrals of Airy functions, and in this latter case the analysis is substantially more difficult than that given in § 23 for an inviscid fluid.

The shear layer defined by equations (29.15) with $k = 1$ has been treated in detail by Esch (1957) and his results are shown in Fig. 4.16. As $\alpha \downarrow 0$ the curve of marginal stability approaches the form described by equation (29.14) with $W_+ = -W_- = -\frac{1}{6}$ and $\alpha \uparrow \alpha_s$ as $R \to \infty$, where $\alpha_s \cong 0.64$ in agreement with § 23.2. Two other simple examples are worth mentioning briefly (for further details, see Drazin 1961). First, for the discontinuous shear layer $U(z) = \operatorname{sgn} z$, we simply obtain the leading term of equation (29.13). This result is to be expected, of course, since any fixed profile $U(z)$ of the shear-layer type approaches sgn z as the characteristic length $L \to 0$. Second, for the broken-line jet,

$$U(z) = \begin{cases} 1 & (|z| < 1), \\ 0 & (|z| > 1), \end{cases} \tag{29.18}$$

a little analysis shows that as $\alpha \downarrow 0$ for fixed values of $X \equiv \alpha R^2$ we have

$$c = \alpha^2 \{ 4(X^{-1} - \tfrac{1}{3}) - 4iX^{-3/2}(\tfrac{4}{45}X^2 - 3X + 4)\alpha^{1/2} + O(\alpha) \}. \tag{29.19}$$

On comparing this result with equation (29.10) it can be seen that the stability characteristics of the broken-line jet as $\alpha \downarrow 0$ are qualitatively similar to those of the smoothly varying Bickley jet. The asymptotes to the curve of marginal stability for this case could also have been obtained directly from equations (29.12).

It must be emphasized, however, that the use of discontinuous profiles to model the stability characteristics of smoothly varying profiles is limited to small values of α and bounded values of αR. The method is therefore not suitable for either bounded or semi-bounded flows, which are known to be stable in that region of the αR-plane.

30 Numerical methods of solution

The first numerical solution of the Orr–Sommerfeld equation was obtained by Thomas (1953) in an effort to resolve the controversies which existed at that time concerning the asymptotic methods of approximation, and his results fully confirmed the conclusions of Heisenberg (1924) and Lin (1945) that plane Poiseuille flow is indeed unstable. Since then a number of numerical methods have been developed for dealing with various aspects of the Orr–Sommerfeld problem, and in this section therefore we shall discuss briefly some of these methods. Among them we can distinguish those which are applicable (a) to the determination of the curve of marginal stability or, more generally, the curves on which $c_i =$

constant, (b) to the determination of the eigenvalue spectrum for fixed values of α and R, and (c) to the determination of the associated eigenfunctions. Numerical methods are particularly helpful in studying the stability characteristics of flows for which the relevant values of the Reynolds number are low and in studying, for example, the dependence of α_c and R_c on other secondary parameters of the problem. A full discussion of the stability of compressible boundary layers lies outside the scope of this book but this important problem is one for which asymptotic methods, similar to those discussed in § 27, have thus far proved to be inadequate (Lees & Reshotko 1962) whereas a direct numerical attack on the problem has been quite successful (Mack 1965). For a discussion of the stability of parallel shear flows in which the role of a computer is emphasized, see the book by Betchov & Criminale (1967).

30.1 *Expansions in orthogonal functions*

In the study of the Bénard and Taylor problems it was found, as described in Chapters 2 and 3, that variational and Galerkin methods could be used to provide extremely accurate approximations to the eigenvalue relation even when only a small number of simple trial functions were used, and it is natural therefore to attempt to apply such methods to the Orr–Sommerfeld problem. In this latter case, however, the required eigenfunction and its derivatives are rapidly varying near the critical point and, for this reason, further difficulties may be expected. In particular, greater care must be taken in the choice of the trial functions and a large number of them must be used. Throughout this discussion we shall suppose that the basic flow is confined to the interval $|z| \leq 1$ and we shall continue to use plane Poiseuille flow for illustrative purposes.

In this approach it is convenient to regard c as the eigenvalue parameter of the problem and to rewrite the Orr–Sommerfeld equation in the form

$$\mathrm{L}\phi + c\mathrm{M}\phi = 0, \tag{30.1}$$

where

and

$$\left. \begin{array}{c} \mathrm{L} = (\mathrm{i}\alpha R)^{-1}(\mathrm{D}^2 - \alpha^2)^2 - \{U(\mathrm{D}^2 - \alpha^2) - U''\} \\ \mathrm{M} = \mathrm{D}^2 - \alpha^2. \end{array} \right\} \tag{30.2}$$

Now let $\{\phi_n(z)\}$ be a complete set of orthonormal functions which satisfy an orthogonality relation of the form

$$(\phi_m, \phi_n) = \delta_{mn}. \tag{30.3}$$

The precise form of this relation will depend, of course, on the functions $\{\phi_n(z)\}$. We then expand $\phi(z)$ in the infinite series

$$\phi(z) = \sum_{n=1}^{\infty} a_n \phi_n(z). \tag{30.4}$$

In the Galerkin method, as described in § 17, this series is substituted into equation (30.1) and the resulting error is required to be orthogonal to $\phi_m(z)$ for $m = 1, 2, \ldots$ This gives

$$\sum_{n=1}^{\infty} (A_{mn} + cB_{mn})a_n = 0 \quad (m = 1, 2, \ldots), \tag{30.5}$$

where

$$A_{mn} = (\phi_m, L\phi_n) \quad \text{and} \quad B_{mn} = (\phi_m, M\phi_n). \tag{30.6}$$

In this procedure it is assumed that the series in equation (30.4) can be differentiated term by term four times. For some choices of trial functions this may not be permissible, but the difficulty can be circumvented, in some cases at least, by performing suitable integration by parts. In practice, of course, the series in equation (30.4) must be truncated after a finite number of terms, N (say), and we then obtain an approximation to the eigenvalue relation of the form

$$\det (A_{mn} + cB_{mn}) = 0 \quad (m, n = 1, 2, \ldots, N). \tag{30.7}$$

The eigenvalues of this system of linear algebraic equations can then be found, for example, by the use of the QR matrix eigenvalue algorithm (Wilkinson 1965, Gary & Helgason 1970).

This method was first used by Dolph & Lewis (1958) who chose the set of functions defined by the equation

$$(D^2 - \alpha^2)^2 \phi_n + \lambda_n (D^2 - \alpha^2)\phi_n = 0 \tag{30.8}$$

and the boundary conditions

$$\phi_n = D\phi_n = 0 \quad \text{at } z = \pm 1. \tag{30.9}$$

For this set of trial functions it can be shown (Orszag 1971) that the error after N terms is of order N^{-4} as $N \to \infty$. The results obtained by Dolph & Lewis for plane Poiseuille flow with $N = 8$ and 20 show that an eight-term approximation is not adequate for quantitative purposes. With $N = 20$, however, they found that $\alpha_c \approx 1$ and $R_c \approx 5800$ and these results are in good agreement with the more accurate values obtained later by Orszag (1971). In the work by Grosch & Salwen (1968) a somewhat simpler set of functions was used. They chose the set of functions defined by the equation

$$(D^2 - \alpha^2)^2 \phi_n - \lambda_n \phi_n = 0 \qquad (30.10)$$

and the boundary conditions (30.9). When $\alpha = 0$ these functions reduce to those used by Chandrasekhar & Reid for the Bénard problem (see Chandrasekhar (1961), appendix V). For both sets of functions the error after N terms is of order N^{-5} as $N \to \infty$ and, on taking up to 50 terms, Grosch & Salwen obtained the values $\alpha_c \approx 1.025$ and $R_c \approx 5750$.

A comprehensive study of the use of expansions in orthogonal functions for the Orr–Sommerfeld problem has been made by Orszag (1971) who has argued that greater accuracy can be achieved by using the Chebyshev polynomials

$$T_n(z) = \cos(n \cos^{-1} z) \quad (n = 0, 1, 2, \ldots). \qquad (30.11)$$

Although the orthogonal functions used by Dolph & Lewis and Grosch & Salwen would appear to be more closely related to the eigenfunctions of the Orr–Sommerfeld problem, the Chebyshev polynomials have the important property that the error after N terms is of infinite order, i.e. it is of smaller order than any power of N^{-1}, as $N \to \infty$. To determine the coefficients a_n in equation (30.4) Orszag used Lanczos's tau method rather than the Galerkin method and, with $N \geq 26$, he obtained the values

$$R_c = 5772.22, \quad \alpha_c = 1.020\,56 \quad \text{and} \quad c = 0.264\,002. \qquad (30.12)$$

These values have been confirmed by A. Davey (unpublished) who obtained

$$R_c = 5772.2218, \quad \alpha_c = 1.020\,547 \quad \text{and} \quad c = 0.264\,000\,3$$

by a shooting method. Orszag also obtained the values of c for the 32 least stable modes for $\alpha = 1$ and $R = 10\,000$ (see Fig. 4.19).

Chebyshev polynomials were also used somewhat earlier by Clenshaw & Elliott (1960) in their study of the stability of the Bickley jet for which $U(z) = \text{sech}^2 z$. On taking $N = 24$, 48 and 96, they obtained good results along the upper branch of the curve of marginal stability for the symmetric mode for which $\alpha \uparrow 2$ as $R \to \infty$ (see Fig. 4.26) but they encountered difficulties for small values of α along the lower branch.

It would appear, therefore, that the method of expansions in orthogonal functions can give accurate results for the Orr–Sommerfeld problem, in some circumstances at least, provided N is taken sufficiently large. This is, of course, due to the rapid variation of the eigenfunctions of the Orr–Sommerfeld problem (and especially of their derivatives) near the critical point and is in marked contrast with the Bénard and Taylor problems for which only a few terms are generally needed.

30.2 *Finite-difference methods*

In the finite-difference or matrix methods the Orr–Sommerfeld equation is replaced by a difference equation using central difference approximations to the derivatives. To reduce the truncation error without increasing the order of the difference equation, Thomas (1953) used the Gauss–Jackson–Noumerov method. To illustrate the essential features of these methods, consider plane Poiseuille flow on the interval $[0, 1]$ and suppose that this interval is divided into a uniform grid of mesh points given by

$$z_n = n/N \quad (n = 0, 1, 2, \ldots, N). \tag{30.13}$$

Use of the boundary conditions then shows that the approximate solution $\boldsymbol{\phi} = (\phi_0, \phi_1, \phi_2, \ldots, \phi_N)^{\mathrm{T}}$, where $\phi_n = \phi(z_n)$, is an eigenvector of a matrix equation of the form

$$[\mathbf{A}(\alpha, R) + c\mathbf{B}(\alpha)]\boldsymbol{\phi} = \mathbf{0}. \tag{30.14}$$

In this equation $\mathbf{A}$ and $\mathbf{B}$ are pentadiagonal matrices and the elements of $\mathbf{A}$ are complex but those of $\mathbf{B}$ are real. The eigenvalue relation is then simply

$$\det [\mathbf{A}(\alpha, R) + c\mathbf{B}(\alpha)] = 0. \tag{30.15}$$

For given values of α and R, this is an eigenvalue problem for c. If we are interested in a particular mode and if an initial estimate for c is known the determinant can be calculated by reducing the matrix $\mathbf{A} + c\mathbf{B}$ to upper triangular form using Gaussian elimination and a more accurate value of c can then be found by using an iterative procedure such as the one due to Muller (1956). This method has been used by Kurtz & Crandall (1962) in their study of the stability of the Blasius boundary layer and the free convection boundary layer on a vertical heated plate, and a modified form of the method has been developed by Osborne (1967) for the Blasius boundary layer.

When initial estimates for the eigenvalues are not available it is necessary to compute them directly by using, for example, the QR matrix eigenvalue algorithm as adapted by Gary & Helgason (1970) for problems of this type. A more accurate value of any particular eigenvalue can then be found, if desired, by using one of the initial-value methods described below and in § 43.

The number of mesh points N which must be used depends not only on the accuracy required but also on the basic flow and the values of the parameters α and R, and no general criterion is available for choosing N. The accuracy of the results can often be assessed, however, by repeating the calculations for several (increasing) values of N and then extrapolating to zero mesh-size. In the case of plane Poiseuille flow, for example, it appears that $N \approx 100$ is adequate near the nose of the curve of marginal stability. For $R \gg R_c$, however, a very large number of mesh points is required and it may then be desirable to use an automatically determined variable mesh as described by Hughes (1972).

30.3 Initial-value methods (shooting)

In describing these methods it is convenient to rewrite the Orr–Sommerfeld equation as a system of first-order equations. Thus, if we let

$$\boldsymbol{\psi} = [\phi, \phi', \phi'' - \alpha^2\phi, \phi''' - \alpha^2\phi']^{\mathrm{T}} \tag{30.16}$$

then the Orr–Sommerfeld equation becomes

$$\boldsymbol{\psi}' = \mathbf{N}\boldsymbol{\psi}, \tag{30.17}$$

where

$$\mathbf{N} = \begin{bmatrix} 0 & 1 & 0 & 0 \\ \alpha^2 & 0 & 1 & 0 \\ 0 & 0 & 0 & 1 \\ -i\alpha R U'' & 0 & \alpha^2 + i\alpha R(U-c) & 0 \end{bmatrix}. \tag{30.18}$$

For illustrative purposes we again consider plane Poiseuille flow on the interval $[0, 1]$ and for the symmetric modes we then have the boundary conditions

$$\phi'(0) = \phi'''(0) = 0 \quad \text{and} \quad \phi(1) = \phi'(1) = 0. \qquad (30.19)$$

Now let ψ_1 and ψ_2 be the two solutions of equation (30.17) which satisfy the initial conditions

$$\psi_1(0) = [1, 0, 0, 0]^{\mathrm{T}} \quad \text{and} \quad \psi_2(0) = [0, 0, 1, 0]^{\mathrm{T}}. \qquad (30.20)$$

The general solution of the Orr–Sommerfeld equation which satisfies the boundary conditions at $z = 0$ must then be of the form

$$\phi(z) = \lambda \phi_1(z) + \mu \phi_2(z). \qquad (30.21)$$

The indexing of the solutions here is *not* related to the indexing used in the asymptotic theory. We are, of course, still free to normalize the solution in any convenient way and a frequent choice is to set $\lambda = 1$. On imposing the boundary conditions at $z = 1$ we have

$$\lambda \phi_1(1) + \mu \phi_2(1) = 0 \quad \text{and} \quad \lambda \phi_1'(1) + \mu \phi_2'(1) = 0, \qquad (30.22)$$

and hence the eigenvalue relation is simply

$$\mathcal{F}(\alpha, c, R) = \phi_1(1)\phi_2'(1) - \phi_1'(1)\phi_2(1) = 0. \qquad (30.23)$$

For fixed values of α and R (say), the eigenvalues c must then be found by searching iteratively for the zeros of $\mathcal{F}$.

Although this procedure is simple and attractive theoretically, its numerical implementation can lead to serious difficulty especially when R is large. This difficulty arises from the fact that although the solutions ψ_1 and ψ_2 are numerically satisfactory near $z = 0$, they both contain some multiple of the rapidly growing viscous solution and this causes a loss of linear independence near $z = 1$. One method of overcoming this difficulty, which was proposed by Nachtsheim (1964), is based on what is now called the *method of matched initial-value problems*. In this method, in addition to the forward integration from $z = 0$, a backward integration is also made from $z = 1$ and the eigenvalue relation is then obtained by matching the results at some interior point of the interval (e.g. the mid-point). Other methods for dealing with this difficulty include *filtering* (Kaplan 1964), *orthonormalization* (Godunov 1961, Conte 1966, Davey 1973b), and *parallel shooting* (Keller 1968). In these methods, the major goal has been to be able to compute two solutions which satisfy the boundary conditions at $z = 0$ (say) and which remain numerically satisfactory over the entire interval. Thus, it is necessary to determine which linear combination of ψ_1 and ψ_2 has an essentially inviscid character over the entire interval. Furthermore, even if one knew the exact initial conditions for such a solution, the *parasitic growth problem* would still be present and this is the basic reason for the development of these methods. For a more detailed discussion of these methods reference may be made to the papers by Lee & Reynolds (1967), Gersting & Jankowski (1972), Davey (1973b), and Ross (1973).

Two rather different methods of overcoming these growth problems have been suggested more recently. One of them, based on the use of a certain Riccati matrix, will be described here, and the other, based on the use of certain compound matrices, will be discussed in § 43. The general theory of the *Riccati method* has been given by Scott (1973) and it has subsequently been applied to the Orr–Sommerfeld problem by Davey (1977), Sloan (1977), and Sloan & Wilks (1977). The basic idea involved in this method is to transform the (linear) eigenvalue problem into a (nonlinear) initial-value problem which does not suffer from the parasitic growth problem.

To illustrate the essential features of the method, consider the system of equations

and
$$\left. \begin{aligned} \mathbf{u}' &= \mathbf{A}\mathbf{u} + \mathbf{B}\mathbf{v} \\ -\mathbf{v}' &= \mathbf{C}\mathbf{u} + \mathbf{D}\mathbf{v}, \end{aligned} \right\} \tag{30.24}$$

where $\mathbf{u}$ and $\mathbf{v}$ are n-vectors and $\mathbf{A}, \mathbf{B}, \mathbf{C}, \mathbf{D}$ are $n \times n$ matrices. For problems with separated boundary conditions, Sloan & Wilks (1976) have shown that there is no loss of generality in supposing that

$$\mathbf{u}(0) = \mathbf{0} \quad \text{and} \quad \alpha\mathbf{u}(1) + \beta\mathbf{v}(1) = \mathbf{0}, \tag{30.25}$$

where α and β are $n \times n$ matrices and $[\alpha\beta]$ is an $n \times 2n$ matrix of rank n. The Riccati matrix $\mathbf{R}$ is then introduced through the transformation

$$\mathbf{u} = \mathbf{R}\mathbf{v}. \tag{30.26}$$

To derive the differential equation satisfied by $\mathbf{R}$ it is necessary to consider the more general system

and
$$\left. \begin{aligned} \mathbf{U}' &= \mathbf{A}\mathbf{U} + \mathbf{B}\mathbf{V} \\ -\mathbf{V}' &= \mathbf{C}\mathbf{U} + \mathbf{D}\mathbf{V} \end{aligned} \right\} \tag{30.27}$$

subject to the initial conditions

$$\mathbf{U}(0) = \mathbf{0} \quad \text{and} \quad \mathbf{V}(0) = \mathbf{I}, \tag{30.28}$$

where $\mathbf{U}$ and $\mathbf{V}$ are $n \times n$ matrices. The transformation (30.26) then becomes

$$\mathbf{U} = \mathbf{RV}. \tag{30.29}$$

On differentiating this equation, using equations (30.27), and observing that $\mathbf{V}$ is non-singular, at least near $z = 0$, it then follows that $\mathbf{R}$ satisfies the first-order nonlinear matrix differential equation

$$\mathbf{R}' = \mathbf{B} + \mathbf{AR} + \mathbf{RD} + \mathbf{RCR} \tag{30.30}$$

with

$$\mathbf{R}(0) = \mathbf{0}. \tag{30.31}$$

To satisfy the boundary conditions at $z = 1$ the eigenvalue parameter must be varied until

$$\det[\boldsymbol{\alpha}\mathbf{R}(1) + \boldsymbol{\beta}] = 0. \tag{30.32}$$

For equations with real coefficients, the Riccati equation (30.30) has at least one singularity in the interval $0 < z < 1$. To deal with this difficulty it is necessary either to consider the equation satisfied by the inverse matrix $\mathbf{S} = \mathbf{R}^{-1}$ as described by Scott (1973) or to deform the path of integration into the complex plane as suggested by Davey (1977).

Consider then the application of this method to the Orr–Sommerfeld problem. For illustrative purposes we consider symmetric flows in a channel and for the even modes we then have the boundary conditions (30.19). Thus, if we let

$$\mathbf{u} = \begin{bmatrix} \phi' \\ \phi''' - \alpha^2 \phi' \end{bmatrix} \quad \text{and} \quad \mathbf{v} = \begin{bmatrix} \phi \\ \phi'' - \alpha^2 \phi \end{bmatrix}, \tag{30.33}$$

then the Orr–Sommerfeld equation can be written in the form described by equations (30.24) where

$$\mathbf{A} = \mathbf{0}, \quad \mathbf{B} = \begin{bmatrix} \alpha^2 & 1 \\ -\mathrm{i}\alpha R U'' & \alpha^2 + \mathrm{i}\alpha R(U - c) \end{bmatrix}, \quad \mathbf{C} = -\mathbf{I}, \quad \mathbf{D} = \mathbf{0}. \tag{30.34}$$

Equation (30.30) then simplifies to

$$\mathbf{R}' + \mathbf{R}^2 = \mathbf{B}, \tag{30.35}$$

where

$$\mathbf{R} = \begin{bmatrix} r_1 & r_2 \\ r_3 & r_4 \end{bmatrix}. \tag{30.36}$$

Equation (30.35) then yields the four scalar equations

$$\left. \begin{aligned} r_1' + r_1^2 + r_2 r_3 &= \alpha^2, \\ r_2' + r_1 r_2 + r_2 r_4 &= 1, \\ r_3' + r_1 r_3 + r_3 r_4 &= -i\alpha R U'', \\ r_4' + r_2 r_3 + r_4^2 &= \alpha^2 + i\alpha R(U - c), \end{aligned} \right\} \tag{30.37}$$

and

and the boundary conditions are

and

$$\left. \begin{aligned} r_1 = r_2 = r_3 = r_4 &= 0 \text{ at } z = 0 \\ r_2 &= 0 \text{ at } z = 1. \end{aligned} \right\} \tag{30.38}$$

In the special case of plane Poiseuille flow, for which $U'' = -2$, the second and third of equations (30.37) shows that $r_3 = 2i\alpha R r_2$ and hence in this case we have a system of three first-order nonlinear equations. This problem has been used by Davey (1977) to provide a comparison between the Riccati method and the method of orthonormalization. From this comparison he concluded that the Riccati method has several advantages over the method of orthonormalization: (a) it is simpler to formulate and program; (b) it provides an easier method of obtaining the eigenfunction; and (c) it is faster by about a factor of two.

31 Stability characteristics of various basic flows

In comparing the stability characteristics of different basic flows, it is helpful to distinguish certain general classes of flows. These include flows in channels, boundary layers, jets and shear layers. Flows with large critical Reynolds numbers, such as channel or boundary-layer flows without inflexion points, have stability characteristics which depend quite sensitively on the form of the basic flow. Flows with an inflexion point, however, have much smaller critical Reynolds numbers and their stability characteristics are much less sensitive to the form of the basic flow. To illustrate some of these variations we now wish to consider some typical examples from each class. These examples are also intended to provide some guidance as to what may be expected in other situations for which detailed calculations are not available.

31.1 *Plane Couette flow*

When $U(z) = z$ with $|z| \leqslant 1$, the Orr–Sommerfeld equation can be substantially simplified. If we let

$$\eta = z - c \quad \text{and} \quad \varepsilon = (i\alpha R)^{-1/3} \text{ with ph } \varepsilon = -\pi/6, \quad (31.1)$$

then the governing equation for plane Couette flow becomes

$$\{\varepsilon^3 (D^2 - \alpha^2) - \eta\}(D^2 - \alpha^2)\phi = 0, \quad (31.2)$$

where D now stands for $d/d\eta$. This equation must be considered together with the boundary conditions

$$\phi = D\phi = 0 \quad \text{at} \quad \eta = \eta_1 \quad \text{and} \quad \eta_2, \quad (31.3)$$

where $\eta_1 = -1 - c$ and $\eta_2 = 1 - c$ are the values of η corresponding to $z = -1$ and 1 respectively.

Much of the work on this problem has been concerned with the question of whether or not plane Couette flow is stable with respect to arbitrary infinitesimal disturbances. For large values of R the problem has been studied by Hopf (1914), who used asymptotic methods of approximation, and for small values of R by Southwell & Chitty (1930), who used power series expansions of the type described in § 26. Although instability might be expected for large values of R, and is certainly observed experimentally when R is sufficiently large (Taylor 1936), in neither of these investigations was any evidence of instability found. The methods employed in these investigations are clearly incapable of providing a general proof of stability. Nevertheless, it was soon believed that plane Couette flow is stable for infinitesimal disturbances but that, as first suggested by Rayleigh (1914), it may be unstable for finite disturbances.

Numerical studies of the linear problem by Grohne (1954) Gallagher & Mercer (1962, 1964), Deardorff (1963), Davey (1973a) and Gallagher (1974) have provided valuable insight into the general modal structure of the problem but, again, no instability was found. A partial proof of stability has been obtained by Diki (1964), who proved that all standing waves, i.e. modes for which

$c_r = 0$, are stable. The first general proof of stability, however, seems to be due to Romanov (1973), who proved that the normal modes of the linear problem are damped for all $\alpha \geqslant 0$ and all $R > 0$. He also considered the nonlinear initial-value problem and proved that it has a unique solution which is asymptotically stable provided the norm of the initial disturbances in the Sobolev space $W_2^{(1)}$ ($|x| \leqslant \infty$, $|y| \leqslant \infty$, $|z| \leqslant 1$) is sufficiently small. This condition on the initial disturbances would appear to be quite severe, especially for large values of R.

There is, however, another important aspect of this problem which concerns the asymptotic theory of equation (31.2) for small values of $|\varepsilon|$ and its relationship to the asymptotic theory of the Orr–Sommerfeld equation for more general velocity distributions. Ideally, one would like to be able to treat plane Couette flow as a special case of a general theory. Unfortunately, however, the heuristic theory which was described in § 27 can only be used for neutrally stable or amplified disturbances and so is not applicable to plane Couette flow for which all the modes are damped.

One method of dealing with equation (31.2), which was first used in the early work by Mises (1912a,b) and Hopf (1914), is based on the observation that it can be reduced to an inhomogeneous second-order equation, the general solution of which can then be obtained by the method of variation of parameters. In this way it is easy to show that the solutions of equation (31.2) which satisfy both of the boundary conditions at $\eta = \eta_1$ can be written in the form

$$\Phi_k(\eta) = \alpha^{-1} \varepsilon^{-2} \int_{\eta_1}^{\eta} \sinh \alpha(\eta - \eta') A_k(\zeta' + \alpha^2 \varepsilon^2) \, d\eta', \quad (31.4)$$

where $\zeta' = \eta'/\varepsilon$ and $A_k(z)$ ($k = 1, 2, 3$) are the Airy functions defined by equations (A1) in the Appendix. These solutions have been normalized, following Grohne (1954), so that

$$(D^2 - \alpha^2)\Phi_k = \varepsilon^{-2} A_k(\zeta + \alpha^2 \varepsilon^2), \quad (31.5)$$

and when $\alpha \to 0$ they reduce to

$$\Phi_k(\eta) = A_k(\zeta, 2) - \zeta A_k(\zeta_1, 1) + A_k'(\zeta_1). \quad (31.6)$$

Any two of the solutions $\Phi_k(\eta)$ are linearly independent. In deriving the eigenvalue relation, however, we choose the solutions with $k = 1$ and 2 to provide a numerically satisfactory pair, and on imposing the boundary conditions at $\eta = \eta_2$ we have

$$\Delta(\eta_1, \eta_2; \alpha, \varepsilon) \equiv \begin{vmatrix} \Phi_1(\eta_2) & \Phi_2(\eta_2) \\ \varepsilon\Phi_1'(\eta_2) & \varepsilon\Phi_2'(\eta_2) \end{vmatrix} = 0. \qquad (31.7)$$

On expansion and simplification this becomes

$$\begin{aligned}
\Delta = \alpha^{-1}\varepsilon^{-3}\Bigg\{ &\int_{\eta_1}^{\eta_2} \cosh \alpha\eta A_1(\zeta + \alpha^2\varepsilon^2)\,d\eta \\
&\times \int_{\eta_1}^{\eta_2} \sinh \alpha\eta A_2(\zeta + \alpha^2\varepsilon^2)\,d\eta \\
&- \int_{\eta_1}^{\eta_2} \sinh \alpha\eta A_1(\zeta + \alpha^2\varepsilon^2)\,d\eta \\
&\times \int_{\eta_1}^{\eta_2} \cosh \alpha\eta A_2(\zeta + \alpha^2\varepsilon^2)\,d\eta \Bigg\}.
\end{aligned} \qquad (31.8)$$

As Synge (1938), p. 262 has remarked, 'the deduction of general results from [this equation] is well-nigh impossible' and some further approximations must obviously be made. In his early work on this problem, Hopf (1914) (see also Grohne (1954)) considered three approximations which lead to an enormous simplification of the problem. Firstly, he effectively approximated the Airy functions $A_k(\zeta + \alpha^2\varepsilon^2)$ by $A_k(\zeta)$; this is clearly permissible since $A_k(\zeta)$ is simply the first term in the uniform expansions of $A_k(\zeta + \alpha^2\varepsilon^2)$ for bounded values of α and $|\eta|$. Secondly, he approximated the Airy functions $A_k(\zeta)$ by their asymptotic expansions for unbounded values of $|\zeta|$; this is also permissible provided $|\zeta|$ is greater than about 2 and the expansions are 'complete' in the sense of Watson (Olver 1974). Hopf's analysis was carried out using the old-fashioned and very cumbersome Hankel function notation and it is remarkable therefore that the expansions which he obtained did satisfy this completeness requirement.† Thirdly, he approximated the hyperbolic functions by the first terms of their power series; this approximation is clearly valid as $\alpha \to 0$ but not otherwise.

The problem of obtaining uniform asymptotic expansions to integrals of the type which appear in equation (31.8) and which are

---

† Apart from some minor technical errors in the analysis which were first noticed by Grohne (1954) and subsequently corrected by Reid (1974a).

valid for bounded values of α and $|\eta|$ has been considered by Reid (1974a), who showed that

$$\int_{\infty_k}^{\eta} f(\eta')A_k(\zeta'+\alpha^2\varepsilon^2)\,d\eta'$$

$$= \varepsilon f(0)A_k(\zeta, 1)+\varepsilon^2\eta^{-1}[f(\eta)-f(0)]A_k'(\zeta)+O(\varepsilon^3), \quad (31.9)$$

where ∞_k denotes a path of integration that tends to infinity in the sector $\mathbf{S}_k$. A first approximation to the eigenvalue relation for $|\varepsilon| \ll 1$ is therefore given by

$$\Delta \sim \{A_1(\zeta_2, 1)-A_1(\zeta_1, 1)\}$$

$$\times \left\{\frac{\sinh \alpha\eta_2}{\alpha\eta_2} A_2'(\zeta_2)-\frac{\sinh \alpha\eta_1}{\alpha\eta_1} A_2'(\zeta_1)\right\}$$

$$-\{A_2(\zeta_2, 1)-A_2(\zeta_1, 1)\}$$

$$\times \left\{\frac{\sinh \alpha\eta_2}{\alpha\eta_2} A_1'(\zeta_2)-\frac{\sinh \alpha\eta_1}{\alpha\eta_1} A_1'(\zeta_1)\right\}. \quad (31.10)$$

When $\alpha \to 0$ with αR finite, this approximation becomes exact. In this special case the eigenvalue relation is a function only of ζ_1 and ζ_2, and further details for this case have been given by Reid (1974a). When $\alpha R \to \infty$ with α bounded, it follows from equation (31.10) that

either

$$\zeta_1 \to z_{\pm s}, \qquad \zeta_2 \to \infty \quad \text{in } \mathbf{S}_1 \text{ and } \quad c_r \downarrow -1$$

or

$$\zeta_2 \to z_{\pm s}\, e^{-2\pi i/3}, \quad \zeta_1 \to \infty \quad \text{in } \mathbf{S}_2 \text{ and } \quad c_r \uparrow +1, \qquad (31.11)$$

where $z_{\pm s}$ $(s = 1, 2, \ldots)$ are the (complex) zeros of $A_1(z, 1)$ with $\text{Im}\,(-z_s) > 0$ and $z_{-s} = z_s^*$. These results can be combined to give

$$(1 \pm c_r+ic_i)(\alpha R)^{1/3} \to -z_{\pm s}\, e^{-\pi i/6}. \quad (31.12)$$

For $s = 1$ we have

$$z_{\pm 1} = -4.1070 \mp 1.1442i, \quad z_1\, e^{-\pi i/6} = -4.1288 + 1.0626i,$$

and

$$z_{-1}\, e^{-\pi i/6} = -2.9847 + 3.0444i. \qquad (31.13)$$

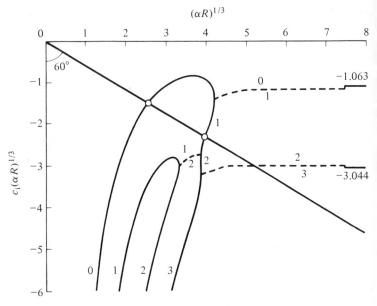

Fig. 4.17. The behaviour of c_i for the first four modes for plane Couette flow. Along the solid curves $c_r = 0$ and along the dashed curves $c_r \neq 0$ with $c_r \to \pm 1$ as $\alpha R \to \infty$. (From Davey & Reid 1977a.)

For larger values of s it is usually sufficient to use the asymptotic approximation

$$(-z_{\pm s})^{3/2} \sim \tfrac{3}{8}\pi(8s-1)$$
$$\pm i\tfrac{3}{2}\cosh^{-1}\{\sqrt{\pi}[\tfrac{3}{8}\pi(8s-1)]^{1/2}\} \quad (s=1,2,\ldots) \qquad (31.14)$$

which is due to Zondek & Thomas (1953). In terms of the notation used in the Appendix we have $z_s = \alpha_s(1)$ (cf. equation (A62)).

The behaviour of c_i for the first four modes for $\alpha = 0$ is shown in Fig. 4.17. The variation of the modal structure with α has also been studied in detail by Gallagher (1974), who considered the first twelve modes for $\alpha = 0, 0.6, 1, 2$ and $0 \leqslant (\alpha R)^{1/3} \leqslant 10$.

31.2 Poiseuille flow in a circular pipe

In discussing the stability of axisymmetric flows we shall use cylindrical polar co-ordinates (x, r, θ) and suppose that the basic flow has velocity components $\{U(r), 0, 0\}$. Although there is no difficulty in deriving the linearized disturbance equations for general non-axisymmetric disturbances, the resulting equations are extremely complicated and they cannot be reduced, by a transformation of Squire's type, to an equivalent axisymmetric problem. For simplicity, therefore, we shall consider only axisymmetric disturbances with velocity components $(u'_x, u'_r, 0)$. It is then convenient to introduce a stream function $\psi'(x, r, t)$ in terms of which we have

$$u'_x = \frac{1}{r}\frac{\partial \psi'}{\partial r} \quad \text{and} \quad u'_r = -\frac{1}{r}\frac{\partial \psi'}{\partial x}. \tag{31.15}$$

On letting $\psi'(x, r, t) = \phi(r) \exp\{i\alpha(x - ct)\}$ as usual, it can then be shown (Sexl 1927) that $\phi(r)$ satisfies the equation

$$(i\alpha R)^{-1}(L - \alpha^2)^2\phi = (U - c)(L - \alpha^2)\phi - r(U'/r)'\phi, \tag{31.16}$$

where

$$L \equiv \frac{d^2}{dr^2} - \frac{1}{r}\frac{d}{dr}. \tag{31.17}$$

This equation bears a striking similarity to the Orr–Sommerfeld equation but there is one important difference: the origin $(r = 0)$ is a regular singular point of equation (31.16). We are thus faced with a singular eigenvalue problem and must require that ϕ/r and ϕ'/r be bounded as $r \to 0$. At a rigid boundary, $r = 1$ (say), $\phi = \phi' = 0$.

The only axisymmetric flow that has been studied in any detail is Poiseuille flow in a circular pipe for which $U(r) = 1 - r^2$. For this flow $(U'/r)' \equiv 0$ and equation (31.16) can then be integrated to give

$$(L - \alpha^2)\phi = f, \tag{31.18a}$$

where

$$(L - \alpha^2)f - i\alpha R(1 - r^2 - c)f = 0. \tag{31.18b}$$

Thus, this problem, like the problem of plane Couette flow, is reducible to an inhomogeneous second-order equation. The inviscid solutions of the problem, which correspond to taking $f = 0$, are

$$\phi_1(r) = rI_1(\alpha r) \quad \text{and} \quad \phi_2(r) = rK_1(\alpha r), \tag{31.19}$$

and the solution that satisfies the boundary conditions at $r = 0$ is then given by

$$\phi(r) = ArI_1(\alpha r)$$

$$+ r \int_0^r [I_1(\alpha r)K_1(\alpha t) - I_1(\alpha t)K_1(\alpha r)]f(t)\, \mathrm{d}t, \qquad (31.20)$$

where $f(r)$ must be chosen as the solution of equation (31.18b) that is regular at the origin. On imposing the boundary conditions at $r = 1$, we find that the eigenvalue relation becomes simply

$$\int_0^1 I_1(\alpha r)f(r)\, \mathrm{d}r = 0. \qquad (31.21)$$

For small values of αR, an approximate solution has been obtained by Pekeris (1948) in the form (cf. equations (26.1))

$$f(r) = \sum_{s=0}^{\infty} (i\alpha R)^s f^{(s)}(r) \quad \text{and} \quad i\alpha Rc = \sum_{s=0}^{\infty} (i\alpha R)^s c^{(s)}. \qquad (31.22)$$

The actual calculations for this problem, however, are substantially more complicated than those for two-dimensional flows, and some further simplification is required. If we suppose that α is small, then the coefficients $c^{(s)}$ can be expanded in powers of α^2 to obtain

$$i\alpha Rc = \kappa_n^2 + \left(\frac{2}{3} + \frac{4}{\kappa_n^2}\right)i\alpha R + O\{\alpha^2, (i\alpha R)^2\}, \qquad (31.23)$$

where $\kappa_n = j_{2,n}$ $(n = 1, 2, \ldots)$. Additional terms in this series have been given by Pekeris (1948). Thus, for the higher modes (i.e. n large), $c_r \to \frac{2}{3}$ as $\alpha R \to 0$, which is simply the mean velocity of the basic flow (cf. equation (26.9)).

More generally, as Pekeris (1948) has shown, the solution of equation (31.18b) which is regular at the origin can be expressed in simple terms of a confluent hypergeometric function. In terms of Kummer's function $M(a, b, z)$ we have

$$f(r) = z\, \mathrm{e}^{-z/2} M(a, 2, z), \qquad (31.24)$$

where

and $\qquad \left. \begin{array}{l} a = 1 + \frac{1}{4}(1 - c)(\alpha R)^{1/2}\, \mathrm{e}^{-\pi i/4} + \frac{1}{4}\alpha^2(\alpha R)^{-1/2}\, \mathrm{e}^{-3\pi i/4} \\ z = (\alpha R)^{1/2}\, \mathrm{e}^{3\pi i/4} r^2. \end{array} \right\} \qquad (31.25)$

The eigenvalue relation (31.21) then becomes

$$\int_0^1 I_1(\alpha r) z e^{-z/2} M(a, 2, z) \, dr = 0. \qquad (31.26)$$

When $|z| \to \infty$ in the sector $-\pi/2 < \mathrm{ph} \, z < 3\pi/2$ we have

$$z \, e^{-z/2} M(a, 2, z) \sim \frac{e^{a\pi i} z^{1-a} \, e^{-z/2}}{\Gamma(2-a)} + \frac{z^{a-1} \, e^{z/2}}{\Gamma(a)}, \qquad (31.27)$$

where the second term is recessive compared to the first when $\mathrm{ph} \, z = 3\pi/4$. The major contribution to the integral (31.26) arises from a small interval near $r = 1$ and the integral is therefore exponentially large unless $1/\Gamma(2-a) = 0$. This gives

$$c = 1 + 4n \, e^{-3\pi i/4} (\alpha R)^{-1/2} + o\{(\alpha R)^{-1/2}\} \quad (n = 1, 2, \ldots) \qquad (31.28)$$

and represents centre modes for which $c_r \uparrow 1$ as $\alpha R \to \infty$. This result is valid for fixed values of n as $\alpha R \to \infty$ but not uniformly so as $n \to \infty$ for fixed values of αR however large.

Another class of modes was found later by Corcos & Sellars (1959) (see also Gill (1965)) and corresponds to wall modes for which $c_r \downarrow 0$ as $\alpha R \to \infty$. For this class of modes they found that

$$c = -2^{2/3} z_{\pm s} \, e^{-\pi i/6} (\alpha R)^{-1/3} + o\{(\alpha R)^{-1/3}\} \quad (s = 1, 2, \ldots), \qquad (31.29)$$

where $z_{\pm s}$ are again the (complex) zeros of $A_1(z, 1)$.

The relationship between these three classes of modes is shown in Fig. 4.18 for $\alpha = 1$ and $R = 5000$. This figure, which is due to Davey & Drazin (1969), also shows a few additional modes which do not belong to any of these classes and for which a theoretical explanation is still lacking.

These results, together with the mathematical similarity between this problem and the problem of plane Couette flow, have led to the belief that Poiseuille flow in a circular pipe is stable with respect to axisymmetric disturbances. There is also increasing evidence that it is stable with respect to non-axisymmetric disturbances (see, for example, Salwen & Grosch (1972)). The instability which is observed experimentally has therefore generally been attributed to finite-amplitude effects. Some suggestions have been made, however, that linear theory may still be important in certain circumstances. Thus, for example, Tatsumi (1952) studied the instability of the boundary layer which forms inside the pipe at the

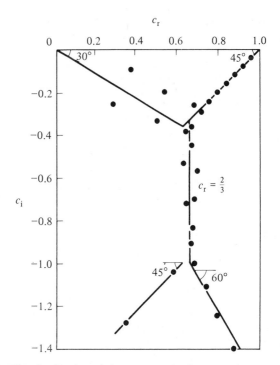

Fig. 4.18. The distribution of eigenvalues of axisymmetric disturbances for Poiseuille flow in a circular pipe at $\alpha = 1$ and $R = 5000$. (From Davey & Drazin 1969.)

inlet and found a critical Reynolds number of nearly 10 000. More recently, Mackrodt (1976) has argued that the observed instability may be due to a very slow rotation of the inlet flow. To model this situation, he considered the stability of circular Poiseuille flow which is in rigid rotation with angular velocity Ω about the x-axis. For this problem it is essential to consider three-dimensional disturbances and, as a result, numerical methods had to be used in the solution of the linearized disturbance equations. For values of the axial Reynolds number R greater than about 500, Mackrodt found that the flow becomes unstable when the azimuthal Reynolds number, $R_\theta \equiv \Omega L^2/\nu > 26.96$, where L is the radius of the pipe. He also found that for values of R_θ greater than about 500, instability occurs when $R > 82.88$. This value of R is the one found previously by Pedley (1969) when $R_\theta \to \infty$ and is close to the energy limit of 81.49 found by Joseph & Carmi (1969).

31.3 *Plane Poiseuille flow*

Some of the stability characteristics of plane Poiseuille flow have already been discussed in § 28. That discussion was concerned primarily with the one mode which becomes unstable when $R > R_c$ and with the curve of marginal stability for that mode. It is also of some interest, however, to consider the behaviour of the higher (damped) modes for this problem. Some results for the first four modes have been obtained by Grohne (1954) (see also Shen (1964), pp. 758–761) for $\alpha = 0.87$ and R in the range 10^3 to 10^7. He also showed that as $\alpha R \to \infty$ the wall modes have the behaviour (cf. equation (31.29))

$$c = -2^{2/3} z_{\pm s} e^{-\pi i/6} (\alpha R)^{-1/3}$$
$$+ O\{(\alpha R)^{-2/3} \ln (\alpha R)\} \quad (s = 1, 2, \ldots), \qquad (31.30)$$

where the roots z_s and z_{-s} are associated with the even and odd modes respectively. For $\alpha = 1$ and $R = 10\,000$, Orszag (1971) has computed the eigenvalues for the first 32 modes, indexed according to decreasing c_i. His results for the even modes have been extended by Mack (1976) and their results are shown in Fig. 4.19. In this figure the modes have been indexed, following Mack, in a manner intended to emphasize the different classes of modes. It should also be remarked that the eigenvalues of the P families for the even and odd modes are very nearly equal but the reasons for this are not fully understood.

The eigenfunction for the first even mode of the A family, for $\alpha = 1$ and $R = 10\,000$ has been computed by Thomas (1953) and is shown in Fig. 4.20. This is a mode which is slightly unstable with eigenvalue $c = 0.2375 + 0.0037i$ and the eigenfunction has been normalized so that $\phi(0) = 1$. From Thomas's results it is then possible to compute the distribution of the Reynolds stress. For this purpose we define the Reynolds stress function (cf. equations (22.35) and (53.6))

$$S(z) \equiv \tfrac{1}{2}i(\phi D\phi^* - \phi^* D\phi)$$
$$= \phi_r D\phi_i - \phi_i D\phi_r \qquad (31.31)$$

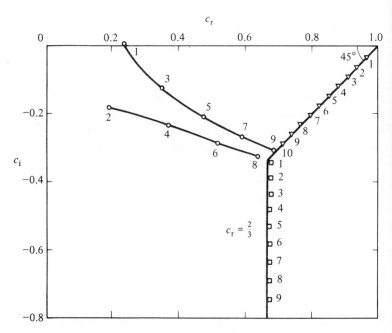

Fig. 4.19. The distribution of eigenvalues of the even modes for plane Poiseuille flow at $\alpha = 1$ and $R = 10\,000$. O, A family; ∇, P family; $\square$, S family. (From Mack 1976.)

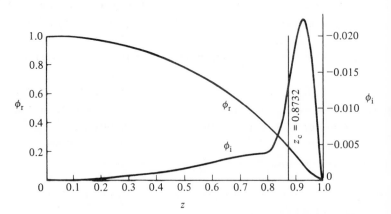

Fig. 4.20. The eigenfunction for the first even mode of the A family for plane Poiseuille flow at $\alpha = 1$ and $R = 10\,000$; $c = 0.2375 + 0.0037i$. (Based on the numerical results of Thomas (1953).)

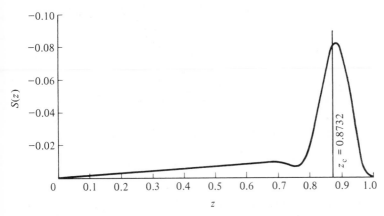

Fig. 4.21. The distribution of the Reynolds stress for plane Poiseuille flow at $\alpha = 1$ and $R = 10\,000$; $c = 0.2375 + 0.0037i$; $S(z) = \phi_r D\phi_i - \phi_i D\phi_r$. (After Stuart 1963.)

and this is shown in Fig. 4.21 which is due to Stuart (1963). Since S is an odd function of z, the product $S\,dU/dz$ is negative everywhere and hence, by equation (53.4), there is a transfer of energy from the basic flow to the disturbance at all values of z. The major contributions to this energy transfer, however, are concentrated near the two critical points.

31.4 *Combined plane Couette and plane Poiseuille flow*

Consider the basic flow

$$U(z) = A(1 - z^2) + Bz \quad (-1 \leqslant z \leqslant 1), \qquad (31.32)$$

where $0 \leqslant A \leqslant 1$ and $0 \leqslant B \leqslant 1$. The choice of a characteristic velocity, however, is still at our disposal. If, for example, we require that $\max U(z) = 1$, then A and B must be related by

$$B = \begin{cases} 1 & (0 \leqslant A \leqslant \tfrac{1}{2}) \\ 2[A(1-A)]^{1/2} & (\tfrac{1}{2} \leqslant A \leqslant 1), \end{cases} \qquad (31.33)$$

and we thus have a one-parameter family of velocity profiles. The stability of these profiles has been studied by Potter (1966), Hains (1967), and Reynolds & Potter (1967) and their results are in substantial agreement. When $A = 1$ and $B = 0$ we have plane Poiseuille flow which is unstable if $R > R_c = 5772$. As B is increased from zero the flow becomes increasingly stable and complete stabilization occurs, i.e. $R_c = \infty$, when

$$A = 0.970, \quad B = 0.341, \quad \alpha = 0, \quad \text{and} \quad c = -0.0331. \quad (31.34)$$

For these values of the parameters there is only one critical point in the interval $-1 \leq z \leq 1$ and it occurs at $z_c = -0.856$. These results show that a relatively small component of plane Couette flow is sufficient to completely stabilize plane Poiseuille flow.

31.5 *The Blasius boundary-layer profile*

The stability of the boundary-layer flow along a semi-infinite flat plate is of great interest both theoretically and experimentally. For this flow $W(x, z)/U(x, z)$ and $\partial U/\partial x$ are both of order $1/R$, where R is a Reynolds number based on the free stream velocity and the thickness of the boundary layer, and it is therefore usually treated as a nearly parallel flow. In this approximation the basic velocity distribution is given by

$$U(z) = f'(z), \quad (31.35)$$

where $z = (U_*/2\nu x_*)^{1/2} z_* = 1.2168 z_*/\delta_*$, δ_* is the displacement thickness of the boundary layer, and $f(z)$ is the Blasius function which satisfies the equation

$$f''' + f f'' = 0 \quad (31.36)$$

and the boundary conditions

$$f(0) = f'(0) = 0 \quad \text{and} \quad f'(z) \to 1 \text{ as } z \to \infty. \quad (31.37)$$

In solving the Orr–Sommerfeld equation on an unbounded interval we must require, as discussed in § 28, that $\phi(z) \sim B e^{-\alpha z}$ as $z \to \infty$. One way of satisfying this requirement is to identify the characteristic length L with the boundary-layer thickness so that $U(z) = 1$ for $z \geq 1$ and then to approximate $U(z)$ by simple polynomials for $0 \leq z \leq 1$. The first calculation of the curve of marginal stability for the Blasius boundary-layer profile was made by Tollmien (1929) who used the approximation

$$U(z) = \begin{cases} 1.70z & (0 \leq z \leq 0.1724) \\ 1 - 1.030(1-z)^2 & (0.1724 \leq z \leq 1) \\ 1 & (z \geq 1) \end{cases} \quad (31.38)$$

Table 4.2. *The values of the parameters associated with the minimum critical Reynolds number for the Blasius boundary-layer profile (based on the displacement thickness δ_*)*

	α_{1c}	c	R_{1c}
Tollmien (1929)	0.34	0.41	420
Schlichting (1933)	0.23	0.42	575
Lin (1945)	0.3718	0.411	421
'Exact' (Jordinson 1970)	0.3012	0.3961	520

for which $\delta_*/L = 0.3416$, where δ_* is the displacement thickness of the boundary layer. He took the eigenvalue relation in a form equivalent to equation (28.33) and used the Tollmien expansions (27.2) and (27.3) for the inviscid solutions. The values of the parameters at the minimum critical Reynolds number which he obtained are given in Table 4.2. He also obtained the asymptotes to both the upper and lower branches of the marginal curve.

A subsequent calculation by Schlichting (1933), which differed from Tollmien's only in the use of the Heisenberg expansions (22.31), led, however, to significantly different results as shown in Table 4.2. To resolve the discrepancy between these two sets of results, Lin (1945) undertook a third calculation. He used an approximation to $U(z)$ which is equivalent to

$$U(z) = \begin{cases} 1.8z - 1.9683z^4 & (0 \leqslant z \leqslant \tfrac{4}{9}) \\ 1 - 0.81(1-z)^2 & (\tfrac{4}{9} \leqslant z \leqslant 1) \\ 1 & (z \geqslant 1) \end{cases} \qquad (31.39)$$

for which $\delta_*/L = 0.3198$. He also took the eigenvalue relation in a form equivalent to equation (28.33) and used the Heisenberg expansions for the inviscid solutions. His results, as shown in Table 4.2, are in close agreement with Tollmien's except for the values of α_c which still differ by about nine per cent.

Nevertheless, except for the values of c, these results are *not* in good agreement with the results of more accurate calculations by Jordinson (1970) and others. In these calculations the basic velocity distribution is defined numerically in terms of the Blasius function.

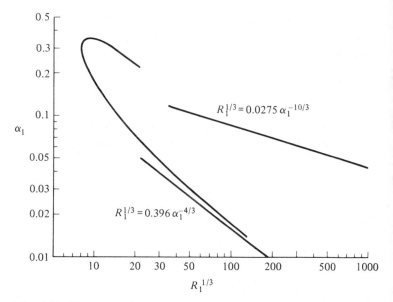

Fig. 4.22. The curve of marginal stability for the Blasius boundary-layer profile. (Based on results by Lin (1945) and Jordinson (1970).)

In the eigenvalue relation, however, the requirement that $\phi(z) \sim Be^{-\alpha z}$ as $z \to \infty$ is replaced by the requirement that $\phi'/\phi = -\alpha$ at some large but finite value of $z = z_2$ (say). Jordinson (1970) chose $z_2 = 7.301$ which corresponds to $z_*/\delta_* = 6$ and this choice would appear to be entirely adequate except for small values of α. More recently, A. Davey (unpublished) has obtained the values

$$R_{1c} = 519.060, \quad \alpha_{1c} = 0.303\,77 \quad \text{and} \quad c = 0.396\,64.$$

The curve of marginal stability for this flow is shown in Fig. 4.22 and the behaviour of c along it is shown in Fig. 4.23.

A comparison between theoretical and experimental results for the Blasius boundary layer will be given in § 32, but there is one important point which should be mentioned here. When the experimental data of Schubauer & Skramstad (1947) first became available it was found to be in good agreement with the theoretical results of Tollmien and Lin. Later, when more accurate calculations led to a larger value of R_c, the agreement with the experimental data became much less satisfactory. More recently, however, it has

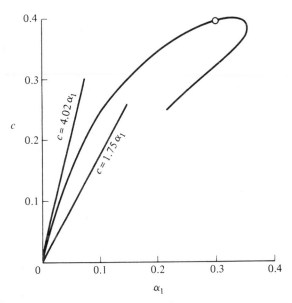

Fig. 4.23. The relationship between the wavenumber α_1 and the wave-speed c along the curve of marginal stability for the Blasius boundary-layer profile. The circled point corresponds to the minimum critical Reynolds number. (Based on results by Lin (1945) and Jordinson (1970).)

been found that agreement can be restored by taking into account the non-parallel character of the basic flow.

31.6 The asymptotic suction boundary-layer profile

The steady flow of a viscous fluid parallel to an infinite flat plate on which there is a prescribed suction velocity W_* leads to an exact solution of the full equations of motion. If we choose the free-stream velocity U_* as our characteristic velocity and the displacement thickness of the boundary layer $\delta_* = -\nu/W_*$ as our characteristic length, then the basic flow can be written in the dimensionless form

$$U = 1 - e^{-z} \quad \text{and} \quad W = -1/R \quad \text{for} \quad 0 \leqslant z < \infty, \quad (31.40)$$

where $R = U_* \delta_*/\nu$. Because of its simple analytical form, this flow permits a more precise discussion of its stability characteristics than most other boundary-layer flows for which the velocity distribution is usually defined only numerically.

The basic flow described by equation (31.40) has usually been treated as a nearly parallel flow (cf. equation (25.4)) for the purpose of studying its stability characteristics. In this approximation the normal component of the basic flow is neglected and the governing equation is again the Orr–Sommerfeld equation. The earlier work on this problem has been reviewed and corrected by Hughes & Reid (1965a), who gave further details. They used an approximation to the eigenvalue relation which is equivalent to equation (28.33). In this approximation it is necessary, as discussed in § 28, to determine the solution of the inviscid equation, Φ (say), that is bounded as $z \to \infty$. Thus, if we let $\Phi = A\phi_1^{(0)} + \phi_2^{(0)}$ as usual, then A must be determined so that $\Phi \sim Be^{-\alpha z}$ as $z \to \infty$. A distinctive feature of this problem, however, is that the exact solution of the inviscid equation can be expressed in terms of the usual hypergeometric function and its analytical continuation (see Problem 4.2). From this exact solution the constants A and B can be determined analytically with the results

and
$$\left.\begin{array}{l} A(\alpha) = 1 - 2\gamma - \alpha - \psi(p+1) - \psi(q+1) \\ B(\alpha, c) = -(1-c)^{-\alpha}\Gamma(p)\Gamma(q)/\Gamma(1+2\alpha), \end{array}\right\} \quad (31.41)$$

where $p = \alpha + (1+\alpha^2)^{1/2}$, $q = \alpha - (1+\alpha^2)^{1/2}$, $\gamma = 0.5772 \ldots$ is Euler's constant, and $\psi(z) = \Gamma'(z)/\Gamma(z)$ is the digamma function. These analytical results also provide a useful check on numerical methods of solving the inviscid equation.

The solution of the eigenvalue relation, in the approximation (28.33), by Hughes & Reid (1965a) led to the following values of the parameters at the minimum critical Reynolds number:

$$R_c = 46\,130, \quad \alpha_c = 0.1600 \quad \text{and} \quad c = 0.1564. \quad (31.42)$$

A more recent numerical solution of the Orr–Sommerfeld equation by Mack (unpublished) has led to the more accurate values

$$R_c = 47\,047, \quad \alpha_c = 0.1630 \quad \text{and} \quad c = 0.1559. \quad (31.43)$$

As expected, these results show that the suction has a strongly stabilizing effect. The good agreement between the two sets of results, however, does *not* confirm, for reasons discussed below, the adequacy of the approximation (28.33).

The basic flow described by equation (31.40) is, of course, not 'strictly parallel' but, since $\partial U/\partial x \equiv 0$ and W is constant, the linearized disturbance equation that governs its stability can be derived with no further approximations beyond the usual linearization. Thus, if the disturbance stream function is taken in the usual form $\phi(z) \exp\{i\alpha(x - ct)\}$, then an easy calculation shows that ϕ must satisfy the equation

$$(i\alpha R)^{-1}\{(D^2 - \alpha^2)^2 + (D^2 - \alpha^2)D\}\phi$$
$$= (U - c)(D^2 - \alpha^2)\phi - U''\phi, \tag{31.44}$$

where we have already substituted for W from equation (31.40). On an exact basis, therefore, the stability of the asymptotic suction profile is governed not by the Orr–Sommerfeld equation but by the modified equation (31.44). A recent numerical solution of this equation by Hocking (1975) led to the following values of the parameters at the minimum critical Reynolds number:

$$R_c = 54\,370, \quad \alpha_c = 0.1555 \quad \text{and} \quad c = 0.150, \tag{31.45}$$

and these results have been confirmed by Mack (unpublished). The change in the value of R_c due to the 'cross flow' terms in equation (31.44) is substantially greater than might have been expected, especially in view of the largeness of R_c, and adds some support to the view that non-parallel effects must be considered when a comparison is made between theoretical and experimental results for boundary-layer flows.

It may be noticed that the inviscid form of equation (31.44) is identical with the inviscid form of the Orr–Sommerfeld equation and, hence, the cross flow terms have no effect on the inviscid solutions, nor do they have any effect on the local turning-point approximation to ϕ_3. Thus, the approximation (28.33) to the eigenvalue relation cannot distinguish between the presence or absence of such terms and the results (31.42) must be interpreted accordingly.

31.7 Boundary layers at separation

An adverse pressure gradient may be expected to have a destabilizing effect on boundary-layer flows owing to the presence of an inflexion point in the basic velocity profile. An interesting example of this effect is provided by the limiting case of a boundary layer *at separation* for which $U'(0) = 0$. Of particular interest is the question of whether the minimum critical Reynolds number is zero or finite when $U'(0) = 0$. This question has been considered by Hughes & Reid (1965b) who chose, for illustrative purposes, to use a simple Pohlhausen fourth-degree polynomial profile. At separation the parameter Λ of the Pohlhausen P4 profiles has the value -12 and the velocity profile is given by (see, for example, Goldstein (1938), p. 158)

$$U(z) = \begin{cases} z^2(6 - 8z + 3z^2) & (0 \leq z < 1) \\ 1 & (z \geq 1). \end{cases} \qquad (31.46)$$

This velocity profile has an inflexion point at $z_s = \frac{1}{3}$ and at that point $c_s = \frac{11}{27}$. In the inviscid limit, the flow is found to be unstable when $0 < \alpha < \alpha_s$, where $\alpha_s \cong 2.0071$. To facilitate the comparison of results for this flow with those for other boundary-layer profiles it is convenient to change the characteristic length from L, which has only been defined implicitly by equation (31.46), to the displacement thickness δ_*. Since $\delta_*/L = \frac{2}{5}$, the wavenumber α_1 and Reynolds number R_1 (based on δ_*) are related to α and R (based on L) by

$$\alpha_1 = \tfrac{2}{5}\alpha \quad \text{and} \quad R_1 = \tfrac{2}{5}R. \qquad (31.47)$$

The calculation of the curve of marginal stability by Hughes & Reid (1965b) was based on the approximation (28.33) to the eigenvalue relation and led to a value of R_{1c} of about 25. A numerical solution of this problem was obtained by Wazzan, Okamura & Smith (1967) who found that $R_{1c} = 112$. Both calculations found that the asymptote to the lower branch of the marginal curve was of the form $R_1^{1/3} \propto \alpha_1^{-7/9}$ but the constants of proportionality differed by a factor of about four. Thus, although the asymptotic results are qualitatively similar to the numerical results, there are substantial quantitative differences between them. These differences can be attributed to at least two causes. One, of course, is the smallness of R_{1c} or, more precisely, of $R_{1c}^{1/3} \cong 4.82$. A more serious limitation of the asymptotic theory, however, is due to the fact that as $c \downarrow 0$ along the lower branch of the marginal curve there are two critical points at $\pm\{2c/U''(0)\}^{1/2}$ which coalesce to form a single turning point of second order and the approximation (27.25) to $\phi_3(z)$ is no longer valid in this limiting situation.

Wazzan *et al.* (1967) also made a calculation of the curve of marginal stability for the Falkner–Skan similarity profile at separation ($\beta = -0.1988$). For this profile they found that $\alpha_{1s} \cong 1.240$, that $R_{1c} \cong 64$, and that $R_1^{1/3} \sim 2.47\alpha_1^{-0.477}$ as $\alpha_1 \downarrow 0$.

These results show that the minimum critical Reynolds numbers for boundary layers at separation are indeed finite but that the stability characteristics of such flows are quite sensitive to the form of the basic velocity distribution.

31.8 *The Falkner–Skan profiles*

The effects of a pressure gradient and surface suction on the stability of laminar boundary layers have been widely studied, and the values of R_c have often been correlated with the shape factor $H = \delta_*/\theta$ (see, for example, Lin (1955), Fig. 6.4, or Stuart (1963), Fig. IX.15). For a given family of velocity profiles there is a unique relationship between R_c and H. For different families of velocity profiles, the correlation between the values of R_c and H is still quite good when the pressure gradient is favourable or when there is surface suction. When the pressure gradient is adverse, however, the values of R_c depend sensitively on the form of the velocity profile and the correlation of R_c with H is then less satisfactory.

To illustrate this general behaviour we shall consider the Falkner–Skan family of boundary-layer profiles which describe the flow past a two-dimensional wedge of (total) angle $\pi\beta$. The basic velocity distribution is again given by $U(z) = f'(z)$, where $f(z)$ satisfies the Falkner–Skan equation

$$f''' + ff'' + \beta(1 - f'^2) = 0 \tag{31.48}$$

and the boundary conditions (31.37). The parameter β provides a measure of the pressure gradient: $\beta = 0$ corresponds to the Blasius flow, $\beta = 1$ corresponds to a two-dimensional stagnation point flow, and $\beta = -0.1988$ corresponds to a boundary layer at separation. When these flows are treated as nearly parallel flows then we obtain the results shown in Fig. 4.24.

For large values of β, Lagerstrom (1964), pp. 125–129 has shown that

$$U(z) \sim 3 \tanh^2 \{(\beta/2)^{1/2} z + \tanh^{-1} (2/3)^{1/2}\} - 2 \text{ as } \beta \to \infty. \tag{31.49}$$

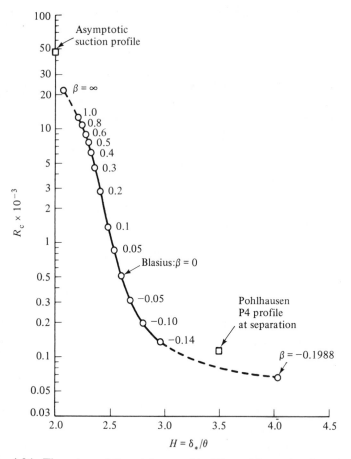

Fig. 4.24. The values of the minimum critical Reynolds number (based on the displacement thickness) for the Falkner–Skan family of boundary-layer profiles. (Based on numerical results given by Obremski, Morkovin & Landahl (1969).)

This result, though not physically relevant to flow past a wedge, does describe the two-dimensional flow in a boundary layer along an infinite plane wall due to a (line) sink at the origin (Schlichting (1968), pp. 152–153). For this flow it is easy to show that

$$\frac{\delta_*}{L} \sim (3\sqrt{2} - 2\sqrt{3})\beta^{-1/2} \quad \text{and} \quad H \sim \tfrac{3}{5}(1 + \sqrt{6}). \qquad (31.50)$$

The stability characteristics of this flow have kindly been computed for us by A. Davey and, in terms of the displacement thickness δ_*, he obtained

$$R_c = 21\,675, \quad \alpha_c = 0.1738 \quad \text{and} \quad c = 0.1841 \quad \text{as} \quad \beta \to \infty.$$
(31.51)

31.9 The Bickley jet

The stability characteristics of two-dimensional jets which are symmetric and have only a single maximum are well illustrated by the Bickley jet for which

$$U = \text{sech}^2 z \quad (0 \leqslant |z| < \infty).$$
(31.52)

In studying the stability of this flow we shall treat it as a nearly parallel flow in the sense of the inequalities (25.4) but the limitations of this approximation, as discussed in § 29, should not be ignored. If we consider the even and odd solutions separately then we can restrict attention to the interval $0 \leqslant z < \infty$ and on this interval the flow has a single inflexion point at

$$z_s = \tanh^{-1}(1/\sqrt{3}) = \tfrac{1}{2}\ln(2+\sqrt{3}) \cong 0.6585.$$
(31.53)

It is also convenient to normalize the even and odd solutions so that $\phi(0) = 1$ and $\phi'(0) = 1$ respectively.

Consider first the marginally stable eigensolutions of the inviscid problem. These have been discussed in some detail by Drazin & Howard (1966) where further references are given. For the even (sinuous) mode we have

$$\phi_s(z) = \text{sech}^2 z, \quad \alpha_s = 2, \quad c_s = \tfrac{2}{3}$$
(31.54)

and for the odd (varicose) mode we have

$$\phi_s(z) = \text{sech}\, z \tanh z, \quad \alpha_s = 1, \quad c_s = \tfrac{2}{3}.$$
(31.55)

Associated with these marginally stable eigensolutions we have even and odd solutions that are unstable when $0 < \alpha < 2$ and $0 < \alpha < 1$ respectively. The growth rates and wave speeds for these unstable eigensolutions have been computed by Drazin & Howard (1966) and their results are shown in Fig. 4.25.

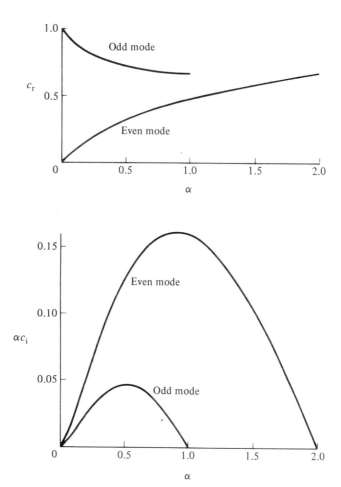

Fig. 4.25. Inviscid stability characteristics of the Bickley jet. (Based on numerical results given by Drazin & Howard (1966).)

Consider next the viscous problem for this profile. From the foregoing inviscid results we would expect to find two curves of marginal stability along which ϕ is either even or odd. We would also expect the even mode to be the least stable because of its larger inviscid growth rate. Calculations for this mode have been made by Curle (1957) and Clenshaw & Elliott (1960) but there is substantial disagreement between their results when α is less than about 0.75. The most accurate and comprehensive results for this flow are those

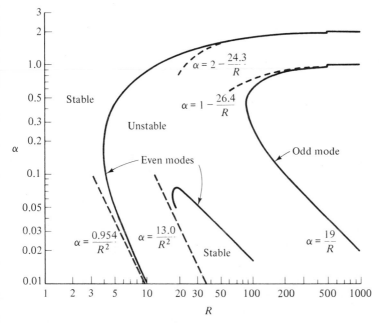

Fig. 4.26. The curves of marginal stability for the Bickley jet. (After Silcock 1975.)

due to Silcock (1975) and they are shown in Fig. 4.26. An interesting feature of these results is the 'minor' curve of marginal stability, the existence of which was first conjectured by Stuart (see Drazin 1961) on the basis of the results of the long-wave approximation given in § 29. The asymptote to the upper branch of this curve is not known analytically nor is the asymptote to the lower branch of the marginal curve on which ϕ is odd. For the odd mode, however, Clenshaw & Elliott (1960) found numerically that $\alpha R \to 19$, approximately, as $\alpha \downarrow 0$.

It is also of some interest to consider the behaviour of the marginal curves as $\alpha \uparrow 2$ for the even mode and $\alpha \uparrow 1$ for the odd mode. For unbounded flows of this type, viscous effects come in through higher approximations to the inviscid solutions and this suggests that we approximate the solution of the Orr–Sommerfeld equation by an expansion of the form

$$\phi = \phi_s + \Phi_1(\alpha - \alpha_s) + \Phi_2(c - c_s) + \cdots + \Phi_3(i\alpha_s R)^{-1} + \cdots. \quad (31.56)$$

It is then easy to show that the Φ_k ($k = 1, 2, 3$) satisfy the inhomogeneous equations

$$\left(D^2 - \alpha_s^2 - \frac{U''}{U - c_s}\right)\Phi_k = f_k \text{ (say)}, \qquad (31.57)$$

where (cf. equations (22.15))

and

$$\left.\begin{aligned} f_1 &= 2\alpha_s\phi_s, \\ f_2 &= (U - c_s)^{-1}(D^2 - \alpha_s^2)\phi_s, \\ f_3 &= (U - c_s)^{-1}(D^2 - \alpha_s^2)^2\phi_s. \end{aligned}\right\} \qquad (31.58)$$

To solve these equations by the method of variation of parameters we need to define a second solution of Rayleigh's equation (with $\alpha = \alpha_s$ and $c = c_s$). For the sinuous mode it is convenient to take this solution in the form

$$\psi_s = \phi_s \int_0^z \phi_s^{-2}\,dz$$

$$= \tfrac{3}{8}z\operatorname{sech}^2 z + (\tfrac{3}{8} + \tfrac{1}{4}\cosh^2 z)\tanh z, \qquad (31.59)$$

where the lower limit of integration has been chosen so that ψ_s is odd and thus contains no multiple of ϕ_s. The Wronskian of ϕ_s and ψ_s is constant with the value $\mathcal{W}(\phi_s, \psi_s) = 1$ and, accordingly, the solutions of equations (31.57) are

$$\Phi_k = \psi_s \int_0^z f_k\phi_s\,dz - \phi_s \int_0^z f_k\psi_s\,dz \quad (k = 1, 2, 3), \quad (31.60)$$

where the lower limits of integration have been chosen so that the Φ_k are even and $\Phi_k(0) = 0$. This last condition preserves the normalization $\phi(0) = \phi_s(0) = 1$ and, without loss of generality, we can now consider the problem on the interval $0 \leqslant z < \infty$. From equations (31.59) and (31.60) we see that the Φ_k become unbounded as $z \to \infty$, the dominant terms being

$$\Phi_k \sim \frac{1}{16}e^{2z}I_k \quad \text{as } z \to \infty, \quad \text{where} \quad I_k = \int_0^\infty f_k\phi_s\,dz. \quad (31.61)$$

The requirement that ϕ be bounded as $z \to \infty$ then leads, in a first approximation, to the eigenvalue relation

$$I_1(\alpha - \alpha_s) + I_2(c - c_s) + I_3(i\alpha_s R)^{-1} = 0. \qquad (31.62)$$

Since the velocity profile is monotone *decreasing* on the interval $0 \leq z < \infty$ we must, in evaluating I_2 and I_3, take a path of integration which lies *above* z_s. In this way we obtain

$$
\left.
\begin{aligned}
I_1 &= \frac{8}{3}, \\[2mm]
I_2 &= -8 - \frac{4}{\sqrt{3}}(2z_s + \pi i) \\[2mm]
I_3 &= \frac{224}{3} + \frac{16}{3\sqrt{3}}(2z_s + \pi i)
\end{aligned}
\right\} \tag{31.63}
$$

and

from which it follows that the asymptote to the upper branch of the marginal curve for this mode is given by

$$
c_s - c \sim 0.224(\alpha_s - \alpha) \quad \text{and} \quad R \sim 24.3(\alpha_s - \alpha)^{-1} \text{ as } \alpha \uparrow \alpha_s. \tag{31.64}
$$

These results were first obtained in a different manner by al-Amir (1968). The inviscid limit for this mode can also be obtained by setting $(i\alpha_s R)^{-1} = 0$ in equation (31.62). This gives

$$
\begin{aligned}
c - c_s &\sim \frac{2}{\sqrt{3}} \frac{2(z_s + \sqrt{3}) - \pi i}{4(z_s + \sqrt{3})^2 + \pi^2}(\alpha - \alpha_s) \quad \text{as} \quad \alpha \to \alpha_s, \\[2mm]
&\cong (0.1687 - 0.1108i)(\alpha - \alpha_s)
\end{aligned} \tag{31.65}
$$

and shows that for α near α_s we have stability for $\alpha > \alpha_s$ and instability for $\alpha < \alpha_s$. The corresponding results for the varicose mode are given in Problem 4.12.

31.10 *The hyperbolic-tangent shear layer*

In §§ 23 and 29 we considered the stability of a piecewise-linear shear layer which exhibits some but not all of the stability characteristics of smoothly varying shear layers. In particular, when $\alpha > \alpha_s$ the eigenvalue relation (23.7) shows that $c_i \equiv 0$ and $c_r \to \pm 1$ as $\alpha \to \infty$ whereas one might have expected that a mode which is unstable for $0 < \alpha < \alpha_s$ would, in the inviscid limit, be damped for $\alpha > \alpha_s$. This expectation has been confirmed by Tatsumi, Gotoh & Ayukawa (1964) and Gotoh (1965) who made a detailed study of the inviscid problem for the velocity profile $U(z) = \tanh z$. Their results, which are shown in Fig. 4.27, are believed to be typical of smoothly varying shear layers. An interesting feature of these

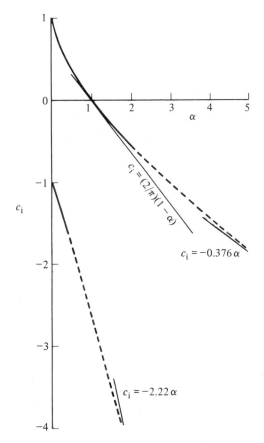

Fig. 4.27. Inviscid stability characteristics of the hyperbolic-tangent shear layer. (From Gotoh 1965.)

results is the existence of a second inviscid mode which is damped for all values of α but the significance of this mode is not fully understood. Some further results for the hyperbolic-tangent profile have also been given by Michalke (1964) who studied the streamline pattern and vorticity distribution of the disturbed flow.

The viscous problem for the hyperbolic-tangent profile has been studied numerically by Betchov & Szewczyk (1963) and their results, which are shown in Fig. 4.28, are in good agreement with the analytical approximations for both large and small values of R. This curve of marginal stability is also believed to be typical of

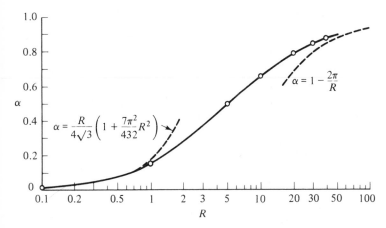

Fig. 4.28. The curve of marginal stability for the hyperbolic-tangent shear layer. (The circled points are from Betchov & Szewczyk (1963).)

smoothly varying profiles; in particular it does not exhibit any of the undulations shown in Fig. 4.16.

32 Experimental results

The study of the stability of laminar boundary layers was originally undertaken in an attempt to account for the phenomenon of transition, and the early work by Tollmien and Schlichting provided detailed theoretical predictions for the growth or decay of small disturbances, now known as *Tollmien–Schlichting waves*, in a Blasius boundary layer. These waves can be seen in the film loops by Brown (FL 1964) and Lippisch (FL 1964). During the 1930s, however, no experimental support for the theory emerged and the prevailing view amongst experimentalists of that period was that transition depended primarily on the magnitude of the disturbances in the boundary layer. When the pressure gradients associated with the disturbances became sufficiently large it was thought that local separation of the boundary layer would then lead to transition. This view is, of course, in marked contrast to the predictions of linear stability theory which show that the stability or instability of small disturbances depends only on their frequency or wavelength and on the Reynolds number.

Experimental confirmation of the theory was finally obtained, however, by Schubauer & Skramstad (1947)† and their work remains one of the great landmarks of the subject. The experimental techniques which they used involved three essential elements. Firstly, the experimental work was conducted in a low-turbulence wind tunnel in which the turbulence level did not exceed 0.0459 per cent at a wind speed of 120 ft s^{-1} (36.6 m s^{-1}); secondly, a vibrating ribbon was used to produce a controlled disturbance of known amplitude and frequency; and, thirdly, hot-wire anemometers of high sensitivity were used to study the growth and decay of the forced oscillations. These techniques have continued to be widely used in experimental work on Tollmien–Schlichting waves and related phenomena.

Until relatively recently the most detailed theoretical results for the Blasius boundary layer were those due to Schlichting (1933) and Shen (1954), who followed Lin's method, and a comparison of their results with the experimental results of Schubauer and Skramstad for the curve of marginal stability is shown in Fig. 4.29. In this figure F $(=\alpha c_r/R)$ is a dimensionless frequency parameter and the displacement thickness of the boundary layer has been used for the characteristic length. On the whole, the agreement with Shen's results is remarkably good.

Nevertheless, 'exact' numerical calculations by Jordinson (1970) and others led to a substantial increase in the value of R_c (see Table 4.2, page 225) and when these more accurate results are compared with the experimental data (see Fig. 4.30 below) the agreement is significantly worsened especially at lower Reynolds numbers. Although the possibility of experimental errors cannot be excluded, Barry & Ross (1970) suggested that the discrepancy is more likely due to the parallel-flow assumption which had been made in all theoretical work up to that time.

Since then a number of attempts have been made to develop weakly non-parallel theories which seek to account for the effects of boundary-layer growth.‡ One approach, which was developed independently by Barry & Ross (1970) and Chen, Sparrow & Tsou (1971), leads to a modified form of the Orr–Sommerfeld equation

† This paper was originally issued in April 1943 by the National Advisory Committee for Aeronautics as an Advance Confidential Report.

‡ An addendum on non-parallel effects may be found on pp. 479–80.

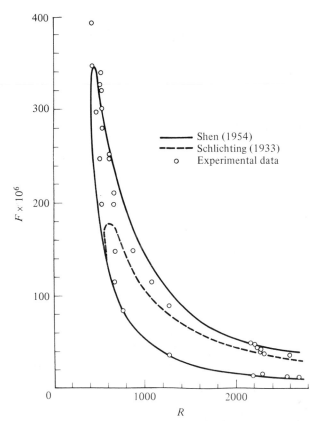

Fig. 4.29. A comparison between the theoretical curves of marginal stability for the Blasius boundary layer and the experimental data of Schubauer & Skramstad (1947). The dimensionless frequency is $F = \alpha c_r / R$. (From Shen 1954.)

which is similar to equation (31.44) and reduces to it for the asymptotic suction profile defined by equation (31.40). This theory leads to a reduction in the value of R_c from 520 to 500 but this is not sufficient to restore agreement with the experimentally observed values of 450 (Schubauer & Skramstad 1947) and 400 (Ross *et al.* 1970). The theories of Ling & Reynolds (1973) and Gaster (1974) are of a quite different type but both lead to very small reductions in the value of R_c and so do not agree well with the experimental data for the high frequency/low Reynolds number part of the marginal curve. The most successful theories thus far would appear to be

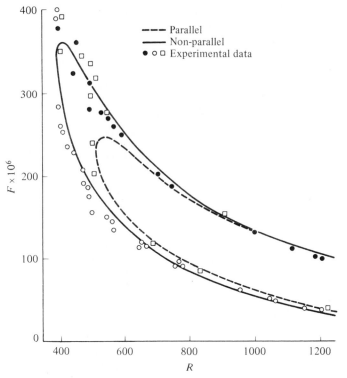

Fig. 4.30. A comparison between the curves of marginal stability for the Blasius boundary layer based on parallel and non-parallel stability theory and experimental data. □ data of Schubauer & Skramstad (1947); ●, ○ upper and lower branch data, respectively, of Ross *et al.* (1970). (From Saric & Nayfeh 1975.)

those due to Bouthier (1972, 1973) and Saric & Nayfeh (1975) which are based largely on the method of multiple scales. Both theories lead to remarkably good agreement with the experimental data as shown, for example, in Fig. 4.30 which is due to Saric & Nayfeh (1975). Although there are some points in these theories which remain obscure or controversial, nevertheless, the results do strongly suggest that the effects of boundary-layer growth must be taken into account when comparing theoretical and experimental results.

Another test of the theory is provided by studying the growth of a disturbance which is propagating downstream at a constant frequency. By perturbing the curve of marginal stability, Schlichting (1933) was able to obtain the local growth rates for the unstable temporal mode (see also Shen (1954)). To relate these results to the spatial growth which is observed in the laboratory, he suggested that the total amplification of a disturbance over a distance $x - x_0$ would then be given by

$$\frac{A(x)}{A(x_0)} = \exp\left(\int_{x_0}^{x} \frac{\beta_i(x)}{\partial \beta_r(x)/\partial \alpha} \, dx\right), \tag{32.1}$$

where $\beta = \alpha c$. A somewhat better description of the growth of such a mode is provided by the spatial theory for which β is real but α is complex, and the total amplification is then given by

$$\frac{A(x)}{A(x_0)} = \exp\left\{-\int_{x_0}^{x} \alpha_i(x) \, dx\right\}. \tag{32.2}$$

For weakly amplified or damped disturbances, Gaster (1962) has shown that

$$\alpha_i(S) = -\frac{\beta_i(T)}{\partial \beta_r/\partial \alpha}, \tag{32.3}$$

where S and T refer to the spatial and temporal theories respectively. This result is based in part on supposing that the maximum value of $\beta_i(T)$ is small. Although $\max\{\beta_i(T)\} = O(10^{-3})$ for both plane Poiseuille flow and the Blasius boundary layer (Shen 1954), the calculations by Jordinson (1970) for the Blasius flow show that equation (32.1) leads to significantly greater amplification than equation (32.2). When comparing such theoretical results with experimental data, it is convenient to rewrite equation (32.2) in the alternative form

$$\frac{A(R)}{A(R_0)} = \exp\left(-\frac{2}{m^2} \int_{R_0}^{R} \alpha_i(R) \, dR\right), \tag{32.4}$$

where R is based on the displacement thickness of the boundary layer and $m = 1.7208$ is the Blasius constant. A comparison between the theoretical results of Jordinson (1970) and the experimental data of Ross et al. (1970) is shown in Fig. 4.31. Although the agreement between theory and experiment seems reasonable, it should be noted that the effects of boundary-layer growth were not taken into account in the theoretical calculations and such effects are difficult to assess.

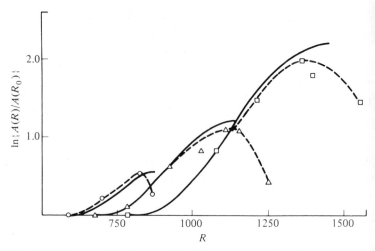

Fig. 4.31. The amplitude of a disturbance propagating downstream at constant values of F. ———, theoretical results of Jordinson (1970); -----, approximate experimental curves. $\square$, $F = 82 \times 10^{-6}$; $\triangle$, $F = 110 \times 10^{-6}$; $\odot$, $F = 157 \times 10^{-6}$. (From Ross *et al.* 1970.)

Experimental confirmation of the linear theory for plane Poiseuille flow has been obtained only recently by Nishioka, Iida & Ichikawa (1975). Their experiments were conducted in a rectangular channel with an aspect ratio of 27.4 in which the background turbulence had been reduced to a level of 0.05 per cent. A vibrating ribbon was used to introduce a sinusoidal disturbance of known frequency and the downstream development of the disturbance was then studied for values of the Reynolds number in the range from 3000 to 7500. Measurements of the spatial amplification or decay rates for several values of R in this range were found to be in good agreement with Itoh's (1974a) theoretical calculations. Good agreement was also obtained with the calculated curve of marginal stability.

One of the important predictions of nonlinear stability theory is that plane Poiseuille flow exhibits subcritical instability near the point (α_c, R_c). This prediction depends on extensive numerical calculations and somewhat different versions of the theory have been given by Reynolds & Potter (1967), Pekeris & Shkoller (1967), Chen & Joseph (1973) and Itoh (1974a). Measurements of the threshold amplitude by Nishioka *et al.* (1975) were found to be in qualitative agreement with Itoh's version of the theory and this provides encouraging support for the basic ideas involved in the nonlinear theory.

Considering the many difficulties involved in the study of the stability of parallel shear flows, it is perhaps not altogether surprising that it has taken so long to obtain agreement between theory and experiment. The agreement is particularly good in the case of the linear theory and future developments may therefore be expected to emphasize nonlinear aspects of the problem.

Problems for chapter 4

4.1. *Derivation of Rayleigh's stability equation from the vorticity equation.* The equation for the stream function ψ which governs two-dimensional motion of an inviscid fluid is

$$\Delta\psi_t + \psi_z\Delta\psi_x - \psi_x\Delta\psi_z = 0,$$

where the subscripts denote partial derivatives. Let $\psi(x, z, t) = \Psi(z) + \psi'(x, z, t)$, where $\Psi(z)$ is the stream function of the basic flow. Show that the linearized equation for ψ' is

$$\Delta\psi'_t + U\Delta\psi'_x - U''\psi'_x = 0$$

and hence derive Rayleigh's stability equation.

*4.2. *The exact solution of Rayleigh's equation for the asymptotic suction boundary-layer profile.* When $U(z) = 1 - e^{-z}$ for $0 \le z < \infty$ show that the solution of equation (21.17) which is bounded as $z \to \infty$ is

$$\Phi(z) = \exp\{-\alpha(z - z_c)\} \frac{F(p, q; r; t)}{F(p, q; r; 1)} \quad \text{for} \quad |t| \le 1, \quad \text{(i)}$$

where $t = \exp\{-(z - z_c)\}$, $F(p, q; r; t)$ is the standard hypergeometric function, and the solution has been normalized so that $\Phi(z_c) = 1$. The parameters in the solution are given by

$$p = \alpha + (1 + \alpha^2)^{1/2}, \quad q = \alpha - (1 + \alpha^2)^{1/2} \quad \text{and} \quad r = 1 + 2\alpha.$$

From equation (i) show that $\Phi(z) \sim Be^{-\alpha z}$ as $z \to \infty$ where $B(\alpha, c)$ is given by equation (31.41).

In approximating the eigenvalue relation for the viscous problem it is necessary to evaluate Φ and Φ' at $z = 0$; hence it is necessary to obtain the analytic continuation of equation (i) into the region $|t| > 1$. By considering the principal branch of $F(p, q; r; t)$, show that the required analytical continuation of equation (i) must contain logarithmic terms and is given by

$$\Phi(z) = \exp\{-\alpha(z - z_c)\}$$
$$\times \left\{1 - (1-t)F(p+1, q+1; 2; 1-t)\ln(1-t) - \sum_{n=0}^{\infty} A_n \frac{(1-t)^{n+1}}{(n+1)!}\right\},$$

(ii)

for $|1 - t| < 1$ and $-\pi < \mathrm{ph}(1-t) < \pi$, where

$$A_n(\alpha) = \frac{\Gamma(p+n+1)\Gamma(q+n+1)}{\Gamma(p+1)\Gamma(q+1)\Gamma(n+1)}$$
$$\times \{\psi(p+n+1) + \psi(q+n+1) - \psi(n+1) - \psi(n+2)\}$$

and $\psi(z) = \Gamma'(z)/\Gamma(z)$ is the digamma function. Now let $\Phi(z) = A\phi_1(z) + \phi_2(z)$, where $\phi_1(z)$ and $\phi_2(z)$ have the forms defined by equations (22.25) and (22.26) respectively. From equation (ii) show that the constant A depends only on α and is given by equation (31.41). [Hughes & Reid (1965a).]

4.3. *Stability characteristics of an inviscid triangular jet.* When $U(z) = 1 - |z|$ for $|z| < 1$ and $U(z) = 0$ for $|z| > 1$ show that the eigenvalue relation for the sinuous mode (ϕ even) is

$$2\alpha^2 c^2 + \alpha(1 - 2\alpha - e^{-2\alpha})c - \{1 - \alpha - (1+\alpha)e^{-2\alpha}\} = 0.$$

Hence show that this mode is unstable for $0 < \alpha < \alpha_s$, where $\alpha_s \cong 1.833$ and $c_s \cong 0.367$. For the varicose mode (ϕ odd) show that

$$c = (2\alpha)^{-1}(1 - e^{-2\alpha}).$$

Compare these results with those obtained in Problem 1.7 for a rectangular jet. [Rayleigh (1894), p. 395.]

4.4. *An initial-value problem for inviscid plane Couette flow.* The solution to the initial-value problem for inviscid plane Couette flow can be expressed in the form

$$\psi'(x, z, t) = \int_{-\infty}^{\infty} d\alpha \int_{-1}^{1} dc\ e^{i\alpha(x-ct)}f(\alpha, c)G(z, c) \quad \text{for } t > 0,$$

where $G(z, c)$ is the Green's function (24.10) and $f(\alpha, c)$ is determined by the initial disturbance $\psi'(x, z, 0)$. If

$$\psi'(x, z, 0) = \delta(x)\psi_0(z) \quad \text{for } -\infty < x < \infty \quad \text{and} \quad -1 < z < 1,$$

show that $f(\alpha, c) = (2\pi)^{-1}\{(D^2 - \alpha^2)\psi_0(z)\}_{z=c}$. [Dikii (1964).]

4.5. *Two-dimensional disturbances for the Orr–Sommerfeld problem.* By taking $\mathbf{u}'$, $p' \propto \exp\{i(\alpha x + \beta y)\}$, by resolving $\mathbf{u}'$ into components parallel and perpendicular to the wavenumber vector, i.e. by writing

$$u_\parallel' = (\alpha u' + \beta v')/\tilde{\alpha} \quad \text{and} \quad u_\perp' = (\beta u' - \alpha v')/\tilde{\alpha},$$

and by using the linearized equations of motion (25.5), show that the equations satisfied by $(u'_\parallel, w', p')$ are independent of $u'_\perp$. Hence deduce Squire's theorem.

Show further that $u'_\perp$ satisfies the inhomogeneous diffusion equation with convection by the basic flow,

$$\left(D^2 - \tilde{\alpha}^2 - i\alpha RU - R\frac{\partial}{\partial t}\right)u'_\perp = R(DU)\beta w'/\alpha,$$

and the no-slip conditions $u'_\perp = 0$ at $z = z_1$ and z_2. Prove that the solution of the equation can be written in the form

$$u'_\perp(\mathbf{x}, t) = \{\hat{u}_\perp(z)\, e^{-i\alpha ct} + f(z, t)\}\, e^{i(\alpha x + \beta y)},$$

where the particular integral has the same value of c as that previously determined for w' (and $u'_\parallel$ and p') but the complementary function satisfies

$$\frac{\partial f}{\partial t} + i\alpha Uf = R^{-1}(D^2 - \tilde{\alpha}^2)f.$$

Suppose now that $\hat{u}_\perp$ and f each vanish at $z = z_1$ and z_2. Hence show that

$$\frac{d}{dt}\int_{z_1}^{z_2}|f|^2 dz = -2R^{-1}\int_{z_1}^{z_2}\{|Df|^2 + \tilde{\alpha}^2|f|^2\}\, dz$$

and therefore that f is stable (in the mean) independently of the values of c determined by the Orr–Sommerfeld problem for $u'_\parallel$, w' and p'.

4.6. *The enstrophy equation and the stability of unbounded plane Poiseuille flow.* By analogy with equations (26.10) and (53.7), show that

$$I_3^2 + 3\alpha^2 I_2^2 + 3\alpha^4 I_1^2 + \alpha^6 I_0^2$$

$$= \tfrac{1}{2}[\phi''\phi^{*\prime\prime\prime} + \phi^{*\prime\prime}\phi''']_{z_1}^{z_2} - \alpha c_i R(I_2^2 + 2\alpha^2 I_1^2 + \alpha^4 I_0^2)$$

$$- \tfrac{1}{2}i\alpha R\int_{z_1}^{z_2} U'''(\phi\phi^{*\prime} - \phi^*\phi')\, dz,$$

where

$$I_n^2 = \int_{z_1}^{z_2}|D^n\phi|^2\, dz \quad (n = 0, 1, 2, 3).$$

This is in fact the equation for the rate of change of twice the mean enstrophy, i.e. the integral of the square of the vorticity, $I_2^2 + 2\alpha^2 I_1^2 + \alpha^4 I_0^2$. Hence show that any discrete mode of unbounded plane Poiseuille (or Couette) flow is stable. [Synge (1938), p. 267.]

4.7. *The second approximation of WKBJ type.* In Chapter 5, in connexion with our discussion of uniform approximations, we will need the second approximation of WKBJ type. Derive the equation satisfied by g_2 and when $g_0 = \mp (U - c)^{1/2}$ show that

$$g_2 = \pm (U-c)^{-1/2} \left\{ \frac{101}{32} \left(\frac{U'}{U-c} \right)^2 - \frac{13}{8} \frac{U''}{U-c} - \frac{1}{2} \alpha^2 \right\},$$

where the signs of g_0 and g_2 are ordered. Hence show that

$$\phi_{3,4}(z) \sim C_{3,4} (U-c)^{-5/4} \exp\left\{ \mp (\alpha R)^{1/2} Q(z) \right\}$$
$$\times [1 \mp (i\alpha R U_c')^{-1/2} G_1(z) + O\{(i\alpha R U_c')^{-1}\}],$$

where

$$G_1(z) = \left(\frac{101}{48} \frac{U'}{U-c} + \frac{23}{24} \frac{U''}{U'} \right) \left(\frac{U_c'}{U-c} \right)^{1/2}$$
$$- \int_{z_c}^{z} \left\{ \frac{23}{24} \left(\frac{U'''}{U'} - \frac{U''^2}{U'^2} \right) - \frac{1}{2} \alpha^2 \right\} \left(\frac{U_c'}{U-c} \right)^{1/2} dz$$

and the normalization of $G_1(z)$ has been fixed by requiring that its expansion about $z = z_c$ contains no constant term. [Eagles (1969), Lakin & Reid (1970), Reid (1974b).]

4.8. *The second inner approximation to $\phi_3(z)$.* From equation (27.45) show that the second approximation of inner type to $\phi_3(z)$ is

$$\chi_3^{(1)}(\xi) = (U_c''/U_c')\{A_1(\xi, 3) - \tfrac{1}{2} A_1(\xi) + \tfrac{1}{10} A_1(\xi, -3)\},$$

where the normalization has been fixed by requiring that $\chi_3^{(1)}$ contains no multiple of $\chi_3^{(0)}$. Show that this result can also be written in the form

$$\chi_3^{(1)}(\xi) = (U_c''/U_c')\{\tfrac{1}{2}\xi^2 \int_{\infty}^{\xi} \mathrm{Ai}(t)\, dt - \tfrac{9}{10}\mathrm{Ai}(\xi) - \tfrac{2}{5}\xi\mathrm{Ai}'(\xi)\}.$$

Hence show that

$$\chi_3^{(1)}(\xi) \sim -(U_c''/U_c')\tfrac{1}{20}\pi^{-1/2}\xi^{5/4} \exp\left(-\tfrac{2}{3}\xi^{3/2}\right)$$

as $|\xi| \to \infty$ in the sector $|\mathrm{ph}\,\xi| < \pi$. Notice that $|\varepsilon\chi_3^{(1)}(\xi)| \ll |\chi_3^{(0)}(\xi)|$ provided $|\xi| \ll |\varepsilon|^{-2/5}$ or $|z - z_c| \ll |\varepsilon|^{3/5}$. [Lock (1955), Reid (1972).]

*4.9. *The heuristic theory for the adjoint Orr–Sommerfeld equation.* Show that the inviscid approximation to the solutions of equation (25.18) are

$$\phi_1^{\dagger(0)}(z) = Q_1(z)$$

and

$$\phi_2^{\dagger(0)}(z) = (z - z_c)^{-1} Q_2(z) + (U_c''/U_c')\phi_1^{\dagger(0)}(z) \ln (z - z_c),$$

where $Q_1(z)$ and $Q_2(z)$ are analytic at z_c with $Q_1(z_c) = Q_2(z_c) = 1$ and, to fix the normalization, $Q_2'(z_c) = 0$. Show further that the improved viscous approximations of Tollmien type are

$$\phi_{3,4}^{\dagger}(z) = \left(\frac{U-c}{U_c'\eta} \right)^{-1/4} A_{1,2}(\zeta).$$

The singularity in $\phi_2^{\dagger(0)}(z)$ is stronger than the one in $\phi_2^{(0)}(z)$ and $\phi_2^{\dagger(0)}(z)$ does not therefore provide an adequate approximation to $\phi_2^{\dagger}(z)$ for use in the eigenvalue relation. By considering the inner approximation to second order show that

$$\phi_2^{\dagger}(z) = \varepsilon^{-1} X_2^{(0)}(\xi) + X_2^{(1)}(\xi) + O(\varepsilon),$$

where

$$X_2^{(0)}(\xi) = -B_3(\xi, 0, 1)$$

and

$$X_2^{(1)}(\xi) = -(U_c''/U_c')\{B_3(\xi, 1, 1) - B_3(\xi, -2, 1) + \tfrac{1}{10} B_3(\xi, -5, 1)$$
$$- [\ln \varepsilon + \psi(1) + 2\pi i]\}.$$

Also discuss the first and second order additive composite approximations to $\phi_2^{\dagger}(z)$. [Reid (1965), Hughes & Reid (1965a).]

*4.10. *The asymptote to the upper branch of the marginal curve for channel flows with an inflexion point.* Let $\phi_s(z)$ denote the marginally stable eigensolution of the inviscid problem. Then $\phi_s(z) = A\phi_1^{(0)}(z) + P_2(z)$, where $U''(z_s) = 0$ and $c_s = U(z_s)$. For symmetrical flows in a channel, the boundary conditions $\phi_s(z_1) = 0$ and $\phi_s'(z_2) = 0$ then determine α_s and A. By expanding ϕ in the form of equation (22.14) and requiring that $\Phi_1'(z_2) = \Phi_2'(z_2) = 0$ show that

$$W(\alpha, c) = -\{U_1'/c_s\phi_s'(z_1)\}\{\Phi_1(z_1)(\alpha - \alpha_s) + \Phi_2(z_1)(c - c_s) + \cdots\},$$

where

$$\Phi_1(z_1) = \frac{2\alpha_s}{\phi_s'(z_1)} \int_{z_1}^{z_2} \phi_s^2 \, dz$$

and

$$\Phi_2(z_1) = \frac{1}{\phi_s'(z_1)} \int_{z_1}^{z_2} \frac{U''}{(U - c_s)^2} \phi_s^2 \, dz,$$

and the path of integration for the second integral must lie below z_s if $U_s' > 0$. (Note the differences in sign with equations (22.18).) By using, for example, the approximation (28.23) to the eigenvalue relation show that

$$c - c_s \sim \frac{\Phi_1(z_1)}{\Phi_{2i}(z_1) - \Phi_{2r}(z_1)} (\alpha - \alpha_s) \quad \text{as } \alpha \to \alpha_s$$

and

$$R \sim \frac{1}{2\pi^2 \alpha_s c_s} \frac{U_s'^4}{U_s'''^2} \left(\frac{\phi_s'(z_1)}{\phi_s(z_s)}\right)^4 (c - c_s)^{-2} \quad \text{as } c \to c_s.$$

Show further that these results remain valid for flows of the boundary-layer type provided that α_s and A are determined by the conditions $\phi_s(z_1) = 0$

and $\phi_s(z)$ is bounded as $z \to \infty$. Discuss the effect of using other approximations to the eigenvalue relation. [Eagles (1969), Hughes & Reid (1965b, 1968).]

4.11. *The asymptote to the upper branch of the marginal curve for the Blasius boundary layer.* If $U_1'' = 0$ but $U'' \neq 0$ in $z_1 < z < z_2$ as occurs, for example, in the case of the Blasius boundary-layer flow along a semi-infinite flat plate, then equation (28.21) must be modified. For such flows the equations of motion show that U_1''' must also vanish. Hence show that

$$R \sim 2\pi^{-2}\{U_1'^{\,19}/(U_1^{iv})^2\}\alpha^{-10}$$

as $\alpha \downarrow 0$ along the marginal curve. [Lin (1945), Reid (1965).]

*4.12. *The asymptote to the upper branch of the marginal curve for the varicose mode of the Bickley jet.* Show that the first approximation to the eigenvalue relation for the varicose mode is also given by an equation of the same form as equation (31.62) where

$$I_1 = \frac{2}{3}, \quad I_2 = 2 - \frac{2}{\sqrt{3}}(2z_s + \pi i) \quad \text{and} \quad I_3 = \frac{16}{3} + \frac{32}{3\sqrt{3}}(2z_s + \pi i).$$

Hence show that the asymptote to the upper branch of the marginal curve for this mode is given by

$$c_s - c \sim 0.141(\alpha_s - \alpha) \quad \text{and} \quad R \sim 26.4(\alpha_s - \alpha)^{-1} \quad \text{as } \alpha \uparrow \alpha_s,$$

where $\alpha_s = 1$ and $c_s = \frac{2}{3}$. Show also that in the inviscid limit we have

$$c - c_s \sim -\frac{1}{\sqrt{3}} \frac{(2z_s - \sqrt{3}) - \pi i}{(2z_s - \sqrt{3})^2 + \pi^2}(\alpha_s - \alpha)$$

$$\cong (0.02386 + 0.1806i)(\alpha_s - \alpha) \quad \text{as } \alpha \uparrow \alpha_s.$$

4.13. *The asymptote to the upper branch of the marginal curve for the hyperbolic-tangent shear layer.* By adapting the analysis given in § 31 for the Bickley jet show that if $U = \tanh z$ for $-\infty < z < \infty$ then as $R \to \infty$ along the marginal curve we have $c = 0$ and $R \sim 2\pi/(1 - \alpha)$. In the inviscid limit show that $c_r = 0$ and $c_i \sim (2/\pi)(1 - \alpha)$ as $\alpha \uparrow 1$. [Drazin & Howard (1962), Tatsumi, Gotoh & Ayukawa (1964).]

4.14. *An application of the Riccati method to the adjoint problem.* The system of equations adjoint to equations (30.24) is

$$\mathbf{u}^{\dagger\prime} = -\mathbf{A}^T\mathbf{u}^\dagger + \mathbf{C}^T\mathbf{v}^\dagger \quad \text{and} \quad -\mathbf{v}^{\dagger\prime} = \mathbf{B}^T\mathbf{u}^\dagger - \mathbf{D}^T\mathbf{v}^\dagger,$$

and the boundary conditions which correspond to (30.25) are

$$\mathbf{v}^\dagger(0) = \mathbf{0} \quad \text{and} \quad \beta\mathbf{u}^\dagger(1) + \alpha\mathbf{v}^\dagger(1) = \mathbf{0}.$$

If the Riccati matrix $\mathbf{R}^\dagger$ for the adjoint problem is introduced through the transformation $\mathbf{v}^\dagger = \mathbf{R}^\dagger\mathbf{u}^\dagger$, show that $\mathbf{R}^\dagger = -(\mathbf{R}^{-1})^T$.

CHAPTER 5

UNIFORM ASYMPTOTIC APPROXIMATIONS

It is in the nature of Applied Mathematics that one should be concerned not only with the application of existing mathematical theories and methods, but also with the stimulation of new mathematical problems, through the study of interesting problems in science, and the attempts to solve these problems. – C. C. Lin (1964)

33 Introduction

The approximations to the solutions of the Orr–Sommerfeld equation which were derived in § 27 suffer from two major defects. First, the approximations (except for the regular inviscid solution) are not uniformly valid in a full neighbourhood of the critical point and, second, the theory does not lead to a systematic method of obtaining higher approximations. In this chapter, therefore, we shall describe some of the attempts which have been made to overcome these deficiencies of the heuristic theory. These improved theories have generally been based on either the comparison-equation method or the method of matched asymptotic expansions. Although neither method is entirely satisfactory by itself, both have played important roles in the development of the subject.

The *comparison-equation method* has been extensively studied by Wasow (1953b), Langer (1957, 1959), Lin (1957a,b, 1958), Lin & Rabenstein (1960, 1969) and others. In all of this work the major aims have been to obtain asymptotic approximations to the solutions of the Orr–Sommerfeld equation which are uniformly valid in a bounded domain containing one simple turning point and to develop an algorithm by which higher approximations can be obtained systematically. Theories of this type are largely based on the idea of generalizing Langer's (1931, 1932, 1949) well-known theory for second-order differential equations with a simple turning point to higher order equations of the Orr–Sommerfeld type. This requires the development of a procedure by which the solutions of the Orr–Sommerfeld equation can be represented asymptotically in terms of the solutions of a suitably chosen comparison equation. The success of this method, however, depends crucially on being able to satisfy two closely related conditions. First, the comparison equation must be sufficiently simple so that its solutions may be considered known, otherwise little would be achieved, and, second, the solutions of the comparison equation must have asymptotic properties which are close to those of the Orr–Sommerfeld equation in order to achieve the desired degree of uniformity in the resulting approximations.

The basic idea in this method, therefore, is to express $\phi(z)$ asymptotically in terms of $u(\eta)$, say, where $u(\eta)$ is a solution of the comparison equation and allowance has been made for a change of both dependent and independent variables. As Lin (1957a,b) has shown, the comparison equation must be of the form

$$\varepsilon^3 u^{\mathrm{iv}} - (\eta u'' + \alpha u' + \beta u) = 0 \tag{33.1}$$

in which α and β must be allowed to have series expansions of the form

$$\alpha(\varepsilon) = \sum_{s=0}^{\infty} \alpha_s \varepsilon^{3s} \quad \text{and} \quad \beta(\varepsilon) = \sum_{s=0}^{\infty} \beta_s \varepsilon^{3s}. \tag{33.2}$$

The coefficients in these asymptotic power series, however, are not known *a priori* but must be determined, in the process of relating $\phi(z)$ to $u(\eta)$, by certain regularity conditions. It is easy to show that $\alpha_0 = 0$ and $\beta_0 = -U_c''/U_c$, but the higher-order coefficients in these series are difficult to determine and only α_1 is known (Lakin & Reid 1970). Furthermore, although it is not difficult, as Rabenstein (1958) has shown, to 'solve' the comparison equation (33.1) by the method of Laplace integrals, the behaviour of the solutions for small values of $|\varepsilon|$ is found to be rather complicated. The integral representations of the solutions can, of course, be re-expanded to yield inner and outer expansions, but they do not lead in any obvious way to uniform expansions. Thus, although the comparison-equation method has perhaps been formally successful, it is technically complicated and difficult to apply in actual stability calculations. It has been used, however, by Lakin & Reid (1970) to obtain first approximations to the Stokes multipliers for the Orr–Sommerfeld equation and hence to obtain outer expansions which are complete in the sense of Watson.

The *method of matched asymptotic expansions* was first applied to the Orr–Sommerfeld equation by Graebel (1966) who obtained results in substantial agreement with those given in § 28. A more systematic application of the method, based on the general theory developed by Fraenkel (1969), was later given by Eagles (1969). The most distinctive feature of Eagles's work lies in its strict adherence to Poincaré's definition of an asymptotic expansion, an important consequence of which is that all recessive series are systematically neglected. As a result he was limited to a discussion of the behaviour of the upper branch of the curve of marginal stability as $\alpha R \to \infty$. A rigorous justification of the results obtained by Eagles has been given by De Villiers (1975).

This method has also been used by Reid (1972) to obtain composite approximations to the solutions of the Orr–Sommerfeld equation which are valid in the complete sense of Watson. The concept of completeness (Olver 1974, p. 543) is based on the observation that it is often more important for numerical purposes to obtain a 'first approximation' which is valid in the complete sense than to obtain the whole of the descending series associated with the dominant term in the expansion. This means that in some circumstances we must retain terms which are exponentially small compared with other terms in the expansion, even though such terms are negligible in the usual Poincaré sense. A further consequence of this concept is that different asymptotic expansions of a given solution must be restricted to non-overlapping sectors, the boundaries of which are Stokes lines, even though the expansions remain valid in the Poincaré sense in larger sectors, the boundaries of which are anti-Stokes lines. The composite approximations obtained by this method are not fully uniform, however, in the sense that they are not valid in a full neighbourhood of the critical point, but they are valid in certain sectorial domains which do not shrink to zero as $|\varepsilon| \to 0$. Nevertheless, it was shown that the inner expansions could be obtained to all orders in terms of the class of generalized Airy functions which are discussed in the Appendix, and it was this result which led to the simple theory of uniform approximations which will be described in § 38.

Asymptotic approximations to the solutions of the Orr–Sommerfeld equation necessarily involve two length scales (z and ξ or η and ζ) and it has sometimes been suggested (see, for example, Lin (1964)) that it should be possible to derive uniform approximations by the *method of multiple scales*. This has been attempted by Tam

(1968) but his results, when compared with those given in § 38, would not appear to be uniform. The difficulties in applying this method to the Orr–Sommerfeld equation are similar to, but more serious than, those encountered by Fowkes (1968) in his study of second-order equations with a simple turning point.

The method of deriving uniform approximations which will be discussed in this chapter was suggested in part by Langer's theory for second-order differential equations with a simple turning point and it may be helpful, therefore, to recall some of the main points of that theory. Consider then the equation

$$w'' = \{\lambda^2 p(z) + q(z)\} w \tag{33.3}$$

in which $|\lambda|$ is large and $p(z)$ and $q(z)$ are analytic functions of z in a domain D containing a simple zero of $p(z)$ at $z = z_0$ (say). Without loss of generality, we can suppose that $p'(z_0) = 1$. To reduce equation (33.3) to standard form we make a preliminary transformation of both dependent and independent variables of the form

$$w(z) = \{\eta'(z)\}^{-1/2} W(\eta), \tag{33.4}$$

where the exponent of $\eta'(z)$ has been chosen so that the transformed equation is in normal form, i.e. so that it does not contain W', and where the Langer variable $\eta(z)$ is defined by

$$\eta(z) = \left[\frac{3}{2} \int_{z_0}^{z} \{p(t)\}^{1/2} \, dt \right]^{2/3}. \tag{33.5}$$

Near $z = z_0$ we have

$$\eta(z) = z - z_0 + \tfrac{1}{10} p''(z_0)(z - z_0)^2 + \cdots \tag{33.6}$$

so that $\eta(z)$ is analytic at $z = z_0$. Under the change of variables given by equation (33.4), equation (33.3) becomes

$$\varepsilon^3 W'' = \{\eta + \varepsilon^3 f(\eta)\} W, \tag{33.7}$$

where

$$f(\eta) = q(z)\eta'^{-2} + \tfrac{1}{2}\eta'''\eta'^{-3} - \tfrac{3}{4}\eta''^2 \eta'^{-4} \tag{33.8}$$

and $f(\eta)$ is analytic in the transformed domain G containing the origin. To emphasize the relation between this theory and the one to be developed for the Orr–Sommerfeld equation we have also let $\lambda^2 = \varepsilon^{-3}$.

According to Langer's theory there exist solutions $W_k(\eta, \varepsilon)$ ($k = 1, 2, 3$) of equation (33.7) such that as $|\varepsilon| \to 0$

$$W_k(\eta, \varepsilon) = \mathscr{A}(\eta, \varepsilon)A_k(\zeta) + \varepsilon^2 \mathscr{B}(\eta, \varepsilon)A_k'(\zeta), \tag{33.9}$$

where $\zeta = \eta/\varepsilon$, $A_k(\zeta)$ are the Airy functions defined by equations (A1), and $\mathscr{A}$ and $\mathscr{B}$ have series expansions of the form

$$\mathscr{A}(\eta, \varepsilon) = \sum_{s=0}^{\infty} \varepsilon^{3s} \mathscr{A}_s(\eta) \tag{33.10}$$

the coefficients of which are required to be analytic at $\eta = 0$. A short calculation then shows that $\mathscr{A}$ and $\mathscr{B}$ satisfy the equations

and
$$\left. \begin{aligned} \mathscr{A}'' - f(\eta)\mathscr{A} + 2\eta\mathscr{B}' + \mathscr{B} &= 0 \\ 2\mathscr{A}' + \varepsilon^3\{\mathscr{B}'' - f(\eta)\mathscr{B}\} &= 0. \end{aligned} \right\} \tag{33.11}$$

When $\mathscr{A}$ and $\mathscr{B}$ have expansions of the form given in equation (33.10), the coefficients $\mathscr{A}_s$ and $\mathscr{B}_s$ are given by the recursion relations

$$\left. \begin{aligned} \mathscr{B}_s(\eta) &= -\tfrac{1}{2}\eta^{-1/2}\int_0^{\eta} t^{-1/2}\{\mathscr{A}_s''(t) - f(t)\mathscr{A}_s(t)\}\,\mathrm{d}t \\ \text{and} \\ \mathscr{A}_{s+1}(\eta) &= -\tfrac{1}{2}\mathscr{B}_s'(\eta) + \tfrac{1}{2}\int f(\eta)\mathscr{B}_s(\eta)\,\mathrm{d}\eta + \text{constant} \end{aligned} \right\} \tag{33.12}$$

for $s = 0, 1, 2, \ldots$ with $\mathscr{A}_0(\eta) = \text{constant} = 1$ (say).

The three solutions $W_k(\eta, \varepsilon)$ cannot, of course, be linearly dependent but must be related by the connexion formula

$$\sum_{k=1}^{3} W_k(\eta, \varepsilon) = 0 \tag{33.13}$$

which follows directly from equations (33.9) and (A2). We may also note that

$$W_k'(\eta, \varepsilon) = (\mathscr{A}' + \eta\mathscr{B})A_k(\zeta) + \varepsilon^{-1}(\mathscr{A} + \varepsilon^3\mathscr{B}')A_k'(\zeta). \tag{33.14}$$

For a general discussion of the errors associated with partial sums of the expansions (33.9) and (33.14), together with a rigorous theory of error bounds, see Olver (1974).

Consider next the generalization of this theory to the Orr–Sommerfeld equation. For this purpose it is desirable to consider first the case of plane Couette flow for which the theory is particularly simple. It is found, for example, that the integral representations of the solutions of dominant-recessive type are reducible to a form to which the method of Chester, Friedman & Ursell (1957) is directly applicable, and this shows that their uniform expansions can be expressed *to all orders* in terms of the Airy functions $A_k(\zeta)$ together with their first derivatives and first integrals. Alternatively, as discussed in § 35, the uniform expansions can also be derived directly from the Orr–Sommerfeld equation without any reference to the integral representations of the solutions.

The structure of the theory for plane Couette flow also suggests how it can be immediately generalized to provide 'first approximations' to the solutions of the Orr–Sommerfeld equation for a more general class of velocity profiles.To this order the general theory is also remarkably simple. It is found, for example, that the slowly varying coefficients in the approximations can be expressed in terms of three well-known quantities: the regular inviscid solution, the regular part of the singular inviscid solution, and a regularized form of one of the coefficients which appears in the outer expansions of dominant-recessive type. The essential elements in the general theory are thus well known from the older heuristic theories, but the final structure of the theory is substantially different. The uniform approximations of this type are also vastly simpler than the composite approximations obtained in § 37 by the method of matched asymptotic expansions.

PLANE COUETTE FLOW

34 The integral representations of the solutions

As discussed in § 31.1, when $U(z) = z$, the Orr–Sommerfeld equation becomes

$$\{\varepsilon^3(D^2 - \alpha^2) - \eta\}(D^2 - \alpha^2)\phi = 0, \qquad (34.1)$$

where

$$\eta = z - c, \quad D = d/d\eta, \quad \text{and} \quad \varepsilon = (i\alpha R)^{-1/3} \quad \text{with} \quad \text{ph}\varepsilon = -\tfrac{1}{6}\pi. \qquad (34.2)$$

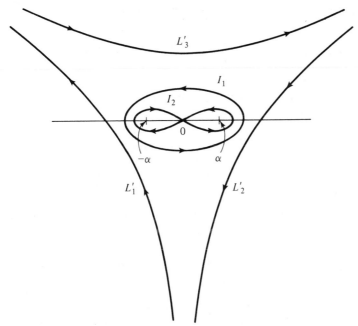

Fig. 5.1. The paths of integration in the t-plane associated with the integrals (34.3). (After Wasow 1953a.)

The solutions of equation (34.1) are of either well-balanced or dominant-recessive type whereas for general velocity profiles the Orr–Sommerfeld equation also has solutions of balanced type. In this respect equation (34.1) fails to exhibit one of the essential features of the general problem but, because of its simplicity, a number of special techniques can be used to treat it, one of which was discussed in § 31.1.

Another method of dealing with equation (34.1), which was first used by Wasow (1953a), exploits the fact that the independent variable occurs linearly in it and hence that one can obtain contour integral representations of the solutions by Laplace's method. A short calculation then shows that contour integrals of the form

$$\int_{\mathscr{C}} (t^2 - \alpha^2)^{-1} \exp\left\{(\eta + \alpha^2 \varepsilon^3)t - \tfrac{1}{3}\varepsilon^3 t^3\right\} dt \qquad (34.3).$$

are solutions of equation (34.1), provided $\mathscr{C}$ is one of the paths shown in Fig. 5.1. The paths I_n $(n = 1, 2)$ lead to solutions of well-balanced type and the paths L'_k $(k = 1, 2, 3)$ lead to solutions of dominant-recessive type.

Consider first the solutions of well-balanced type which can conveniently be defined by

$$U_n(\eta) = \frac{\alpha}{2\pi i} \int_{I_n} (t^2 - \alpha^2)^{-1} \exp\{(\eta + \alpha^2 \varepsilon^3)t - \tfrac{1}{3}\varepsilon^3 t^3\} \, dt$$

$$(n = 1, 2). \tag{34.4}$$

A simple residue calculation then shows that

$$U_1(\eta) = \sinh(\alpha\eta + \tfrac{2}{3}\alpha^3 \varepsilon^3) \quad \text{and} \quad U_2(\eta) = \cosh(\alpha\eta + \tfrac{2}{3}\alpha^3 \varepsilon^3), \tag{34.5}$$

and their Wronskian has the value $\mathscr{W}(U_1, U_2) = -\alpha$. Alternatively, observe that equation (34.1) admits two exact solutions for which $(D^2 - \alpha^2)\phi = 0$, and the solutions of this equation are clearly related to the 'inviscid' solutions of the Orr–Sommerfeld equation. To emphasize this relationship let

$$\phi_1(\eta) = \alpha^{-1} \sinh \alpha\eta \quad \text{and} \quad \phi_2(\eta) = \cosh \alpha\eta. \tag{34.6}$$

Then $\phi_1(\eta)$ is precisely the regular inviscid solution with the usual normalization and $\phi_2(\eta)$ can be identified with the regular part of the singular inviscid solution. The solutions $U_1(\eta)$ and $U_2(\eta)$ are thus linear combinations of the solutions $\phi_1(\eta)$ and $\phi_2(\eta)$.

Consider next the solutions of dominant-recessive type which are defined by

$$V_k(\eta) = \frac{\varepsilon^{-1}}{2\pi i} \int_{L'_k} (t^2 - \alpha^2)^{-1} \exp\{(\eta + \alpha^2 \varepsilon^3)t - \tfrac{1}{3}\varepsilon^3 t^3\} \, dt$$

$$(k = 1, 2, 3), \tag{34.7}$$

where the factor ε^{-1} has been introduced to provide a convenient scaling of the solutions. By considering the paths shown in Fig. 5.1 we immediately obtain the connexion formula

$$\sum_{k=1}^{3} V_k(\eta) = -(\alpha\varepsilon)^{-1} U_1(\eta). \tag{34.8}$$

This result can also be written in the form

$$\sum_{k=1}^{3} V_k(\eta) = -\varepsilon^{-1}\{C_1(\varepsilon)\phi_1(\eta) + C_2(\varepsilon)\phi_2(\eta)\}, \tag{34.9}$$

where

$$C_1(\varepsilon) = \cosh \tfrac{2}{3}\alpha^3\varepsilon^3 \quad \text{and} \quad C_2(\varepsilon) = \alpha^{-1}\sinh \tfrac{2}{3}\alpha^3\varepsilon^3.$$

$$(34.10)$$

Our immediate goal then is to obtain asymptotic expansions of the solutions $V_k(\eta)$ which are uniformly valid in a bounded domain of the (complex) η-plane containing the origin. Before deriving the uniform expansions, however, let us consider the inner and outer expansions of the solutions. To obtain the inner expansions let

$$V_k(\eta, \varepsilon) = \tilde{V}_k(\zeta, \varepsilon), \text{ where } \zeta = \eta/\varepsilon. \qquad (34.11)$$

On letting $\tau = \varepsilon t$ in equation (34.7) we have

$$\tilde{V}_k(\zeta, \varepsilon) = \frac{1}{2\pi i}\int_{L_k} (\tau^2 - \alpha^2\varepsilon^2)^{-1}\exp\{(\zeta + \alpha^2\varepsilon^2)\tau - \tfrac{1}{3}\tau^3\}\,d\tau, \quad (34.12)$$

where L_k are the usual Airy contours shown in Fig. A2. Expansion of the factor $(\tau^2 - \alpha^2\varepsilon^2)^{-1}\exp(\alpha^2\varepsilon^2\tau)$ in powers of $(\alpha\varepsilon)^2$ then yields the inner expansions in the form

$$\tilde{V}_k(\zeta, \varepsilon) = \sum_{n=0}^{\infty} (\alpha\varepsilon)^{2n}\tilde{v}_k^{(2n)}(\zeta), \qquad (34.13)$$

where

$$\tilde{v}_k^{(2n)}(\zeta) = \sum_{m=0}^{n} \frac{1}{m!}A_k(\zeta, 2+2n-3m). \qquad (34.14)$$

Thus, the inner expansions of the solutions $V_k(\eta, \varepsilon)$ can be expressed *to all orders* in terms of the generalized Airy functions $A_k(\zeta, p)$ with $p = 0, \pm 1, \pm 2, \dots$. The series (34.13) is convergent for bounded values of $|\zeta|$ and truncation of the series after a finite number of terms thus provides an asymptotic approximation for bounded values of $|\zeta|$. To obtain an approximation which is uniformly valid for bounded values of $|\eta|$ one could, at least in principle, use the recursion formula (A9) to express all of the Airy functions in terms of three of them, with $p = 2, 1, 0$ (say). This procedure would lead to an expression of the form

$$V_k(\eta) = a(\eta, \varepsilon)A_k(\zeta, 2) + \varepsilon^2 b(\eta, \varepsilon)A_k(\zeta, 1) + \varepsilon c(\eta, \varepsilon)A_k(\zeta, 0),$$

$$(34.15)$$

where the slowly varying coefficients a, b and c all have expansions of the form

$$a \equiv a(\eta, \varepsilon) = \sum_{s=0}^{\infty} a_s(\eta)\varepsilon^{3s}. \qquad (34.16)$$

Although this procedure does lead to the correct *form* for the uniform expansions of $V_k(\eta)$, it does not provide a useful algorithm for the determination of the coefficients $a_s(\eta)$, $b_s(\eta)$ and $c_s(\eta)$.

One method of determining the slowly varying coefficients in the uniform expansions in equation (34.15) is based on a knowledge of the outer expansions of the solutions $V_k(\eta)$. The required outer expansions can be obtained either by re-expansion of the inner expansions in equation (34.13) or by an application of the method of steepest descents to the integrals in equation (34.7). If we let $\mathbf{T}'_k$ denote the Stokes sectors in the η-plane and restrict ph η to the range $-\frac{3}{2}\pi < \mathrm{ph}\,\eta < \frac{1}{2}\pi$, then we have

$$\left.\begin{aligned}
V_1(\eta) &\sim \bar{v}_-(\eta) \quad (\eta \in \mathbf{T}'_2 \cup \mathbf{T}'_3), \\
V_2(\eta) &\sim i\bar{v}_+(\eta) \quad (\eta \in \mathbf{T}'_3 \cup \mathbf{T}'_1), \\
\text{and} \qquad\qquad & \\
V_3(\eta) &\sim \begin{cases} -\bar{v}_-(\eta) & (\eta \in \mathbf{T}'_1) \\ -i\bar{v}_+(\eta) & (\eta \in \mathbf{T}'_2), \end{cases}
\end{aligned}\right\} \qquad (34.17)$$

where

$$\bar{v}_\pm(\eta) = \tfrac{1}{2}\pi^{-1/2}\varepsilon^{5/4}\eta^{-5/4}\exp\left(\pm\tfrac{2}{3}\varepsilon^{-3/2}\eta^{3/2}\right)\sum_{s=0}^{\infty}(\pm\varepsilon^{3/2})^s H_s(\eta),$$

$$(34.18)$$

$$H_0(\eta) = 1, \; H_1(\eta) = \tfrac{101}{48}\eta^{-3/2} + a^2\eta^{1/2}, \ldots, \qquad (34.19)$$

and the sectors of validity have been restricted to insure completeness. The outer expansions of the solutions $V_k(\eta)$ in the sectors $\mathbf{T}'_k$ can then be obtained by use of the connexion formula (34.8) but they are not required for the present purposes.

These inner and outer expansions could be combined, as described in § 37, to form composite approximations of either the additive or multiplicative type. This is particularly simple in the case of plane Couette flow since the inner expansions (34.13) are 'self-composite' in the sectors $\mathbf{I} - \mathbf{T}'_k$ with respect to either additive or multiplicative composition, i.e. they contain their outer expansions in the sectors $\mathbf{I} - \mathbf{T}'_k$. Nevertheless, composite approximations of this type are not fully uniform, since they are not valid in a full neighbourhood of the turning point.

To obtain a uniform 'first approximation' to $V_k(\eta)$ it is necessary to determine the three coefficients $a_0(\eta)$, $b_0(\eta)$, and $c_0(\eta)$ in equation (34.15) and this can now easily be done in the following manner. Observe, first, from equations (34.15) and (34.8) that

$$\sum_{k=1}^{3} V_k(\eta) = -\varepsilon^{-1}\{\eta \, a(\eta, \varepsilon) + \varepsilon^3 b(\eta, \varepsilon)\}$$

$$= -(\alpha\varepsilon)^{-1} U_1(\eta), \tag{34.20}$$

and hence

$$a_0(\eta) = (\alpha\eta)^{-1} \sinh \alpha\eta. \tag{34.21}$$

Consider next the outer expansion of $V_1(\eta)$ (say). For $\eta \in \mathbf{T}'_2 \cup \mathbf{T}'_3$ and $|\eta| \geq |\eta_0| > 0$, equation (34.15) gives

$$V_1(\eta) = \tfrac{1}{2}\pi^{-1/2}\varepsilon^{5/4}\eta^{-5/4}\exp\left(-\tfrac{2}{3}\varepsilon^{-3/2}\eta^{3/2}\right)\{a_0 + \eta \, c_0$$
$$- \varepsilon^{3/2}(\tfrac{101}{48}\eta^{-3/2}a_0 + \eta^{1/2}b_0 + \tfrac{5}{48}\eta^{-1/2}c_0) + O(\varepsilon^3)\}. \tag{34.22}$$

Since this expansion must be identical to the one given by equation (34.17), we see immediately that

and
$$\left.\begin{array}{l} a_0 + \eta \, c_0 = H_0 \\ \tfrac{101}{48}\eta^{-3/2}a_0 + \eta^{1/2}b_0 + \tfrac{5}{48}\eta^{-1/2}c_0 = H_1. \end{array}\right\} \tag{34.23}$$

A short calculation then gives

$$b_0(\eta) = \frac{2}{\eta^2}\left(1 - \frac{\sinh \alpha\eta}{\alpha\eta}\right) + \alpha^2 \quad \text{and} \quad c_0(\eta) = \frac{1}{\eta}\left(1 - \frac{\sinh \alpha\eta}{\alpha\eta}\right), \tag{34.24}$$

and it is evident that b_0 and c_0 are both analytic at $\eta = 0$. This method of determining b_0 and c_0 is of particular importance because it can easily be generalized so as to apply to problems for which integral representations of the solutions do not exist.

It is clearly of some interest, however, to show how the uniform expansions (34.15) and the slowly varying coefficients in them can be obtained directly from the integral representations given by equation (34.7). For this purpose consider the integrals

$$I_k(\eta, \alpha) = \frac{1}{2\pi i} \int_{L'_k} (t - \alpha)^{-1} \exp\{(\eta + \alpha^2 \varepsilon^3)t - \tfrac{1}{3}\varepsilon^3 t^3\} \, dt$$

$$(34.25)$$

in terms of which we have

$$V_k(\eta) = \tfrac{1}{2}(\alpha\varepsilon)^{-1}\{I_k(\eta, \alpha) - I_k(\eta, -\alpha)\}. \qquad (34.26)$$

On making the change of variable $t \to \varepsilon^{-3/2} t + \alpha$ in equation (34.25) we find after a little reduction that

$$I_k(\eta, \alpha) = \exp(\alpha\eta + \tfrac{2}{3}\alpha^3 \varepsilon^3)\{A_k(\zeta, 1) + J_k(\eta, \alpha)\}, \quad (34.27)$$

where

$$J_k(\eta, \alpha) = \frac{1}{2\pi i} \int_{L''_k} g(t, \alpha) \exp\{\varepsilon^{-3/2}(\eta t - \tfrac{1}{3}t^3)\} \, dt, \quad (34.28)$$

$$g(t, \alpha) = t^{-1}\{\exp(-\alpha t^2) - 1\}, \qquad (34.29)$$

and the paths L''_k are obtained from the paths L'_k by a rotation through an angle of $-\tfrac{1}{4}\pi$. The problem has thus been reduced to a consideration of the integrals $J_k(\eta, \alpha)$ and these integrals are of a type which can be treated by the method of Chester, Friedman & Ursell (1957). In applying that method to the present problem, some simplification is possible since $g(t, \alpha)$ is an odd function of t and, for bounded values of α and $|\eta|$, it then follows that the integrals $J_k(\eta, \alpha)$ have uniform asymptotic expansions of the form

$$J_k(\eta, \alpha) = \varepsilon^2 A_k(\zeta) \sum_{s=0}^{\infty} \varepsilon^{3s} p_s(\eta, \alpha) + \varepsilon A'_k(\zeta) \sum_{s=0}^{\infty} \varepsilon^{3s} q_s(\eta, \alpha).$$

$$(34.30)$$

The slowly varying coefficients $p_s(\eta, \alpha)$ and $q_s(\eta, \alpha)$ can be obtained recursively (Bleistein & Handelsman 1975, p. 374) but, except for $q_0(\eta, \alpha)$, their actual determination is complicated. In the case of $q_0(\eta, \alpha)$, however, it is easy to show that

$$q_0(\eta, \alpha) = \eta^{-1/2} g(\eta^{1/2}, \alpha) = \eta^{-1}(e^{-\alpha\eta} - 1) \qquad (34.31)$$

and this leads to a determination of $a_0(\eta)$ and $c_0(\eta)$ which is in agreement with equations (34.21) and (34.24).

On collecting together these results and then using the recursion formula (A9), we see that the uniform expansions of the solutions

$V_k(\eta)$ are indeed of the form (34.15) as inferred earlier from the inner expansions of the solutions. Although the treatment of this problem by the method of Chester, Friedman & Ursell is attractive on general theoretical grounds, a reduction of the problem to integrals of the type (34.28) is, of course, very special. In most problems we do not have integral representations for the solutions and it is important, therefore, to devise methods for obtaining the uniform expansions which do not depend in any way on the existence of integral representations of the solutions.

35 The differential equation method

Once the general form of the uniform expansions has been determined, by considering the inner expansions or otherwise, then the slowly varying coefficients which appear in them can be obtained by deriving and then solving the differential equations which they satisfy. In the case of plane Couette flow they can also be determined by considering a certain second-order inhomogeneous differential equation (see Problem 5.2) but this is not typical of the situation for general velocity profiles. Consider then the problem of determining the slowly varying coefficients in the expansion (34.15) directly from the differential equation (34.1). As in the case of Langer's theory for second-order equations with a simple turning point, we simply substitute the expansion (34.15) into equation (34.1) and require that the coefficients of $A_k(\zeta, p)$ for $p = 2, 1, 0$ all vanish. After a straightforward but somewhat lengthy calculation in which the recursion formula (A9) is used repeatedly, we find, without any approximation, that equation (34.1) is formally satisfied provided the coefficients a, ℓ and c satisfy the equations

$$\eta a'' + 4a' - \alpha^2 \eta a + 2\eta c' + 2c$$
$$- \varepsilon^3 \{(D^2 - \alpha^2)^2 a - 2(3D^2 - \alpha^2)\ell - 4(D^2 - \alpha^2)Dc\} = 0, \quad (35.1)$$
$$2\eta D(a + \eta c) + \varepsilon^3 \{4(D^2 - \alpha^2)Da$$
$$+ \eta(5D^2 - \alpha^2)\ell + 4\eta(D^2 - \alpha^2)Dc\} + \varepsilon^6(D^2 - \alpha^2)^2 \ell = 0, \quad (35.2)$$

and

$$2(3D^2 - \alpha^2)a + 2\eta\ell' + \ell + 5\eta c'' + 4c' - \alpha^2 \eta c$$
$$+ \varepsilon^3 \{(3D^2 - 4\alpha^2)D\ell + (D^2 - \alpha^2)^2 c\} = 0. \quad (35.3)$$

These equations have a somewhat complicated structure and we will therefore not attempt to derive general recursion formulae for the coefficients a_s, b_s and c_s.

One general result, however, is worth noting. On eliminating c between equations (35.1) and (35.2) we obtain

$$\{\varepsilon^3(D^2 - \alpha^2) - \eta\}(D^2 - \alpha^2)(\eta a + \varepsilon^3 b) = 0. \tag{35.4}$$

Thus the quantity $\eta a + \varepsilon^3 b$ satisfies the original differential equation and it would appear therefore that no simplification has been achieved. But if we recall that a, b, and c must have expansions in powers of ε^3 of Poincaré type (cf. equation (34.16)) then we must have

$$(D^2 - \alpha^2)(\eta a + \varepsilon^3 b) = 0. \tag{35.5}$$

A result of this type is not unexpected and is closely related to the connexion formula (34.20). The general solution of equation (35.5) can conveniently be taken in the form

$$\eta a + \varepsilon^3 b = p(\varepsilon) \sinh \alpha \eta + q(\varepsilon) \cosh \alpha \eta, \tag{35.6}$$

where p and q both have expansions of the form

$$p(\varepsilon) = \sum_{s=0}^{\infty} \varepsilon^{3s} p_s. \tag{35.7}$$

The requirement that $a(\eta, \varepsilon)$ be analytic at $\eta = 0$ then shows that we must set $q_0 = 0$ and $q_{s+1} = b_s(0)$ for $s \geq 0$ but the p_s are arbitrary and can be chosen in any convenient way. In particular, to achieve agreement with the integral representation (34.7) we would have to choose

$$p(\varepsilon) = \cosh \tfrac{2}{3}\alpha^3 \varepsilon^3 \quad \text{and} \quad q(\varepsilon) = \alpha^{-1} \sinh \tfrac{2}{3}\alpha^3 \varepsilon^3, \tag{35.8}$$

but this choice is certainly not necessary and one could, for example, choose $p_0 = 1$ and $p_s = 0$ for $s \geq 1$.

Consider next the problem of determining a 'first approximation' to the solutions $V_k(\eta)$. For this purpose we must determine a_0, b_0 and c_0 which, from equations (35.1)–(35.3), satisfy the equations

$$\left. \begin{aligned} \eta a_0'' + 4a_0' - \alpha^2 \eta a_0 + 2\eta c_0' + 2c_0 &= 0, \\ 2\eta D(a_0 + \eta c_0) &= 0 \end{aligned} \right\} \tag{35.9}$$

and

$$2(3D^2 - \alpha^2)a_0 + 2\eta b_0' + b_0 + 5\eta c_0'' + 4c_0' - \alpha^2 \eta c_0 = 0.$$

The second of these equations shows that $a_0 + \eta c_0 =$ constant and if we fix the normalization in a natural way by requiring that $a_0(0) = 1$ then we have

$$a_0(\eta) + \eta c_0(\eta) = 1. \tag{35.10}$$

On using this relation in the first of equations (35.9) we find that a_0 satisfies the equation (cf. equation (35.5))

$$\eta(D^2 - \alpha^2)(\eta a_0) = 0. \tag{35.11}$$

Since a_0 must be analytic at $\eta = 0$, ηa_0 must be proportional to $\sinh \alpha \eta$ and, with the normalization we have adopted, we then recover equation (34.21). Finally, the third of equations (35.9) provides a first-order inhomogeneous equation for the determination of ℓ_0, and the solution of this equation which is analytic at $\eta = 0$ is given by

$$\ell_0(\eta) = 2\eta^{-1} c_0(\eta) + \alpha^2. \tag{35.12}$$

These results for ℓ_0 and c_0 are in complete agreement with equations (34.24).

The asymptotic theory of the Orr–Sommerfeld equation is thus reasonably complete in the case of plane Couette flow and we now wish to consider how the theory can be generalized so as to apply to more general velocity profiles.

GENERAL VELOCITY PROFILES

36 A preliminary transformation

In the derivation of uniform approximations to the solutions of the Orr–Sommerfeld equation, it is customary to make a preliminary transformation of both dependent and independent variables of the form

$$\phi(z) = \{\eta'(z)\}^m \chi(\eta), \tag{36.1}$$

where $\eta(z)$ is the usual Langer variable

$$\eta(z) = \left[\frac{3}{2} \int_{z_c}^{z} \left(\frac{U - c}{U_c'} \right)^{1/2} dz \right]^{2/3}. \tag{36.2}$$

The exponent m in equation (36.1) is often chosen so that the transformed equation is in normal form, i.e. so that it does not contain χ''', and this condition requires that $m = -\tfrac{3}{2}$. Clearly the final results must be independent of m and, especially when dealing with equations such as equation (31.44) which are not initially in normal form, a more convenient choice would appear to be $m = 0$. Thus, with $m = 0$ in equation (36.1) we find that the Orr–Sommerfeld equation becomes

$$\varepsilon^3(\chi^{iv} + f_0\chi''') - (\eta + \varepsilon^3 f_1)\chi'' - (g_0 + \varepsilon^3 g_1)\chi' - (h_0 + \varepsilon^3 h_1)\chi = 0,$$

$$(36.3)$$

where

$$\left.\begin{aligned}
f_0(\eta) &= 6\gamma, \\
f_1(\eta) &= -(4\gamma' + 11\gamma^2 - 2\alpha^2\eta'^{-2}), \\
g_0(\eta) &= \eta\gamma, \\
g_1(\eta) &= -(\gamma'' + 7\gamma'\gamma + 6\gamma^3 - 2\alpha^2\gamma\eta'^{-2}), \\
h_0(\eta) &= -(2\eta\gamma' + 6\eta\gamma^2 + 5\gamma + \alpha^2\eta\eta'^{-2}), \\
h_1(\eta) &= -\alpha^4\eta'^{-4},
\end{aligned}\right\} \qquad (36.4)$$

and

$$\gamma(\eta) = \eta''/\eta'^2. \qquad (36.5)$$

It would, of course, have been possible to develop the present theory in a more general form by assuming only that the coefficients $f_0, f_1, \ldots, h_1$ which appear in equation (36.3) are analytic functions of η in some neighbourhood of $\eta = 0$. The general structure of the theory then depends in a crucial way on the value of $g_0(0)$ and to a lesser extent on the value of $h_0(0)$. As Lin & Rabenstein (1960) have shown, there are then three cases which must be considered depending on the nature of the solutions of the reduced equation $\eta\chi'' + g_0\chi' + h_0\chi = 0$. To treat each of these cases in full detail would require a lengthy discussion and tend to obscure the other important elements of the theory. We shall suppose therefore that $f_0, f_1, \ldots, h_1$ have the forms given by equations (36.4) and it is then easily seen that

$$g_0(0) = 0 \quad \text{and} \quad h_0(0) = -5\gamma(0) = -U_c''/U_c'. \qquad (36.6)$$

We are thus dealing with an example of case II in Lin & Rabenstein's classification for which $g_0(0)$ is an integer (positive, negative or zero) and $h_0(0) \neq 0$.

In discussing the solutions of equation (36.3) we wish to exploit so far as possible certain symmetries which they exhibit in the complex plane, and for this purpose it is convenient to consider seven solutions which will be denoted by $U_0(\eta)$, $U_k(\eta)$ and $V_k(\eta)$ ($k = 1, 2, 3$). Although we do not, in general, have exact representations for these solutions, they can be uniquely defined (to within multiplicative factors and modulo an arbitrary additive multiple of U_0 in the case of the U_k) in terms of their asymptotic properties. Thus, we require that U_0 be well balanced, that U_k be (purely) balanced in $\mathbf{T}'_k$, and that V_k be recessive in $\mathbf{S}'_k$, where $\mathbf{T}'_k$ and $\mathbf{S}'_k$ are the Stokes and anti-Stokes sectors respectively in the η-plane. The existence of solutions having these asymptotic properties has been proved by Lin & Rabenstein (1969) but we shall make no essential use of this fact. These seven solutions cannot, of course, be linearly independent but must be related by three connexion formulae. Approximations to these connexion formulae can be obtained either from the inner expansions of the solutions or from the uniform approximations, but they will not be needed for the present purposes.

These solutions of the transformed equation are related to the four standard solutions of the Orr–Sommerfeld equation in the manner

$$\left. \begin{array}{ll} \phi_1(z) \equiv C_1 U_0(\eta), & \phi_2(z) \equiv C_0 U_0(\eta) + C_2 U_3(\eta), \\ \phi_3(z) \equiv C_3 V_1(\eta), & \text{and} \quad \phi_4(z) \equiv C_4 V_2(\eta), \end{array} \right\} \quad (36.7)$$

where the values of the constants $C_0, C_1, \ldots, C_4$ depend only on the way in which the solutions have been normalized. Without loss of generality, however, we can fix the normalization of the solutions so that

$$C_0 = 0 \quad \text{and} \quad C_1 = C_2 = C_3 = C_4 = 1, \quad (36.8)$$

and we will suppose that this has been done.

37 The inner and outer expansions

The application of the method of matched asymptotic expansions to the Orr–Sommerfeld equation was first studied by Graebel (1966), who obtained results essentially similar to the heuristic approximations discussed in §§ 27 and 28. A more systematic application of this method was later made by Eagles (1969) and Reid (1972), who were primarily concerned with the derivation of composite approximations which, though not uniformly valid in a full neighbourhood of the critical point, are valid in certain deleted neighbourhoods which do not, however, shrink to zero as $|\varepsilon| \to 0$.

The theory developed by Eagles for the Orr–Sommerfeld equation was based on a strict adherence to Poincaré's definition of an asymptotic expansion whereas Reid first made the preliminary transformation (36.1) with $m = 0$ and then required that all asymptotic expansions be complete in the sense of Watson. Although the introduction of a Langer variable is clearly not essential when using the method of matched asymptotic expansions, it has the important consequence of bringing the Stokes and anti-Stokes lines associated with the inner and outer expansions into coincidence and this, in turn, substantially simplifies the subsequent solution of the central matching problem and the formation of composite approximations.

Our interest here is not in composite approximations *per se* but rather in using the inner expansions to infer the general structure of the uniform expansions. The outer expansions can then be used to provide a simple method of determining some of the slowly varying coefficients in the uniform expansions.

37.1 *The inner expansions*

To derive the required inner expansions it is convenient to let

$$\chi(\eta, \varepsilon) = \tilde{\chi}(\zeta, \varepsilon), \quad \text{where } \zeta = \eta/\varepsilon. \tag{37.1}$$

Equation (36.3) can then be rewritten in the form

$$A D^2 \tilde{\chi} = (g_0 + \varepsilon^3 g_1) \tilde{\chi}' + \varepsilon (h_0 + \varepsilon^3 h_1) \tilde{\chi} - \varepsilon f_0 \tilde{\chi}''' + \varepsilon^2 f_1 \tilde{\chi}'', \tag{37.2}$$

where

$$D = d/d\zeta \quad \text{and} \quad A = D^2 - \zeta. \tag{37.3}$$

The solutions of equation (37.2) are then expanded in series of the form

$$\tilde{\chi}(\zeta, \varepsilon) = \sum_{s=0}^{\infty} \varepsilon^s \tilde{\chi}^{(s)}(\zeta) \tag{37.4}$$

which are convergent for bounded values of $|\zeta|$. We cannot, of course, compute the whole of these series, but their partial sums provide us with the required inner expansions. On substituting equation (37.4) into equation (37.2) and using the fact that $g_0(0) = 0$, we find that the $\tilde{\chi}^{(s)}$ for $s = 0, 1, 2$ must satisfy the equations

$$AD^2\tilde{\chi}^{(0)} = 0, \qquad (37.5a)$$

$$AD^2\tilde{\chi}^{(1)} = M_0\tilde{\chi}^{(0)}, \qquad (37.5b)$$

and

$$AD^2\tilde{\chi}^{(2)} = M_0\tilde{\chi}^{(1)} + M_1\tilde{\chi}^{(0)}, \qquad (37.5c)$$

where

$$M_0 = -f_0(0)D^3 + g_0'(0)\zeta D + h_0(0) \qquad (37.6a)$$

and

$$M_1 = -f_0'(0)\zeta D^3 + \tfrac{1}{2}g_0''(0)\zeta^2 D + h_0'(0)\zeta + f_1(0)D^2. \qquad (37.6b)$$

If $g_0(0) \neq 0$ then the operator AD^2 which appears in equations (37.5) must be replaced by $\{AD - g_0(0)\}D$ and this shows how the general structure of the inner expansions depends on the value of $g_0(0)$.

We next define seven inner expansions $(\tilde{u}_0, \tilde{u}_k, \tilde{v}_k)$ in such a way that their asymptotic behaviours for unbounded values of $|\zeta|$ are identical to those prescribed for the solutions (U_0, U_k, V_k). Thus, for unbounded values of $|\zeta|$ we require that $\tilde{u}_0$ be well balanced, that $\tilde{u}_k$ be (purely) balanced in T_k, and that $\tilde{v}_k$ be recessive in S_k. These conditions serve to define $\tilde{u}_0$ and $\tilde{v}_k$ uniquely to within multiplicative factors but, since an arbitrary multiple of $\tilde{u}_0$ can be added to the $\tilde{u}_k$ without affecting their asymptotic behaviour, some additional conditions must be imposed on the $\tilde{u}_k$ to fix their normalization.

Consider first the inner expansion of well-balanced type. From equation (37.5a) we see immediately that $\tilde{u}_0^{(0)}$ must be of the form $\tilde{a}_0 + \tilde{b}_0\zeta$, where $\tilde{a}_0$ and $\tilde{b}_0$ are arbitrary constants. If $\tilde{a}_0 \neq 0$, however, then $\tilde{u}_0^{(1)}$ will (unless $\gamma(0) = 0$, i.e. $U_c'' = 0$) contain a multiple of $B_k(\zeta, 2, 1)$ with $k = 1, 2$ or 3 and hence cannot be well balanced. Accordingly, we must set $\tilde{a}_0 = 0$. The normalization is then fixed by setting $\tilde{b}_0 = 1$ and by requiring that $\tilde{u}_0^{(s)}$ for $s \geq 1$ contain no multiple of $\tilde{u}_0^{(0)}$. This does not imply, however, that $\tilde{u}_0^{(s)}$ for $s \geq 1$ cannot contain a constant term but rather that the constant term, $\tilde{a}_s$ (say), in $\tilde{u}_0^{(s)}$ must be chosen so that $\tilde{u}_0^{(s+1)}$ is well balanced, i.e. $\tilde{a}_s$ must be chosen so that there is no constant term in the inhomogeneous part of the equation which determines $\tilde{u}_0^{(s+1)}$. It can then be shown that $\tilde{a}_{3s} = \tilde{a}_{3s+1} = 0$ but that $\tilde{a}_{3s+2} \neq 0$ $(s = 0, 1, 2, \ldots)$. Thus the inner expansion of well-balanced type can be found to all orders in terms of the polynomials $B_0(\zeta, p)$ which are defined in the Appendix or, alternatively, in terms of the powers of ζ. In this way we obtain

$$\tilde{u}_0^{(0)}(\zeta) = \zeta, \tag{37.7a}$$

$$\tilde{u}_0^{(1)}(\zeta) = 4\gamma_0 \frac{\zeta^2}{2!}, \tag{37.7b}$$

and

$$\tilde{u}_0^{(2)}(\zeta) = (6\gamma_0' + 12\gamma_0^2 + \alpha^2)\frac{\zeta^3}{3!} + \tilde{a}_2, \tag{37.7c}$$

where

$$\tilde{a}_2 = -\left(\frac{U_c^{iv}}{U_c''} - \frac{2}{3}a^2\right) \tag{37.8}$$

and $\gamma_0 = \gamma(0)$, $\gamma_0' = \gamma'(0), \ldots$.

Consider next the inner expansions of balanced type. From equation $(37.5a)$ we see that $\tilde{u}_k^{(0)}$ must also be of the form $\tilde{a}_0 + \tilde{b}_0\zeta$, but if we require that $\tilde{u}_k^{(0)}$ contain no multiple of $\tilde{u}_0^{(0)}$ then, as is usually done, we must set $\tilde{b}_0 = 0$. The fixing of the normalization of $\tilde{u}_k$, however, requires some care. Observe first that $\tilde{u}_k^{(s)}$ for $s \geq 1$ is composed of two parts: a well-balanced part which is expressible in terms of the polynomials $B_0(\zeta, p)$ or the powers of ζ and a balanced part which is expressible to all orders in terms of $B_k(\zeta, p, q)$. We shall fix the normalization therefore by setting $\tilde{a}_0 = 1$ and by requiring that the well balanced part of $\tilde{u}_k^{(s)}$ for $s \geq 1$ contains no constant or linear term. The solutions which satisfy these conditions are then found to be

$$\tilde{u}_k^{(0)}(\zeta) = 1, \tag{37.9a}$$

$$\tilde{u}_k^{(1)}(\zeta) = -5\gamma_0 B_k(\zeta, 2, 1), \tag{37.9b}$$

$$\tilde{u}_k^{(2)}(\zeta) = (7\gamma_0' - 19\gamma_0^2 + \alpha^2)\frac{\zeta^2}{2!}$$

$$- 5\gamma_0\{4\gamma_0 B_k(\zeta, 3, 1) - \tfrac{5}{2}\gamma_0 B_k(\zeta, 0, 1)\}, \tag{37.9c}$$

and

$$\tilde{u}_k^{(3)}(\zeta) = (\tfrac{9}{2}\gamma_0'' + \tfrac{35}{2}\gamma_0'\gamma_0 - \tfrac{107}{2}\gamma_0^3 - \tfrac{1}{2}\alpha^2\gamma_0)\frac{\zeta^3}{3!}$$
$$- 5\gamma_0\{(6\gamma_0' + 12\gamma_0^2 + \alpha^2)B_k(\zeta, 4, 1)$$
$$- (5\gamma_0' + 22\gamma_0^2 - \alpha^2)B_k(\zeta, 1, 1)$$
$$- (\tfrac{5}{4}\gamma_0' - \tfrac{25}{8}\gamma_0^2)B_k(\zeta, -2, 1)\}. \tag{37.9d}$$

The inner expansions of dominant-recessive type are uniquely defined, to within multiplicative factors, by the requirement that they be recessive in $\mathbf{S}_k$. From equation $(37.5a)$ we see that $\tilde{v}_k^{(0)}$ must be a multiple of $A_k(\zeta, 2)$. The normalization is then fixed by setting $\tilde{v}_k^{(0)}(\zeta) = A_k(\zeta, 2)$ and by requiring that $\tilde{v}_k^{(s)}$ for $s \geq 1$ contain no multiple of $\tilde{v}_k^{(0)}$. We then easily find that

$$\tilde{v}_k^{(0)}(\zeta) = A_k(\zeta, 2), \tag{37.10a}$$

$$\tilde{v}_k^{(1)}(\zeta) = 4\gamma_0 A_k(\zeta, 3) - \tfrac{5}{2}\gamma_0 A_k(\zeta, 0), \tag{37.10b}$$

and

$$\tilde{v}_k^{(2)}(\zeta) = (6\gamma_0' + 12\gamma_0^2 + \alpha^2)A_k(\zeta, 4)$$
$$- (5\gamma_0' + 22\gamma_0^2 - \alpha^2)A_k(\zeta, 1) - (\tfrac{5}{4}\gamma_0' - \tfrac{25}{8}\gamma_0^2)A_k(\zeta, -2). \tag{37.10c}$$

To this order, the coefficients in the inner expansions are expressible in terms of $A_k(\zeta, p) \equiv A_k(\zeta, p, 0)$; for $s = 3, 4, 5$, however, we also need the functions $A_k(\zeta, p, 1)$; for $s = 6, 7, 8$ we also need $A_k(\zeta, p, 2)$; and so on. Similar remarks apply to the inner expansions of balanced type.

37.2 *The outer expansions*

In discussing the outer expansions to the solutions of equation (36.3) it is again convenient to define seven expansions of this type. These expansions, which will be denoted by $\bar{u}_0$, $\bar{u}_k$ and $\bar{v}_k$, are required to be well balanced, balanced in $\mathbf{T}_k'$, and recessive in $\mathbf{S}_k'$, respectively.

To derive the outer expansions of well-balanced and balanced type, which are often described as being of 'inviscid' type, it is convenient to rewrite equation (36.3) in the form

$$R_2\chi = \varepsilon^3 R_4\chi, \tag{37.11}$$

where

$$R_2 = \eta D^2 + g_0 D + h_0 \qquad (37.12)$$

and

$$R_4 = D^4 + f_0 D^3 - f_1 D^2 - g_1 D - h_1. \qquad (37.13)$$

We then consider a formal expansion in powers of ε^3 of the form

$$\bar{\chi}(\eta, \varepsilon) = \sum_{s=0}^{\infty} \varepsilon^{3s} \bar{\chi}^{(s)}(\eta), \qquad (37.14)$$

where

$$R_2 \bar{\chi}^{(0)} = 0 \qquad (37.15)$$

and

$$R_2 \bar{\chi}^{(s)} = R_4 \bar{\chi}^{(s-1)} \quad (s \geq 1). \qquad (37.16)$$

Equation (37.15) is, of course, simply Rayleigh's equation written in terms of the Langer variable. One solution of equation (37.15) is analytic at $\eta = 0$ and it can therefore be identified with $\bar{u}_0^{(0)}$. It is also convenient to fix the normalization so that

$$\bar{u}_0^{(0)}(\eta) \equiv \phi_1^{(0)}(z) \qquad (37.17)$$

and we then have (cf. equation (22.25))

$$\bar{u}_0^{(0)}(\eta) = \eta Q_1(\eta)$$
$$= \eta + 4\gamma_0 \frac{\eta^2}{2!} + (6\gamma_0' + 12\gamma_0^2 + \alpha^2) \frac{\eta^3}{3!} + O(\eta^4). \qquad (37.18)$$

If we further require that $\bar{u}_0^{(s)}$ for $s \geq 1$ contains no multiple of $\bar{u}_0^{(0)}$ then in general we have

$$\bar{u}_0^{(s)}(\eta) = \bar{a}_s + O(\eta^2) \quad (s \geq 1). \qquad (37.19)$$

As will be seen shortly, $\bar{u}_0(\eta)$ and $\varepsilon \tilde{u}_0(\zeta)$ are merely different representations of the same solution and hence $\bar{a}_s \equiv \tilde{a}_{3s-1}(s \geq 1)$.

The other solution of equation (37.15), which will be denoted simply by $\bar{u}^{(0)}$, must necessarily have a logarithmic branch point at $\eta = 0$ (provided $U_c'' \neq 0$), and, if we normalize $\bar{u}^{(0)}$ so that

$$\bar{u}^{(0)}(\eta) \equiv \phi_2^{(0)}(z), \qquad (37.20)$$

then we have (cf. equation (22.26))

$$\bar{u}^{(0)}(\eta) = Q_2(\eta) + (U_c''/U_c')\bar{u}_0^{(0)}(\eta) \ln \eta, \qquad (37.21)$$

where

$$Q_2(\eta) = 1 + (7\gamma_0' - 29\gamma_0^2 + \alpha^2)\frac{\eta^2}{2!}$$
$$+ \left(\frac{9}{2}\gamma_0'' - \frac{15}{2}\gamma_0'\gamma_0 - \frac{207}{2}\gamma_0^3 - \frac{14}{3}\alpha^2\gamma_0\right)\frac{\eta^3}{3!} + O(\eta^4).$$

$$(37.22)$$

From equation (37.20) it follows that

$$Q_2(\eta) \equiv P_2(z) - (U_c''/U_c')\phi_1^{(0)}(z)\ln[\eta/(z-z_c)], \quad (37.23)$$

where $P_2(z)$ is the regular part of $\phi_2^{(0)}(z)$. In § 38 it will be found that the quantity $Q_2(\eta)$ plays an important role in the determination of the uniform approximations to the solutions of balanced type and for purposes of that discussion it is helpful to note that $Q_2(\eta)$ can also be defined as the solution of the inhomogeneous equation (cf. equation (22.28))

$$R_2 Q_2 = -(U_c''/U_c')\{2D - \eta^{-1}(1-g_0)\}\bar{u}_0^{(0)} \quad (37.24)$$

that satisfies the initial conditions $Q_2(0) = 1$ and $Q_2'(0) = 0$. Since $\mathcal{W}(\phi_1^{(0)}, \phi_2^{(0)})(z) = -1$, we see that

$$\mathcal{W}(\bar{u}_0^{(0)}, \bar{u}^{(0)})(\eta) = \eta'^{-1}. \quad (37.25)$$

The higher approximations to $\bar{u}(\eta)$ are not needed in the present theory and we shall therefore only note, because of its relevance to the central matching problem, that $\bar{u}^{(s)}$ for $s \geq 1$ has a pole of order $3s - 1$ at $\eta = 0$ and that

$$\lim_{\eta \to 0}\{\eta^{3s-1}\bar{u}^{(s)}(\eta)\} = \frac{(3s-2)!}{3^s s!}\frac{U_c''}{U_c'} \quad (s \geq 1). \quad (37.26)$$

The branch of $\bar{u}(\eta)$ can be chosen arbitrarily but we shall suppose for convenience, and without loss of generality, that ph η is restricted to the range $-\frac{3}{2}\pi < \text{ph } \eta < \frac{1}{2}\pi$, and we then regard $\bar{u}(\eta)$ as defined for all values of ph η in this range. This convention regarding the definition of $\bar{u}(\eta)$ does not imply the existence of a solution of equation (36.3) which is asymptotic to $\bar{u}(\eta)$ throughout this open sector of angle 2π; even in the Poincaré sense that would be false. There do exist, however, three solutions of equation (36.3) which are asymptotic (in the complete sense of Watson) to $\bar{u}(\eta)$ in non-overlapping sectors of angle $\frac{2}{3}\pi$ and this suggests that we define three outer expansions of balanced type according to the scheme

$$\bar{u}_k(\eta) = \bar{u}(\eta) \quad (\eta \in \mathbf{T}'_k). \tag{37.27}$$

The outer expansions of dominant-recessive type, which are often described as being of 'viscous' type, can easily be derived by the WKBJ method in which, as discussed in § 27, equation (36.3) is first transformed into a nonlinear third-order equation, the solutions of which are then formally expanded in powers of $\varepsilon^{3/2}$. A preliminary study of the general structure of these expansions, however, suggests that when $f_0 = 6\gamma$ a more systematic derivation can be obtained by introducing the functions

$$\bar{v}_\pm(\eta) = \tfrac{1}{2}\pi^{-1/2}\varepsilon^{5/4}(\eta\eta'^2)^{-5/4} \exp\left(\pm\tfrac{2}{3}\varepsilon^{-3/2}\eta^{3/2}\right)H(\eta), \tag{37.28}$$

where $H(\eta)$ will subsequently be expanded in an asymptotic power series the leading term of which is a constant. On substituting these expressions into equation (36.3) we find that H must satisfy a *linear* equation of the form

$$H' \pm \varepsilon^{3/2}\mathcal{H}_1(H) + \varepsilon^3\mathcal{H}_2(H) \pm \varepsilon^{9/2}\mathcal{H}_3(H) = 0. \tag{37.29}$$

When $f_1, g_0, \ldots, h_1$ have the forms given by equations (36.4) we find that

$$\begin{aligned}\mathcal{H}_1(H) = \eta^{-1/2}\{\tfrac{5}{2}H'' - (\tfrac{13}{4}\eta^{-1} + 4\gamma)H' \\ + (\tfrac{101}{32}\eta^{-2} + \tfrac{9}{2}\gamma\eta^{-1} - \tfrac{13}{4}\gamma' + \tfrac{23}{8}\gamma^2 - \tfrac{1}{2}\alpha^2\eta'^2)H\};\end{aligned} \tag{37.30}$$

the detailed forms of $\mathcal{H}_2(H)$ and $\mathcal{H}_3(H)$ are somewhat lengthy and are not needed for the present purposes. If we now expand H in the form

$$H(\eta, \varepsilon) = \sum_{s=0}^{\infty} (\pm 1)^s \varepsilon^{3s/2} H_s(\eta), \tag{37.31}$$

then

$$H'_0 = 0, \quad H'_1 = -\mathcal{H}_1(H_0), \ldots. \tag{37.32}$$

To this order the normalization of this expansion will be fixed by setting $H_0 = 1$ and by requiring that there is no constant term in the expansion of H_1 about $\eta = 0$. Thus we obtain

$$H_0(\eta) = 1$$

and

$$\left. \begin{array}{l} H_1(\eta) = \tfrac{101}{48}\eta^{-3/2} + 9\gamma\eta^{-1/2} \\[2mm] \qquad - \displaystyle\int_0^\eta (\tfrac{23}{4}\gamma' + \tfrac{23}{8}\gamma^2 - \tfrac12\alpha^2\eta'^{-2})\eta^{-1/2}\,\mathrm{d}\eta. \end{array} \right\} \quad (37.33)$$

To fix the branch of $\bar{v}_\pm(\eta)$ we shall continue to suppose that ph η is restricted to the range $-\tfrac32\pi < \mathrm{ph}\ \eta < \tfrac12\pi$ and consider $\bar{v}_\pm(\eta)$ as being defined for all values of ph η in this range. The outer expansions of dominant-recessive type can then be defined according to the scheme (cf. equations (A14))

$$\left. \begin{array}{ll} \bar{v}_1(\eta) = \bar{v}_-(\eta) & (\eta \in \mathbf{T}_2' \cup \mathbf{T}_3'), \\[1mm] \bar{v}_2(\eta) = i\bar{v}_+(\eta) & (\eta \in \mathbf{T}_3' \cup \mathbf{T}_1'), \\[1mm] \bar{v}_3(\eta) = -i\bar{v}_+(\eta) & (\eta \in \mathbf{T}_2'). \end{array} \right\} \quad (37.34)$$

On crossing the branch cut in the positive sense from $\mathbf{T}_2'$ to $\mathbf{T}_1'$ the value, though not the form, of $\bar{v}_3(\eta)$ must be preserved. Because of the presence of logarithms in the coefficients $H_s(\eta)$ for $s = 2, 4, \ldots$, some care is needed in specifying $\bar{v}_3(\eta)$ when $\eta \in \mathbf{T}_1'$. To this order, however, it is found that

$$\bar{v}_3(\eta) = -\bar{v}_-(\eta)\{1 + O(\varepsilon^3)\} \quad (37.35)$$

and this reflects the asymmetry of the problem which first becomes evident at order ε^3.

For purposes of evaluating $H_1(\eta)$ at the lower boundary η_1 ($\equiv \eta(z_1)$), it is more convenient to express it as a function of z. This can easily be done by comparing $\bar{v}_1(\eta)$ or $\bar{v}_2(\eta)$ with the results given in Problem 4.7 and this shows that

$$H_1(\eta) \equiv G_1(z), \quad (37.36)$$

where

$$G_1(z) = \left(\frac{101}{48}\frac{U'}{U-c} + \frac{23}{24}\frac{U''}{U'} \right)\left(\frac{U_c'}{U-c} \right)^{1/2}$$
$$\qquad - \int_{z_c}^z \left\{ \frac{23}{24}\left(\frac{U'''}{U'} - \frac{U''^2}{U'^2} \right) - \frac12\alpha^2 \right\}\left(\frac{U_c'}{U-c} \right)^{1/2}\,\mathrm{d}z.$$
$$\qquad (37.37)$$

37.3 *The central matching problem*

Thus far we have fixed the normalization of the inner and outer expansions independently, and we now wish to consider the matching of these expansions so that they provide approximations to the solutions (U_0, U_k, V_k). To simplify the present discussion we shall consider the four solutions U_0, U_3, V_1 and V_2 which form a 'numerically satisfactory' set when one boundary point lies in $\mathbf{T}_1$ and the other lies in $\mathbf{T}_2$ as occurs, for example, in the case of marginal stability. Let us first fix the normalization of these solutions in terms of their outer expansions. This agrees with the normalization of the uniform approximations given in § 38 and is more convenient than the normalization, in terms of the inner expansions, which was originally adopted by Reid (1972). Thus, for $|\eta| \gg |\varepsilon|$ we require that

$$\left.\begin{aligned}
U_0(\eta) &\sim \bar{u}_0(\eta) &&(\eta \in \mathbf{I}), \\
U_3(\eta) &\sim \bar{u}_3(\eta) &&(\eta \in \mathbf{T}'_3), \\
V_1(\eta) &\sim \bar{v}_1(\eta) &&(\eta \in \mathbf{I} - \mathbf{T}'_1),
\end{aligned}\right\} \tag{37.38}$$

and

$$V_2(\eta) \sim \bar{v}_2(\eta) \quad (\eta \in \mathbf{I} - \mathbf{T}'_2),$$

where the sectors of validity have again been restricted to insure completeness in the sense of Watson.

The matching of $\tilde{u}_0(\zeta)$ and $\bar{u}_0(\eta)$ is somewhat special since both expansions are convergent and, since they have been normalized in the same way, they are merely different representations of the same solution, i.e.

$$U_0(\eta) = \bar{u}_0(\eta) = \varepsilon \tilde{u}_0(\zeta) \quad (\eta \in \mathbf{I}). \tag{37.39}$$

The other central matching coefficients can now be determined by the use of the 'asymptotic matching principle' (Van Dyke 1975, Fraenkel 1969). For this purpose it is convenient to use the inner and outer expansion operators introduced by Fraenkel. Thus we let $H_p f$ and $E_p f$ denote the inner and outer expansions of f respectively up to and including terms of order ε^p. To describe the domains in which the matching takes place it is also convenient, following Fraenkel, to use Hardy's notation for relative orders of magnitude. Thus, if $f(\varepsilon) = o\{g(\varepsilon)\}$ as $\varepsilon \to 0$, then we write $f < g$ which may be read as 'f is of order less than g'; similarly, if $f(\varepsilon) = O\{g(\varepsilon)\}$ as $\varepsilon \to 0$, then we write $f \leqslant g$.

Because of the way in which we have defined the inner and outer expansions it is clear that for bounded values of $|\zeta|$ we must have

$$U_3(\eta) = \varepsilon a_3(\varepsilon)\tilde{u}_0(\zeta) + b_3(\varepsilon)\tilde{u}_3(\zeta), \left.\begin{array}{l} \\ \\ \end{array}\right\}$$
$$V_1(\eta) = c_1(\varepsilon)\tilde{v}_1(\zeta) \text{ and } V_2(\eta) = c_2(\varepsilon)\tilde{v}_2(\zeta), \quad (37.40)$$

where $a_3(\varepsilon)$, $b_3(\varepsilon)$, $c_1(\varepsilon)$, and $c_2(\varepsilon)$ are the central matching coefficients which must now be determined by matching the inner and outer expansions in their common domains of validity.

Consider first the determination of $a_3(\varepsilon)$ and $b_3(\varepsilon)$. When η lies in the domain $\mathbf{T}_3' \cap (|\varepsilon| < |\eta| < 1)$ the matching principle requires that

$$E_p H_q\{\varepsilon a_3(\varepsilon)\tilde{u}_0(\zeta) + b_3(\varepsilon)\tilde{u}_3(\zeta)\} - H_q E_p \bar{u}_3(\eta) = 0 \quad (37.41)$$

for suitably chosen values of p and q. A preliminary investigation shows that $a_3(\varepsilon)$ and $b_3(\varepsilon)$ can be expressed in terms of the asymptotic sequence

$$\ln \varepsilon, 1, \varepsilon^3 \ln^2 \varepsilon, \varepsilon^3 \ln \varepsilon, \varepsilon^3, \ldots \quad (37.42)$$

and, with $p = 2$ and $q = 3$, we then obtain

and
$$a_3(\varepsilon) = (U_c''/U_c')[\ln \varepsilon + \psi(2) + 2\pi i] + O(\varepsilon^3 \ln^2 \varepsilon) \left.\begin{array}{l} \\ \\ \end{array}\right\}$$
$$b_3(\varepsilon) = 1 + O(\varepsilon^3 \ln \varepsilon), \quad (37.43)$$

where $\psi(z) = \Gamma'(z)/\Gamma(z)$ is the digamma function.

Consider next the determination of the coefficients $c_1(\varepsilon)$ and $c_2(\varepsilon)$. In matching the expansions of dominant-recessive type it is necessary to make a minor modification in the outer expansion operator E_p to deal with the exponential factors which appear in $\bar{v}_k$ and in the asymptotic expansions of $\tilde{v}_k$. For this purpose let

$$G^{\pm} = \tfrac{1}{2}\pi^{-1/2}\zeta^{-5/4} \exp (\pm \tfrac{2}{3}\zeta^{3/2})$$
$$\equiv \tfrac{1}{2}\pi^{-1/2}\varepsilon^{5/4}\eta^{-5/4} \exp (\pm \tfrac{2}{3}\varepsilon^{-3/2}\eta^{3/2}), \quad (37.44)$$

where $-\tfrac{3}{2}\pi < \mathrm{ph}\, \eta < \tfrac{1}{2}\pi$, and then define the modified outer expansion operators $E_p^{\pm}$ by

$$E_p^{\pm} = E_p(1/G^{\pm}). \quad (37.45)$$

The coefficient $c_1(\varepsilon)$ can then be found by applying the matching principle in the form

$$E_p^- H_q \{c_1(\varepsilon) \tilde{v}_1(\varepsilon)\} - H_q E_p^- \bar{v}_1(\eta) = 0, \tag{37.46}$$

where η lies in the domain $(\mathbf{I} - \mathbf{T}_1') \cap (|\varepsilon| < |\eta| < 1)$, with a similar expression for the determination of $c_2(\varepsilon)$. A preliminary investigation shows that these coefficients can be expressed in terms of the asymptotic sequence

$$1, \varepsilon^3 \ln \varepsilon, \varepsilon^3, \varepsilon^6 \ln^2 \varepsilon, \varepsilon^6 \ln \varepsilon, \varepsilon^6, \ldots \tag{37.47}$$

and, with $p = \frac{3}{2}$ and $q = 2$, we then obtain

$$c_1(\varepsilon) = 1 + O(\varepsilon^3 \ln \varepsilon) \quad \text{and} \quad c_2(\varepsilon) = 1 + O(\varepsilon^3 \ln \varepsilon). \tag{37.48}$$

Thus, to this order, the inner and outer expansions of dominant-recessive type are automatically matched; this is, of course, simply a consequence of the way in which the two expansions were normalized.

37.4 Composite approximations

Approximations of this type can be obtained by combining the inner and outer expansions according to various rules (Van Dyke 1975). We shall denote the approximations, obtained by additive or multiplicative composition, by (cf. equation (27.56))

$$\mathbf{C}_{pq}^+ f = (E_p + H_q - E_p H_q) f \tag{37.49}$$

and

$$\mathbf{C}_{pq}^\times f = (E_p f)(H_q f)/E_p H_q f, \tag{37.50}$$

respectively. When dealing with the solutions of dominant-recessive type, however, the outer expansion operator E_p must be replaced by $G^\pm E_p^\pm$ with an appropriate choice of signs.

Consider first the well-balanced solution U_0. If $q \leq p$ it is easy to see that $(H_q - E_p H_q) \tilde{u}_0 = 0$ and hence that

$$\mathbf{C}_{pq}^+ U_0 = \mathbf{C}_{pq}^\times U_0 = E_p \bar{u}_0 \quad (q \leq p). \tag{37.51}$$

The outer expansion of U_0 is thus self-composite in the sense that it automatically contains the inner expansion of U_0.

Consider next the solution U_3 for which the rule for additive composition gives

$$\mathbf{C}_{00}^+ U_3 = \bar{u}^{(0)}(\eta) \tag{37.52}$$

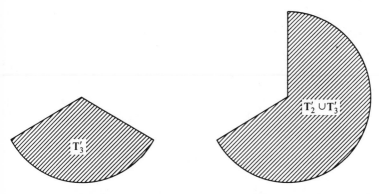

Fig. 5.2. The domains of validity of the composite approximations to U_3 (left) and V_1 (right) in the η-plane. (After Reid 1972.)

and

$$C_{11}^+ U_3 = \bar{u}^{(0)}(\eta) - (U_c''/U_c')\eta \ln \eta$$
$$- \varepsilon (U_c''/U_c')\{B_3(\zeta, 2, 1) - [\ln \varepsilon + \psi(2) + 2\pi i]\zeta\}. \quad (37.53)$$

These approximations are valid in the complete sense in the domain $\mathbf{T}_3'$ shown in Fig. 5.2 and the errors associated with them are $O(\varepsilon \ln \varepsilon)$ and $O(\varepsilon^2 \ln \varepsilon)$ respectively. The approximation (37.53) can also be written in the form

$$C_{11}^+ U_3 = Q_2(\eta) + \varepsilon \chi_2^{(1)}(\zeta) + (U_c''/U_c')\{Q_1(\eta) - 1\}\eta \ln \eta,$$
$$(37.54)$$

where $\chi_2^{(1)}(\zeta)$ is given by equation (27.52) with ξ replaced by ζ, and it is seen to have the same general structure as the approximation (27.57).

In the case of V_1, however, the composite approximations of additive type which have been given by Eagles (1969) and Reid (1972) cannot be used in the inner region $|\zeta| \leq K < \infty$ (see also De Villiers (1975)). For the solutions of dominant-recessive type, it seems preferable to use multiplicative composition and this gives

$$C_{00}^\times V_1 = \eta'^{-5/2} A_1(\zeta, 2) \quad (37.55)$$

which is seen to have precisely the same form as Tollmien's improved viscous approximation (27.42). From the present point of view, however, we would expect this approximation to be valid in the complete sense only in the domain $\mathbf{T}_2' \cup \mathbf{T}_3'$ shown in Fig. 5.2.

The method of matched asymptotic expansions thus provides a simple and systematic method of deriving composite approximations to the solutions of the Orr–Sommerfeld equation. But the composite approximations which are obtained by this method are not fully uniform, in the sense that they are not valid in a full neighbourhood of the critical point. They are valid, however, in certain sectorial neighbourhoods of the critical point which do not shrink to zero as $|\varepsilon| \to 0$.

38 Uniform approximations

We now wish to consider the derivation of 'first approximations' to the solutions of the Orr–Sommerfeld equation which are uniformly valid in a full neighbourhood of the critical point. The theory to be described here (Reid 1974b) was suggested in part by the results given in § 35 for plane Couette flow and in part by the structure of the inner expansions given in § 37.

38.1 *The solution of well-balanced type*

The required first approximation to U_0 is simply the first term in its outer expansion. Thus we have

$$U_0(\eta) = \bar{u}_0^{(0)}(\eta) + O(\varepsilon^3), \tag{38.1}$$

where $\bar{u}_0^{(0)}(\eta) \equiv \phi_1^{(0)}(z)$ is the regular solution of Rayleigh's equation.

38.2 *The solutions of balanced type*

To infer the general structure of the expansions for the solutions of balanced type we first recall from § 37 that the inner expansions of the solutions of this type can be expressed *to all orders* in terms of the generalized Airy functions $B_k(\zeta, p, q)$ ($p = 0, \pm1, \pm2, \ldots; q = 0, 1, 2, \ldots$). Alternatively, since these functions also satisfy the recursion formula (A32), it is sufficient to let p take on the values 2, 1, 0. The structure of the inner expansions then suggests that the uniform expansions must be of the form

$$U_k(\eta) = \mathscr{G}(\eta, \varepsilon) - \{\varepsilon\mathscr{A}(\eta, \varepsilon)B_k(\zeta, 2, 1) + \varepsilon^3\mathscr{B}(\eta, \varepsilon)B_k(\zeta, 1, 1)$$
$$+ \varepsilon^2\mathscr{C}(\eta, \varepsilon)B_k(\zeta, 0, 1)\}$$
$$- \varepsilon^3\{\varepsilon\mathscr{D}(\eta, \varepsilon)B_k(\zeta, 2, 2) + \varepsilon^3\mathscr{E}(\eta, \varepsilon)B_k(\zeta, 1, 2)$$
$$+ \varepsilon^2\mathscr{F}(\eta, \varepsilon)B_k(\zeta, 0, 2)\} - \cdots, \tag{38.2}$$

where all of the slowly varying coefficients have expansions of the form

$$\mathcal{A}(\eta, \varepsilon) = \sum_{s=0}^{\infty} \mathcal{A}_s(\eta)\varepsilon^{3s}. \tag{38.3}$$

In a 'first approximation' to $U_k(\eta)$ we have simply

$$U_k(\eta) \sim \mathcal{G}(\eta) - \{\varepsilon\mathcal{A}(\eta)B_k(\zeta, 2, 1) + \varepsilon^3\mathcal{B}(\eta)B_k(\zeta, 1, 1)$$
$$+ \varepsilon^2\mathcal{C}(\eta)B_k(\zeta, 0, 1)\}, \tag{38.4}$$

where we have now dropped the subscripts on the slowly varying coefficients. The corresponding approximation to $U'_k(\eta)$ is given by

$$U'_k(\eta) \sim \mathcal{G}' - \eta\mathcal{C} - \{\varepsilon(\mathcal{A}' - \mathcal{C})B_k(\zeta, 2, 1) + (\mathcal{A} + \eta\mathcal{C})B_k(\zeta, 1, 1)$$
$$+ \varepsilon^2(\mathcal{B} + \mathcal{C}')B_k(\zeta, 0, 1)\}. \tag{38.5}$$

In some circumstances the term involving $\varepsilon^3\mathcal{B}(\eta)$ in equation (38.4) can be omitted and the errors associated with these approximations are then nominally of order ε^3. The problem has been reduced therefore to the determination of the four slowly varying coefficients $\mathcal{A}$, $\mathcal{B}$, $\mathcal{C}$, and $\mathcal{G}$.

On substituting the expansion (38.2) into equation (36.3) we find, after a straightforward but rather lengthy calculation, that the differential equation is satisfied provided

$$R_2\mathcal{G} + 4\eta\mathcal{A}' + (1 + \eta f_0)\mathcal{A} + (2\eta D + \eta f_0 - g_0)(\eta\mathcal{C}) = 0, \tag{38.6}$$

$$\eta\mathcal{A}'' + (4 + g_0)\mathcal{A}' + (f_0 + h_0)\mathcal{A} + 2\eta\mathcal{C}' + (2 + \eta f_0 - g_0)\mathcal{C} = 0, \tag{38.7}$$

$$(2\eta D + \eta f_0 - g_0)(\mathcal{A} + \eta\mathcal{C}) = 0 \tag{38.8}$$

and

$$6\mathcal{A}'' + 3f_0\mathcal{A}' - f_1\mathcal{A} + 2\eta\mathcal{B}' + (1 + \eta f_0 - g_0)\mathcal{B}$$
$$+ 5\eta\mathcal{C}'' + (4 + 3\eta f_0 - g_0)\mathcal{C}' + (f_0 - \eta f_1 - h_0)\mathcal{C} = 0. \tag{38.9}$$

In spite of the complicated appearance of these equations, they are surprisingly easy to solve. Consider first equation (38.8) which can be integrated immediately to give

$$\mathcal{A}(\eta) + \eta\mathcal{C}(\eta) = \mathcal{A}(0)\eta'^{-5/2}. \tag{38.10}$$

Next, on eliminating $\mathscr{C}$ between equations (38.7) and (38.8), we have

$$R_2(\eta\mathscr{A}) = 0 \tag{38.11}$$

and the requirement that $\mathscr{A}$ be analytic at $\eta = 0$ implies that

$$\mathscr{A}(\eta) = \mathscr{A}(0)\eta^{-1}\bar{u}_0^{(0)}(\eta). \tag{38.12}$$

Equation (38.6) can then be simplified to

$$R_2\mathscr{G} = -\mathscr{A}(0)\{2D - \eta^{-1}(1-g_0)\}\bar{u}_0^{(0)}. \tag{38.13}$$

It is easy to show that the solution of this equation will be analytic at $\eta = 0$ if and only if $\mathscr{A}(0) = U_c''/U_c'$ and, on comparing equation (38.13) with equation (37.24), we then see that the general solution of equation (38.13) which is analytic at $\eta = 0$ must be of the form

$$\mathscr{G}(\eta) = Q_2(\eta) + C\bar{u}_0^{(0)}(\eta), \tag{38.14}$$

where C is an arbitrary constant. Finally, $\mathscr{B}(\eta)$ can be found from equation (38.9) by quadrature but the calculations involved are somewhat lengthy and not very illuminating. Alternatively, it can be found by a simple matching technique which was first suggested by Lin (1957b) in a slightly different context and which will be discussed in connexion with the solutions of dominant-recessive type.

The normalization of these approximations is completely fixed once we have specified the value of the constant C which appears in equation (38.14). The normalization, in terms of the inner expansions, adopted by Reid (1972) corresponds to setting $C = 0$. For the present purposes, however, a different choice for C is preferable. Thus, consider the solution U_3 which will eventually be used in the eigenvalue relation. When η lies in the domain $\mathbf{T}_3' \cap (|\varepsilon| < |\eta| \leqslant 1)$ we have the outer expansion

$$\begin{aligned}
E_2\{U_3(\eta)\} &= \mathscr{G} + \eta\mathscr{A}\ln\eta - \eta\mathscr{A}[\ln\varepsilon + \psi(2) + 2\pi i] \\
&\equiv \bar{u}^{(0)}(\eta) + \{C - (U_c''/U_c')[\ln\varepsilon + \psi(2) + 2\pi i]\}\bar{u}_0^{(0)}(\eta)
\end{aligned} \tag{38.15}$$

and if we require that this expansion contain no multiple of $\bar{u}_0^{(0)}$ then we must have

$$C = (U_c''/U_c')[\ln\varepsilon + \psi(2) + 2\pi i]. \tag{38.16}$$

This last condition could also be used, once $\mathscr{A}$ is known, to provide an alternate derivation of $\mathscr{G}$.

38.3 *The solutions of dominant-recessive type*

Approximations to the solutions of dominant-recessive type can be obtained directly from the corresponding approximations to the solutions of balanced type by using the connexion formulae (A43). Although this approach is attractive from a theoretical viewpoint, it has two minor disadvantages. First, it leads to a somewhat unnatural normalization of the approximations and, second, it is not applicable to plane Couette flow which has no solutions of balanced type. Both of these difficulties can be avoided, however, by starting with a 'first approximation' of the form (cf. equation (34.15))

$$V_k(\eta) \sim a(\eta)A_k(\zeta, 2) + \varepsilon^2 b(\eta)A_k(\zeta, 1) + \varepsilon\, c(\eta)A_k(\zeta, 0).$$
(38.17)

The corresponding approximation to V'_k is given by

$$V'_k(\eta) \sim (a' - c)A_k(\zeta, 2) + \varepsilon^{-1}(a + \eta c)A_k(\zeta, 1) + \varepsilon(b + c')A_k(\zeta, 0).$$
(38.18)

Again, for some purposes, the term involving $\varepsilon^2 b(\eta)$ in equation (38.17) can be omitted.

On substituting equation (38.17) into equation (36.3) we find that the coefficients a, b and c satisfy precisely the same differential equations as $\mathscr{A}$, $\mathscr{B}$ and $\mathscr{C}$. Thus we have immediately

$$a(\eta) = a(0)\eta^{-1}\bar{u}_0^{(0)}(\eta) \quad \text{and} \quad a(\eta) + \eta c(\eta) = a(0)\eta'^{-5/2},$$
(38.19)

and we now wish to consider an alternate method of determining b which avoids having to solve equation (38.9) and which provides greater insight into the structure and role of this coefficient. For this purpose consider the outer expansion of V_1 (say). When η lies in the domain $(\mathbf{I} - \mathbf{T}'_1) \cap (|\varepsilon| < |\eta| \leqslant 1)$ we have

$$V_1(\eta) \sim \tfrac{1}{2}\pi^{-1/2}\varepsilon^{5/4}\eta^{-5/4} \exp\left(-\tfrac{2}{3}\varepsilon^{-3/2}\eta^{3/2}\right)\{a + \eta c$$
$$- \varepsilon^{3/2}(\tfrac{101}{48}\eta^{-3/2}a + \eta^{1/2}b + \tfrac{5}{48}\eta^{-1/2}c) + O(\varepsilon^3)\}.$$
(38.20)

On the other hand, if we identify the outer expansion of V_1 with $\bar{v}_1$, thereby fixing the normalization of V_1, then we have

$$V_1(\eta) \sim \tfrac{1}{2}\pi^{-1/2}\varepsilon^{5/4}(\eta\eta'^2)^{-5/4} \exp\left(-\tfrac{2}{3}\varepsilon^{-3/2}\eta^{3/2}\right)$$
$$\times \{1 - \varepsilon^{3/2}H_1(\eta) + O(\varepsilon^3)\},$$
(38.21)

where H_1 is given by equations (37.33) and (37.36). Since these two expressions for the outer expansion of V_1 must be identical, we have

$$a(\eta) + \eta\, c(\eta) = \eta'^{-5/2} \tag{38.22}$$

which shows that $a(0) = 1$. Similarly, we have

$$\tfrac{101}{48}\eta^{-3/2} a(\eta) + \eta^{1/2}\ell(\eta) + \tfrac{5}{48}\eta^{-1/2} c(\eta) = \eta'^{-5/2} H_1(\eta), \tag{38.23}$$

which gives

$$\ell(\eta) = 2\eta^{-1} c(\eta) + \eta'^{-5/2}\eta^{-1/2}\{H_1(\eta) - \tfrac{101}{48}\eta^{-3/2}\}, \tag{38.24}$$

and it is not difficult to verify that ℓ is analytic at $\eta = 0$. Thus, ℓ is essentially a regularized form of the coefficient $H_1(\eta)$ which appears in the outer expansions of dominant-recessive type.

From these results we see that

$$(\mathscr{A}, \mathscr{B}, \mathscr{C}) = (U_c''/U_c')(a, \ell, c). \tag{38.25}$$

Hence, on using the connexion formulae satisfied by the generalized Airy functions we immediately obtain the following first approximations to the connexion formulae satisfied by the solutions:

$$\left.\begin{aligned}
U_1 - U_2 &= -2\pi i\varepsilon\,(U_c''/U_c')\,V_3, \\
U_2 - U_3 &= -2\pi i\varepsilon\,(U_c''/U_c')\,V_1, \\
\sum_{k=1}^{3} V_k &= -\varepsilon^{-1}\bar{u}_0^{(0)}.
\end{aligned}\right\} \tag{38.26}$$

and

Thus, so long as we consider only 'first approximations', the theory is particularly simple. To this order the slowly varying coefficients in the approximations can be expressed in terms of $\phi_1^{(0)}(z)$, the regular part of $\phi_2^{(0)}(z)$, and a regularized form of $G_1(z)$, and the rapidly varying terms in the approximations can be expressed in terms of the solutions of the equations $u''' - \zeta u' = 0$ and $u''' - \zeta u' = 1$. Most of the essential elements in the theory are thus well known from the older heuristic theories but the general structure of the uniform approximations is substantially different.

39 A comparison with Lin's theory

Uniform approximations to the solutions of equation (36.3) have been derived by Lin (1957a,b, 1958) by a method which would appear to be quite different from the one described in § 38, and it is of some interest, therefore, to consider the relationship between these two approaches. Lin supposed that the transformed equation was in normal form, but that is not essential for the present discussion. According to Lin's theory, asymptotic solutions of equation (36.3) can be found in the form

$$\chi = Au + Bu' + \varepsilon^3 (Cu'' + Du'''), \tag{39.1}$$

where u is a solution of the comparison equation

$$\varepsilon^3 u^{iv} - (\eta u'' + \alpha u' + \beta u) = 0. \tag{39.2}$$

In equation (39.1), the slowly varying coefficients A, B, C and D all have expansions of the form

$$A \equiv A(\eta, \varepsilon) = \sum_{s=0}^{\infty} A_s(\eta)\varepsilon^{3s}. \tag{39.3}$$

Similarly, in equation (39.2), α and β must also have expansions of the form

$$\alpha \equiv \alpha(\varepsilon) = \sum_{s=0}^{\infty} \alpha_s \varepsilon^{3s}. \tag{39.4}$$

The coefficients A_s, B_s, C_s and D_s can be determined by deriving and then solving the differential equations which they satisfy. In doing so, however, the constants α_s and β_s must be chosen so that the solutions are all analytic at $\eta = 0$. Alternatively, as Lin (1957b) has shown, these difficulties can be avoided by the use of a simple 'matching' technique. (The matching here is not of the term-by-term type used in the usual method of matched asymptotic expansions.) In this latter approach, outer expansions to the solutions of the comparison equation are substituted into the expansion (39.1) and the results are then matched to the corresponding outer expansions to the solutions of equation (36.3). Although this method has been formally successful, nevertheless there are two remaining difficulties which have prevented it from being used in actual calculations. One of these difficulties is concerned with the truncation of the expansion (39.1) and the other with the approximation of the solutions of the comparison equation (39.2).

Rabenstein (1958) has shown that the solutions of the comparison equation (39.2) can be expressed in terms of Laplace integrals, but these are of too complicated a form to be used directly in the expansion (39.1). From the integral representation of the solutions he derived inner and outer expansions which were, of course, automatically matched. Neither expansion, however, is uniformly valid and hence neither expansion is suitable for use in equation (39.1). What are needed here are uniform approximations to the solutions of the comparison equation, and these provide the necessary link between Lin's method and the method, based on the use of generalized Airy functions, which was described in § 38.

For the present purposes it is sufficient to consider three solutions of equation (39.2) which will be denoted by u_0, u_3, and v_1 where, as the notation suggests, u_0 is well balanced, u_3 is balanced in $\mathbf{T}_3'$, and v_1 is recessive in $\mathbf{S}_1'$. In deriving 'first approximations' to these solutions it is necessary only to determine the values of α_0 and β_0. A comparison of equation (39.2) with equation (36.3) suggests that

$$\alpha_0 = g_0(0) = 0 \quad \text{and} \quad \beta_0 = h_0(0) = -U_c''/U_c'. \quad (39.5)$$

These are, in fact, the correct choices for α_0 and β_0 (Lin 1957b) and, to simplify the present discussion, we shall adopt these values at the outset.

Consider then the outer expansions of the solutions which can be obtained either from their integral representations (see Problem 5.4) or by the formal methods described in § 37. Thus, if we let

$$u_0(\eta) = u_0^{(0)}(\eta) + O(\varepsilon^3) \quad (\eta \in \mathbf{I}) \quad (39.6)$$

and

$$u_3(\eta) = u_3^{(0)}(\eta) + O(\varepsilon^3) \quad (\eta \in \mathbf{T}_3'), \quad (39.7)$$

then $u_0^{(0)}$ and $u_3^{(0)}$ satisfy the reduced form of the comparison equation, i.e. $\eta u'' + \beta_0 u = 0$, and if they are normalized in the same way as equations (37.18) and (37.21) then we have

$$u_0^{(0)}(\eta) = \eta q_1(\eta) \quad (39.8)$$

and

$$u_3^{(0)}(\eta) = q_2(\eta) + (U_c''/U_c')u_0^{(0)}(\eta)\ln\eta, \qquad (39.9)$$

where

$$q_1(\eta) = (\beta_0\eta)^{-1/2}J_1(2\beta_0^{1/2}\eta^{1/2})$$
$$= \sum_{s=0}^{\infty} \frac{(-1)^s}{s!\,(s+1)!}(\beta_0\eta)^s \qquad (39.10)$$

and

$$q_2(\eta) = (\beta_0\eta)^{1/2}\{[2\gamma-1+\ln(\beta_0\eta)]J_1(2\beta_0^{1/2}\eta^{1/2})$$
$$- \pi Y_1(2\beta_0^{1/2}\eta^{1/2})\}$$
$$= 1 + \sum_{s=1}^{\infty} \frac{(-1)^s}{s!\,(s+1)!}\left\{2[\psi(s+1)+\gamma]-\frac{s}{s+1}\right\}(\beta_0\eta)^{s+1}.$$
$$(39.11)$$

We also note that $\mathcal{W}(u_0^{(0)}, u_3^{(0)}) = -1$ and that the regular part of $u_3^{(0)}$ contains no multiple of $u_0^{(0)}$. From these results we can now determine A_0 and B_0 by matching. Thus, by considering the solutions U_0 and U_3 we have

$$E_2\{U_0(\eta)\} = \bar{u}_0^{(0)}(\eta)$$
$$= A_0 u_0^{(0)}(\eta) + B_0 u_0^{(0)\prime}(\eta) \quad (\eta\in\mathbf{I}) \quad (39.12)$$

and

$$E_2\{U_3(\eta)\} = \bar{u}_3^{(0)}(\eta)$$
$$= A_0 u_3^{(0)}(\eta) + B_0 u_3^{(0)\prime}(\eta) \quad (\eta\in\mathbf{T}_3') \quad (39.13)$$

from which we obtain

$$A_0(\eta) = \bar{u}_3^{(0)}(\eta)u_0^{(0)\prime}(\eta) - \bar{u}_0^{(0)}(\eta)u_3^{(0)\prime}(\eta) \qquad (39.14)$$

and

$$B_0(\eta) = \bar{u}_0^{(0)}(\eta)u_3^{(0)}(\eta) - \bar{u}_3^{(0)}(\eta)u_0^{(0)}(\eta). \qquad (39.15)$$

The results can also be expressed in the alternate forms

$$A_0(\eta) = Q_2(\eta)\{\eta q_1'(\eta)+q_1(\eta)\} - \eta Q_1(\eta)\{q_2'(\eta)-\beta_0 q_1(\eta)\}$$
$$= 1 + O(\eta^2) \qquad (39.16)$$

and

$$B_0(\eta) = \eta\{Q_1(\eta)q_2(\eta) - Q_2(\eta)q_1(\eta)\}$$
$$= \tfrac{1}{10}\beta_0\eta^2 + O(\eta^3) \tag{39.17}$$

which show explicitly that A_0 and B_0 are analytic at $\eta = 0$; this is, of course, a direct consequence of equations (39.5).

Consider next the outer expansion of v_1 which, with the usual normalization, is given by

$$v_1(\eta) = \tfrac{1}{2}\pi^{-1/2}\varepsilon^{5/4}\eta^{-5/4}\exp\left(-\tfrac{2}{3}\varepsilon^{-3/2}\eta^{3/2}\right)\sum_{s=0}^{\infty}(-1)^s\varepsilon^{3s/2}h_s(\eta),$$
$$\tag{39.18}$$

where

$$h_0(\eta) = 1, \quad h_1(\eta) = \tfrac{101}{48}\eta^{-3/2} - \beta_0\eta^{-1/2},\ldots. \tag{39.19}$$

If this expansion for v_1 and the corresponding expansions for its derivatives are substituted into equation (39.1) then we obtain

$$V_1(\eta) = \tfrac{1}{2}\pi^{-1/2}\varepsilon^{5/4}\eta^{-5/4}\exp\left(-\tfrac{2}{3}\varepsilon^{-3/2}\eta^{3/2}\right)\{-\varepsilon^{-3/2}\eta^{1/2}(B_0 + \eta D_0)$$
$$+ A_0 - \tfrac{5}{4}\eta^{-1}B_0 + \eta C_0 - \tfrac{9}{4}D_0 + \eta^{1/2}h_1(B_0 + \eta D_0)$$
$$- \varepsilon^{3/2}[\eta^{1/2}(B_1 + \eta D_1) - 2\eta^{-1/2}C_0 + \tfrac{101}{16}\eta^{-3/2}D_0$$
$$+ h_1(A_0 - \tfrac{5}{4}\eta^{-1}B_0 + \eta C_0 - \tfrac{9}{4}D_0)$$
$$+ h_1'(B_0 + 3\eta D_0) + h_2\eta^{1/2}(B_0 + \eta D_0)]$$
$$+ O(\varepsilon^3)\} \quad (\eta \in \mathbf{I} - \mathbf{T}_1'). \tag{39.20}$$

This expression for $V_1(\eta)$ must now be matched to the outer expansion $\bar{v}_-(\eta)$ given by equation (37.28). To lowest order this gives

$$B_0 + \eta D_0 = 0, \tag{39.21}$$

so that the terms involving h_2 drop out and, on using the explicit form for h_1, equation (39.20) can be simplified to

$$V_1(\eta) = \tfrac{1}{2}\pi^{-1/2}\varepsilon^{5/4}\eta^{-5/4}\exp\left(-\tfrac{2}{3}\varepsilon^{-3/2}\eta^{3/2}\right)[A_0 + \eta C_0 - D_0$$
$$- \varepsilon^{3/2}\{\tfrac{101}{48}\eta^{-3/2}(A_0 + \eta C_0 - D_0)$$
$$- \eta^{-1/2}[\beta_0(A_0 + \eta C_0 - 2D_0) + 2C_0] + \eta^{1/2}(B_1 + \eta D_1)\}$$
$$+ O(\varepsilon^3)] \quad (\eta \in \mathbf{I} - \mathbf{T}_1'). \tag{39.22}$$

Continuing the matching then gives

$$C_0 = \eta^{-1}(\eta'^{-5/2} - A_0 + D_0) \tag{39.23}$$

and

$$B_1 + \eta D_1 = \eta^{-1}\{\beta_0(\eta'^{-5/2} - D_0) + 2C_0\}$$
$$+ \eta'^{-5/2}\eta^{-1/2}\{H_1(\eta) - \tfrac{101}{48}\eta^{-3/2}\}, \tag{39.24}$$

and it can easily be verified that these expressions are both analytic at $\eta = 0$.

Consider next the uniform approximations to the solutions u_3 and v_1 of equation (39.2). According to the theory described in the previous section, these solutions must have uniform first approximations of the form

$$u_3(\eta) \sim g(\eta) + \beta_0\{\varepsilon a(\eta)B_3(\zeta, 2, 1) + \varepsilon^3 b(\eta)B_3(\zeta, 1, 1)$$
$$+ \varepsilon^2 c(\eta)B_3(\zeta, 0, 1)\} \tag{39.25}$$

and

$$v_1(\eta) \sim a(\eta)A_1(\zeta, 2) + \varepsilon^2 b(\eta)A_1(\zeta, 1) + \varepsilon c(\eta)A_1(\zeta, 0), \tag{39.26}$$

and a short calculation then shows that the slowly varying coefficients in these approximations are given by

$$\left.\begin{array}{c} a(\eta) = q_1(\eta), \quad a(\eta) + \eta c(\eta) = 1, \\ b(\eta) = \eta^{-1}[2c(\eta) - \beta_0] \\ g(\eta) = q_2(\eta) - \beta_0\eta q_1(\eta)[\ln \varepsilon + \psi(2) + 2\pi i], \end{array}\right\} \tag{39.27}$$

and

where $q_1(\eta)$ and $q_2(\eta)$ are given by equations (39.10) and (39.11).

If, for example, the approximation (39.26) to $v_1(\eta)$ and the corresponding approximations to its derivatives are substituted into the expansion (39.1) then, in a first approximation to $V_1(\eta)$, we obtain

$$V_1(\eta) \sim \{aA_0 + (a' - c)B_0 - (a + \eta c)D_0\}A_1(\zeta, 2)$$
$$+ \varepsilon^2\{bA_0 + b'B_0 + (2a' + \eta b + 2\eta c')C_0$$
$$+ (3a'' + 3\eta b' + 3\eta c'')D_0 + (a + \eta c)(B_1 + \eta D_1)\}A_1(\zeta, 1)$$
$$+ \varepsilon\{cA_0 + (b + c')B_0 + (a + \eta c)C_0 + (3a' + \eta b$$
$$+ 3\eta c' + c)D_0\}A_1(\zeta, 0). \tag{39.28}$$

To simplify this result we use the fact that $\mathscr{W}(u_0^{(0)}, u_3^{(0)}) = -1$ or, alternatively, that

$$\eta \mathscr{W}(q_1, q_2) - q_1 q_2 - \beta_0 \eta q_1^2 = -1. \tag{39.29}$$

It is then found, after a somewhat lengthy calculation, that equation (39.28) reduces to equation (38.17) with $k = 1$. Similarly, if the approximation (39.25) to $u_3(\eta)$ and the corresponding approximations to its derivatives are used in the expansion (39.1) then we recover equation (38.4) with $k = 3$.

In this way, therefore, we can pass directly from the uniform approximations to the solutions of the comparison equation to the corresponding uniform approximations to the solutions of the Orr–Sommerfeld equation. Thus, we can conclude that Lin's theory leads to uniform 'first approximations' which are identical to those derived in § 38 by a simpler and more direct method. A similar conclusion would be expected to hold for higher approximations though it is doubtful that they would ever be needed.

40 Preliminary simplification of the eigenvalue relation

In § 28 we considered a number of heuristic approximations to the eigenvalue relation and we now wish to consider the derivation of a 'first approximation' to the eigenvalue relation which is uniformly valid along the entire curve of marginal stability. Before doing so, however, it is helpful to make certain preliminary approximations which greatly simplify the subsequent analysis.

The precise form of the eigenvalue relation depends, of course, on the class of basic flows considered and, for simplicity, we shall consider only symmetrical flows in a channel. We will suppose further that $U'(z) > 0$ on the interval (z_1, z_2) and, without loss of generality, that $U(z_1) = 0$ and $U(z_2) = 1$. An important flow of this type is plane Poiseuille flow for which $U(z) = 1 - z^2$ with $z_1 = -1$ and $z_2 = 0$, and we shall continue to use this flow for illustrative purposes. As in § 28, we will consider only the even solutions which must then satisfy the boundary conditions

$$\phi = D\phi = 0 \text{ at } z = z_1 \quad \text{and} \quad D\phi = D^3\phi = 0 \text{ at } z = z_2. \tag{40.1}$$

For flows of the boundary-layer type, however, the approximations derived in § 38 must be modified so as to remain uniformly valid as $z \to \infty$.

Now let $\phi_i(z)$ $(i = 1, 2, 3, 4)$ denote a linearly independent set of solutions of the Orr–Sommerfeld equation. The general solution can then be written in the form

$$\phi(z) = \sum_{i=1}^{4} C_i \phi_i(z). \tag{40.2}$$

On imposing the boundary conditions (40.1), we obtain a system of four linear homogeneous equations for the constants C_i. For a non-trivial solution, the determinant of the coefficients must vanish and this leads to the eigenvalue relation

$$\mathscr{F}(\alpha, c, \varepsilon) = \begin{vmatrix} \phi_{11} & \phi_{21} & \phi_{31} & \phi_{41} \\ \phi'_{11} & \phi'_{21} & \phi'_{31} & \phi'_{41} \\ \phi_{12} & \phi_{22} & \phi_{32} & \phi_{42} \\ \phi'''_{12} & \phi'''_{22} & \phi'''_{32} & \phi'''_{42} \end{vmatrix} = 0, \tag{40.3}$$

where $\phi_{i\nu} = \phi_i(z_\nu)$ $(i = 1, 2, 3, 4; \nu = 1, 2)$. If we now let $\mathscr{W}(u, v)(z)$ denote the usual second-order Wronskian of $u(z)$ and $v(z)$ then, by Laplace's expansion of a determinant by complementary minors, we obtain

$$\mathscr{F}(\alpha, c, \varepsilon) = \mathscr{W}(\phi_2, \phi_3)(z_1) \frac{\mathrm{d}}{\mathrm{d}z} \mathscr{W}(\phi'_1, \phi'_4)(z_2)$$

$$- \mathscr{W}(\phi_1, \phi_3)(z_1) \frac{\mathrm{d}}{\mathrm{d}z} \mathscr{W}(\phi'_2, \phi'_4)(z_2)$$

$$+ \mathscr{W}(\phi_1, \phi_2)(z_1) \frac{\mathrm{d}}{\mathrm{d}z} \mathscr{W}(\phi'_3, \phi'_4)(z_2)$$

$$+ \mathscr{W}(\phi_3, \phi_4)(z_1) \frac{\mathrm{d}}{\mathrm{d}z} \mathscr{W}(\phi'_1, \phi'_2)(z_2)$$

$$- \mathscr{W}(\phi_2, \phi_4)(z_1) \frac{\mathrm{d}}{\mathrm{d}z} \mathscr{W}(\phi'_1, \phi'_3)(z_2)$$

$$+ \mathscr{W}(\phi_1, \phi_4)(z_1) \frac{\mathrm{d}}{\mathrm{d}z} \mathscr{W}(\phi'_2, \phi'_3)(z_2) = 0. \tag{40.4}$$

The importance of being able to express the eigenvalue relation in terms of these second-order Wronskians lies in the fact that if $u(z)$

and $v(z)$ denote *any* two solutions of the Orr–Sommerfeld equation then it is possible, as shown by Lakin, Ng & Reid (1978), to derive a pair of coupled third-order equations for $\mathcal{W}(u, v)(z)$ and $\mathcal{W}(u', v')(z)$. From these equations it would then be possible, at least in principle, to derive approximations to the Wronskians and hence to the eigenvalue relation without having to obtain approximations to the solutions themselves.

Consider then the possibility of simplifying the eigenvalue relation (40.4). For this purpose it necessary to characterize the asymptotic behaviour of the solutions as $\varepsilon \to 0$ and, as in § 27, we shall suppose that ϕ_1 is well balanced, that ϕ_2 is balanced at z_1 and z_2, and that ϕ_3 and ϕ_4 are recessive at z_2 and z_1 respectively. The first simplification of equation (40.4) we wish to consider is based on the observation, which follows from equation (27.13), that ϕ_3 and ϕ_4 are necessarily dominant at z_1 and z_2 respectively, and hence the first two terms in equation (40.4) are also dominant, the second two terms are balanced and the last two terms are recessive. Thus, with an exponentially small error, we can approximate the eigenvalue relation by

$$\mathcal{G}(\alpha, c, \varepsilon) = \mathcal{W}(\phi_2, \phi_3)(z_1) \frac{\mathrm{d}}{\mathrm{d}z} \mathcal{W}(\phi_1', \phi_4')(z_2)$$

$$- \mathcal{W}(\phi_1, \phi_3)(z_1) \frac{\mathrm{d}}{\mathrm{d}z} \mathcal{W}(\phi_2', \phi_4')(z_2) = 0. \quad (40.5)$$

This approximation is equivalent, as Lin (1955), pp. 35–36 has shown, to neglecting ϕ_{32}' and ϕ_{32}''' compared to ϕ_{31} and ϕ_{31}' in equation (40.3) and similarly neglecting ϕ_{41} and ϕ_{41}' compared to ϕ_{42}' and ϕ_{42}'''.

To obtain a more precise estimate of the error associated with this approximation we first note that the largest error arises from the neglect of ϕ_{32}''' compared to ϕ_{31}'. Thus, consider the ratio ϕ_{32}'''/ϕ_{31}' which can easily be estimated by using the WKBJ approximation (27.13) to ϕ_3. If $0 < c < 1$, with c bounded away from both 0 and 1, then a simple calculation gives

$$|\phi_{32}'''/\phi_{31}'| \sim (1-c)^{1/4} c^{3/4} \alpha R \exp\{-(\tfrac{1}{2}\alpha R)^{1/2} \lambda(c)\}, \quad (40.6)$$

where

$$\lambda(c) = \int_{z_1}^{z_2} |U - c|^{1/2} \,\mathrm{d}z. \quad (40.7)$$

To obtain a numerical estimate of equation (40.6) consider plane Poiseuille flow for which

$$\lambda(c) = \tfrac{1}{2}\sqrt{c} - \tfrac{1}{2}(1-c)\tanh^{-1}\sqrt{c} + \tfrac{1}{4}\pi(1-c). \qquad (40.8)$$

For values of α, c, and R near the nose of the marginal curve, equation (40.6) then gives $|\phi_{32}'''/\phi_{31}'| \approx 3 \times 10^{-12}$ which is many orders of magnitude smaller than any of the terms retained in the subsequent analysis.

On the lower branch of the curve of marginal stability α and c both tend to zero as $R \to \infty$ and in this case it is necessary to approximate ϕ_{31}' by the first term of its inner expansion. A short calculation then gives

$$|\phi_{32}'''/\phi_{31}'| \sim \text{constant} \times R^{9/14} \exp(-\text{constant} \times R^{3/7})$$
$$(40.9)$$

as $R \to \infty$ along the lower branch of the curve of marginal stability. The constants appearing in this equation are both positive and independent of R. They can be evaluated in terms of

$$U'(z_1), \quad U''(z_1) \quad \text{and} \quad \int_{z_1}^{z_2} U^2 \, dz,$$

but the results are somewhat complicated and are not needed for the present purposes. For flows without inflexion points, α and c also tend to zero as $R \to \infty$ on the upper branch of the curve of marginal stability. In this case, however, equation (40.6) remains valid and leads to the estimate

$$|\phi_{32}'''/\phi_{31}'| \sim \text{constant} \times R^{17/22} \exp(-\text{constant} \times R^{5/11})$$
$$(40.10)$$

as $R \to \infty$ along the upper branch of the curve of marginal stability.

Thus, when $0 \leqslant c < 1$, the error associated with the approximation (40.5) is exponentially small as $R \to \infty$ and it is very small numerically for values of R near the nose of the marginal curve. When $c \uparrow 1$, however, we have a situation where two simple turning points coalesce to form a single turning point of second order at $z = z_2$. This situation lies outside the scope of the present theory and does not occur anywhere on the curve of marginal stability.

In discussing the next approximation it is convenient to rewrite equation (40.5) in a somewhat different form. For this purpose let

$$\Phi = A\phi_1 + \phi_2 \quad \text{and} \quad \hat{\Phi} = B\phi_1 + \phi_2, \qquad (40.11)$$

where

and
$$\left. \begin{aligned} A(\alpha, c, \varepsilon) &= -\phi'_{22}/\phi'_{12} \\ B(\alpha, c, \varepsilon) &= -\phi'''_{22}/\phi'''_{12}. \end{aligned} \right\} \qquad (40.12)$$

Equation (40.5) can then be written in the equivalent form

$$\mathcal{H}(\alpha, c, \varepsilon) = \mathcal{W}(\Phi, \phi_3)(z_1) - \frac{\phi'''_{12}}{\phi'_{12}} \frac{\phi'_{42}}{\phi'''_{42}} \mathcal{W}(\hat{\Phi}, \phi_3)(z_1) = 0. \qquad (40.13)$$

We will now assume that z_c is closer to z_1 than to z_2, as it happens to be in most problems. This assumption then permits the neglect of those terms in the asymptotic expansions of ϕ_2 and ϕ_4 near z_2 which are recessive on a length scale $z_2 - z_c$. The distinction here is again between expansions which are valid in the usual Poincaré sense as opposed to being valid in the complete sense of Watson (Olver 1974, p. 543). The outer expansions of ϕ_1 and ϕ_4 then show that

and
$$\left. \begin{aligned} \frac{\phi'''_{12}}{\phi'_{12}} &= \alpha^2 + \frac{U''(z_2)}{1-c} + O(\varepsilon^3) \\ \frac{\phi'_{42}}{\phi'''_{42}} &= \frac{U'_c}{1-c} \varepsilon^3 + O(\varepsilon^{9/2}). \end{aligned} \right\} \qquad (40.14)$$

We also see that the quantities A and B defined by equations (40.12) must have expansions of the form

and
$$\left. \begin{aligned} A(\alpha, c, \varepsilon) &= \sum_{s=0}^{\infty} \varepsilon^{3s} A^{(s)}(\alpha, c) \\ B(\alpha, c, \varepsilon) &= \sum_{s=0}^{\infty} \varepsilon^{3s} B^{(s)}(\alpha, c), \end{aligned} \right\} \qquad (40.15)$$

where

$$A^{(0)}(\alpha, c) = B^{(0)}(\alpha, c) = -\phi_{22}^{(0)'}/\phi_{12}^{(0)'}. \qquad (40.16)$$

This last result shows that $\mathcal{W}(\hat{\Phi}, \phi_3)(z_1) = \mathcal{W}(\Phi, \phi_3)(z_1) + O(\varepsilon^3)$ and hence, with a relative error of order ε^6, we can approximate the eigenvalue relation by

$$\Delta(\alpha, c, \varepsilon) = \mathcal{W}(\Phi, \varepsilon\phi_3)(z_1) = 0, \qquad (40.17)$$

where the factor ε has been introduced for scaling purposes. This form of the eigenvalue relation is exact for flows of the boundary-layer type provided A is determined from the condition that Φ is bounded as $z \to \infty$.

In the subsequent discussion of the various approximations to equation (40.17) it is convenient to let

$$\Delta(z) \equiv \Delta(\alpha, c, \varepsilon; z) = \mathscr{W}(\Phi, \varepsilon\phi_3)(z) \qquad (40.18)$$

so that the eigenvalue relation in this approximation becomes simply $\Delta(z_1) = 0$. Our problem then is to derive a 'first approximation' to $\Delta(z)$ which is uniformly valid in a domain of the (complex) z-plane containing z_1.

41 The uniform approximation to the eigenvalue relation

We now wish to derive a first approximation to $\Delta(\alpha, c, \varepsilon; z)$ which is uniformly valid in a full neighbourhood of z_c containing z_1. For that purpose it is convenient to rewrite equation (40.18) in the equivalent form

$$\Delta(z) = \eta'(z)\mathscr{W}(\Psi, \varepsilon V_1)(\eta), \qquad (41.1)$$

where

$$\Psi(\eta) = AU_0(\eta) + U_3(\eta) \qquad (41.2)$$

and $A(\alpha, c, \varepsilon)$ is still given by equation (40.12).

In the derivation (§ 38) of the uniform approximations to the solutions of the Orr–Sommerfeld equation it was assumed that α and c were bounded away from zero. As $R \to \infty$ along the upper and lower branches of the curve of marginal stability, however, they both tend to zero and in these limits $\eta(z_1)$ is of order $\varepsilon^{3/5}$ and ε respectively. Ideally, of course, we would like to have error bounds for the approximations which are valid in a full neighbourhood of the turning point, but that would require the development of a theory of error bounds similar to the one which Olver (1974) has developed for second-order equations with a simple turning point. We can, however, as Lakin, Ng & Reid (1978) have shown, obtain simple error estimates for the inner, intermediate and outer expansions of the solutions which correspond to η (or z) being of order

Table 5.1. *Orders of magnitude for symmetrical flows in a channel without inflexion points*

	z & η	ξ & ζ	α	c	$A^{(0)}$	$\eta A^{(0)}$
Inner	ε	1	$\varepsilon^{1/2}$	ε	ε^{-1}	1
Intermediate	$\varepsilon^{3/5}$	$\varepsilon^{-2/5}$	$\varepsilon^{3/10}$	$\varepsilon^{3/5}$	$\varepsilon^{-3/5}$	1
Outer	1	ε^{-1}	1	1	1	1

ε, $\varepsilon^{3/5}$, and 1 respectively, and we then have the orders of magnitude shown in Table 5.1.

Consider, for example, the solution of well-balanced type. It has a uniform expansion of the form

$$U_0(\eta) = \sum_{s=0}^{\infty} \varepsilon^{3s} U_0^{(s)}(\eta), \qquad (41.3)$$

where $U_0^{(0)}(\eta) = O(\eta)$ as $\eta \to 0$ and $U_0^{(1)}(0) = \frac{2}{3}\alpha^2 - (U_c^{iv}/U_c'')$. Thus, if $U_c^{iv} \neq 0$ then $U_0^{(0)}(\eta)$ provides an approximation to the inner, intermediate and outer expansions of $U_0(\eta)$ with *relative* errors of order ε^2, $\varepsilon^{12/5}$ and ε^3, respectively, and we shall denote such error estimates by $O(\varepsilon^2, \varepsilon^{12/5}, \varepsilon^3)$. If $U_c^{iv} = 0$, as it does for plane Poiseuille flow, then the error is of order ε^3 for all η. Error estimates for the approximations to the solutions of balanced and dominant-recessive type can be obtained in a similar manner and reference may be made to Lakin, Ng & Reid (1978) where a full discussion is given.

We also need to consider, however, the errors involved in approximating $A(\alpha, c, \varepsilon)$ by $A^{(0)}(\alpha, c)$. For symmetrical flows in a channel we have (Nield (1972), cf. also equation (28.8))

$$A^{(0)}(\alpha, c) = U_c'^2 \alpha^{-2} \left\{ \int_{z_c}^{z_2} (U-c)^2 \, dz \right\}^{-1} + O(1) \text{ as } \alpha \downarrow 0 \qquad (41.4)$$

and it is not difficult to show that $A^{(1)}(\alpha, c) = O(\alpha^{-4})$ as $\alpha \downarrow 0$. Accordingly, $A^{(0)}(\alpha, c)$ provides an approximation to $A(\alpha, c, \varepsilon)$ with relative errors $O(\varepsilon^2, \varepsilon^{12/5}, \varepsilon^3)$.

In deriving a consistent first approximation to $\Delta(z)$ it is found that the term involving $\theta(\eta)$ in equation (38.17) must be omitted. This is not unexpected, however, since Lakin & Reid (1970) have observed that when outer expansions are used to approximate the eigenvalue relation then it is necessary to retain two terms in the outer expansion of ϕ_3' but only one term in the outer expansion of ϕ_3. It is also found that the final form of the approximation can be greatly simplified if it is expressed in terms of generalized Airy functions with $p = 0, \pm 1$ rather than $p = 0, 1, 2$. Thus on omitting the term involving $\ell(\eta)$ in equation (38.17) and then using the recursion formula (A9) we obtain

$$\varepsilon V_k(\eta) \sim \mathscr{L}A_k(\zeta, 1)$$
$$\equiv \eta\, a(\eta)A_k(\zeta, 1) - \varepsilon\, a(\eta)A_k(\zeta, -1) + \varepsilon^2 c(\eta)A_k(\zeta, 0)$$
$$(41.5)$$

and

$$\varepsilon V_k'(\eta) \sim \mathscr{M}A_k(\zeta, 1)$$
$$\equiv (\eta\, a)'A_k(\zeta, 1) - \varepsilon(a' - c)A_k(\zeta, -1) + \varepsilon^2(\ell + c')A_k(\zeta, 0),$$
$$(41.6)$$

and the corresponding approximations to the solutions of balanced type are

$$U_k(\eta) \sim \mathscr{G}(\eta) - \eta\mathscr{A}(\eta) - (U_c''/U_c')\mathscr{L}B_k(\zeta, 1, 1) \qquad (41.7)$$

and

$$U_k'(\eta) \sim \mathscr{G}'(\eta) - \eta\mathscr{A}'(\eta) - (U_c''/U_c')\mathscr{M}B_k(\zeta, 1, 1). \qquad (41.8)$$

Suppose now that we approximate V_1 and V_1' by equations (38.17) and (41.6) but temporarily regard the remaining terms in equation (41.1) as exact. Then we obtain

$$\Delta(z) \sim \eta'\{[\Psi(\eta\, a)' - \Psi'(\eta\, a + \varepsilon^3\ell)]A_1(\zeta, 1)$$
$$- \varepsilon[\Psi(a' - c) - \Psi'a]A_1(\zeta, -1)$$
$$+ \varepsilon^2[\Psi(\ell + c') - \Psi'c]A_1(\zeta, 0)\}. \qquad (41.9)$$

Clearly we can neglect $\varepsilon^3\ell$ compared to $\eta\, a$ with relative errors $O(\varepsilon^2, \varepsilon^{12/5}, \varepsilon^3)$ and this is equivalent to approximating V_1 by equation (41.5) rather than equation (38.17). A similar argument shows that in a first approximation to Δ we must also approximate

U_3 by equation (41.7) rather than equation (38.4). Thus, on approximating A, U_0 and U_3 in equation (41.9) we obtain

$$
\begin{aligned}
\Delta(z) \sim \eta'\{&[(\mathcal{G}-\eta\mathscr{A})(\eta\,a)' - (\mathcal{G}'-\eta\mathscr{A}')\eta\,a]A_1(\zeta, 1)\\
&-\varepsilon[(A^{(0)}U_0^{(0)} + \mathcal{G}-\eta\mathscr{A})(a'-c)\\
&-(A^{(0)}U_0^{(0)'} + \mathcal{G}'-\eta\mathscr{A}')a]A_1(\zeta, -1)\\
&+\varepsilon^2[(A^{(0)}U_0^{(0)} + \mathcal{G}-\eta\mathscr{A})(\mathcal{b}+c')\\
&-(A^{(0)}U_0^{(0)'} + \mathcal{G}'-\eta\mathscr{A}')c]A_1(\zeta, 0)\\
&+\varepsilon\mathscr{A}(a+\eta\,c)\frac{\mathrm{d}}{\mathrm{d}\zeta}\,W[A_1(\zeta, 1), B_3(\zeta, 1, 1)]\\
&-\varepsilon^2[\mathscr{C}(\eta\,a)' - (\mathscr{B}+\mathscr{C}')\eta\,a]W[A_1(\zeta, 1), B_3(\zeta, 1, 1)]\\
&+\varepsilon^3[\mathscr{A}(\mathcal{b}+c') - (\mathscr{A}'-\mathscr{C})c]W[A_1(\zeta, 0), B_3(\zeta, 0, 1)]\}.
\end{aligned}
$$
(41.10)

A crucial step in the simplification of this result is the recognition of the fact that it is possible, by the use of equations (A54), (A55) and (A56), to eliminate the B-type Airy functions from equation (41.10). This shows immediately that the terms in equation (41.10) which are formally of order ε^3 can also be neglected in a first approximation to Δ. The coefficient of $A_1(\zeta, 1)$ can then be substantially simplified by noting that

$$
(\mathcal{G}-\eta\mathscr{A})(\eta\,a)' - (\mathcal{G}'-\eta\mathscr{A}')\eta\,a \equiv -\eta'^{-1}W(\phi_1^{(0)}, \phi_2^{(0)})(z) = \eta'^{-1}.
$$
(41.11)

It is also convenient, for later computational purposes, to express the remaining coefficients, so far as possible, in terms of z. If, as usual, we let

$$
\Phi^{(0)}(z) = A^{(0)}(\alpha, c)\phi_1^{(0)}(z) + \phi_2^{(0)}(z),
$$
(41.12)

then, after some reduction, we obtain

$$
\begin{aligned}
\Delta(z) \sim &A_1(\zeta, 1)\\
&+\varepsilon\eta^{-1}\eta'^{-3/2}\{[\Phi^{(0)}(z) - \eta'^{3/2}]A_1(\zeta, -1)\\
&\quad+(U_c''/U_c')\phi_1^{(0)}(z)[2A_1(\zeta, -1, 1) - (\ln\zeta + 2\pi\mathrm{i})A_1(\zeta, -1)]\}\\
&+\varepsilon^2\{[\eta'\Phi^{(0)}(z)(\mathcal{b}+c') - \Phi^{(0)'}(z)c]A_1(\zeta, 0)\\
&\quad+(U_c''/U_c')\{\eta'\phi_1^{(0)}(z)(\mathcal{b}+c') - \phi_1^{(0)'}(z)c][2A_1(\zeta, 0, 1)\\
&\qquad\qquad\qquad - (\ln\zeta + 2\pi\mathrm{i})A_1(\zeta, 0)]\}.
\end{aligned}
$$
(41.13)

We also note that for any function $f(z)$ we have

$$\eta' f(z)(\mathscr{E} + c') - f'(z)c = \eta^{-2} \mathscr{W}(\phi_1^{(0)}, f)(z)$$

$$- \eta^{1/4} \left(\frac{U-c}{U_c'} \right)^{-3/4} \left\{ \left(\frac{U-c}{U_c'} \right)^{-1/2} [f'(z) + \frac{5}{4} \frac{U'}{U-c} f(z)] \right.$$

$$\left. - G_1(z)f(z) - \frac{7}{48} \eta^{-3/2} f(z) \right\} \tag{41.14}$$

The slowly varying coefficients in equation (41.13) can therefore all be expressed in terms of the Langer variable $\eta(z)$, the basic velocity distribution $U(z)$, the solutions of the inviscid equation $\phi_1^{(0)}(z)$ and $\phi_2^{(0)}(z)$, and the coefficient $G_1(z)$ which appears in the outer expansion of $\phi_3(z)$.

Equation (41.13) thus provides a 'first approximation' to $\Delta(z)$ with relative errors $O(\varepsilon^2 \ln \varepsilon, \varepsilon^{12/5}, \varepsilon^3)$, and further simplification is not possible without destroying the uniformity of the approximation. The form in which equation (41.13) has been written shows that the terms

$$2A_1(\zeta, p, 1) - (\ln \zeta + 2\pi i)A_1(\zeta, p) \quad (p = -1, 0) \tag{41.15}$$

play a role very similar to the 'viscous corrections' introduced by Tollmien (1947) in his discussion of the singular inviscid solution.

41.1 A computational form of the first approximation to the eigenvalue relation

The first approximation to the eigenvalue relation can now be obtained by simply setting $z = z_1$ in equation (41.13) and then equating the right-hand side to zero. The Airy functions which appear in this approximation are all rapidly varying functions of $\zeta_1 = \eta(z_1)/\varepsilon$ and this behaviour can cause numerical difficulties. Such difficulties can be entirely avoided, however, by dividing equation (41.13) through by $A_1(\zeta, 0)$ (say). The resulting ratios of Airy functions can then be expressed in terms of certain Tietjens type functions which are easy to compute. Consider then the generalized Tietjens function (Hughes & Reid 1968)

$$F(Z, p) = \frac{A_1(\zeta_1, p)}{\zeta_1 A_1(\zeta_1, p-1)} \tag{41.16}$$

where (cf. equation (28.45))

$$\zeta_1 = Z\, e^{-5\pi i/6} \quad \text{and} \quad Z = \left[\frac{3}{2}\int_{z_1}^{z_c} (c-U)^{1/2}\, dz\right]^{2/3} (\alpha R)^{1/3}.$$
(41.17)

When $p = 2$, this gives the ordinary Tietjens function defined by equation (28.34). For the present purposes, however, it is more convenient to let

$$H(Z, p) = ZF(Z, p)$$

$$= e^{5\pi i/6}\frac{A_1(\zeta_1, p)}{A_1(\zeta_1, p-1)}$$
(41.18)

and to define the related functions

$$K(Z, p) = \frac{A_1(\zeta_1, p, 1)}{A_1(\zeta_1, p)}.$$
(41.19)

The first approximation to the eigenvalue relation can then be written in the form

$$\frac{e^{5\pi i/6}}{A_1(\zeta_1, 0)}\Delta(z_1) \sim H(Z, 1) + \eta'^{-3/2}\{\Phi^{(0)}(z_1) - \eta'^{3/2}$$

$$+ (U_c''/U_c')\phi_1^{(0)}(z_1)$$

$$\times[2K(Z, -1) - (\ln Z + \tfrac{7}{6}\pi i)]\}H(Z, -1)$$

$$+ \varepsilon^2 e^{5\pi i/6}\{\eta'\Phi^{(0)}(z_1)(\ell + c') - \Phi^{(0)'}(z_1)c$$

$$+ (U_c''/U_c')[\eta'\phi_1^{(0)}(z_1)(\ell + c') - \phi_1^{(0)'}(z_1)c]$$

$$\times[2K(Z, 0) - (\ln Z + \tfrac{7}{6}\pi i)]\} = 0,$$
(41.20)

where it is understood that η' is to be evaluated at z_1 and ℓ, c and c' are to be evaluated at $\eta_1 \equiv \eta(z_1)$.

The Tietjens type functions which appear in equation (41.20) are particularly easy to compute. We first note that

$$H(Z, -1) = i/\{ZH(Z, 0)\}$$
(41.21)

and that $H(Z, 0)$ satisfies the first-order *nonlinear* equation

$$H'(Z, 0) - iZ\{H(Z, 0)\}^2 = 1$$
(41.22)

and the initial condition

$$H(0, 0) = 3^{-1/3}\{\Gamma(\tfrac{1}{3})/\Gamma(\tfrac{2}{3})\}\, e^{-\pi i/6}.$$
(41.23)

Once $H(Z, 0)$ has been computed, we can obtain $H(Z, 1)$ as the solution of the first-order *linear* equation

$$H'(Z, 1) + \{H(Z, 0)\}^{-1} H(Z, 1) = 1 \qquad (41.24)$$

which satisfies the initial conditions

$$H(0, 1) = 3^{-1/3} \Gamma(\tfrac{2}{3}) \, e^{-\pi i/6}. \qquad (41.25)$$

It is also easy to show that

$$K(Z, 0) = -\frac{1}{2} i \int_0^Z \{H(Z', 1)\}^2 \, dZ' + \frac{1}{3} \left(-\gamma + \frac{\pi}{2\sqrt{3}} - \frac{1}{2} \ln 3 + 3\pi i\right) \qquad (41.26)$$

and

$$K(Z, -1) = K(Z, 0) - \tfrac{1}{2} i H(Z, 0) \{H(Z, 1)\}^2. \qquad (41.27)$$

Thus, the computation of the four Tietjens type functions which appear in equation (41.20) require only the integration of two first-order equations and one quadrature. The general behaviour of these functions is shown in Figs. 5.3 and 5.4. We also note that

$$H(Z, p) \sim e^{\pi i/4} Z^{-1/2} \{1 + \tfrac{1}{4}(2p + 1) \, e^{\pi i/4} Z^{-3/2} + O(Z^{-3})\} \qquad (41.28)$$

and

$$K(Z, p) \sim \tfrac{1}{2} \ln Z + \tfrac{7}{12} \pi i - \tfrac{1}{48}(24p - 17) \, e^{\pi i/4} Z^{-3/2} + O(Z^{-3}). \qquad (41.29)$$

These expansions are valid in the complete sense of Watson in the sector $\pi/6 < \mathrm{ph}\, Z < 3\pi/2$ but they remain valid in the sense of Poincaré in the larger sector $-\pi/6 < \mathrm{ph}\, Z < 11\pi/6$.

41.2 *Results for plane Poiseuille flow*

To illustrate the accuracy of the approximation (41.20), Lakin, Ng & Reid (1978) made a calculation of the curve of marginal stability for plane Poiseuille flow. For this flow we have $U(z) = 1 - z^2$, $z_1 = -1$, $z_c = -(1-c)^{1/2}$, and $z_2 = 0$. We also have (cf. equation (28.51))

$$Z = \{\tfrac{3}{4}[\sqrt{c} - (1 - c) \tanh^{-1} \sqrt{c}]\}^{2/3} (\alpha R)^{1/3} \qquad (41.30)$$

and

$$G_1(z_1) = -i\sqrt{2}(1 - c)^{1/4} \left(\frac{101}{24} c^{-3/2} + \frac{23}{24} c^{-1/2} + \frac{23}{24} \frac{c^{1/2}}{1 - c}\right.$$

$$\left. + \tfrac{1}{2} \alpha^2 \tanh^{-1} \sqrt{c}\right). \qquad (41.31)$$

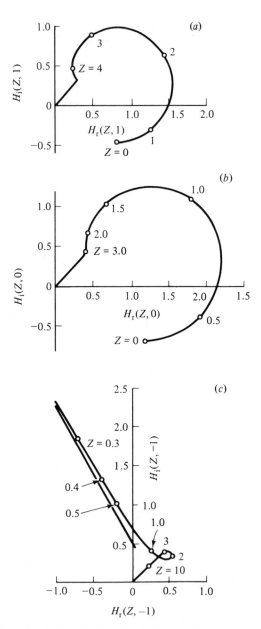

Fig. 5.3. The behaviour of $H(Z, p)$ for (a) $p = 1$, (b) $p = 0$, and (c) $p = -1$. In part (c) the asymptote as $Z \to 0$ is $\sqrt{3}H_r(Z, -1) + H_i(Z, -1) \cong 0.531$. (From Lakin, Ng & Reid 1978.)

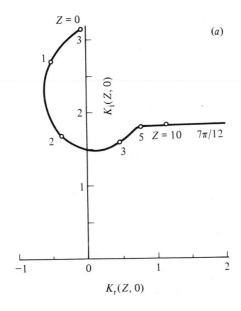

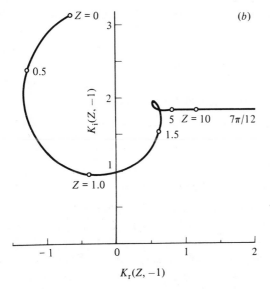

Fig. 5.4. The behaviour of $K(Z, p)$ for (a) $p = 0$, and (b) $p = -1$. (From Lakin, Ng & Reid 1978.)

Table 5.2. *A comparison with the results of Reynolds & Potter* (1967)

R		Uniform approximation α	c	Reynolds & Potter α	c
7500	Upper branch	1.0944	0.2597	1.094	0.2597
	Lower branch	0.8750	0.2345	0.875	0.2344
9000	Upper branch	1.0971	0.2515	1.097	0.2515
	Lower branch	0.8233	0.2203	0.823	0.2203

The computational procedure used to solve equation (41.20) was similar to the one described by Hughes & Reid (1968). In calculating the curve of marginal stability, it is found to be convenient to fix Z and then iterate for α, c and R. Thus, the calculations were made for $Z = 2.325(0.025)2.925(0.0025)3.275(0.025)8$ which corresponds to $R_c \leqslant R \leqslant 6 \times 10^8$ along the upper branch and $R_c \leqslant R \leqslant 4 \times 10^7$ along the lower branch. These remarks also apply to the results described in § 42. Some further details of the computational procedure and tables of the results have been given by Ng (1977).

The values of the parameters associated with the minimum critical Reynolds number were found to be

$$R_c = 5769.7, \quad \alpha_c = 1.0207 \quad \text{and} \quad c = 0.2640, \quad (41.32)$$

and these results should be compared with the 'exact' numerical values

$$R_c = 5772.22, \quad \alpha_c = 1.0205 \quad \text{and} \quad c = 0.2640, \quad (41.33)$$

which were obtained by Orszag (1971) and subsequently confirmed by A. Davey (unpublished). Accurate values of α and c for $R = 7500$ and 9000 have also been computed numerically by Reynolds & Potter (1967) and a comparison with their results is given in Table 5.2. Near the nose of the curve of marginal stability the error associated with the approximation (41.20) would be expected to be of the order of ε^3. The worst error in the present results is in the value of R_c for which the actual error is about 0.044 per cent compared with an expected error of about 0.030 per cent.

It is easy to verify that equation (41.20) gives the correct behaviour of the asymptotes to the upper and lower branches of the curve of marginal stability and that the errors are then of order $\varepsilon^{12/5}$ and $\varepsilon^2 \ln \varepsilon$, respectively. Thus, equation (41.20) provides a 'first approximation' to the eigenvalue relation which is uniformly valid along the entire curve of marginal stability.

42 A comparison with the heuristic approximations to the eigenvalue relation

The approximations (41.13) and (41.20) have very different structures from the heuristic approximations which were discussed in § 28 and it is of some interest, therefore, to compare the results obtained by using the heuristic approximations to the eigenvalue relation with those obtained by using the fully uniform approximation. In the heuristic approach to the eigenvalue problem, ϕ_1 and ϕ_2 are approximated by the first terms in their outer expansions and various approximations to ϕ_3 are then considered. Thus, for purposes of the present discussion, we approximate equation (40.18) at the outset by

$$\Delta(z) \approx \Delta^{(0)}(z) \equiv \mathcal{W}(\Phi^{(0)}, \varepsilon\phi_3)(z). \tag{42.1}$$

42.1 The local turning-point approximation to $\phi_3(z)$

In this approximation ϕ_3 is given by (cf. equation (27.25))

$$\phi_3(z) \sim A_1(\xi, 2), \quad \text{where } \xi = (z - z_c)/\varepsilon, \tag{42.2}$$

and equation (42.1) then becomes

$$\Delta^{(0)}(z) \sim [\Phi^{(0)}(z) - (z - z_c)\Phi^{(0)'}(z)]A_1(\xi, 1) + \varepsilon\Phi^{(0)'}(z)A_1(\xi, -1). \tag{42.3}$$

For comparison with equation (41.20) this can be written in the form

$$\frac{e^{5\pi i/6}}{A_1(\xi_1, 0)} \Delta^{(0)}(z_1) \sim [\Phi^{(0)}(z_1) - (z_1 - z_c)\Phi^{(0)'}(z_1)]H(Y, 1)$$

$$+ (z_1 - z_c)\Phi^{(0)'}(z_1)H(Y, -1) = 0, \tag{42.4}$$

Table 5.3 *Results for plane Poiseuille flow based on the heuristic approximations to the eigenvalue relation*

Approximation to $\phi_3(z)$	Equation	R_c	α_c	c
Local turning point	(42.4)	5397.1	1.022	0.2672
Tollmien	(42.8)	5697.3	1.010	0.2607
Uniform (truncated equation)	(42.13)	4880.9	1.034	0.2721
Uniform (Orr–Sommerfeld equation)	(42.15)	6052.1	1.020	0.2621
'Exact' values (Orszag 1971)		5772.2	1.021	0.2640

where $\xi_1 = \xi(z_1)$ and Y is defined by equation (28.31). This result is equivalent to equation (28.33) since

$$F(Y) \equiv F(Y, 2) = 1 - \{H(Y, -1)/H(Y, 1)\}. \qquad (42.5)$$

The results of calculations based on this approximation to the eigenvalue relation have been given in § 28 and they are also given here in Table 5.3 and Fig. 5.5. As $R \to \infty$ along the upper and lower branches of the curve of marginal stability, the errors associated with this approximation are of order $\varepsilon^{6/5}$ and $\varepsilon \ln \varepsilon$ respectively. The approximation is clearly not uniform, however, since it involves the outer expansions of ϕ_1 and ϕ_2 combined with the inner expansion of ϕ_3.

42.2 Tollmien's improved approximation to $\phi_3(z)$

From equation (27.42) we have

$$\phi_3(z) \sim \eta'^{-5/2} A_1(\zeta, 2) \qquad (42.6)$$

and equation (42.1) then becomes

$$\Delta^{(0)}(z) \sim \eta'^{-5/2}\{[\eta'\Phi^{(0)}(z) - \eta\Phi^{(0)'}(z)]A_1(\zeta, 1) + \varepsilon\Phi^{(0)'}(z)A_1(\zeta, -1)\}. \qquad (42.7)$$

In this approximation the eigenvalue relation is given by

$$\frac{e^{5\pi i/6}}{A_1(\zeta_1, 0)} \Delta^{(0)}(z_1) \sim \eta'^{-5/2}\{[\eta'\Phi^{(0)}(z_1) - \eta\Phi^{(0)'}(z_1)]H(Z, 1)$$
$$+ \eta\Phi^{(0)'}(z_1)H(Z, -1)\} = 0, \qquad (42.8)$$

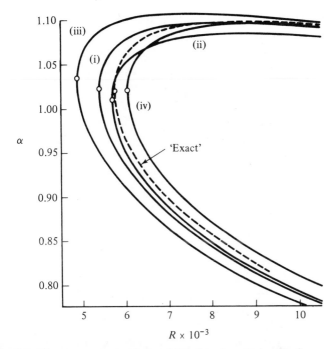

Fig. 5.5. The curves of marginal stability for plane Poiseuille flow according to the heuristic approximations to the eigenvalue relation. The approximations used for $\phi_3(z)$ are (i) the local turning point approximation, (ii) Tollmien's improved viscous approximation, (iii) uniform approximation (truncated equation), and (iv) uniform approximation (Orr–Sommerfeld equation). The circled points correspond to the values of α_c and R_c. The dashed curve is based on 'exact' numerical values obtained by Reynolds & Potter (1967), Orszag (1971), and T. H. Hughes (unpublished); on this scale it is indistinguishable from the results obtained by using the fully uniform approximation (41.20). (From Lakin, Ng & Reid 1978.)

where it is understood that η and η' are to be evaluated at z_1. This approximation to the eigenvalue relation is equivalent to equation (28.47) and leads to the results given in Table 5.3 and Fig. 5.5.

To assess the errors associated with Tollmien's approximation to $\phi_3(z)$, consider equation (38.17) which, with $k = 1$, can be written in the alternate form

$$\phi_3(z) \sim \eta'^{-5/2} A_1(\zeta, 2) - \varepsilon (2c - \eta \ell) A_1(\zeta, 3) - 3\varepsilon^2 \ell A_1(\zeta, 4).$$

$$(42.9)$$

This shows that when $\eta \in \mathbf{T}_2' \cup \mathbf{T}_3'$, the relative errors associated with the approximation (42.6) are $O(\varepsilon, \varepsilon^{6/5}, \varepsilon^{3/2})$. When $\eta \in \mathbf{T}_1'$, i.e. in the sector containing η_1, equations (38.17) and (42.6) can each be written as the sum of dominant, balanced and recessive terms. The relative errors associated with the dominant and recessive terms in equation (42.6) remain unchanged, as would be expected, but those associated with the balanced term are $O(\varepsilon, \varepsilon^{3/5}, 1)$. Thus, equation (42.6) does not provide a uniform approximation to the solution of either the truncated equation (27.34) or the Orr–Sommerfeld equation (25.12).

42.3 *The uniform approximation to $\phi_3(z)$ based on the truncated equation*

To derive uniform approximations to the solutions of the truncated equation (27.34) we again make the preliminary transformation (36.1) with $m = 0$. This leads to an equation of the form of equation (36.3) where the coefficients are now given by

$$\left.\begin{aligned}
&f_0(\eta) = 6\gamma, &&f_1(\eta) = -(4\gamma' + 11\gamma^2), \\
&g_0(\eta) = \eta\gamma, &&g_1(\eta) = -(\gamma'' + 7\gamma'\gamma + 6\gamma^3), \\
&h_0(0) = 0, &&h_1(\eta) = 0.
\end{aligned}\right\} \quad (42.10)$$

Let $\hat{V}_k(\eta)$ denote the solutions of (the transformed version of) the truncated equation which are of dominant-recessive type. Then they must also have uniform approximations of the form of equation (38.17) and a short calculation shows that the slowly varying coefficients in the approximations are given by

$$\text{and} \quad \left.\begin{aligned}
&a(\eta) = (z - z_c)/\eta, \quad a(\eta) + \eta c(\eta) = \eta'^{-5/2} \\
&b(\eta) = 2\eta^{-1}c(\eta) + \eta'^{-5/2}\eta^{-1/2}\{G_1(z) - \tfrac{101}{48}\eta^{-3/2}\},
\end{aligned}\right\} \quad (42.11)$$

where $G_1(z)$ is now given by (cf. equation (37.37) and Problem 4.7)

$$G_1(z) = \left(\frac{101}{48}\frac{U'}{U-c} - \frac{1}{24}\frac{U''}{U'}\right)\left(\frac{U_c'}{U-c}\right)^{1/2}$$
$$+ \frac{1}{24}\int_{z_c}^{z}\left(\frac{U'''}{U'} - \frac{U''^2}{U'^2}\right)\left(\frac{U_c'}{U-c}\right)^{1/2}\,\mathrm{d}z. \quad (42.12)$$

If we now approximate $\varepsilon\hat{V}_1$ and $\varepsilon\hat{V}_1'$ by $\mathscr{L}A_1(\zeta, 1)$ and $\mathscr{M}A_1(\zeta, 1)$ respectively, where a, b and c are given by equations (42.11), then equation (42.1) becomes

$$
\begin{aligned}
\Delta^{(0)}(z) \sim &[\Phi^{(0)}(z) - (z - z_c)\Phi^{(0)\prime}(z)]A_1(\zeta, 1) \\
&+ \varepsilon\eta^{-1}\eta^{\prime-3/2}\{\Phi^{(0)}(z) \\
&- \eta^{\prime 3/2}[\Phi^{(0)}(z) - (z - z_c)\Phi^{(0)\prime}(z)]\}A_1(\zeta, -1) \\
&+ \varepsilon^2[\eta^{\prime}\Phi^{(0)}(z)(\ell + c^{\prime}) - \Phi^{(0)\prime}(z)c]A_1(\zeta, 0). \quad (42.13)
\end{aligned}
$$

Although the structure of this approximation to $\Delta^{(0)}(z)$ is similar to that of equation (41.13) it leads, as Table 5.3 and Fig. 5.5 clearly show, to surprisingly poor results. The approximation (42.13) is, of course, defective in two respects since the approximation to ϕ_2 is of outer type and hence not uniform and the approximation to ϕ_3 is based on the truncated equation.

42.4 The uniform approximation to $\phi_3(z)$ based on the Orr–Sommerfeld equation

To assess the relative effects of these two defects, consider now an approximation to $\Delta^{(0)}(z)$ in which $\varepsilon\phi_3$ and $\varepsilon\phi_3^{\prime}$ are approximated by $\mathscr{L}A_1(\zeta, 1)$ and $\eta^{\prime}\mathscr{M}A_1(\zeta, 1)$ respectively, where a, ℓ and c are given by equations (38.19) and (38.24). This leads immediately to

$$
\begin{aligned}
\Delta^{(0)}(z) \sim &A_1(\zeta, 1) + \varepsilon\eta^{-1}\eta^{\prime-3/2}[\Phi^{(0)}(z) - \eta^{\prime 3/2}]A_1(\zeta, -1) \\
&+ \varepsilon^2[\eta^{\prime}\Phi^{(0)}(z)(\ell + c^{\prime}) - \Phi^{(0)\prime}(z)c]A_1(\zeta, 0). \quad (42.14)
\end{aligned}
$$

For computational purposes this can be written in the form

$$
\begin{aligned}
\frac{e^{5\pi i/6}}{A_1(\zeta_1, 0)}\Delta^{(0)}(z_1) \sim &H(Z, 1) + \eta^{\prime-3/2}[\Phi^{(0)}(z_1) - \eta^{\prime 3/2}]H(Z, -1) \\
&+ \varepsilon^2 e^{5\pi i/6}[\eta^{\prime}\Phi^{(0)}(z_1)(\ell + c^{\prime}) - \Phi^{(0)\prime}(z_1)c] = 0. \quad (42.15)
\end{aligned}
$$

This approximation to the eigenvalue relation differs from the fully uniform approximation (41.20) only in the absence of the 'viscous corrections' $K(Z, p) - (\frac{1}{2}\ln Z + \frac{7}{12}\pi i)$ whose behaviour is shown in Fig. 5.6 for $p = 0$ and -1. Although the differences between the approximations (41.20) and (42.15) might appear to be small, they obviously are of crucial importance numerically as the results given in Table 5.3 and Fig. 5.5 clearly show.

When $\eta \in \mathbf{T}_2^{\prime} \cup \mathbf{T}_3^{\prime}$, the relative errors associated with the approximation (42.14) are $O(\varepsilon, \varepsilon^{9/5}, \varepsilon^3)$ and these relative errors

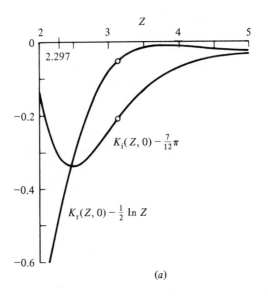

(a)

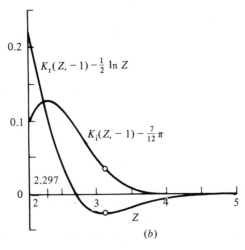

(b)

Fig. 5.6. The behaviour of the 'viscous corrections' $K(Z, p) -$ $(\frac{1}{2}\ln Z + \frac{7}{12}\pi i)$ for (a) $p = 0$, and (b) $p = -1$. The circled points correspond to the minimum critical Reynolds number and $Z = 2.297$ corresponds to the asymptote to the lower branch of the marginal curve. (From Lakin, Ng & Reid 1978.)

remain unchanged when $\eta \in \mathbf{T}_1'$ and equation (42.14) is written as the sum of dominant, balanced and recessive terms. There is thus an intrinsic limitation to the accuracy which can be achieved by starting with equation (42.1). It would appear, therefore, that equation (41.13) provides a 'first' fully uniform approximation to the eigenvalue relation for the Orr–Sommerfeld problem and that further simplification is not possible without destroying the uniformity of the approximation.

43 A numerical treatment of the Orr–Sommerfeld problem using compound matrices

In § 30 we considered some of the methods which have been developed for the numerical solution of the Orr–Sommerfeld problem, and in this section we wish to consider one additional method which is based on the use of certain compound matrices. This method has been used by Gilbert & Backus (1966) in their discussion of elastic wave problems and it has also been used by Lakin, Ng & Reid (1978) to provide an alternate derivation of the uniform approximation (41.13) to the eigenvalue relation. The use of compound matrices for the numerical treatment of eigenvalue problems has been discussed by Ng & Reid (1979) and the present section is largely based on their work.

To illustrate the basic ideas involved in this method, consider the linear fourth-order equation

$$\phi^{\mathrm{iv}} - a_1\phi''' - a_2\phi'' - a_3\phi' - a_4\phi = 0, \qquad (43.1)$$

where a_1, a_2, a_3 and a_4 are functions of z and $z_1 \leqslant z \leqslant z_2$. To be definite, we shall also suppose that the boundary conditions at $z = z_1$ are $\phi = \phi' = 0$. The boundary conditions at $z = z_2$, however, need not be specified until later.

For the present purposes it is convenient to rewrite equation (43.1) as a system of first-order equations. Thus, if we let $\boldsymbol{\phi} = [\phi, \phi', \phi'', \phi''']^{\mathrm{T}}$ then equation (43.1) becomes

$$\boldsymbol{\phi}' = \mathbf{A}(z)\boldsymbol{\phi}, \qquad (43.2)$$

where

$$\mathbf{A}(z) = \begin{bmatrix} 0 & 1 & 0 & 0 \\ 0 & 0 & 1 & 0 \\ 0 & 0 & 0 & 1 \\ a_4 & a_3 & a_2 & a_1 \end{bmatrix}. \qquad (43.3)$$

Now let $\boldsymbol{\phi}_1$ and $\boldsymbol{\phi}_2$ be two solutions of equation (43.1) which satisfy the initial conditions

$$\boldsymbol{\phi}_1(z_1) = [0, 0, 1, 0]^T \quad \text{and} \quad \boldsymbol{\phi}_2(z_1) = [0, 0, 0, 1]^T, \qquad (43.4)$$

and consider the 4×2 solution matrix

$$\boldsymbol{\Phi}(z) = \begin{bmatrix} \phi_1 & \phi_2 \\ \phi_1' & \phi_2' \\ \phi_1'' & \phi_2'' \\ \phi_1''' & \phi_2''' \end{bmatrix}. \qquad (43.5)$$

The 2×2 minors of the matrix $\boldsymbol{\Phi}$ are

$$\left. \begin{aligned} y_1 &= \phi_1\phi_2' - \phi_1'\phi_2, & y_4 &= \phi_1'\phi_2'' - \phi_1''\phi_2', \\ y_2 &= \phi_1\phi_2'' - \phi_1''\phi_2, & y_5 &= \phi_1'\phi_2''' - \phi_1'''\phi_2', \\ y_3 &= \phi_1\phi_2''' - \phi_1'''\phi_2, & y_6 &= \phi_1''\phi_2''' - \phi_1'''\phi_2'', \end{aligned} \right\} \qquad (43.6)$$

and they satisfy the quadratic identity

$$y_1 y_6 - y_2 y_5 + y_3 y_4 = 0. \qquad (43.7)$$

On differentiating equations (43.6) and then using equation (43.1) to eliminate ϕ_1^{iv} and ϕ_2^{iv}, it is easy to show that $\mathbf{y} = [y_1, \ldots, y_6]^T$ satisfies the equation

$$\mathbf{y}' = \mathbf{B}(z)\mathbf{y}, \qquad (43.8)$$

where

$$\mathbf{B}(z) = \begin{bmatrix} 0 & 1 & 0 & 0 & 0 & 0 \\ 0 & 0 & 1 & 1 & 0 & 0 \\ a_3 & a_2 & a_1 & 0 & 1 & 0 \\ 0 & 0 & 0 & 0 & 1 & 0 \\ -a_4 & 0 & 0 & a_2 & a_1 & 1 \\ 0 & -a_4 & 0 & -a_3 & 0 & a_1 \end{bmatrix}. \qquad (43.9)$$

Thus $\mathbf{y}$ is the second compound of $\boldsymbol{\Phi}$ and it satisfies the initial conditions

$$\mathbf{y}(z_1) = [0, 0, 0, 0, 0, 1]^\mathrm{T}. \tag{43.10}$$

The boundary conditions on ϕ at $z = z_2$ will imply that some element of $\mathbf{y}$ or, more generally, a linear combination of the elements of $\mathbf{y}$ must vanish there and this condition will provide the required eigenvalue relation. In actual calculations, of course, some iterative procedure must be used to vary the eigenvalue parameter until this condition is satisfied to some prescribed degree of accuracy.

It should be observed, however, that the initial conditions (43.4) play no essential role in the derivation of equation (43.8) and it is evident therefore that if ϕ_1 and ϕ_2 are *any* two solutions of equation (43.1) then $\mathbf{y}$ still satisfies equation (43.8). Furthermore, as Sloan & Wilks (1976) have shown, the case of general separated boundary conditions can be reduced, without loss of generality, to requiring that any two components of $\boldsymbol{\phi}$ vanish at $z = z_1$ (say) and

$$\mathbf{Q}\boldsymbol{\phi}(z_2) = \mathbf{0}, \quad \text{where } \mathbf{Q} = \begin{bmatrix} q_{11} & q_{12} & q_{13} & q_{14} \\ q_{21} & q_{22} & q_{23} & q_{24} \end{bmatrix} \tag{43.11}$$

and $\mathbf{Q}$ is of rank 2. The eigenvalue relation in this more general case then becomes

$$(q_{11}q_{22} - q_{12}q_{21})y_1(z_2) + (q_{11}q_{23} - q_{13}q_{21})y_2(z_2)$$
$$+ (q_{11}q_{24} - q_{14}q_{21})y_3(z_2) + (q_{12}q_{23} - q_{13}q_{22})y_4(z_2)$$
$$+ (q_{12}q_{24} - q_{14}q_{22})y_5(z_2) + (q_{13}q_{24} - q_{14}q_{23})y_6(z_2) = 0. \tag{43.12}$$

Once the required eigenvalue has been obtained by the method just described, we can then proceed to the determination of the corresponding eigenfunction ϕ (say). To be definite, suppose that the boundary conditions are $\phi(z_1) = \phi'(z_1) = 0$ and $\phi'(z_2) = \phi'''(z_2) = 0$. Then $\mathbf{y}$ is found by integrating equation (43.8) subject to the initial conditions (43.10) and by requiring that $y_5(z_2) = 0$. Clearly there must exist constants C_1 and C_2 such that

$$\left. \begin{array}{ll} \phi = C_1\phi_1 + C_2\phi_2, & \phi' = C_1\phi_1' + C_2\phi_2', \\ \phi'' = C_1\phi_1'' + C_2\phi_2'', & \phi''' = C_1\phi_1''' + C_2\phi_2'''. \end{array} \right\} \tag{43.13}$$

The constants C_1 and C_2 can be eliminated from these equations in four different ways and if this is done then we obtain

$$y_1\phi'' - y_2\phi' + y_4\phi = 0, \qquad (43.14a)$$

$$y_1\phi''' - y_3\phi' + y_5\phi = 0, \qquad (43.14b)$$

$$y_2\phi''' - y_3\phi' + y_6\phi = 0, \qquad (43.14c)$$

and

$$y_4\phi''' - y_5\phi'' + y_6\phi' = 0. \qquad (43.14d)$$

Thus we have four possible equations for the determination of the eigenfunction ϕ and it is certainly not obvious which one is best for numerical purposes.

Consider then the behaviour of the solutions of equations (43.14) near $z = z_1$. For this purpose we observe that as $z \to z_1$ we have

$$\left.\begin{aligned}
&y_1 \sim \tfrac{1}{12}(z - z_1)^4, \quad y_2 \sim \tfrac{1}{3}(z - z_1)^2, \quad y_3 \sim \tfrac{1}{2}(z - z_1)^2, \\
&y_4 \sim \tfrac{1}{2}(z - z_1)^2, \quad y_5 \sim z - z_1, \quad y_6 \sim 1,
\end{aligned}\right\} \qquad (43.15)$$

and this limiting behaviour is seen to be independent of the coefficients in equation (43.1). The point $z = z_1$ is therefore a regular singular point of equations (43.14) and at that point they have exponents $(2, 3)$, $(-2, 2, 3)$, $(-\tfrac{1}{2}, 2, 3)$, and $(0, 2, 3)$, respectively. It is easy to show, however, as a consequence of equation (43.8), that none of the solutions contains logarithmic terms. Accordingly, near $z = z_1$ the solution of equations (43.14) that satisfies the boundary conditions must be of the form

$$\phi(z) = \sum_{s=0}^{\infty} b_s(z - z_1)^{s+2}, \qquad (43.16)$$

where b_0 and b_1 are arbitrary. When equation (43.1) is even moderately stiff, however, and this is certainly the case for the Orr–Sommerfeld equation when αR is moderately large, forward integration of equations (43.14) from $z = z_1$ to z_2 leads, as might have been expected, to a serious growth problem.

To avoid this growth problem, consider the possibility of determining ϕ by integrating one of equations (43.14) backwards from $z = z_2$ to z_1. In all cases it is convenient to fix the normalization of the solution so that $\phi(z_2) = 1$ and we shall also assume that $y_1(z)$ does not vanish anywhere in the interval $z_1 < z \le z_2$. For equation (43.14a) the initial conditions are $\phi(z_2) = 1$ and $\phi'(z_2) = 0$, and equation (43.14a) then shows that $\phi'''(z_2)$ vanishes automatically. On integrating equation (43.14a) from $z = z_2$ to z_1, we see that ϕ must necessarily satisfy the boundary conditions at $z = z_1$ since the exponents of equation (43.14a) at $z = z_1$ are 2 and 3. Thus we have a simple marching problem for the determination of the eigen-function ϕ. This procedure, however, will fail to yield the final values at $z = z_1$ since equation (43.14a) is singular there but this is only a very minor limitation. It can further be shown (Ng & Reid 1979) that the solution of equation (43.14a) obtained in this way is also a solution of equation (43.1).

For equations (43.14b) to (43.14d) the initial conditions are $\phi(z_2) = 1$, $\phi'(z_2) = 0$ and $\phi''(z_2) = -y_4(z_2)/y_1(z_2)$. On integrating these equations from $z = z_2$ to z_1, however, some numerical difficulties would be expected owing to the exponents -2, $-\frac{1}{2}$, and 0 of the equations respectively at $z = z_1$, and these difficulties are particularly severe in the case of equation (43.14b).

Consider now the application of this method to the Orr–Sommerfeld equation for which the coefficients in equation (43.1) are

$$a_1 = 0, \quad a_2 = 2\alpha^2 + i\alpha R(U - c),$$
$$a_3 = 0, \quad a_4 = -\{\alpha^4 + i\alpha R[\alpha^2(U - c) + U'']\}. \qquad (43.17)$$

Equation (43.8) must therefore be solved subject to appropriate initial and end conditions. These conditions depend on the form of the basic velocity profile and two cases will be considered to illustrate the main principles involved.

43.1 Symmetrical flows in a channel

As in the asymptotic theory, we shall choose $z_1 = -1$ and $z_2 = 0$. The initial conditions are then given by equation (43.10) and the eigen-value relation for an even or odd mode is simply $y_5(0) = 0$ or $y_2(0) = 0$ respectively. As an example of this procedure, consider the unstable (even) mode for plane Poiseuille flow with $\alpha = 1$ and $R = 10\,000$. To obtain the eigenvalue c correct to 5D (say), it is found that about 800 (equally spaced) steps are required for the numerical integration. The behaviour of the components of $\mathbf{y}$ for this problem is shown in Fig. 5.7, and, in particular, it is seen that none of them vanishes in the interval $(0, 1)$. Their rapid variation near $z = 0$ is, of course, a direct consequence of the largeness of αR.

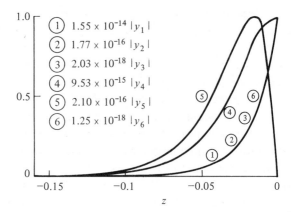

Fig. 5.7. The behaviour of the components of **y** for plane Poiseuille flow with $\alpha = 1$ and $R = 10\,000$. When normalized to unity, the graphs of $|y_1|$, $|y_2|$, $|y_3|$, and $|y_6|$, though not identical, are indistinguishable on this scale. (After Ng & Reid 1979.)

43.2 *Boundary-layer flows*

For unbounded flows of this type we choose $z_1 = 0$ and z_2 must then be chosen sufficiently large so that $U(z_2) = 1$ and $U''(z_2) = 0$ to some prescribed degree of accuracy, i.e. the basic velocity profile is given by

$$\left. \begin{array}{ll} U(z) & (0 \leqslant z \leqslant z_2) \\ 1 & (z_2 \leqslant z < \infty). \end{array} \right\} \tag{43.18}$$

For numerical purposes, of course, we wish to impose boundary conditions at $z = 0$ and z_2. To derive the appropriate boundary conditions at $z = z_2$, we observe that for $z > z_2$ the solution must be of the form

$$\phi(z) = C_3 \, \mathrm{e}^{-\alpha z} + C_4 \, \mathrm{e}^{-\beta z}, \tag{43.19}$$

where $\beta = [\alpha^2 + i\alpha R(1 - c)]^{1/2}$ and $\mathrm{Re}\,(\beta) > 0$. If we now require, following Itoh (1974a), that ϕ, ϕ', ϕ'' and ϕ''' are continuous then a short calculation leads to the required two boundary conditions

and
$$\left. \begin{array}{l} \phi''(z_2) + (\alpha + \beta)\phi'(z_2) + \alpha\beta\phi(z_2) = 0 \\ \phi'''(z_2) + (\alpha + \beta)\phi''(z_2) + \alpha\beta\phi'(z_2) = 0. \end{array} \right\} \tag{43.20}$$

These conditions are equivalent to the ones given by Keller (1976), p. 58 and, in the notation of equations (43.11), they can be written in the form

$$\mathbf{Q}\boldsymbol{\phi}(z_2) = \mathbf{0}, \quad \text{where} \quad \mathbf{Q} = \begin{bmatrix} \alpha\beta & \alpha+\beta & 1 & 0 \\ 0 & \alpha\beta & \alpha+\beta & 1 \end{bmatrix}.$$

(43.21)

Thus, to solve the eigenvalue problem by forward integration from $z = 0$ to z_2, equation (43.8) must be integrated subject to the initial conditions (43.10), and the eigenvalue relation (43.12) then becomes

$$\alpha^2\beta^2 y_1 + \alpha\beta(\alpha+\beta)y_2 + \alpha\beta y_3 + (\alpha^2 + \alpha\beta + \beta^2)y_4$$
$$+ (\alpha+\beta)y_5 + y_6 = 0 \quad \text{at } z = z_2.$$

(43.21)

It is also of some interest to see how the eigenvalue problem can be solved by *backward* integration from $z = z_2$ to 0. For this purpose we now define $\boldsymbol{\phi}_1$ and $\boldsymbol{\phi}_2$ such that for $z > z_2$

$$\text{and} \qquad \begin{aligned} \boldsymbol{\phi}_1(z) &= [1, -\alpha, \alpha^2, -\alpha^3]^{\mathrm{T}} \, e^{-\alpha z} \\ \boldsymbol{\phi}_2(z) &= [1, -\beta, \beta^2, -\beta^3]^{\mathrm{T}} \, e^{-\beta z}. \end{aligned} \right\}$$

(43.22)

The initial conditions for $\mathbf{y}$ at $z = z_2$ then follow immediately from equations (43.6) and, on discarding common factors, we obtain

$$\mathbf{y}(z_2) = [1, -(\alpha+\beta), \alpha^2 + \alpha\beta + \beta^2, \alpha\beta, -\alpha\beta(\alpha+\beta), \alpha^2\beta^2]^{\mathrm{T}}.$$

(43.23)

Thus, the eigenvalue problem can also be solved by backward integration of equation (43.8) subject to the initial conditions (43.23) and the eigenvalue relation then becomes simply $y_1(0) = 0$.

Problems for chapter 5

5.1 *The adjoint problem for plane Couette flow.* By considering the integral representation of the solutions of the adjoint equation

$$(\mathrm{D}^2 - \alpha^2)\{\varepsilon^3(\mathrm{D}^2 - \alpha^2) - \eta\}\phi^\dagger = 0,$$

show that it has a solution of well-balanced type which is given by

$$U_0^\dagger(\eta) = \alpha^{-1} \int_0^\alpha \cosh\{\eta t + \varepsilon^3(\alpha^2 t - \tfrac{1}{3}t^3)\} \, \mathrm{d}t$$

$$= (\alpha\eta)^{-1} \sinh \alpha\eta + O(\varepsilon^3),$$

that it has three solutions of balanced type which are given by

$$U_k^\dagger(\eta) = G^\dagger(\eta) - \varepsilon^{-1} B_k(\zeta + \alpha^2 \varepsilon^2, 0),$$

where

$$G^\dagger(\eta) = \int_0^\alpha \sinh\{\eta t + \varepsilon^3(\alpha^2 t - \tfrac{1}{3}t^3)\} \, dt$$

$$= \eta^{-1}(\cosh \alpha\eta - 1) + O(\varepsilon^3),$$

and that it has three solutions of dominant-recessive type which are given by

$$V_k^\dagger(\eta) = A_k(\zeta + \alpha^2 \varepsilon^2, 0).$$

Derive the three connexion formulae satisfied by these seven solutions. [Reid (1974a).]

5.2 *The second-order inhomogeneous problem for plane Couette flow.* Equation (34.1) can be integrated to give

$$(D^2 - \alpha^2)\phi = CA_k(\zeta + \alpha^2 \varepsilon^2),$$

where C is independent of η but could depend on α and/or ε. If $C = 0$ then we have the two well-balanced solutions given by equation (34.6). If $C \neq 0$, however, then the solutions of this inhomogeneous equation are clearly of dominant-recessive type. Show that a natural choice for C is ε^{-2} and that the solutions of this inhomogeneous equation then have uniform expansions of the form of equation (34.15) with a_0, b_0 and c_0 given by equations (34.21) and (34.24). [Reid (1974a).]

5.3 *The exact solution for plane Couette flow.* In Problem 5.2 let $C = \varepsilon^{-1}$ and write the solutions of dominant-recessive type in the form

$$V_k(\eta) = A(\eta, \varepsilon)A_k(\zeta + \alpha^2 \varepsilon^2, 1) + \varepsilon^2 B(\eta, \varepsilon)A_k(\zeta + \alpha^2 \varepsilon^2, 0)$$
$$+ \varepsilon C(\eta, \varepsilon)A_k(\zeta + \alpha^2 \varepsilon^2, -1).$$

Show that the coefficients A, B and C must satisfy the equations

$$A'' - \alpha^2 A = 0,$$

$$2A' + \eta B + 2\eta C' + C + \varepsilon^3(B'' + 2\alpha^2 C') = 1,$$

and

$$A + \eta C + \varepsilon^3(2B' + C'') = 0.$$

Hence *verify* that the exact solution is given by

$$A(\eta, \varepsilon) = \alpha^{-1} \sinh(\alpha\eta + \tfrac{2}{3}\alpha^3 \varepsilon^3),$$

$$B(\eta, \varepsilon) = -\alpha^{-1} \int_0^\alpha (\alpha - \tau) \sinh\{\eta\tau + \varepsilon^3(\alpha\tau^2 - \tfrac{1}{3}\tau^3)\} \, d\tau,$$

and

$$C(\eta, \varepsilon) = -\alpha^{-1} \int_0^\alpha \cosh\{\eta\tau + \varepsilon^3(\alpha\tau^2 - \tfrac{1}{3}\tau^3)\}\, d\tau.$$

Note that, for bounded values of α, A, B and C have *convergent* expansions in powers of ε^3. Discuss the relationship between this exact solution and the uniform expansions (34.15). [For a *derivation* of these results, see Reid (1979).]

*5.4 *The integral representations of the solutions of the comparison equation.* Show that the solutions of equation (39.2) can be expressed in terms of Laplace contour integrals of the form

$$\int_\mathscr{C} t^{\alpha-2} \exp(\eta t - \beta t^{-1} - \tfrac{1}{3}\varepsilon^3 t^3)\, dt,$$

where the path of integration $\mathscr{C}$ must be chosen so that

$$[t^\alpha \exp(\eta t - \beta t^{-1} - \tfrac{1}{3}\varepsilon^3 t^3)]_\mathscr{C} = 0.$$

By investigating various choices of $\mathscr{C}$, show that there exist solutions $u_0(\eta)$ and $u_3(\eta)$ of equation (39.2) which have outer expansions given by equations (39.6) and (39.7) respectively. [Lakin & Reid (1970).]

5.5 *The relationship between the Riccati method and the compound matrix method.* Consider equation (43.1) on the interval $z_1 \leqslant z \leqslant z_2$ and suppose that $\phi(z_1) = \phi'(z_1) = 0$. In the notation of § 30, let $\mathbf{u} = [\phi, \phi']^\mathrm{T}$ and $\mathbf{v} = [\phi'', \phi''']^\mathrm{T}$ so that the Riccati matrix is given by (cf. equation (30.29))

$$\mathbf{R} \equiv \begin{bmatrix} r_1 & r_2 \\ r_3 & r_4 \end{bmatrix} = \begin{bmatrix} \phi_1 & \phi_2 \\ \phi_1' & \phi_2' \end{bmatrix} \begin{bmatrix} \phi_1'' & \phi_2'' \\ \phi_1''' & \phi_2''' \end{bmatrix}^{-1}.$$

Hence show that

$$r_1 = y_3/y_6, \quad r_2 = -y_2/y_6, \quad r_3 = y_5/y_6, \quad r_4 = -y_4/y_6.$$

Also derive the differential equations satisfied by the elements of $\mathbf{R}$. [Davey (1979).]

Note, however, that if $\phi'(z_1) = \phi'''(z_1) = 0$ and if $\mathbf{u}$ and $\mathbf{v}$ are defined by equation (30.33) then[1]

$$r_1 = (y_4 + \alpha^2 y_1)/y_2, \quad r_2 = y_1/y_2,$$

$$r_3 = -(y_6 + \alpha^2 y_4 - \alpha^2 y_3 + \alpha^4 y_1)/y_2, \quad r_4 = (y_3 - \alpha^2 y_1)/y_2.$$

In the case of plane Poiseuille flow, show that

$$(\alpha^4 + 2i\alpha R)y_1 - \alpha^2 y_3 + \alpha^2 y_4 + y_6 = 0.$$

CHAPTER 6

ADDITIONAL TOPICS IN LINEAR STABILITY THEORY

There can never be a last word in regard to the axioms of any physical theory. All we can ask of them is that they lead to conclusions in agreement with observation. Sooner or later more refined observations will find the weak point in any set of physical axioms. Nature is far too complicated to be completely described in a few equations.

— J. L. Synge (1938)

Of the problems in the linear theory of hydrodynamic stability which have been solved, only a few have been presented in the previous chapters. These few were selected both to include the major results and to illustrate the fundamental ideas of the theory. A few more problems are treated briefly in this chapter, some to indicate the variety of the applications of the theory and others to cover more advanced topics.

44 Instability of parallel flow of a stratified fluid

44.1 *Introduction*

In Chapter 4 we developed the theory of stability of parallel flow of an inviscid fluid because it has direct applications and because it is fundamental to the theory of a viscous fluid. However, it is fundamental also to the theory of stability of parallel flow of an inviscid fluid under the action of external forces. There are many force fields important in diverse applications, but lack of space obliges us to confine our attention to a single external force, and we have chosen the force of buoyancy because it is as important as any for both theoretical developments and practical applications, notably to meteorology and oceanography.

The interplay of two major themes appears in this section. We introduced these themes in § 4 in our discussion of Kelvin–Helmholtz instability, where we described how two layered fluids at rest are stable to internal gravity waves and how a vortex sheet of homogeneous fluid is unstable. We found that for a wave of given length on a vortex sheet between two layered fluids, either the buoyancy is strong enough to overcome the tendency of shear instability and render the wave stable, or it is not and the wave is unstable. Here we shall consider this interplay of the stabilizing influence of gravity on a continuously stratified fluid and of the destabilizing influence of basic shear in a generalized form of the Kelvin–Helmholtz instability of § 4, although this generalization is sometimes just called Kelvin–Helmholtz instability. We shall see that the intuitions that heavy fluid below light fluid has a stabilizing influence, that strong shear has a destabilizing influence and that light fluid below heavy fluid renders any basic flow unstable are usually correct.

Following the analysis of §§ 4 and 21, we model the problem by taking a basic state in dynamic equilibrium, with velocity, density and pressure given by

$$\mathbf{u}_* = U_*(z_*)\mathbf{i}, \quad \rho_* = \bar{\rho}_*(z_*), \quad p_* = p_{0*} - g \int^{z_*} \bar{\rho}_*(z'_*)\, dz'_*$$

$$\text{for } z_{1*} \leqslant z_* \leqslant z_{2*} \qquad (44.1)$$

respectively, where z_* is the height and g the acceleration due to gravity, and where each of the horizontal planes at $z_* = z_{1*}$ and z_{2*} is taken to be rigid. We take scales L of length and V of velocity characteristic of the basic velocity distribution $U_*(z_*)$ and ρ_0 characteristic of the basic density $\rho_*(z_*)$. We assume first that the fluid is inviscid and incompressible, density being convected but not diffused. Then the equations of motion, incompressibility and continuity in dimensionless form give

$$\left.\begin{aligned} \rho\left(\frac{\partial \mathbf{u}}{\partial t} + \mathbf{u} \cdot \boldsymbol{\nabla} \mathbf{u}\right) &= -\boldsymbol{\nabla} p - F^{-2}\rho \mathbf{k}, \\ \boldsymbol{\nabla} \cdot \mathbf{u} &= 0 \end{aligned}\right\} \qquad (44.2)$$

and

$$\frac{\partial \rho}{\partial t} + \mathbf{u} \cdot \boldsymbol{\nabla}\rho = 0;$$

where the *Froude number* is defined by

$$F = V/\sqrt{gL}. \qquad (44.3)$$

It can be verified that the basic state satisfies the equations for *arbitrary* distributions $U(z)$ and $\bar{\rho}(z)$. To study its stability we put

$$
\left.
\begin{aligned}
\mathbf{u}(\mathbf{x}, t) &= U(z)\mathbf{i} + \mathbf{u}'(\mathbf{x}, t), \\
p(\mathbf{x}, t) &= p_0 - F^{-2} \int^z \bar{\rho}(z')\, dz' + p'(\mathbf{x}, t),
\end{aligned}
\right\}
\tag{44.4}
$$

and

$$
\rho(\mathbf{x}, t) = \bar{\rho}(z) + \rho'(\mathbf{x}, t),
$$

substitute these expressions into equations (44.2), and neglect quadratic terms in the small primed quantities to derive the linearized equations for the disturbance. We also take normal modes of the form

$$
\{\mathbf{u}'(\mathbf{x}, t), p'(\mathbf{x}, t), \rho'(\mathbf{x}, t)\} = \{\hat{\mathbf{u}}(z), \hat{p}(z), \hat{\rho}(z)\} \exp\{i(\alpha x + \beta y - \alpha c t)\},
\tag{44.5}
$$

in the manner of § 21. Thus equations (44.2) give

$$
\left.
\begin{aligned}
i\alpha\bar{\rho}(U - c)\hat{u} + \bar{\rho}U'\hat{w} &= -i\alpha\hat{p}, \\
i\alpha\bar{\rho}(U - c)\hat{v} &= -i\beta\hat{p}, \\
i\alpha\bar{\rho}(U - c)\hat{w} &= -D\hat{p} - F^{-2}\hat{\rho}, \\
i\alpha\hat{u} + i\beta\hat{v} + D\hat{w} &= 0,
\end{aligned}
\right\}
\tag{44.6}
$$

and

$$
i\alpha(U - c)\hat{\rho} + \bar{\rho}'\hat{w} = 0,
$$

where differentiation with respect to z of a basic quantity is denoted by a prime and of a perturbed quantity by D. One may eliminate $\hat{u}$ and $\hat{v}$ from the first two of equations (44.6) and from the fourth; then one may eliminate $\hat{p}$ and $\hat{\rho}$ in turn with the aid of the third and fifth of equations (44.6) to find

$$
(U - c)\{D^2\hat{w} - (\alpha^2 + \beta^2)\hat{w}\} - U''\hat{w}
$$
$$
- \frac{(\alpha^2 + \beta^2)\bar{\rho}'}{\alpha^2 F^2 (U - c)\bar{\rho}}\,\hat{w} + \frac{\bar{\rho}'}{\bar{\rho}}\{(U - c)D\hat{w} - U'\hat{w}\} = 0.
\tag{44.7}
$$

The conditions at the rigid boundaries give

$$
\hat{w} = 0 \quad \text{at } z = z_1, z_2.
\tag{44.8}
$$

Yih (1955) applied Squire's transformation to this system. It can be seen that the characteristics of two-dimensional waves are simply related to those of three-dimensional ones, for each three-dimensional wave with numbers (α, β) there being a two-dimensional one with the same value of the complex velocity c but with wavenumbers $((\alpha^2 + \beta^2)^{1/2}, 0)$ and Froude number $\alpha F/(\alpha^2 + \beta^2)^{1/2}$. Thus two-dimensional waves effectively have reduced gravity and magnified relative growth rate $(\alpha^2 + \beta^2)^{1/2}c_i$, and they are usually found to be the most unstable waves. For these reasons we shall henceforth consider only two-dimensional waves.

We used F^{-2} as a dimensionless measure of gravity because it was the first one at hand. However, it can now be seen from equation (44.7) that F^{-2} arises only in a product with $-\bar{\rho}'/\bar{\rho}$. Further, the physical effects of gravity on the modes will be seen to create internal rather than surface gravity waves (if the upper boundary is rigid). So we shall use the overall *Richardson number J*, defined as a characteristic value of

$$-\frac{\bar{\rho}'}{F^2\bar{\rho}} = -\frac{gL^2}{V^2}\frac{d\bar{\rho}_*}{\bar{\rho}_* dz_*},$$

rather than the Froude number. It is also convenient to define the *Brunt–Väisälä frequency* (or the *buoyancy frequency*) N_* by

$$N_*^2(z_*) = -g\frac{d\bar{\rho}_*}{dz_*}\Big/ \bar{\rho}_* = JN^2(z)V^2/L^2. \qquad (44.9)$$

Thus

$$JN^2/U'^2 = -g\frac{d\bar{\rho}_*}{dz_*}\Big/ \bar{\rho}_*\left(\frac{dU_*}{dz_*}\right)^2$$

is the *local Richardson number* of the flow at each height z_*, and we shall identify $J^{1/2}N$ as the dimensionless frequency of short internal gravity waves in the case for which $U \equiv 0$ and N is constant (see equation (44.17)).

In many applications of this theory it happens that $\bar{\rho}_*(z_*)$ varies much more slowly with height than $U_*(z_*)$, so that $-\bar{\rho}'/\bar{\rho} \ll 1$, yet J is nonetheless of order of magnitude unity because $F \ll 1$; in this approximation, which resembles the Boussinesq approximation, we neglect the last two terms of equation (44.7). Thus the effects of variation of density are neglected in the inertia but retained in the buoyancy.

Considering only two-dimensional disturbances, using the Richardson number instead of the Froude number, and neglecting the inertial effects of the variation of density, we can reduce the system described by equations (44.7) and (44.8) to the form

$$(U-c)(D^2-\alpha^2)\phi - U''\phi + JN^2\phi/(U-c) = 0, \quad (44.10)$$

$$\alpha\phi = 0 \quad \text{at } z = z_1 \text{ and } z_2, \quad (44.11)$$

where

$$u' = \partial\psi'/\partial z, \quad w' = -\partial\psi'/\partial x, \quad (44.12)$$

and

$$\psi' = \phi(z) \exp\{i\alpha(x-ct)\}. \quad (44.13)$$

Equation (44.10) is called the *Taylor–Goldstein equation* in honour of its derivation and exploitation by Taylor (1931) and Goldstein (1931), although a more general form of the equation was independently published by Haurwitz (1931) in the same year.

44.2 *Internal gravity waves and Rayleigh–Taylor instability*

The important special case of internal gravity waves or Rayleigh–Taylor instability arises when

$$U_* \equiv 0. \quad (44.14)$$

Of course this is equivalent to the case when U_* has any constant value, by Galilean transformation. Here there is no scale of the basic velocity, so we use dimensional variables, for which the Taylor–Goldstein problem reduces to

$$c_*^2(D_*^2 - \alpha_*^2)\phi + N_*^2\phi = 0 \quad (44.15)$$

and

$$\alpha_*\phi = 0 \quad \text{at } z_* = z_{1*}, z_{2*}, \quad (44.16)$$

a problem originally due to Rayleigh (1883b).

The problem has no solution in finite terms for a general function $N_*^2(z_*)$, but there are a few simple solutions known for particular functions $N_*^2(z_*)$. For the simplest, we follow Rayleigh (1883b) and suppose that $\bar\rho_* = \rho_0 \exp(-z_*/H)$ so that $N_*^2 = g/H$ is constant. We deduce at once that

$$c_*^2 = \{\alpha_*^2 + n^2\pi^2/(z_{2*}-z_{1*})^2\}^{-1}N_*^2,$$
$$\phi = \sin\{n\pi(z_*-z_{1*})/(z_{2*}-z_{1*})\} \quad \text{for } n = 1, 2, \ldots . \quad (44.17)$$

This gives a discrete spectrum of internal gravity waves, stable or unstable according as N_*^2 is positive or negative respectively, with a complete set of eigenfunctions.

A regular Sturm–Liouville problem with eigenvalue c_*^{-2} is given by equations (44.15) and (44.16). Detailed properties of its solutions, for both general and particular density distributions, are described by Yih (1965), Chapter 2 and Krauss (1966). They also treat cases when the upper boundary is a free surface, when the inertial terms due to the variation of density are not negligible, and when the fluid is compressible.

Rayleigh (1883b) himself proved the outstanding general property, namely that there is instability if and only if light fluid is locally below heavier fluid, i.e. there is instability if and only if N_*^2 is negative somewhere. His proof runs as follows. Multiply equation (44.15) by the complex conjugate ϕ^* and integrate from z_{1*} to z_{2*} to deduce that

$$c_*^2 \int_{z_{1*}}^{z_{2*}} \{ |D_* \phi|^2 + \alpha_*^2 |\phi|^2 \} \, dz_* = \int_{z_{1*}}^{z_{2*}} N_*^2 |\phi|^2 \, dz_*, \quad (44.18)$$

on integration by parts and use of boundary conditions (44.16). It follows that c_*^2 and therefore ϕ is real, and that c_* is real if $N_*^2 > 0$ everywhere. Thus there is stability if $N_*^2 > 0$ everywhere. To prove the converse, Rayleigh noted that the variational principle associated with the Sturm–Liouville problem described by equations (44.15) and (44.16) gives c_*^2 as the minimum of

$$\int_{z_{1*}}^{z_{2*}} N_*^2 f^2 \, dz_*$$

over the class of functions $f(z_*)$ with square-integrable derivatives such that

$$\int_{z_{1*}}^{z_{2*}} \{ (D_* f)^2 + \alpha_*^2 f^2 \} \, dz_* = 1.$$

It follows at once by the calculus of variations that $c_*^2 < 0$ if $N_*^2(z_*) < 0$ anywhere.

44.3 Kelvin–Helmholtz instability

The interplay of the effects of basic shear and buoyancy is seen in the eigensolutions of the Taylor–Goldstein problem described by equations (44.10) and (44.11). We shall show how, for a given flow and wavenumber, the modes may be divided into five classes, some of which may be empty:

(i) There is a finite class of non-singular unstable modes. These, giving instability, are the most important and have been given most attention in the literature. The class is empty certainly if the local Richardson number is everywhere greater than or equal to a quarter, the flow then being stable to all waves. These modes are in general the modifications by buoyancy of the unstable modes of shear instability (as described in Chapter 4), but, exceptionally, buoyancy with $N^2 > 0$ everywhere may render unstable a wave of given number which is stable when $N^2 \equiv 0$ (cf. Howard & Maslowe 1973).

(ii) There is an equal number of conjugate damped modes, forming a finite class of non-singular asymptotically stable modes.

(iii) The marginally stable modes form a finite class of singular neutral modes, each having a branch point at its critical layer.

(iv) There is a continuous spectrum of singular neutral modes, each eigenfunction having a singularity no worse than a discontinuous derivative at its critical layer. These are associated with algebraic rather than exponential decay of the disturbance.

(v) The internal gravity waves modified by the basic shear form a discrete class of stable modes when $N^2 > 0$ everywhere. There is a finite number or countable infinity of these waves, for which $c > U_{\max}$ or $c < U_{\min}$ and the eigenfunctions are non-singular. There are similar unstable modes when $N^2 < 0$ somewhere.

To discuss the properties of these classes of modes, first note, as in § 21, that we may take $\alpha \geqslant 0$ without loss of generality and that to each unstable mode there corresponds a conjugate stable mode.

The essential mechanism of the instability is the conversion of the available kinetic energy of relative motion of layers of the basic flow into kinetic energy of the disturbance, overcoming the potential energy needed to raise or lower fluid when $d\bar{\rho}_* / dz_* < 0$ everywhere. Thus shear tends to destabilize and buoyancy to stabilize the flow. To quantify these tendencies, suppose that two neighbouring fluid particles of equal volumes, at heights z_* and $z_* + \delta z_*$, are interchanged. Then the work δW per unit volume needed to overcome gravity and effect this interchange is given by

$$\delta W = -g \delta \bar{\rho}_* \delta z_*,$$

where $\delta \bar{\rho}_* = (\mathrm{d}\bar{\rho}_*/\mathrm{d}z_*)\delta z_*$. In order that the horizontal momentum of the inviscid fluid is conserved in the interchange, the particle originally at height z_* will plausibly have final velocity $(U_* + k\delta U_*)\mathbf{i}$ and the other particle $\{U_* + (1-k)\delta U_*\}\mathbf{i}$, where $\delta U_* = (\mathrm{d}U_*/\mathrm{d}z_*)\delta z_*$ and k is some number between zero and one. Then the kinetic energy δT per unit volume released by the basic flow in this way is given by

$$\begin{aligned}
\delta T &= \tfrac{1}{2}\bar{\rho}_* U_*^2 + \tfrac{1}{2}(\bar{\rho}_* + \delta\bar{\rho}_*)(U_* + \delta U_*)^2 - \tfrac{1}{2}\bar{\rho}_*(U_* + k\delta U_*)^2 \\
&\quad - \tfrac{1}{2}(\bar{\rho}_* + \delta\bar{\rho}_*)\{U_* + (1-k)\delta U_*\}^2 \\
&= k(1-k)\bar{\rho}_*(\delta U_*)^2 + U_*\delta U_*\delta\bar{\rho}_* \\
&\leqslant \tfrac{1}{4}\bar{\rho}_*(\delta U_*)^2 + U_*\delta U_*\delta\bar{\rho}_*,
\end{aligned}$$

the equality holding for $k = \tfrac{1}{2}$. Now a *necessary* condition for this interchange, and thus for instability, is that $\delta W \leqslant \delta T$ and therefore that somewhere in the field of flow

$$-g\frac{\mathrm{d}\bar{\rho}_*}{\mathrm{d}z_*} \leqslant \frac{1}{4}\bar{\rho}_*\left(\frac{\mathrm{d}U_*}{\mathrm{d}z_*}\right)^2 + U_*\frac{\mathrm{d}U_*}{\mathrm{d}z_*}\frac{\mathrm{d}\bar{\rho}_*}{\mathrm{d}z_*}, \qquad (44.19)$$

i.e.

$$-g\frac{\mathrm{d}\bar{\rho}_*}{\mathrm{d}z_*} \Big/ \bar{\rho}_*\left(\frac{\mathrm{d}U_*}{\mathrm{d}z_*}\right)^2 \leqslant \frac{1}{4} \qquad (44.20)$$

if the inertial effects of the variation of density are negligible. The essential idea of this argument is due to Richardson, who in 1920 applied it to turbulence. However, it has been recast by Prandtl, Taylor and many others since. The above form of the argument is essentially that of Chandrasekhar (1961), p. 491. The argument is heuristic in the sense that only energetics are considered, the detailed kinematics and dynamics of the interchange of the particles being ignored.

A rigorous form of the argument comes on assuming $c_i \neq 0$, defining H by

$$H = \phi/(U - c)^{1/2}, \qquad (44.21)$$

and substituting H for ϕ in the Taylor–Goldstein equation. This gives

$$D\{(U-c)DH\} - \{\alpha^2(U-c) + \tfrac{1}{2}U'' + (\tfrac{1}{4}U'^2 - JN^2)/(U-c)\}H = 0.$$

$$(44.22)$$

Multiplying this equation by H^* and integrating, we find

$$\int_{z_1}^{z_2} \left\{ (U-c)\{|DH|^2 + \alpha^2|H|^2\} + \tfrac{1}{2}U''|H|^2 + \frac{\tfrac{1}{4}U'^2 - JN^2}{U-c}|H|^2 \right\} \, dz = 0.$$

$$(44.23)$$

The imaginary part of this equation gives

$$-c_i \int_{z_1}^{z_2} \left\{ |DH|^2 + \alpha^2|H|^2 + (JN^2 - \tfrac{1}{4}U'^2)|H|^2/|U-c|^2 \right\} \, dz = 0.$$

$$(44.24)$$

Therefore

$$0 > -\int_{z_1}^{z_2} |DH|^2 \, dz$$

$$= \int_{z_1}^{z_2} \{(JN^2 - \tfrac{1}{4}U'^2) + \alpha^2|U-c|^2\}|H|^2/|U-c|^2 \, dz \quad (44.25)$$

if $c_i \neq 0$. Therefore the local Richardson number satisfies the inequality

$$JN^2/U'^2 < \tfrac{1}{4} \qquad (44.26)$$

somewhere in the field of flow. This is Howard's (1961) general proof of the necessary condition for instability which Miles (1961) had proved for a special class of flows. It can be stated in the form that *there is stability if the local Richardson number is everywhere greater than or equal to one quarter.* Howard (1961) also showed that the inequality (44.25) gives

$$\alpha^2 c_i^2 \leq \max_{z_1 \leq z \leq z_2} (\tfrac{1}{4}U'^2 - JN^2). \qquad (44.27)$$

The methods used in § 22 to prove Rayleigh's inflexion-point theorem can be shown to give the condition that instability implies

$$U'' = 2(U-c_r)JN^2/\{(U-c_r)^2 + c_i^2\} \qquad (44.28)$$

somewhere in the field of flow (Synge 1933). Unfortunately this condition involves the unknowns c_r and c_i, and so does not give a simple criterion like Rayleigh's. However, it implies that if $U'' \neq 0$ then

$$c_i \leqslant \max_{z_1 \leqslant z \leqslant z_2} |2JN^2/U''|, \tag{44.29}$$

because $c_i \leqslant |U - c| = |2(U - c_r)JN^2/U''(U - c)|$ somewhere.

Howard's semicircle theorem given by equation (22.39) can be shown to apply to the Taylor–Goldstein problem by the same method as before provided that $N^2 \geqslant 0$ everywhere in the field of flow. The method also shows that if $N^2 \leqslant 0$ everywhere then no non-singular neutral mode exists; unfortunately this does not imply instability because a continuous spectrum of singular stable modes with velocity c within the range of $U(z)$ may exist (cf. Case 1960b).

For any given basic flow the marginally stable modes are modifications of the s-modes of § 22 for the case $J = 0$, although the significance of a point of inflexion is lost when $J \neq 0$. For illustration, take the example

$$U = \tanh z, \quad N^2 = \operatorname{sech}^2 z \quad \text{for } -\infty < z < \infty. \tag{44.30}$$

Note that this Brunt–Väisälä frequency comes from taking $\ln \bar{\rho} \propto -\tanh z$, so that $\bar{\rho}(-\infty) - \bar{\rho}(0) \cong \bar{\rho}(0) - \bar{\rho}(\infty)$ and $\bar{\rho}(\infty)\bar{\rho}(-\infty) = 1$. Holmboe (cf. Miles 1963) verified that then a neutral eigensolution of the system described by equations (44.10) and (44.11) is given by

$$c = 0, \quad J = \alpha(1 - \alpha), \quad \phi = (\operatorname{sech} z)^\alpha (\tanh z)^{1-\alpha} \quad \text{for } 0 \leqslant \alpha \leqslant 1. \tag{44.31}$$

Miles (1963) examined the branch point at the critical layer where $z = 0$ in the limit as $c_i \downarrow 0$, finding that the solution (44.31) is the unique marginally stable solution provided that one interprets $(\tanh z)^{1-\alpha} > 0$ for $z > 0$ but

$$(\tanh z)^{1-\alpha} = e^{-i\pi(1-\alpha)}|\tanh z|^{1-\alpha} \quad \text{for } z < 0.$$

Hazel (1972) computed the unstable mode for various values of α and J. These results are shown in Fig. 6.1. It can be seen that $\alpha c_i \leqslant (\frac{1}{4} - J)^{1/2}$ in accord with inequality (44.27), equality being attained only when $J = \frac{1}{4}$ and $\alpha = \frac{1}{2}$. Thus the condition $J \geqslant \frac{1}{4}$ everywhere happens to be both necessary and sufficient for stability of the

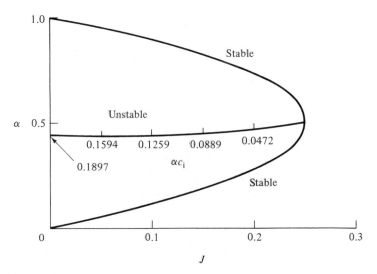

Fig. 6.1. Stability boundary and curve of maximum relative growth rate αc_i for the basic shear layer with $U = \tanh z$ and $N^2 = \text{sech}^2 z$ for $-\infty < z < \infty$. (After Hazel 1972.)

flow described by equations (44.30). The semicircle theorem is satisfied, it being found that $c_r = 0$ and $c_i \leqslant 1$ for the unstable mode (with $c_i = 1$ for $\alpha = 0$, $J = 0$). The s-solution can be seen to arise from equation (44.31) when $J = 0$ and $\alpha = \alpha_s = 1$.

Howard (1963a) showed that if the marginal curve for a given class of flows has equation

$$J = J_s(\alpha) \tag{44.32}$$

in the $J\alpha$-plane, then the generalization of formula (22.20) gives

$$\left(\frac{\partial c}{\partial \alpha^2}\right)_J = \lim_{c_i \downarrow 0}\left[\int_{z_1}^{z_2}\phi^2\,dz\bigg/\int_{z_1}^{z_2}\{2JN^2(U-c)^{-1} - U''\}\right.$$

$$\left. \times \phi^2/(U-c)^2\,dz\right]. \tag{44.33}$$

It also follows by partial differentiation that

$$\left(\frac{\partial c}{\partial J}\right)_\alpha = -\left(\frac{\partial c}{\partial \alpha}\right)_J\frac{dJ_s(\alpha_s)}{d\alpha_s} \tag{44.34}$$

at each value α_s of α on the marginal curve defined by equation (44.32). Care is needed to evaluate the integrals (44.33) in the limit as $c_i \downarrow 0$. Incorrect results may arise when a critical layer is at a boundary or when one of the integrals in equation (44.33) vanishes in the limit as $c_i \downarrow 0$ (Huppert 1973, Banks & Drazin 1973, Banks, Drazin & Zaturska 1976). Nonetheless Howard successfully applied his formulae (44.33) and (44.34) to Holmboe's solution (44.31), finding that on the inside of the marginal curve

$$\left(\frac{\partial c}{\partial \alpha}\right)_J = \frac{i \cos \pi \alpha_s}{\pi \alpha_s} B\left(\tfrac{3}{2} - \alpha_s, \alpha_s\right) \quad \text{and} \quad \left(\frac{\partial c}{\partial J}\right)_\alpha = -\frac{i}{\pi \alpha_s} B\left(\alpha_s, \tfrac{1}{2}\right),$$

in terms of the beta function.

If U, U' or $\bar{\rho}$ is discontinuous, at $z = z_0$ say, the arguments of § 23 show that

$$\Delta[\phi/(U-c)] = 0 \quad \text{and} \quad \Delta[\bar{\rho}\{(U-c)\mathrm{D}\phi - U'\phi - \phi/F^2(U-c)\}] = 0,$$
$$(44.35)$$

in order that the normal velocity and the pressure are continuous at the interface with mean position $z = z_0$.

These conditions are useful in working out simple examples for which the Taylor–Goldstein equation can be solved piecewise in terms of elementary functions. Here we take the example with the basic flow

$$U = \begin{cases} 1 \\ z \\ -1 \end{cases} \quad \text{and} \quad \bar{\rho} = \begin{cases} \rho_\infty & \text{for } z > 1 \\ \tfrac{1}{2}(\rho_\infty + \rho_{-\infty}) & \text{for } -1 < z < 1, \\ \rho_{-\infty} & \text{for } z < -1 \end{cases} \quad (44.36)$$

after Taylor (1931), § 3 and Goldstein (1931), §§ 3, 5. (Note that $N^2 = 0$ for $z \neq \pm 1$.) They found that there is instability if and only if

$$2\alpha/(1 + e^{-2\alpha}) - 1 < J < 2\alpha/(1 - e^{-2\alpha}) - 1, \quad (44.37)$$

where $J = gL(\rho_{-\infty} - \rho_\infty)/V^2(\rho_{-\infty} + \rho_\infty)$ and $(\rho_{-\infty} - \rho_\infty) \ll (\rho_{-\infty} + \rho_\infty)$, as shown in Fig. 6.2. When $J = 0$ this problem reduces to one of Rayleigh's (§ 23.2). It can be seen that for a given value of $\alpha > \alpha_s \cong 0.64$ a wave is stable when $J = 0$, unstable for a narrow band of positive values of J and stable again for large values of J. Thus buoyancy, which *seems* to have a stabilizing influence, can render unstable a wave that would otherwise be stable. Taylor (1931), p. 500 remarked that the unstable band centres round the particular wave whose length is such that a backward-moving internal gravity wave at $z = 1$, regarded as existing at a discontinuity of density in

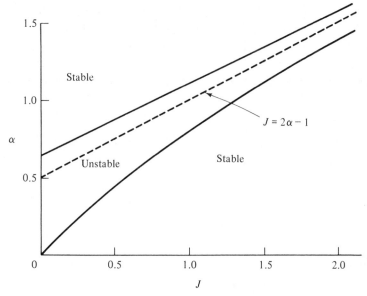

Fig. 6.2. The region of instability defined by equation (44.37) for the piecewise-linear basic shear layer defined by equations (44.36).

the fluid, all of which moves with velocity $U = 1$, has the same velocity as a similar forward-moving wave at $z = -1$, and therefore the instability may be regarded as due to a kind of resonance between these backward-moving and forward-moving internal gravity waves. Howard & Maslowe (1973) have discussed this phenomenon, and also how a flow which is stable when $J = 0$ may be unstable for $J > 0$.

Viscous fluid. The instability of plane parallel flow of a *viscous* stratified fluid leads to a sixth-order eigenvalue problem, it being desirable to allow for the diffusion of density as well as momentum in order to model a real fluid consistently. The problem thus depends upon the Reynolds and Prandtl numbers (if the density variation is due to temperature variation) as well as the Richardson number for each class of dynamically similar basic flows. Flows with points of inflexion, such as unbounded flows, are the most unstable at large values of the Reynolds number; thus the inviscid theory gives a useful criterion of their overall stability. Maslowe & Thompson (1971) found numerically some detailed stability characteristics of a hyperbolic-tangent shear layer, which exemplify this. Gage & Reid (1968) found the characteristics for plane Poiseuille flow of stratified viscous fluid with Prandtl number $Pr = 1$ by asymptotic methods for large Reynolds numbers. Gage (1971) generalized these methods to find a universal criterion for stability of flows without a point of inflexion when $Pr = 1$, namely that

$$JN^2/U'^2 > 0.0554 \qquad (44.38)$$

everywhere in the field of flow. Thus, for example, a boundary layer is stable at all values of the Reynolds number if this criterion is satisfied.

Viscous damping of instability caused by hydrostatically stable stratification (i.e. with $J > 0$) was investigated by Davey & Reid (1977b). They treated plane Couette flow with $U = z$ and $N^2 = z^2$ for $-1 \leqslant z \leqslant 1$. Now plane Couette flow of a homogeneous fluid ($J = 0$) is stable for all values of the Reynolds number R. Yet plane Couette flow of an inviscid fluid is *un*stable for all $J > \frac{1}{4}$ (Huppert 1973). Davey & Reid found that this instability of a viscous fluid with $Pr = 1$ occurs only for $R > R_c = 183$.

Observational work. Taylor's (1931) paper contained work originally written for his Adams Prize essay of 1915, but he delayed publication for 16 years in the vain hope of performing experiments to confirm his theory. Even today the technical difficulties of controlled experiments on the flow of stratified fluids are formidable. They have prevented close quantitative comparison with the theory, but Thorpe (1971) and Scotti & Corcos (1972) have found encouraging agreement between theoretical results and their experiments. The importance of this kind of instability in clear-air turbulence and in generating internal gravity waves in the atmosphere has also led meteorologists to relate their observations to the theory (see, for example, Gossard & Hooke (1975)).

For further details of generalized Kelvin–Helmholtz instability, the reader may consult the surveys by Drazin & Howard (1966) and by Howard & Maslowe (1973), and the book by Turner (1973).

45 Baroclinic instability

An important problem in meteorology is the large-scale instability of the westerly winds in mid-latitudes. In the laboratory this is

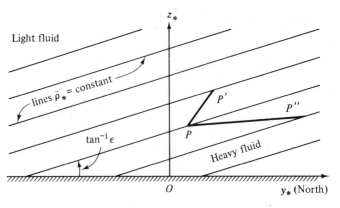

Fig. 6.3. Sketch to illustrate the physical argument of baroclinic instability.

modelled by the instability of shear flow in a differentially heated rotating annulus (see, for example, Fultz, Kaylor & Hide (FL 1965)). It is called *baroclinic instability* because it depends essentially upon the difference between the surfaces of constant density and of constant pressure in the fluid. To illustrate the essential physical mechanism, consider the basic density distribution $\bar{\rho}_* = \rho_0\{1 - k(z_* - \varepsilon y_*)\}$, where Oz_* is the upward vertical and Oy_* the northward direction. The slope ε of the surfaces of constant density in dynamic equilibrium is due to the rotation of the system, and is associated with an eastward (i.e. westerly) zonal flow. Now suppose that two fluid particles, P and P', are interchanged, P' coming from a level higher than P. We can plausibly assume that P keeps its original density, as in an adiabatic exchange. Then P will begin to sink back towards its original level if P is denser than P'. Therefore P will begin to sink back if the vector PP' lies outside the angle $\tan^{-1}\varepsilon$ shown in Fig. 6.3. Similarly P' will be lighter than the ambient fluid in its new position, and will tend to rise. If, however, particles P and P'' are interchanged, where P'' comes from a level higher than P but is denser, then P will tend to rise and P'' to fall, so that P and P'' will separate further; this is instability. In fact the mechanism we have just described heuristically is limited by kinematic and dynamic realities of particle exchange, which give rise to a criterion for instability, not instability in all circumstances as the description may seem to imply.

The background and literature of baroclinic instability have been surveyed by Drazin (1978) and Hart (1979). Scores of papers date from the work of Charney (1947). To introduce the topic here, we shall state and briefly solve a fundamental problem due to Eady (1949).

Consider then an inviscid non-conducting Boussinesq fluid in a frame of reference rotating with steady angular velocity $\Omega\mathbf{k}$ about the vertical. It follows that the governing equations are

$$\partial\mathbf{u}_*/\partial t_* + \mathbf{u}_* \cdot \boldsymbol{\nabla}_*\mathbf{u}_* + 2\Omega\mathbf{k}\times\mathbf{u}_* = -\rho_0^{-1}\boldsymbol{\nabla}_*p_* + \alpha g(\theta_* - \theta_0)\mathbf{k}, \quad (45.1)$$

$$\partial\theta_*/\partial t_* + \mathbf{u}_* \cdot \boldsymbol{\nabla}_*\theta_* = 0, \quad (45.2)$$

and

$$\boldsymbol{\nabla}_* \cdot \mathbf{u}_* = 0; \quad (45.3)$$

where $\mathbf{u}_*$ is the velocity relative to the rotating frame, $\rho_* = \rho_0\{1 - \alpha(\theta_* - \theta_0)\}$ the density, p_* the relative pressure and θ_* the temperature of the fluid.

We consider flow in a long rectangular channel with rigid walls at $y_* = 0, L$ and $z_* = 0, H$, so that

$$v_* = 0 \quad \text{at } y_* = 0, L, \quad (45.4)$$

and

$$w_* = 0 \quad \text{at } z_* = 0, H. \quad (45.5)$$

Suppose that the basic state is given by

$$\left.\begin{array}{l}
\mathbf{U}_* = VH^{-1}z_*\mathbf{i}, \quad \Theta_* = \theta_0 + (\Delta\theta)H^{-1}z_* - 2\Omega V(\alpha gH)^{-1}y_*, \\
\text{and} \\
P_* = \rho_0\{\tfrac{1}{2}\alpha g(\Delta\theta)H^{-1}z_*^2 - 2\Omega VH^{-1}y_*z_*\} \\
\quad \text{for } -\infty < x_* < \infty, 0 \leqslant y_* \leqslant L, 0 \leqslant z_* \leqslant H;
\end{array}\right\} \quad (45.6)$$

where $\Delta\theta$ is a constant scale of basic vertical temperature difference. (Note that this gives the basic density,

$$\bar{\rho}_* = \rho_0\{1 - \alpha(\Delta\theta)H^{-1}(z_* - 2\Omega Vy_*/\alpha g\Delta\theta)\},$$

of the form taken in the physical argument above.) Then it is convenient to define the dimensionless variables

$$\left.\begin{aligned}
(x, y) &= (x_*, y_*)/L, \quad z = z_*/H, \quad t = Vt_*/L, \\
(u, v) &= (u_*, v_*)/V, \quad w = Lw_*/VHRo, \\
\theta &= \alpha g H\{\theta_* - \theta_0 - (\Delta\theta)z_*/H\}/2\Omega VL, \\
p &= \{p_* - \tfrac{1}{2}\alpha\rho_0 g(\Delta\theta)H^{-1}z_*^2\}/2\Omega VL\rho_0;
\end{aligned}\right\} \qquad (45.7)$$

where the *Rossby number*, a characteristic ratio of the inertia to the Coriolis force, is given by

$$Ro = V/2\Omega L. \qquad (45.8)$$

With dimensionless basic flow

$$\mathbf{U} = z\mathbf{i}, \quad \Theta = -y, \quad P = -yz \quad \text{for } -\infty < x < \infty, \quad 0 \leqslant y, z \leqslant 1, \qquad (45.9)$$

and total flow

$$\mathbf{u} = \mathbf{U} + \mathbf{u}', \quad \theta = \Theta + \theta', \quad p = P + p', \qquad (45.10)$$

it follows that the governing equations for the perturbation become

$$Ro\left(\frac{\partial u'}{\partial t} + z\frac{\partial u'}{\partial x} + u'\frac{\partial u'}{\partial x} + v'\frac{\partial u'}{\partial y} + Ro\, w'\frac{\partial u'}{\partial z} + Ro\, w'\right)$$
$$- v' = -\frac{\partial p'}{\partial x}, \qquad (45.11)$$

$$Ro\left(\frac{\partial v'}{\partial t} + z\frac{\partial v'}{\partial x} + u'\frac{\partial v'}{\partial x} + v'\frac{\partial v'}{\partial y} + Ro\, w'\frac{\partial v'}{\partial z}\right) + u' = -\frac{\partial p'}{\partial y}, \qquad (45.12)$$

$$Ro^2 H^2 L^{-2}\left(\frac{\partial w'}{\partial t} + z\frac{\partial w'}{\partial x} + u'\frac{\partial w'}{\partial x} + v'\frac{\partial w'}{\partial y} + Ro\, w'\frac{\partial w'}{\partial z}\right) = -\frac{\partial p'}{\partial z} + \theta', \qquad (45.13)$$

$$\frac{\partial \theta'}{\partial t} + z\frac{\partial \theta'}{\partial x} + u'\frac{\partial \theta'}{\partial x} + v'\frac{\partial \theta'}{\partial y} + Ro\, w'\frac{\partial \theta'}{\partial z} - v' + Bw' = 0, \qquad (45.14)$$

$$\frac{\partial u'}{\partial x} + \frac{\partial v'}{\partial y} + Ro\,\frac{\partial w'}{\partial z} = 0; \qquad (45.15)$$

where the *Burger number* is given by

$$B = \alpha g H \Delta\theta/4\Omega^2 L^2. \qquad (45.16)$$

Note that $B = J Ro^2$, where the overall Richardson number is given by

$$J = \alpha g H \Delta\theta / V^2 \cong -gH^2 V^{-2} (\mathrm{d}\bar{\rho}_* / \bar{\rho}_* \mathrm{d}z_*). \qquad (45.17)$$

For the westerlies in the atmosphere it is found that $Ro \ll 1$, $H \ll L$ and $B \approx 1$, and therefore that the slope of the surfaces of constant density $\varepsilon = 2\Omega V / \alpha g \Delta\theta \ll 1$.

The boundary conditions (45.4) and (45.5) give

$$v' = 0 \quad \text{at } y = 0, 1 \qquad (45.18)$$

and

$$w' = 0 \quad \text{at } z = 0, 1. \qquad (45.19)$$

Linearizing the perturbation equations (45.11)–(45.14), we find

$$Ro\left(\frac{\partial u'}{\partial t} + z\frac{\partial u'}{\partial x} + Ro\, w'\right) - v' = -\frac{\partial p'}{\partial x}, \qquad (45.20)$$

$$Ro\left(\frac{\partial v'}{\partial t} + z\frac{\partial v'}{\partial x}\right) + u' = -\frac{\partial p'}{\partial y}, \qquad (45.21)$$

$$Ro^2 H^2 L^{-2}\left(\frac{\partial w'}{\partial t} + z\frac{\partial w'}{\partial x}\right) = -\frac{\partial p'}{\partial z} + \theta', \qquad (45.22)$$

and

$$\frac{\partial\theta'}{\partial t} + z\frac{\partial\theta'}{\partial x} - v' + Bw' = 0. \qquad (45.23)$$

Elimination of p' from equations (45.20) and (45.21) gives the vorticity equation,

$$Ro\left(\frac{\partial}{\partial t} + z\frac{\partial}{\partial x}\right)\left(\frac{\partial v'}{\partial x} - \frac{\partial u'}{\partial y}\right) - Ro^2\frac{\partial w'}{\partial y} + \frac{\partial u'}{\partial x} + \frac{\partial v'}{\partial y} = 0,$$

and thence

$$\left(\frac{\partial}{\partial t} + z\frac{\partial}{\partial x}\right)\left(\frac{\partial v'}{\partial x} - \frac{\partial u'}{\partial y}\right) - Ro\frac{\partial w'}{\partial y} - \frac{\partial w'}{\partial z} = 0 \qquad (45.24)$$

on use of equation (45.15). In the *geostrophic* limit as $Ro \to 0$, equations (45.21), (45.20) and (45.22) become

$$u' = -\frac{\partial p'}{\partial y}, \quad v' = \frac{\partial p'}{\partial x}, \quad \theta' = \frac{\partial p'}{\partial z}, \qquad (45.25)$$

and therefore equation (45.24) becomes

$$\left(\frac{\partial}{\partial t}+z\frac{\partial}{\partial x}\right)\left(\frac{\partial^2 p'}{\partial x^2}+\frac{\partial^2 p'}{\partial y^2}\right)=\frac{\partial w'}{\partial z}. \tag{45.26}$$

Similarly equation (45.23) becomes

$$\left(\frac{\partial}{\partial t}+z\frac{\partial}{\partial x}\right)\frac{\partial p'}{\partial z}-\frac{\partial p'}{\partial x}+Bw'=0. \tag{45.27}$$

Elimination of w' from equations (45.26) and (45.27) finally gives

$$\left(\frac{\partial}{\partial t}+z\frac{\partial}{\partial x}\right)\left\{\frac{\partial^2 p'}{\partial z^2}+B\left(\frac{\partial^2 p'}{\partial x^2}+\frac{\partial^2 p'}{\partial y^2}\right)\right\}=0. \tag{45.28}$$

The boundary conditions (45.18) and (45.19) give

$$\partial p'/\partial x = 0 \quad \text{at } y = 0, 1, \tag{45.29}$$

and

$$\left(\frac{\partial}{\partial t}+z\frac{\partial}{\partial x}\right)\frac{\partial p'}{\partial z}-\frac{\partial p'}{\partial x}=0 \quad \text{at } z=0, 1. \tag{45.30}$$

It may be noted that the eigenvalue problem defined by equations (45.28)–(45.30) is *mathematically* equivalent to the problem governing the stability of plane Couette flow of an inviscid homogeneous incompressible fluid in an inertial frame with constant pressure on the bounding planes (see equations (24.2) and (23.2)).

Take normal modes with

$$p' = \hat{p}(z) \exp\{i\alpha(x-ct)\} \sin\beta y. \tag{45.31}$$

Now $\beta = n\pi$ for some positive integer n in order to satisfy the boundary condition (45.29) on the vertical walls. Equation (45.28) gives

$$\frac{d^2\hat{p}}{dz^2}-B(\alpha^2+\beta^2)\hat{p}=0.$$

Therefore

$$\hat{p}=R\cosh 2qz + S\sinh 2qz; \tag{45.32}$$

where

$$q=\tfrac{1}{2}\{B(\alpha^2+n^2\pi^2)\}^{1/2} \text{ for } n=1,2,\ldots, \tag{45.33}$$

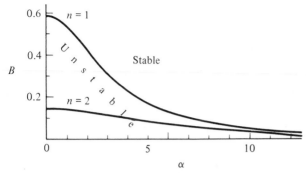

Fig. 6.4. The curves of marginal stability $B = 4q_c^2/(\alpha^2 + n^2\pi^2)$ for the first two modes of the Eady problem of baroclinic instability.

and R and S are some constants. Finally conditions (45.30) on the horizontal walls give the eigenvalue relation on elimination of R and S:

$$(4q^2 \sinh 2q)c^2 - (4q^2 \sinh 2q)c - \sinh 2q + 2q \cosh 2q = 0,$$

i.e.

$$c = \tfrac{1}{2} \pm \{(q \coth q - 1)(q \tanh q - 1)\}^{1/2}/2q. \qquad (45.34)$$

It follows that the modes are stable ($\alpha c_i \leqslant 0$) if and only if $q \geqslant q_c \cong 1.2$, where q_c is defined as the positive root of the transcendental equation $q \tanh q = 1$. Therefore the modes are stable if and only if

$$B \geqslant B_c(\alpha, n) = 4q_c^2/(\alpha^2 + n^2\pi^2) \qquad (45.35)$$

for given real α and positive integer n. Therefore the flow is stable if and only if

$$B \geqslant \max B_c(\alpha, n) \cong 0.58, \qquad (45.36)$$

the maximum occurring for $\alpha = 0$ and $n = 1$. The curves of marginal stability for the modes with $n = 1, 2$ are shown in Fig. 6.4. Note that the longest waves are unstable for the largest values of B but that their growth rates are not the largest because $\alpha c_i \to 0$ as $\alpha \to 0$.

46 Instability of the pinch

The confinement of a fluid in equilibrium by a magnetic field is a well-known topic in the subject of *magnetohydrodynamics*. Understanding of stability is very important in order to exploit such a

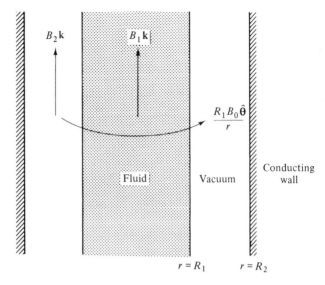

Fig. 6.5. The configuration of the linear pinch.

configuration to produce thermonuclear energy, and is introduced here by one problem of the *linear pinch*. Some aspects of this type of stability are shown in the film loop by Dattner (FL 1963). The wider background to magnetohydrodynamics, or *hydromagnetics*, may be followed up in many textbooks, e.g. Roberts (1967).

The problem of the stability of the linear pinch was solved by various authors independently (see, for example, Chandrasekhar (1961), pp. 575–576), but we shall adopt the approach of Tayler (1957a). Consider then the configuration of the basic state shown in Fig. 6.5. We suppose that an incompressible perfectly conducting fluid is confined in a cylinder by the basic magnetic field,

$$\mathbf{B} = \begin{cases} B_1\mathbf{k} & \text{for } 0 \leq r \leq R_1 \\ R_1 B_0\hat{\boldsymbol{\theta}}/r + B_2\mathbf{k} & \text{for } R_1 \leq r \leq R_2, \end{cases} \quad (46.1)$$

where cylindrical polar coordinates (r, θ, z) are used, $\mathbf{k}$ is the unit vector in the z-direction, $\hat{\boldsymbol{\theta}}$ is the unit vector in the θ-direction, and R_1, R_2, B_0, B_1 and B_2 are constants. We suppose that the fluid is at rest in the cylinder $0 \leq r \leq R_1$ and is surrounded by a vacuum, and that there is a perfectly conducting wall at $r = R_2$. To maintain the discontinuous field in the perfectly conducting fluid there must be a basic surface current at $r = R_1$, of strength

$$\mathbf{J} = \Delta[(\hat{\mathbf{r}} \times \mathbf{B})/\mu]$$

$$= \left(\frac{B_1}{\mu} - \frac{B_2}{\mu_0}\right)\hat{\boldsymbol{\theta}} + \frac{B_0}{\mu_0}\mathbf{k}, \tag{46.2}$$

where $\hat{\mathbf{r}}$ is the unit vector in the r-direction and μ is the magnetic permeability of the fluid. Also continuity of normal stress across the surface of the fluid implies that the uniform basic pressure inside is

$$p_0 = \frac{1}{2}\left(\frac{B_0^2}{\mu_0} + \frac{B_2^2}{\mu_0} - \frac{B_1^2}{\mu}\right); \tag{46.3}$$

note that p_0 must be positive on physical grounds.

It can be shown that the equations of motion of the fluid in this problem are as follows (see, for example, Roberts (1967)):

$$\frac{\partial \mathbf{u}}{\partial t} + \mathbf{u} \cdot \nabla \mathbf{u} = -\nabla\left(\frac{p}{\rho} + \frac{B^2}{2\mu\rho}\right) + \frac{1}{\mu\rho}\mathbf{B} \cdot \nabla \mathbf{B}, \tag{46.4}$$

$$\frac{\partial \mathbf{B}}{\partial t} = \nabla \times (\mathbf{u} \times \mathbf{B}), \tag{46.5}$$

$$\nabla \cdot \mathbf{u} = 0 \quad \text{and} \quad \nabla \cdot \mathbf{B} = 0. \tag{46.6}$$

The boundary conditions are that
  (i) a particle on the disturbed surface of the fluid remains there;
  (ii) the radial component B_r of the magnetic field vanishes on the conducting wall ($r = R_2$);
  (iii) the component of the magnetic field normal to the disturbed surface of the fluid is continuous;
  (iv) the normal component of the stress there is continuous.
Let the perturbed fields be

$$\mathbf{u} = \mathbf{u}', \quad \mathbf{B} = B_1\mathbf{k} + \mathbf{B}' \quad \text{and} \quad p = p_0 + p' \tag{46.7}$$

within the fluid, and

$$\mathbf{B} = R_1 B_0\hat{\boldsymbol{\theta}}/r + B_2\mathbf{k} + \mathbf{B}'' \tag{46.8}$$

in the vacuum between the surface of the fluid and the outer cylinder ($r = R_2$). Then the linearized equations for the disturbances are

$$\frac{\partial \mathbf{u}'}{\partial t} = -\boldsymbol{\nabla}\left(\frac{p'}{\rho} + \frac{B_1 B_z'}{\mu\rho}\right) + \frac{B_1}{\mu\rho}\frac{\partial \mathbf{B}'}{\partial z}, \tag{46.9}$$

$$\frac{\partial \mathbf{B}'}{\partial t} = B_1 \frac{\partial \mathbf{u}'}{\partial z}, \tag{46.10}$$

$$\boldsymbol{\nabla}\cdot\mathbf{u}' = 0 \quad \text{and} \quad \boldsymbol{\nabla}\cdot\mathbf{B}' = 0 \tag{46.11}$$

within the fluid; and

$$\boldsymbol{\nabla}\cdot\mathbf{B}'' = 0 \quad \text{and} \quad \boldsymbol{\nabla}\times\mathbf{B}'' = 0 \tag{46.12}$$

in the vacuum. Boundary condition (i) on the fluid surface, with equation $r = R_1 + \zeta'(\theta, z, t)$, say, gives

$$u_r' = \frac{\partial \zeta'}{\partial t} + u_\theta'\frac{\partial \zeta'}{r\partial \theta} + u_z'\frac{\partial \zeta'}{\partial z} \quad \text{at } r = R_1 + \zeta', \tag{46.13}$$

and, on linearization,

$$u_r' = \frac{\partial \zeta'}{\partial t} \quad \text{at } r = R_1. \tag{46.14}$$

Condition (ii) gives exactly

$$B_r'' = 0 \quad \text{at } r = R_2. \tag{46.15}$$

Condition (iii) gives

$$(B_1\mathbf{k} + \mathbf{B}')\cdot\mathbf{n} = (R_1 B_0\hat{\boldsymbol{\theta}}/r + B_2\mathbf{k} + \mathbf{B}'')\cdot\mathbf{n} \quad \text{at } r = R_1 + \zeta', \tag{46.16}$$

where the unit normal out of the fluid surface is

$$\mathbf{n} = \left(1, -\frac{\partial \zeta'}{r\partial\theta}, -\frac{\partial \zeta'}{\partial z}\right)\bigg/\left\{1 + \left(\frac{\partial \zeta'}{r\partial\theta}\right)^2 + \left(\frac{\partial \zeta'}{\partial z}\right)^2\right\}^{1/2}.$$

On linearization this gives

$$B_r' - B_1\frac{\partial \zeta'}{\partial z} = -\frac{B_0}{R_1}\frac{\partial \zeta'}{\partial\theta} - B_2\frac{\partial \zeta'}{\partial z} + B_r'' \quad \text{at } r = R_1. \tag{46.17}$$

Condition (iv) gives

$$p_0 + p' + (B_1\mathbf{k} + \mathbf{B}')^2/2\mu = (R_1 B_0\hat{\boldsymbol{\theta}}/r + B_2\mathbf{k} + \mathbf{B}'')^2/2\mu_0$$
$$\text{at } r = R_1 + \zeta', \tag{46.18}$$

and, on linearization, we have

$$p' + B_1 B_z'/\mu = (B_0 B_\theta'' + B_2 B_z'' - B_0^2\zeta'/R_1)/\mu_0 \quad \text{at } r = R_1. \tag{46.19}$$

This linearized system has coefficients dependent only on r, so we take normal modes with

$$p', \mathbf{B}', \mathbf{B}'' \propto \exp\{st + i(n\theta + kz)\},$$

where k is a real and n an integral wavenumber. Further, let

$$\Pi' = \frac{p'}{\rho} + \frac{B_1 B_z'}{\mu\rho}. \tag{46.20}$$

Now the divergence of equation (46.9), with equations (46.11), gives

$$\Delta\Pi' = 0, \tag{46.21}$$

where $\Delta = \partial^2/\partial r^2 + \partial^2/r^2\partial\theta^2 + \partial^2/\partial z^2$ is the Laplacian here. Therefore

$$\Pi' = AI_n(kr)\exp\{st + i(n\theta + kz)\} \tag{46.22}$$

for some complex constant A, where I_n is the modified Bessel function of order n. The other solution, K_n, is rejected because the pressure is bounded at $r = 0$.

Also equation (46.10) gives

$$\mathbf{B}' = ikB_1\mathbf{u}'/s. \tag{46.23}$$

Whence it follows that equation (46.9) gives

$$(s^2 + k^2 B_1^2/\mu\rho)\mathbf{u}' = -s\boldsymbol{\nabla}\Pi'. \tag{46.24}$$

Equations (46.12) give the potential field in the vacuum,

$$\mathbf{B}'' = \boldsymbol{\nabla}[\{CI_n(kr) + DK_n(kr)\}\exp\{st + i(n\theta + kz)\}], \tag{46.25}$$

for some constants C and D.

Boundary condition (46.14), with equations (46.22) and (46.24), gives

$$\zeta' = -\frac{kAI_n'(kR_1)}{s^2 + k^2 B_1^2/\mu\rho}\exp\{st + i(n\theta + kz)\}; \tag{46.26}$$

and equation (46.15) gives

$$CI_n'(kR_2) + DK_n'(kR_2) = 0. \tag{46.27}$$

Then condition (46.17) gives

$$\frac{ikB_0(n + kR_1B_2/B_0)AI_n'(kR_1)}{R_1(s^2 + k^2 B_1^2/\mu\rho)}$$
$$+ k\{CI_n'(kR_1) + DK_n'(kR_1)\} = 0; \tag{46.28}$$

and equation (46.19) gives

$$\left\{ R_1 \mu_0 \rho I_n(kR_1) - \frac{kB_0^2 I_n'(kR_1)}{s^2 + k^2 B_1^2/\mu\rho} \right\} A$$

$$= iB_0(n + kR_1 B_2/B_0)\{CI_n(kR_1) + DK_n(kR_1)\}. \quad (46.29)$$

Elimination of A, C and D from equations (46.27)–(46.29) gives the eigenvalue relation

$$\frac{R_1^2 \mu_0 \rho}{B_0^2} \frac{I_n(kR_1)}{I_n'(kR_1)} (s^2 + k^2 B_1^2/\mu\rho)$$

$$= kR_1 + (n + kR_1 B_2/B_0)^2$$

$$\times \frac{I_n(kR_1)K_n'(kR_2) - K_n(kR_1)I_n'(kR_2)}{I_n'(kR_1)K_n'(kR_2) - K_n'(kR_1)I_n'(kR_2)}, \quad (46.30)$$

i.e.

$$\frac{I_n(\alpha\eta)}{I_n'(\alpha\eta)} (R_1^2 \mu_0 \rho s^2/B_0^2 + \alpha^2 \beta_1^2 \eta^2 \mu_0/\mu)$$

$$= \alpha\eta + (n + \alpha\beta_2\eta)^2 \frac{I_n(\alpha\eta)K_n'(\alpha) - K_n(\alpha\eta)I_n'(\alpha)}{I_n'(\alpha\eta)K_n'(\alpha) - K_n'(\alpha\eta)I_n'(\alpha)}, \quad (46.31)$$

where

$$\alpha = kR_2, \quad \eta = R_1/R_2, \quad \beta_1 = B_1/B_0 \quad \text{and} \quad \beta_2 = B_2/B_0. \quad (46.32)$$

It can be seen that s^2 is real, and therefore that the principle of exchange of stabilities is valid, and the pinch is stable if and only if

$$\alpha\eta + (n + \alpha\beta_2\eta)^2 \frac{I_n(\alpha\eta)K_n'(\alpha) - K_n(\alpha\eta)I_n'(\alpha)}{I_n'(\alpha\eta)K_n'(\alpha) - K_n'(\alpha\eta)I_n'(\alpha)}$$

$$\leqslant \frac{\alpha^2 \beta_1^2 \eta^2 \mu_0 I_n(\alpha\eta)}{\mu I_n'(\alpha\eta)} \quad (46.33)$$

for all real k and integral n. Also s^2 is an even function of k and is unchanged when both n and β_2 change sign.

If $B_1 = 0$ we see that the inequality (46.33) gives instability because then $\beta_1 = 0$ and we may consider $\alpha = -n/\beta_2 > 0$.

If $n = 0$, then the inequality (46.33) becomes

$$1 + \alpha\beta_2^2\eta \frac{I_0(\alpha\eta)K_1(\alpha) + K_0(\alpha\eta)I_1(\alpha)}{I_1(\alpha\eta)K_1(\alpha) - K_1(\alpha\eta)I_1(\alpha)} \leqslant \frac{\alpha\beta_1^2\eta\mu_0 I_0(\alpha\eta)}{\mu I_1(\alpha\eta)}. \quad (46.34)$$

By the use of known properties of the Bessel functions, it can be shown (Tayler 1957a) that if the inequality (46.34) holds for some $\alpha \geq 0$ then it holds for each greater value of α. Now the inequality (46.34) is even in α, so a necessary and sufficient condition for stability to axisymmetric disturbances comes in the limit as $\alpha \to 0$. The condition is therefore that

$$\beta_1^2 \geq \tfrac{1}{2} - \beta_2^2 \eta^2 / (1 - \eta^2). \tag{46.35}$$

Further discussion of the complicated stability condition (46.33) needs a little numerical analysis, given by Tayler (1957a). It is found that if there is no outside wall ($\eta = 0$) then all modes other than some with $n = -1$ may be stabilized by suitable choice of β_1 and β_2. However, even if $n = -1$, all modes may be stabilized when η is greater than some critical value.

47 Development of linear instability in time and space

At various points, in particular § 24, we have remarked upon the relationship between the normal modes and the development of a localized disturbance in time and space. These remarks are amplified with a few technical details in this section.

47.1 *Initial-value problems*

For any steady basic flow in a bounded domain the normal modes are wholly discrete, and there exists in general just one mode whose relative growth rate, σ_m say, is distinctly greater than that of each of the countable infinity of the other modes. This is the most unstable, or least stable, mode. After some time of exponential growth or damping of components of a general initial disturbance, the most unstable mode outstrips the others and dominates the instability. So the separation of modes gives a reliable view of linear instability, the instability having the character of a single mode, namely the most unstable one. Even in this case of discrete modes it must be remembered that nonlinearity is likely in practice to modify an unstable disturbance after a short time of exponential growth.

However, for a basic flow in an unbounded domain the normal modes vary continuously with respect to one or more wavenumbers, and the most unstable mode is merely first among equals because neighbouring modes have relative growth rates arbitrarily close to σ_m. Therefore an unstable disturbance, so long as it is governed by the linear approximation, has the character of a group of waves similar to the most unstable mode, that is to say the character of a wavepacket rather than a single wave. This packet moves at its group velocity (possibly zero) and spreads while it grows.

These ideas can be illustrated by an initial-value problem for Tollmien–Schlichting waves, which we describe below after Benjamin (1961). In the theory of the Orr–Sommerfeld problem (Chapter 4) we found a spectrum of normal modes, which by integration with respect to the wavenumbers α and β can be superposed to give the velocity component

$$w'(\mathbf{x}, t) = \mathrm{Re}\Bigg[\int_{-\infty}^{\infty} \int_{-\infty}^{\infty} \sum_{j=1}^{\infty} \frac{g_j(\alpha, \beta)\hat{w}_j(z)}{\hat{w}_j(0)}$$

$$\times \exp\{\mathrm{i}(\alpha x + \beta y) + s_j t\}\, \mathrm{d}\alpha\, \mathrm{d}\beta\Bigg], \qquad (47.1)$$

where the jth eigenfunction $\hat{w}_j(z) = \hat{w}_j(z, \alpha, \beta, R)$ has eigenvalue $s = -\mathrm{i}\alpha c = \sigma + \mathrm{i}\omega = s_j(\alpha, \beta, R)$, and $g_j(\alpha, \beta)$ is the generalized Fourier transform of the initial distribution $w'(\mathbf{x}, 0)$. (There are similar expressions for the other velocity perturbations u' and v', and for the pressure perturbation p'.) In the ranges of α, β and R of chief interest to us, that is to say in the ranges where there is instability, it seems that the same mode j is the most unstable. So we may simplify the discussion by ignoring other values of the integer j. Accordingly, we shall henceforth omit the summation over j, the subscripts j, etc., considering only the value of j which corresponds to the most unstable mode.

Expression (47.1) can be evaluated asymptotically as $t \to +\infty$ by Laplace's method (cf. Olver 1974, pp. 80–96), on the basis that the most unstable mode and its neighbouring modes dominate the others after some time of exponential growth. We must look at a few of the details to see how to apply Laplace's method here. First note that Squire's transformation gives $c = f(\tilde{\alpha}\,\mathrm{sgn}\,\alpha, \tilde{R})$ for some function $f = f_r + \mathrm{i}f_i$, where $\tilde{\alpha} = (\alpha^2 + \beta^2)^{1/2}$ and $\tilde{R} = |\alpha|R/\tilde{\alpha}$. Also $f(-\alpha, R) = f^*(\alpha, R)$, $\hat{w}(z, -\alpha, \beta, R) = w^*(z, \alpha, \beta, R)$. From this it can be shown (Watson 1960a) that the most unstable mode is two-dimensional, at least for a substantial range of $R > R_c$ if not for all R. So we may take $\alpha = \alpha_m > 0$ and $\beta = 0$ for a most unstable mode in the range of Reynolds numbers of chief interest. There is also a conjugate most unstable mode with $\sigma = \sigma_m$, $\alpha = -\alpha_m$ and $\beta = 0$ and complex conjugate eigenfunction. Expanding the relative growth rate about its maximum value, we have

$$\sigma = \sigma_m - p(\alpha - \alpha_m)^2 - q\beta^2 + \cdots, \tag{47.2}$$

where

$$p = -\frac{1}{2}\left(\frac{\partial^2 \sigma}{\partial \alpha^2}\right)_m = -\frac{1}{2}\alpha_m\left(\frac{\partial^2 f_i}{\partial \tilde{\alpha}^2}\right)_m + \frac{1}{\alpha_m}(f_i)_m > 0, \tag{47.3}$$

and

$$q = -\frac{1}{2}\left(\frac{\partial^2 \sigma}{\partial \beta^2}\right)_m = \frac{1}{2\alpha_m}\left\{(f_i)_m + R\left(\frac{\partial f_i}{\partial \tilde{R}}\right)_m\right\} > 0. \tag{47.4}$$

Similarly, the expansion of the angular frequency is

$$\omega = \omega_m - U_m(\alpha - \alpha_m) + \kappa(\alpha - \alpha_m)^2 + \mu\beta^2 + \cdots, \tag{47.5}$$

where

$$\omega_m = -\alpha_m(c_r)_m, \quad \kappa = \frac{1}{2}\left(\frac{\partial^2 \omega}{\partial \alpha^2}\right)_m, \quad \mu = \frac{1}{2}\left(\frac{\partial^2 \omega}{\partial \beta^2}\right)_m \tag{47.6}$$

and the group velocity of the most unstable wave is

$$U_m = -\left(\frac{\partial \omega}{\partial \alpha}\right)_m = (f_r)_m + \alpha_m\left(\frac{\partial f_r}{\partial \tilde{\alpha}}\right)_m. \tag{47.7}$$

(It seems that $U_m > 0$ for flows of interest.) It is also useful to define

$$X = x - U_m t, \quad V = U_m - (c_r)_m = \alpha_m\left(\frac{\partial f_r}{\partial \tilde{\alpha}}\right)_m \quad \text{and} \quad \zeta = \alpha - \alpha_m. \tag{47.8}$$

Now the imaginary part of the exponent of the integrand of equation (47.1) may be arranged in the form

$$\text{Im}\{i(\alpha x + \beta y) + st\}$$
$$= \alpha_m(X + Vt) + \zeta X + \beta y + \kappa\zeta^2 t + \mu\beta^2 t + \text{cubic terms in } \beta, \zeta \tag{47.9}$$

for small β and ζ. It can thence be shown that contributions from the conjugate pair of most unstable modes in expression (47.1) give

$$w'(\mathbf{x}, t) \sim g(\alpha_m, 0)I + g(-\alpha_m, 0)I^* \quad \text{as } t \to \infty, \tag{47.10}$$

where

$$I = \frac{\hat{w}(z, \alpha_m, 0, R)}{\hat{w}(0, \alpha_m, 0, R)} \exp\{\sigma_m t + i\alpha_m(X + Vt)\}$$

$$\times \int_{-\infty}^{\infty} \exp\{i\zeta X + (-p + i\kappa)\zeta^2 t\}\, d\zeta$$

$$\times \int_{-\infty}^{\infty} \exp\{i\beta y + (-q + i\mu)\beta^2 t\}\, d\beta. \qquad (47.11)$$

This gives

$$w' \propto t^{-1} \exp\left\{\sigma_m t + i\alpha_m(X + Vt) - \frac{p + i\kappa}{4(p^2 + \kappa^2)t} X^2 - \frac{q + i\mu}{4(q^2 + \mu^2)t} y^2\right\}$$

$$\text{as } t \to \infty. \quad (47.12)$$

The other perturbation quantities can be treated similarly.

It can be seen from equation (47.12) that the amplitude of the wavepacket grows like $t^{-1} \exp(\sigma_m t)$, *not* like an exponential as does the amplitude of a single mode. The factor $1/t$ arises from interference of components of the packet. The extent of the disturbance in the xy-plane may be said to be confined within an ellipse similar to the one with equation

$$\frac{pX^2}{p^2 + \kappa^2} + \frac{qy^2}{q^2 + \mu^2} = 4(\ln 10)t, \qquad (47.13)$$

in the sense that outside this ellipse the magnitude of the disturbance is less than the magnitude at the centre by a factor of more than 10. Now the centre of the ellipse has equations $X = 0$, i.e. $x = U_m t$, and $y = 0$, and the area is proportional to t. So there is an expanding elliptically shaped disturbance travelling downstream at the group velocity U_m. In addition, the disturbance varies in the z-direction like a linear combination of the conjugate pair of eigenfunctions for the most unstable modes.

Benjamin (1961), Fig. 4, described his experiments on a slightly unstable film flowing down an inclined plane, which vividly illustrated his theory.

In independent work, Criminale & Kovasznay (1962) analysed numerically an initial-value problem for a boundary layer. Like Benjamin, they considered the development of a pulse, i.e. of a disturbance forced only at an initial instant. Gaster (1968) studied the development of a pulse in a boundary layer and Gaster & Davey (1968) that in a wake.

The development of a pulse may be regarded as a model of the early stages of a turbulent spot moving downstream. However, instability may arise from local roughness on a wall, and in many controlled experiments instability is continuously triggered at a fixed point by a vibrating ribbon or microphone. Accordingly Gaster (1965, 1975) studied the forced oscillation of a slightly unstable boundary layer due to a vibrating source of fixed angular frequency ω at a fixed point on the wall. He supposed that the source was turned on at an initial instant and led to two-dimensional flow. His asymptotic analysis of the disturbance as $t \to \infty$ indicates that the disturbance is ultimately only downstream of the source, where it varies like $\exp\{i(\alpha x + \omega t)\}$, α being a *complex* function of ω whose imaginary part α_i is negative. Thus, in the limit as $t \to \infty$ for a fixed x, the disturbance grows exponentially with x downstream. The disturbance is nonetheless localized at each instant so that it vanishes in the limit as $x \to \infty$ for fixed t. (If the group velocity were negative, the disturbance would grow exponentially *up*stream of the source.) This result is used to support the ideas of the next subsection. Further details of the initial-value problem are given by Craik (1981).

47.2 *Spatially growing modes*

Landau & Lifshitz (1959), § 29 were the first to suggest the use of modes which oscillate sinusoidally in time but are damped or amplified exponentially in distance downstream, although such modes had been used by plasma physicists since the work of Landau in 1946. Thus $s = i\omega$ for a *real* frequency ω and *complex* wavenumber $\alpha = \alpha_r + i\alpha_i$, where the eigenvalue problem determines the relation $\alpha = \alpha(\omega)$. Such a disturbance, varying like $\exp\{i(\alpha_r x + \omega t) - \alpha_i x\}$, is called a *spatially growing mode* or a *spatial mode* to distinguish it from an orthodox *temporal* mode with real wavenumber and complex frequency. Spatial modes are inappropriate for stationary instability, such as Bénard convection or instability of Couette flow. They have, however, the obvious advantage for parallel flows that an unstable disturbance is observed to propagate downstream while it grows.

Watson (1962) used spatial modes for a weakly nonlinear theory of plane Poiseuille flow, and Gaster (1962) examined the relationship between spatial and temporal modes. Since then spatial modes have been used in the theory of instability of many parallel flows, and they seem in general to be better than temporal modes as models of instability observed in experiments.

Nonetheless a stronger justification of the use of spatial modes is desirable. Gaster's (1965) work on the development of a disturbance due to a continuously oscillating source which is fixed in space and turned on at $t = 0$ (say), the work we have just discussed, suggests a general basis for the justification of the use of spatial modes, but we shall examine the matter further here. To simplify our examination a little, we shall consider only two-dimensional flow, so that $\beta = 0$. Then the orthodox eigenvalue problem of finding a complex frequency αc (given a real wavenumber α) is no better posed than the problem of finding a complex wavenumber (given a real frequency), because the eigenvalue relation of the Orr–Sommerfeld problem has the form

$$\mathscr{F}(\alpha, s) = 0, \qquad (47.14)$$

where $\mathscr{F}$ is some integral function of α and $s = -i\alpha c$ (and of R). However, asymptotic solutions of the Orr–Sommerfeld equation for large R are more easily used and numerical solutions are slightly easier to find when α is real. Also the introduction of temporal modes arises from the Fourier transforms in space and Laplace transform in time of an initial-value problem for the linearized equations and boundary conditions in a convincing way, even though a rigorous justification may be lacking. This is in contrast to the use of spatial modes in an unbounded fluid, because these modes do not satisfy the boundary conditions both far upstream and downstream at any instant and therefore are mathematically suspect. This suspicion may be dispelled by solution of an initial-value problem with a continuously oscillating source, because the unboundedness of the disturbance as $x \to +\infty$ may arise only *after* taking the limit as $t \to \infty$ for *fixed* x, and, indeed, Gaster (1965) showed this for one case. In applications of spatial modes to hydrodynamic stability it is accordingly the custom to assume that the flow is stable if $\alpha_i \geqslant 0$ for all real ω.

That this assumption is not always true can be seen by use of a simple example, the Korteweg–de Vries equation in the form

$$\frac{\partial u}{\partial t}+(1+u)\frac{\partial u}{\partial x}+\frac{\partial^3 u}{\partial x^3}=0. \qquad (47.15)$$

On linearization one finds

$$\frac{\partial u'}{\partial t}+\frac{\partial u'}{\partial x}+\frac{\partial^3 u'}{\partial x^3}=0, \qquad (47.16)$$

and on taking normal modes with $u'=\exp(ikx+st)$, we have

$$s=i(k^3-k), \qquad (47.17)$$

which is purely imaginary for all real k. Therefore the 'flow' is stable. Note that the angular frequency $\omega=\operatorname{Im} s$ and the group velocity

$$c_g=-\frac{d\omega}{dk}=1-3k^2.$$

Therefore $c_g=0$ if $k=\pm 1/\sqrt{3}$, for which values $\omega=\mp 2/3\sqrt{3}$ respectively. If, however, one considers spatial modes one finds $k=k(\omega)$ by solving the cubic

$$k^3-k-\omega=0.$$

If $\omega^2=4/27$, i.e. $\omega=\pm 2/3\sqrt{3}$, then the roots of the cubic are $\mp 1/\sqrt{3}$, $\mp 1/\sqrt{3}$ and $\pm 2/\sqrt{3}$ respectively. The cubic has three real roots if $\omega^2<4/27$, but one real root and two complex conjugate ones if $\omega^2>4/27$. It is nonetheless false to assume that the 'flow' is unstable. Indeed, solution of the initial-value problem confirms that the spatial modes due to a continuously oscillating source occur in the direction indicated by their group velocity (i.e. downstream of the source if the group velocity is positive and upstream if negative) and decay exponentially away from the source (Drazin 1977a). Problem 6.12 is another example of how naive use of spatial modes may indicate that a stable flow is unstable.

This difficulty has been recognized for a long time by plasma physicists, and is discussed by Clemmow & Dougherty (1969), §6.1. They consider the solutions of some simple initial-value problems in the form of Fourier integrals, and show that if the real k-axis may be smoothly transformed into the real ω-axis then the behaviour of the spatial modes correctly indicates stability or instability. Although little is known of the analytical properties of the function $s(k)$ for most problems of hydrodynamic stability, it seems that spatial modes do correctly indicate the stability or instability of channel flows, boundary layers, jets, free shear layers and pipe flows.

In the most important case, that of marginal stability, the spatial and temporal modes coincide, because then $s = i\omega = -i\alpha c = -i\alpha c_r$ is purely imaginary. In considering slightly unstable temporal modes Schlichting (1933), p. 199 recognized the importance of group velocity. In the present context of the relationship of spatial to temporal modes, Gaster (1962) examined the mathematical implications of relation (47.14) near marginal stability and was led to the group velocity again. On assuming that equation (47.14) gives s as an analytic function of α, the Cauchy–Riemann relations follow:

$$\frac{\partial \sigma}{\partial \alpha_r} = \frac{\partial \omega}{\partial \alpha_i}, \quad \frac{\partial \sigma}{\partial \alpha_i} = -\frac{\partial \omega}{\partial \alpha_r}. \qquad (47.18)$$

The former relation gives the same angular frequency ω of spatial and temporal modes in the limit as $\sigma \to 0$, i.e. as marginal stability is approached. The wavenumber α_r is likewise seen to be the same for spatial and temporal modes near marginal stability, by regarding α as an analytic function of s. The latter Cauchy–Riemann relation (47.18) gives

$$\alpha_i \sim -\sigma \Big/ \left(\frac{\partial \omega}{\partial \alpha_r}\right)_{\sigma=0} = -\sigma/c_g \quad \text{as } \sigma \to 0, \qquad (47.19)$$

where c_g is the group velocity. (This is the basis of equation (32.3).) It is important to note that *small* spatial growth of a mode is related to the temporal growth by the *group* not the phase velocity.

Also three-dimensional spatial modes should be considered. Now Squire's transformation gives $\omega = \alpha f(\tilde{\alpha} \text{ sgn } \tilde{\alpha}, \alpha R/\tilde{\alpha})$, and we must not put $\beta = 0$ and $\tilde{\alpha} = \alpha$ to determine $\alpha = \alpha_r + i\alpha_i$ if the source of frequency ω is not two-dimensional. The problem to determine α and β from ω is indeterminate unless an extra complex equation is given, and this can only be given by the nature of the source and of the initial-value problem. In fact, of all the unstable three-dimensional spatial modes of given ω and R, the two-dimensional one with $\beta = 0$ is not necessarily the most rapidly growing.

In summary, spatial modes seem to describe observed weak instability of parallel flows more faithfully than temporal modes, although their use has not been justified mathematically in an entirely convincing or complete way as yet. Spatial modes may emerge as $t \to \infty$ on solution of the linearized initial-value problem, in which the functional relationship between the complex variables s and α is used to evaluate a contour integral. Even when unstable spatial modes can be shown to emerge from the initial-value problem, it must be remembered that nonlinearity will become significant as the disturbance grows downstream and that therefore the linearization of the problem becomes invalid. So, while awaiting further work to clarify the matter, one may say that spatial modes seem a valuable tool in studying weak linear instability of flows forced at a fixed frequency.

48 Instability of unsteady flows

48.1 *Introduction*

Each of the basic flows we have considered is steady, yet in Nature and in the laboratory laminar flows are never perfectly steady and are often very unsteady. For example, to investigate Bénard convection in the laboratory, the heating of the lower plate must be switched on at some moment and may be given the desired flux almost at once or may be gradually intensified so that the basic temperature distribution varies with time, albeit slowly. This is typical of the setting up of any steady basic flow, and raises the question whether a nearly steady flow has stability characteristics close to those of the steady flow it approximates. Again, many of the exact solutions of the Navier–Stokes equations which are known depend upon time. The flow due to an infinite flat plate which, starting at $t = 0$, is moved in its own plane with constant velocity through a viscous fluid initially at rest is a classic example of an unsteady similarity solution. The flow of a Stokes layer, due to an infinite plate moving in its own plane with a sinusoidal variation of velocity, is a classic example of a periodic solution. It is natural to ask whether and how these solutions are unstable.

Stability of unsteady flows is a relatively new and little studied topic in the theory of hydrodynamic stability. The mathematics is more difficult, because the method of normal modes is not applicable: the linearized partial differential equations have coefficients which vary with time so that the exponential time dependence of the perturbation is not separable. Even the meaning of instability is not clear when the magnitude of the basic flow changes substantially with time. This can be quickly seen by considering an example in which the basic velocity $\mathbf{U}$ is proportional to e^{at}, say, yet the fastest-growing disturbance is of order $e^{\sigma t}$, where $a < \sigma < 0$; here any disturbance is exponentially damped, yet it may dominate the decelerating basic flow after sufficient time. In this example intuition suggests that we should deem the flow unstable although $\sigma < 0$. To deal with these conceptual difficulties it is wiser merely to describe how disturbances evolve than to get embroiled in a discussion of whether they are stable or unstable. However, when the basic flow is periodic in time or otherwise does not change its magnitude much after a long time, our previous notions of the meaning of stability suffice, even though the method of normal modes is not applicable.

Unsteadiness of the basic flow introduces new mechanisms of instability. The best understood of these, called *parametric instability* or *parametric resonance*, occurs when the basic flow is periodic in time. Parametric instability is better known in particle dynamics, e.g. when the point of suspension of a simple pendulum oscillates in a vertical line (cf. Stoker 1950, p. 189). It is also well known in the *Floquet theory* of ordinary differential equations with periodic time dependence, e.g. in Mathieu's equation. The essential mechanism is a resonance between the forced oscillation of the basic flow and the free oscillation of stable perturbations of the time-averaged basic flow when the frequency of a free oscillation is half or an integral multiple of the frequency of the forced oscillation. The detailed properties of parametric instability may be exploited to create instability or stability by controlling the frequency and amplitude of the forced oscillations of the basic flow. Parametric instability is treated in greater detail in the next sub-section and the instability of aperiodic unsteady basic flows is sketched in the one after.

48.2 *Instability of periodic flows*

The first problem of hydrodynamic stability of a periodic basic flow was discovered by Faraday (1831) and finally solved by Benjamin &

Ursell (1954). Following investigations by 'Oersted, Wheatstone and probably others' on Chladni figures, Faraday covered a horizontal plate with water and vibrated the plate. He noted that 'the water usually presents a beautifully crispated appearance in the neighbourhood of the centres of vibration'. He used the word 'crispation' to describe a pattern of standing waves of water. Faraday seems to be the first to have noted the important property that the frequency of the waves was *half* the frequency of the forced vibration of the plate. Later Matthiessen (1868, 1870) found in similar experiments that the standing waves had the same frequency as the vibration. Rayleigh (1883a,c) suggested a qualitative explanation and made further experiments. A more complete theory, adequately explaining all these experiments, was given by Benjamin & Ursell (1954) and is described below.

Consider an incompressible inviscid liquid of depth h in a rectangular tank of sides of lengths a and b, the tank being oscillated in the vertical with acceleration $f \cos \omega t$. We suppose that there is surface tension γ at the liquid's free surface. Then choose Cartesian coordinates so that the average surface of the liquid is the plane $z = 0$, the horizontal bottom of the tank is $z = -h$, and the vertical sides are $x = 0, a$ and $y = 0, b$. By the principle of dynamical equivalence, the flow is as if the tank were at rest and the gravitational acceleration were replaced by $g - f \cos \omega t$. So we may take irrotational flow and linearize the boundary conditions as in the theory of Kelvin–Helmholtz instability. Thus if $z = \zeta(x, y, t)$ is the equation of the liquid surface then it is easily shown that the linearized system for the velocity potential ϕ of the perturbation of relative rest is as follows:

$$\Delta\phi = 0; \tag{48.1}$$

$$\frac{\partial \zeta}{\partial t} = \frac{\partial \phi}{\partial z} \quad \text{at } z = 0, \tag{48.2}$$

$$\frac{\partial \phi}{\partial t} + (g - f \cos \omega t)\zeta - \frac{\gamma}{\rho}\left(\frac{\partial^2 \zeta}{\partial x^2} + \frac{\partial^2 \zeta}{\partial y^2}\right) = 0 \quad \text{at } z = 0, \tag{48.3}$$

$$\frac{\partial \phi}{\partial x} = 0 \quad \text{at } x = 0, a, \tag{48.4}$$

$$\frac{\partial \phi}{\partial y} = 0 \quad \text{at } y = 0, b, \tag{48.5}$$

and

$$\frac{\partial \phi}{\partial z} = 0 \quad \text{at } z = -h. \tag{48.6}$$

To solve this problem, one can expand ζ as a series in the complete set of orthogonal eigenfunctions $S_{lm}(x, y)$, where

$$(\Delta_1 + k_{lm}^2)S_{lm} = 0, \tag{48.7}$$

$$\frac{\partial S_{lm}}{\partial n} = 0 \tag{48.8}$$

on the perimeter of the rectangle, and Δ_1 is the horizontal Laplacian. This gives

$$\zeta = \sum_{l,m=0}^{\infty} A_{lm}(t)S_{lm}(x, y), \tag{48.9}$$

where

$$k_{lm}^2 = \pi^2\left(\frac{l^2}{a^2} + \frac{m^2}{b^2}\right) \quad \text{and} \quad S_{lm}(x, y) = \cos(l\pi x/a)\cos(m\pi y/b). \tag{48.10}$$

Then equations (48.1), (48.2) and (48.6) give

$$\phi(\mathbf{x}, t) = \sum_{l,m=0}^{\infty} \frac{dA_{lm}}{dt} \frac{\cosh\{k_{lm}(h+z)\}}{k_{lm}\sinh(k_{lm}h)} S_{lm} + G(t) \tag{48.11}$$

for some function $G(t)$ of integration. The total volume of fluid is independent of time, and therefore $A_{00} \equiv 0$. Then it can be proved that equation (48.3) gives $G \equiv 0$ and

$$\sum_{l,m=0}^{\infty} \left[\frac{d^2 A_{lm}}{dt^2} + k_{lm}\left\{\tanh(k_{lm}h)\left(\frac{k_{lm}^2\gamma}{\rho} + g - f\cos\omega t\right)\right\}A_{lm}\right]$$

$$\times \frac{S_{lm}}{k_{lm}\tanh(k_{lm}h)} = 0. \tag{48.12}$$

Therefore

$$\frac{d^2 A_{lm}}{dt^2} + k_{lm}\left\{\tanh(k_{lm}h)\left(\frac{k_{lm}^2\gamma}{\rho} + g - f\cos\omega t\right)\right\}A_{lm} = 0. \tag{48.13}$$

This is Mathieu's equation, which can be transformed to the standard form (Abramowitz & Stegun 1964, p. 722),

$$\frac{d^2 A}{du^2} + (p - 2q\cos 2u)A = 0, \tag{48.14}$$

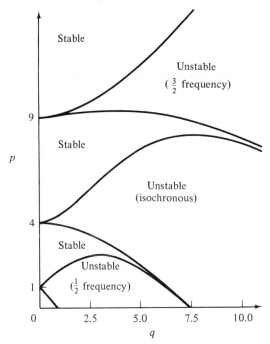

Fig. 6.6. The regions of stability and instability in the qp-plane for the Mathieu equation (48.14). (After Abramowitz & Stegun (1964), Fig. 20.1.)

where

and

$$u = \tfrac{1}{2}\omega t, \quad p = 4\omega^{-2}k_{lm} \tanh(k_{lm}h)(g + k_{lm}^2\gamma/\rho)$$
$$q = 2f\omega^{-2}k_{lm} \tanh(k_{lm}h). \tag{48.15}$$

The properties of Mathieu's equation have been investigated in great detail, but the properties relevant to the present problem can be briefly summarized as follows. The stability diagram in the qp-plane is shown in part in Fig. 6.6. The diagram is symmetric in q and $-q$ and we want $p > 0$, so only the first quadrant is shown. The p-axis corresponds to the stable free oscillations of water waves in the unaccelerated tank, with angular frequencies

$$\omega_{lm} = \{k_{lm}(k_{lm}^2\gamma/\rho + g) \tanh(k_{lm}h)\}^{1/2}. \tag{48.16}$$

The curves of marginal stability correspond to solutions $A(t)$ of period $2\pi/\omega$ or π/ω, i.e. of the frequency of the forced oscillation or half that frequency, as indicated in Fig. 6.6. The unstable solutions oscillate with these periods, whereas their amplitudes increase exponentially. It can be seen that parametric resonance between frequencies of free and forced oscillations occurs for small values of q when $\omega/2\omega_{lm}$ is close to an integer. The stable solutions are oscillatory but not periodic for $p \neq 0$.

One may note that for rapid oscillations the flow has the same stability characteristics as the average basic state (of rest), because $p \to 0$, $q \to 0$ and $A \sim \exp(\pm ip^{1/2}u)$ as $\omega \to \infty$.

The theory may be simply extended when the horizontal cross-section of the tank is not rectangular or when the vertical acceleration of the tank is some periodic function of time other than $f \cos \omega t$. In the former extension the eigenvalue k_{lm} and its eigenfunction S_{lm} are modified by the configuration, and in the latter, equation (48.14) is replaced by a *Hill equation*.

In a given run of an experiment, f and ω are fixed, so that a series of values of p and q are specified by $l, m = 0, 1, 2, \ldots$. At least one of those points (q, p) is likely to lie in a region of instability, so one might think that the liquid is always unstable. However, in practice higher-order modes are damped by viscosity or dissipative contact effects at the edge of the liquid surface and an appreciable disturbance of the free surface is observed only when one of the lower-order modes is unstable. It can now be seen how Faraday (1831) could find standing waves of period π/ω and Matthiessen (1868, 1870) of period $2\pi/\omega$. Benjamin & Ursell (1954) ran some experiments with water in an oscillating tank of circular cross-section and found convincing agreement with their theory.

There is an important assumption implicit in the theory. Equations (48.8) and (48.9) imply that $\partial\zeta/\partial n = 0$ at the wall, i.e. the angle of contact of the liquid is 90°. This means that viscous and capillary contact effects between the liquid and the walls have been neglected. Those surface effects seem to be more important than viscous dissipation within the liquid and so are thought to account for some minor discrepancies between experiment and theory (Benjamin & Ursell 1954, Miles 1967).

A conceptual difficulty occurs in the above problem of the stability of an unsteady flow when ω is very small and $f > g$. Then for a long time during each period the tank has downward acceleration greater than gravity. Because ω is small, $f \cos \omega t$ changes very slowly and one may may use a quasi-steady approximation to predict that Rayleigh–Taylor instability of all modes occurs. Yet the parameters p and q may correspond to a stable point in Fig. 6.6. The resolution of this paradox comes from considering the order of the limits as $\omega \to 0$ and $A \to 0$. In practice, disturbances may have amplitudes A large enough so that nonlinearity becomes significant after rapid growth in that part of one period $2\pi/\omega$ for which $f \cos \omega t > g$; indeed, the liquid might splash right out of the tank before the first period is complete! In theory, if $A \to 0$ first then disturbances may be stable, in the sense that they remain small over an indefinite number of periods yet grow fast for part of each period. This point is elaborated by Rosenblat & Herbert (1970), who considered a problem of low-frequency modulation of Bénard convection. The point leads one to question the suitability of the usual definition of stability.

The ideas of Benjamin & Ursell (1954) have been applied to the stability of other unsteady basic flows. Each physical quantity may be expanded in some complete set of functions of **x** which satisfy the boundary conditions, e.g.

$$\mathbf{u}'(\mathbf{x}, t) = \sum_{j=1}^{\infty} A_j(t) f_j(\mathbf{x}). \tag{48.17}$$

Substitution of these expansions into the linearized partial differential equations of motion then yields a linear system of ordinary differential equations for the amplitudes A_j. This system is not autonomous but, like the basic flow, depends explicitly upon t. The system may be plausibly truncated by supposing that $A_j \equiv 0$ for all $j > N$ and by ignoring all but the first N equations of the system, where the integer N is chosen to be as large as is practical. Special simplifications arise in the important case when the basic flow, and thence the truncated system, is periodic in time, because a system of linear ordinary differential equations whose explicit dependence on time is periodic, called a *Floquet system*, has many well-known properties, which can be found in advanced textbooks on ordinary differential equations (for example, Coddington & Levinson (1955), pp. 78–81, 218–220 and Magnus & Winkler (1966)). Mathieu's equation is the best known example of a Floquet system, but a general system has solutions which are products of periodic and exponential functions of time and may exhibit parametric resonance.

Donnelly (1964) initiated some experiments on instability between coaxial cylinders with an oscillatory rotation of the inner cylinder. His work seemed to point to a method of stabilizing an unstable steady basic flow by modulating the amplitude and frequency of a small oscillation of the basic flow. It can be seen, however, from the variegated regions of stability and instability in a typical stability diagram (like Fig. 6.6) for a Floquet system that if one mode with a given set of wavenumbers is stabilized by the oscillation then some other modes are likely to be destabilized. So it seems unlikely that the instability of a given steady basic flow can be reduced much by oscillation of the flow. In spite of this difficulty in controlling hydrodynamic instability by modulation of the oscillation of a basic flow, Donnelly's work has stimulated a lot of further work.

Several problems of instability of other oscillatory flows have been investigated theoretically and experimentally. They include Stokes layers, Poiseuille flow in a pipe with an oscillatory pressure gradient, oscillatory plane Couette flow, plane Poiseuille flow with an oscillatory pressure gradient, an oscillating boundary layer and thermal instability with oscillatory basic temperature. Many details and references of this work are given by Davis (1976).

In addition, finite-amplitude waves comprise an important class of flows which are periodic in time (and space), although a mono-chromatic plane wave is a steady flow relative to an inertial frame moving at the phase velocity of the wave. The linear instability of waves in fluids is accordingly governed by Floquet theory. Another theory which is used to investigate the stability of waves, the theory of resonant wave interactions, is described in § 52. This is a theory of weakly nonlinear interaction of waves, which enables one to treat the stability of waves of only small finite amplitudes to perturbations of a similar nature. It may be shown (Drazin 1977b) that the condition for a resonant wave interaction corresponds to the condition for parametric resonance. An important example is the instability of a gravity wave of finite amplitude in deep water. Longuet-Higgins (1978) treated this problem by what is essentially Floquet theory. He followed Benjamin & Feir (1967), who did experiments and treated the problem by what is essentially the theory of resonant wave interactions. The limitations of this weakly nonlinear theory prevented Benjamin & Feir from considering the mode of instability of steep waves which Longuet-Higgins discovered.

48.3 Instability of other unsteady basic flows

The structure of the analysis of the previous sub-section is largely independent of the periodicity of the basic flow. The spectral expansion given by equation (48.17) transforms the linearized problem from one of partial to ordinary differential equations, but does no more. However, the substantial results of Floquet theory do depend upon the periodicity of the basic flow. The stability characteristics of aperiodic flows are relatively obscure.

Some of the difficulties can be seen from Shen's (1961) work. He noted that for a plane parallel flow of an inviscid incompressible fluid with basic velocity $\mathbf{U} = U(z, t)\mathbf{i}$ the linearized vorticity equation gives

$$\left(\frac{\partial}{\partial t} + i\alpha U\right)\left(\frac{\partial^2 \psi'}{\partial z^2} - \alpha^2 \psi'\right) - i\alpha \frac{\partial^2 U}{\partial z^2}\psi' = 0, \qquad (48.18)$$

where the perturbation of the stream function is $\psi' = \hat{\psi}(z, t)\, e^{i\alpha x}$. In common with many linearized equations of stability of unsteady basic flows, this equation is of the same operational form as the equation for a steady basic flow (namely, the Rayleigh stability equation in this case), although the time variable is not separable. Shen also supposed that there are rigid plane boundaries at $z = z_1, z_2$.

For profiles of the rather special form $U(z, t) = T(t)Z(z)$ Shen showed that

$$\psi' = \phi(z) \exp\{i\alpha(x - c\tau)\}, \qquad (48.19)$$

where $\tau = \int_0^t T(t')\mathrm{d}t'$ and ϕ is the eigenfunction belonging to the eigenvalue c of the Rayleigh stability problem for the *steady* basic velocity Z, i.e.

$$(Z - c)(\phi'' - \alpha^2 \phi) - Z''\phi = 0 \qquad (48.20)$$

and

$$\phi = 0 \quad \text{at } z = z_1, z_2. \qquad (48.21)$$

It follows that if the steady flow $\mathbf{U} = Z(z)\mathbf{i}$ is stable then c is real and therefore each disturbance of the unsteady flow $\mathbf{U} = T(t)Z(z)\mathbf{i}$ is bounded as $t \to \infty$. Again, even if $\alpha c_i \neq 0$ then ψ' is bounded for all $t > 0$ provided that $\int_0^\infty T(t)\,dt < \infty$. Therefore if either the steady flow $\mathbf{U} = Z(z)\mathbf{i}$ is stable or $\int_0^t T(t')\,dt'$ is bounded for all $t > 0$ then one might be tempted to say that the unsteady flow $\mathbf{U} = Z(z)T(t)\mathbf{i}$ is stable. It must be remembered, however, that $T(t)$ may decrease as t increases so that a relatively small disturbance ψ' at $t = 0$ may remain absolutely small yet dominate the decelerating flow as $t \to \infty$. Shen argued that the criterion of stability should be

$$\frac{1}{E}\frac{dE}{dt} \leq 0 \quad \text{for all sufficiently large } t, \tag{48.22}$$

where

$$E(t) = \int_{\mathcal{V}} \mathbf{u}'^2\,d\mathbf{x} \Big/ \int_{\mathcal{V}} \mathbf{U}^2\,d\mathbf{x} \tag{48.23}$$

is the ratio of the kinetic energy of any disturbance to that of the basic flow within the domain $\mathcal{V}$ of flow. In the present case of an infinite domain of flow $\mathcal{V}$ may be taken as the rectangle $0 \leq x \leq 2\pi/\alpha$, $z_1 \leq z \leq z_2$ of a wavelength downstream between the walls. Now

$$\mathbf{u}'^2 = \left(\mathrm{Re}\left[\frac{d\phi}{dz}\exp\{i\alpha(x - c\tau)\}\right]\right)^2 + (\mathrm{Re}\,[-i\alpha\phi\,\exp\{i\alpha(x - c\tau)\}])^2$$

so that

$$E = \frac{\int_{z_1}^{z_2}(|\phi'|^2 + \alpha^2|\phi|^2)\,dz}{\int_{z_1}^{z_2} Z^2\,dz}\,\frac{\exp\{2\alpha c_i \int_0^t T\,dt\}}{T^2}.$$

Therefore

$$\frac{1}{E}\frac{dE}{dt} = 2\alpha c_i T - \frac{2}{T}\frac{dT}{dt}, \tag{48.24}$$

and Shen's criterion gives instability of $\mathbf{U} = Z T\mathbf{i}$ if

$$\alpha c_i > \frac{dT}{T^2 dt} \quad \text{for all large } t. \tag{48.25}$$

Thus accelerating flows with $dT/dt > 0$ are more stable than decelerating ones with $dT/dt < 0$.

Conrad & Criminale (1965a,b) seem to be the first to have applied the energy method to the stability of unsteady flows. The energy method, introduced in § 26.2 and developed in § 53.1, is one of the few known methods applicable to unsteady flows. It gives sufficient conditions for stability and bounds on critical values of parameters such as the Reynolds number. However, its very generality, which allows its application to nonlinear disturbances as well as to unsteady basic flows, sometimes leads to sufficient conditions which are far from necessary and to bounds which are far from the exact values.

Bouthier (1972) developed an asymptotic method to find the stability characteristics of a basic flow varying slowly in time or space.

Benjamin (1972) has used conservation relations and the calculus of variations to prove that the solitary-wave solution of the Korteweg–de Vries equation is stable; this is an original and profound study but does not seem susceptible of application to other flows.

In summary, one may say that at this stage satisfactory methods of finding the stability characteristics of general unsteady basic flows have still to be found.

Problems for chapter 6

6.1. *Internal gravity waves in an unbounded inviscid fluid.* Taking the basic state

$$U_* \equiv 0 \quad \text{and} \quad \bar{\rho}_* = \begin{cases} \rho_{\infty*} & \text{for } z_* > L, \\ \frac{1}{2}(\rho_{\infty*} + \rho_{-\infty*}) & \text{for } -L < z_* < L, \\ \rho_{-\infty*} & \text{for } z_* < -L, \end{cases}$$

and approximating $(\rho_{-\infty*} - \rho_{\infty*}) \ll (\rho_{-\infty*} + \rho_{\infty*})$, show that

$$\frac{\rho_{-\infty*} + \rho_{\infty*}}{\rho_{-\infty*} - \rho_{\infty*}} \frac{c_*^2}{gL} = \frac{1}{\alpha(1 + \tanh \alpha)} \quad \text{or} \quad \frac{1}{\alpha(1 + \coth \alpha)}$$

for the sinuous or varicose mode respectively of wavenumber k, where $\alpha = kL$. [Rayleigh (1883b).]

6.2. *The instability of a vortex sheet in a stratified fluid.* Taking the basic state

$$U = \begin{cases} 1 & \text{for } z > 0 \\ -1 & \text{for } z < 0 \end{cases} \quad \text{and} \quad N^2 = 1,$$

show that the eigenvalue relation of the Taylor–Goldstein problem is

$$(1-c)^2\{1-J/\alpha^2(1-c)^2\}^{1/2} + (1+c)^2\{1-J/\alpha^2(1+c)^2\}^{1/2} = 0.$$

Deduce that

$$c^2 = -1 + J/2\alpha^2,$$

and thence that the vortex sheet is stable to a given wave if and only if $J \geqslant 2\alpha^2$. [Haurwitz (1931), Bjerknes *et al.* (1933), Alterman (1961).]

6.3. *The instability of a jet of stratified fluid.* Given the basic state

$$U = N^2 = \operatorname{sech}^2 z \quad \text{for} \quad -\infty < z < \infty,$$

verify that a neutral eigensolution for the sinuous mode of the Taylor–Goldstein problem is given by

$$c = (6+\alpha^2)/15, \quad J = \alpha^2(4-\alpha^2)(9-\alpha^2)/225$$

and

$$\phi = (U-c)^k U^m \text{ for } 0 \leqslant \alpha \leqslant 2,$$

where $k = 3(4-\alpha^2)/2(6+\alpha^2)$ and $m = 1-k$. [Drazin & Howard (1966), pp. 74–75. Hazel's (1972) numerical results show that this solution in fact gives the stability boundary.]

*6.4. *Long-wave theory.* Generalizing the method used at the end of § 22, show that, for an unbounded shear layer in a stratified inviscid fluid with $U_\infty = -U_{-\infty} = 1$,

$$c^2 = \frac{J}{\alpha} - 1 + O(\alpha) \quad \text{as } \alpha \to 0 \text{ for fixed } J/\alpha,$$

where $(\rho_{-\infty} - \rho_\infty) \ll (\rho_{-\infty} + \rho_\infty)$ and $J = gL(\rho_{-\infty} - \rho_\infty)/V^2(\rho_{-\infty} + \rho_\infty)$. State why and how this approximate eigenvalue relation is given by equation (4.20) for the vortex sheet defined by equation (4.1). Similarly show that for an unbounded jet with $U_\infty = U_{-\infty} = 0$,

$$c^2 = \frac{J}{\alpha} - \frac{1}{2}\alpha \int_{-\infty}^{\infty} U^2 \, dz + O(\alpha^2) \quad \text{as } \alpha \to 0 \text{ for fixed } J/\alpha^2.$$

[Drazin & Howard (1966), pp. 65–66.]

6.5. *The instability of a channel flow.* Given the basic state

$$U = \sin z \quad \text{and} \quad N^2 = 1 \quad \text{for} \quad -\pi \leqslant z \leqslant \pi,$$

verify that neutral eigensolutions of the Taylor–Goldstein problem are given by

$$c = 0, \quad \alpha = \tfrac{1}{2}\sqrt{3}, \quad \phi = (\cos\tfrac{1}{2}z)^{1/2+\nu}(\sin\tfrac{1}{2}z)^{1/2-\nu} \quad \text{for } 0 \leqslant J \leqslant \tfrac{1}{4}$$

and

$$c = 0, \quad J = (1-\alpha^2)^{1/2} - 1 + \alpha^2, \quad \phi = (\sin z)^m \quad \text{for } 0 \leqslant \alpha \leqslant \tfrac{1}{2}\sqrt{3};$$

where $\nu = (\frac{1}{4} - J)^{1/2}$ and $m = (1 - \alpha^2)^{1/2}$. [Drazin & Howard (1966), p. 69. Hazel's (1972) numerical results show that these solutions in fact give the stability boundary.]

6.6. *An energy integral for baroclinic instability.* Show that the perturbation equations (45.20)–(45.23) and (45.15), together with the boundary conditions (45.18) and (45.19), imply the energy equation of the Eady problem

$$\frac{d}{dt} \int_{\mathcal{V}} \frac{1}{2}(u'^2 + v'^2 + Ro^2 H^2 L^{-2} w'^2 + B^{-1}\theta'^2)\, d\mathbf{x}$$

$$= -\int_{\mathcal{V}} \left(B^{-1}\theta'v'\frac{d\Theta}{dy} + Ro\frac{dU}{dz}u'w' \right) d\mathbf{x},$$

where $\mathcal{V}$ is the volume $0 \le x \le 2\pi/\alpha, 0 \le y, z \le 1$. Discuss the energy balance represented by the various terms in the energy equation.

6.7. *The Eady problem for a slightly viscous fluid.* The effect of a little viscosity on baroclinic instability may be represented approximately by taking thin viscous layers (i.e. *Ekman layers*) on the horizontal walls. It may be shown that this leaves the Eady problem unchanged except for the replacement of boundary conditions (45.19) by

$$Ro\, w' = \pm (\tfrac{1}{2}E)^{1/2}(\partial v'/\partial x - \partial u'/\partial y) \text{ at } z = 0, 1$$

respectively, where the *Ekman number* $E = \nu/2\Omega H^2$ and $Ro^2 \ll E^{1/2} \ll 1$. Show that this condition and equations (45.25), (45.28) and (45.29) give

$$c = \tfrac{1}{2} - i\lambda (1 + T^2)/4qT \pm \{4T(q - T)(qT - 1) - \lambda^2(1 - T^2)^2\}^{1/2}/4qT,$$

where $T \equiv \tanh q$ and $\lambda = B(\tfrac{1}{2}E)^{1/2}(\alpha^2 + n^2\pi^2)/\alpha Ro$. Deduce that for each $q < q_c$ when $\lambda = \{(q - T)(1 - qT)/T\}^{1/2}$ there is marginal stability and $c = \tfrac{1}{2}$ and when $\lambda < \{(q - T)(1 - qT)/T\}^{1/2}$ there is instability. [Barcilon (1964).]

6.8. *The Charney–Stern problem.* If, in the model of § 45, the basic state has the dimensional form $\mathbf{U} = U(y, z)\mathbf{i}$, $\Theta = \Theta(y, z)$ and $\mathbf{\Omega} = \Omega(y)\mathbf{k}$, then show that the basic velocity satisfies the *thermal wind equation*, $2\Omega\partial U/\partial z = -\alpha g\partial\Theta/\partial y$. (The variation of Ω with latitude is used to represent approximately the dynamical effects of the rotation of a sphere on the motion of a thin surface layer of fluid.) Further, show that small perturbations are governed by the equation,

$$\frac{\partial q'}{\partial t} + U\frac{\partial q'}{\partial x} + \frac{\partial Q}{\partial y}v' = 0,$$

in the limit as $Ro \to 0$, where

$$q' = \frac{\partial v'}{\partial x} - \frac{\partial u'}{\partial y} + 2\Omega\frac{\partial}{\partial z}\left(\frac{1}{\rho_0 N^2}\frac{\partial p'}{\partial z}\right),$$

$$\frac{\partial Q}{\partial y} = 2\frac{\partial\Omega}{\partial y} - \frac{\partial^2 U}{\partial y^2} - \frac{\partial}{\partial z}\left(\frac{4\Omega^2}{N^2}\frac{\partial U}{\partial z}\right),$$

and $N^2 = \alpha g \partial \Theta / \partial z$. Deduce that normal modes $p' = \hat{p}(y, z) \exp\{ik(x - ct)\}$ are governed by

$$(U - c)\left\{\frac{\partial}{\partial z}\left(\frac{4\Omega^2}{N^2}\frac{\partial \hat{p}}{\partial z}\right) + \frac{\partial^2 \hat{p}}{\partial y^2} - k^2 \hat{p}\right\} + \frac{\partial Q}{\partial y}\hat{p} = 0,$$

$$\hat{p} = 0 \quad \text{at } y = 0, L,$$

and

$$(U - c)\frac{\partial \hat{p}}{\partial z} - \frac{\partial U}{\partial z}\hat{p} = 0 \quad \text{at } z = 0, H.$$

Hence prove that if $c_i \neq 0$ then

$$\int_0^L \left\{\int_0^H \frac{\partial Q/\partial y}{|U - c|^2}|\hat{p}|^2 \, \mathrm{d}z\right\} \mathrm{d}y = -\int_0^L \left[\frac{4\Omega^2}{N^2}\frac{\partial U}{\partial z}\frac{|\hat{p}|^2}{|U - c|^2}\right]_{z=0}^H \mathrm{d}y.$$

Use this to show that if the flow is unstable and $\partial U / \partial z = 0$ at $z = 0, H$ then $\partial Q / \partial y = 0$ somewhere in the field of flow. [Charney & Stern (1962).]

6.9. *Energy equation for stability of the pinch.* Denoting the Lagrangian displacement of the fluid particle at $\mathbf{x}$ at time t by $\boldsymbol{\xi}(\mathbf{x}, t) = (\xi_r, \xi_\theta, \xi_z)$, so that $\partial \boldsymbol{\xi}/\partial t = \mathbf{u}'$, show that equations (46.9)–(46.12) give

$$\frac{\partial^2 \boldsymbol{\xi}}{\partial t^2} - \frac{B_1^2}{\mu\rho}\frac{\partial^2 \boldsymbol{\xi}}{\partial z^2} = -\boldsymbol{\nabla}\Pi',$$

$$\mathbf{B}' = B_1\frac{\partial \boldsymbol{\xi}}{\partial z}, \quad \boldsymbol{\nabla}\cdot\boldsymbol{\xi} = 0,$$

$$\mathbf{B}'' = \boldsymbol{\nabla}\Omega'' \quad \text{and} \quad \Delta\Omega'' = 0.$$

Also show that boundary conditions (46.14), (46.19), (46.17) and (46.15) give

$$\Pi' = \frac{1}{\mu_0\rho}\left(\frac{B_0}{R_1}\frac{\partial \Omega''}{\partial \theta} + B_2\frac{\partial \Omega''}{\partial z} - \frac{B_0^2 \xi_r}{R_1}\right) \quad \text{at } r = R_1,$$

$$\frac{\partial \Omega''}{\partial r} = \frac{B_0}{R_1}\frac{\partial \xi_r}{\partial \theta} + B_2\frac{\partial \xi_r}{\partial z} \quad \text{at } r = R_1,$$

and

$$\frac{\partial \Omega''}{\partial r} = 0 \quad \text{at } r = R_2.$$

Hence deduce the energy equation

$$\frac{\mathrm{d}}{\mathrm{d}t}\left[\int_{\mathscr{V}'}\left\{\frac{1}{2}\rho\left(\frac{\partial \boldsymbol{\xi}}{\partial t}\right)^2 + \frac{B_1^2}{2\mu}\left(\frac{\partial \boldsymbol{\xi}}{\partial z}\right)^2\right\}\mathrm{d}\mathbf{x} + \int_{\mathscr{V}''}\frac{1}{2\mu_0}(\boldsymbol{\nabla}\Omega'')^2\,\mathrm{d}\mathbf{x} - \int_{\mathscr{S}}\frac{B_0^2\xi_r^2}{2\mu_0 R_1}\,\mathrm{d}S\right] = 0,$$

where $\mathscr{V}'$ is the infinite cylinder $r < R_1$, $\mathscr{V}''$ is the infinite annular cylinder $R_1 < r < R_2$, and $\mathscr{S}$ is their common boundary $r = R_1$, $0 \leqslant \theta \leqslant 2\pi$, $-\infty <$

$z < \infty$. [Form the scalar product of $\partial \boldsymbol{\xi}/\partial t$ and the first displayed equation of this problem, integrate over $\mathcal{V}'$, and apply Gauss's divergence theorem. Similarly show that $\partial_{\frac{1}{2}}(\nabla \Omega'')^2/\partial t = \nabla \cdot (\Omega'' \nabla \partial \Omega''/\partial t)$ and integrate over $\mathcal{V}''$. Finally use the boundary conditions.]

6.10. *The principle of exchange of stabilities.* Taking normal modes with $\boldsymbol{\xi}, \Omega'' \propto e^{st}$ in Problem 6.9, prove that if $\boldsymbol{\xi}_1, \Omega_1$ are eigenfunctions belonging to an eigenvalue s_1 and $\boldsymbol{\xi}_2, \Omega_2$ to s_2 then

$$s_1^2 \int_{\mathcal{V}''} \rho \boldsymbol{\xi}_1 \cdot \boldsymbol{\xi}_2 \, \mathrm{d}\mathbf{x} = - \int_{\mathcal{V}''} \frac{B_1^2}{\mu} \frac{\partial \boldsymbol{\xi}_1}{\partial z} \cdot \frac{\partial \boldsymbol{\xi}_2}{\partial z} \, \mathrm{d}\mathbf{x}$$

$$- \int_{\mathcal{V}''} \frac{1}{\mu_0} \nabla \Omega_1 \cdot \nabla \Omega_2 \, \mathrm{d}\mathbf{x} + \int_{\mathcal{S}} \frac{B_0^2}{\mu_0 R_1} \xi_{1r} \xi_{2r} \, \mathrm{d}S.$$

Hence prove that

$$\int_{\mathcal{V}''} \boldsymbol{\xi}_1 \cdot \boldsymbol{\xi}_2 \, \mathrm{d}\mathbf{x} = 0 \quad \text{if } s_2^2 \neq s_1^2,$$

that the principle of exchange of stabilities is valid, and that a mode is stable if and only if

$$W = \int_{\mathcal{V}''} \frac{B_1^2}{2\mu} \left(\frac{\partial \boldsymbol{\xi}}{\partial z} \right)^2 \mathrm{d}\mathbf{x} + \int_{\mathcal{V}''} \frac{1}{2\mu_0} (\nabla \Omega'')^2 \, \mathrm{d}\mathbf{x} - \int_{\mathcal{S}} \frac{B_0^2}{2\mu_0 R_1} \xi_r^2 \, \mathrm{d}S \geq 0.$$

6.11. *A variational principle for the equations of marginal stability.* Consider the reformulation of Problem 6.9 for steady displacements as follows, with a view to application of the Rayleigh–Ritz method. *Assuming* that

$$W = \int_{\mathcal{V}''} \frac{B_1^2}{2\mu} \left(\frac{\partial \boldsymbol{\xi}}{\partial z} \right)^2 \mathrm{d}\mathbf{x} + \int_{\mathcal{V}''} \frac{1}{2\mu_0} (\nabla \times \mathbf{A}'')^2 \, \mathrm{d}\mathbf{x} - \int_{\mathcal{S}} \frac{B_0^2}{2\mu_0 R_1} \xi_r^2 \, \mathrm{d}S$$

is stationary with respect to variations of the vector functions $\boldsymbol{\xi}$ and $\mathbf{A}''$ which satisfy $\nabla \cdot \boldsymbol{\xi} = 0$ and the boundary conditions

$$A_\theta'' = -B_2 \xi_r, \quad A_z'' = B_0 \xi_r \quad \text{at } r = R_1,$$

and

$$A_\theta'' = 0, \quad A_z'' = 0 \quad \text{at } r = R_2;$$

deduce by the calculus of variations that

$$\frac{B_1^2}{\mu \rho} \frac{\partial^2 \boldsymbol{\xi}}{\partial z^2} = \nabla \Pi' \quad \text{and} \quad \nabla \times \mathbf{B}'' = 0,$$

where $\mathbf{B}''$ is defined by $\mathbf{B}'' = \nabla \times \mathbf{A}''$, and

$$\Pi' = \frac{1}{\mu_0 \rho} \left(B_0 B_\theta'' + B_2 B_z'' - \frac{B_0^2 \xi_r}{R_1} \right) \quad \text{at } r = R_1.$$

[Introduce the function Π' as a Lagrangian multiplier associated with the constraint $\nabla \cdot \boldsymbol{\xi} = 0$, and show that

$$\delta\{W - \int \rho\Pi'(\nabla \cdot \boldsymbol{\xi})\mathrm{d}\mathbf{x}\} = \int_{\gamma'} \left(\frac{B_1^2}{\mu}\frac{\partial^2 \boldsymbol{\xi}}{\partial z^2} - \rho\nabla\Pi'\right) \cdot \delta\boldsymbol{\xi}\,\mathrm{d}\mathbf{x}$$

$$- \int_{\gamma''} (\nabla \times \mathbf{B}'') \cdot \delta\mathbf{A}''\,\mathrm{d}\mathbf{x}$$

$$+ \int_{\mathscr{S}} \frac{1}{\mu_0}\left(\rho\mu_0\Pi' - B_0 B_\theta'' - B_\theta B_z'' + \frac{B_0^2 \xi_r}{R_1}\right)\delta\xi_r\,\mathrm{d}S.$$

Note that the boundary conditions imply that $\mathbf{B} \cdot \mathbf{n}$ is continuous at $r = R_1$ and zero at $r = R_2$. For a more general discussion of the derivation and use of such variational principles see Bernstein *et al.* (1958).]

6.12. *The spatial modes for inviscid plane Couette flow.* Considering the Rayleigh stability equation (21.17) with $U = z$ and boundary conditions (21.18) at $z = \pm 1$, show that the spatial modes for plane Couette flow of an inviscid incompressible fluid are given by the complete set

$$\phi_n(z) = \sin\{\tfrac{1}{2}n\pi(1 + z)\}, \quad \alpha = \pm\tfrac{1}{2}n\pi\mathrm{i} \quad \text{for } n = 1, 2, \ldots$$

and any value of ω. Discuss how the flow may be stable yet these spatial modes may grow in space like $\exp(\mp\tfrac{1}{2}n\pi x)$.

6.13. *Vertical oscillation of a circular tank of liquid.* If the horizontal cross-section of the tank in § 48.2 were the circle $r = a$ instead of the rectangle, show that

$$S_{lm} = J_l(k_{lm}r) \begin{cases} \sin l\theta \\ \cos l\theta \end{cases},$$

where J_l is the Bessel function of order l and k_{lm} is the mth positive zero of $J_l'(k_{lm}a)$.

If, moreover, the vertical acceleration of the tank were the general function F, show that

$$\frac{\mathrm{d}^2 A_{lm}}{\mathrm{d}t^2} + k_{lm}\left[\tanh(k_{lm}h)\left\{\frac{k_{lm}^2\gamma}{\rho} + g - F(t)\right\}\right]A_{lm} = 0.$$

If F is a periodic function, this is a *Hill equation*.

6.14. *Kelvin–Helmholtz instability of an unsteady flow.* If $U_1 = U_1(t)$ and $U_2 = U_2(t)$ in the problem of § 4, show that one may take $\zeta(x, y, t) \propto \exp\{\mathrm{i}(kx + ly)\}$ where

$$\frac{\mathrm{d}^2\zeta}{\mathrm{d}t^2} + \frac{2\mathrm{i}k(\rho_1 U_1 + \rho_2 U_2)}{\rho_1 + \rho_2}\frac{\mathrm{d}\zeta}{\mathrm{d}t}$$

$$+ \left\{g\tilde{k}\frac{\rho_1 - \rho_2}{\rho_1 + \rho_2} - k^2\frac{\rho_1 U_1^2 + \rho_2 U_2^2}{\rho_1 + \rho_2} + \mathrm{i}k\left(\rho_1\frac{\mathrm{d}U_1}{\mathrm{d}t} + \rho_2\frac{\mathrm{d}U_2}{\mathrm{d}t}\right)\right\}\zeta = 0.$$

Putting

$$Z = \zeta \exp\left\{ ik \int \frac{\rho_1 U_1 + \rho_2 U_2}{\rho_1 + \rho_2} \, dt \right\},$$

deduce that

$$\frac{d^2 Z}{dt^2} + \left\{ g\tilde{k} \frac{\rho_1 - \rho_2}{\rho_1 + \rho_2} - \frac{k^2 \rho_1 \rho_2 (U_1 - U_2)^2}{(\rho_1 + \rho_2)^2} \right\} Z = 0.$$

[Kelly (1965).]

6.15. *Instability of a linear pinch in an alternating magnetic field.* In the problem of § 46, suppose that the components B_0 and B_2 of the vacuum field are functions of time. Then, taking $\zeta' = \hat{\zeta}(t) \exp\{i(n\theta + kz)\}$, show that the pinch is stable if and only if the solution $\hat{\zeta} \equiv 0$ of the equation,

$$\frac{d}{dt}\left(\frac{d^2 \hat{\zeta}}{dt^2} + f\hat{\zeta} \right) = 0$$

is stable where

$$f = \frac{k^2 B_1^2}{\mu\rho} - \frac{Y_n}{\mu_0 \rho R_1^2} \{ kR_1 B_0^2 + Z_n (nB_0 + kR_1 B_2)^2 \},$$

$$Y_n = \frac{I_n'(kR_1)}{I_n(kR_1)} \quad \text{and} \quad Z_n = \frac{I_n(kR_1)K_n'(kR_2) - K_n(kR_1)I_n'(kR_2)}{I_n'(kR_1)K_n'(kR_2) - K_n'(kR_1)I_n'(kR_2)}.$$

Hence show that if $B_0 = b_0(1 + \varepsilon \cos 2\omega t)$, but B_1 and B_2 are constants, then stability is governed by the equation

$$\frac{d^2 \hat{\zeta}}{dt^2} + (f_0 + f_1 \cos 2\omega t + f_2 \cos 4\omega t)\hat{\zeta} = 0,$$

where

$$f_0 = \frac{k^2 B_1^2}{\mu\rho} - \frac{Y_n}{\mu_0 \rho R_1^2} \{ kR_1 b_0^2 (1 + \tfrac{1}{2}\varepsilon^2) + (nb_0 + kR_1 B_2)^2 Z_n + \tfrac{1}{2} n^2 b_0^2 \varepsilon^2 Z_n \},$$

$$f_1 = -\frac{2\varepsilon b_0 Y_n}{\mu_0 \rho R_1^2} \{ kR_1 b_0 + n(nb_0 + kR_1 B_2) Z_n \},$$

and

$$f_2 = -\frac{\varepsilon^2 b_0^2 Y_n}{2\mu\rho R_1^2} (kR_1 + n^2 Z_n).$$

This equation for $\hat{\zeta}$ is a special case of a Hill equation called a *Whittaker equation*. [Tayler (1957b), Magnus & Winkler (1966).]

CHAPTER 7

NONLINEAR STABILITY

Nonlinear hydrodynamic stability theory is really concerned, ultimately, with phenomena such as transition to turbulence. In practice,
however, that phenomenon is so complex as to defy rational understanding at the present time. A more limited objective is that of gaining
some understanding of nonlinear processes in fluid mechanics, perhaps
with reference to the early, relatively-simple stages of the evolution of
laminar flow to turbulence. Even then, the mathematical problems
posed are challenging enough.   – J. T. Stuart (1977)

49 Introduction

49.1 *Landau's theory*

Although Reynolds (1883) appreciated the importance of
nonlinear disturbances of Poiseuille flow in a pipe, and Bohr (1909),
Noether (1921) and Heisenberg (1924) treated them theoretically
for other special problems, it may be said that the foundations of the
theory of nonlinear hydrodynamic stability were laid by Landau in
1944 (see Landau & Lifshitz (1959), § 27 for a similar account). In
a prophetic essay, Landau outlined the development of linear
instability towards the onset of turbulence. His ideas have required
substantiation or qualification, but his overall vision has been
confirmed by subsequent work. His theory is described below from
a modern viewpoint, with some of the details filled in.

The linear theory of stability of a steady basic flow quite generally
gives a spectrum of independent modes, each with velocity perturbation of the form

$$\mathbf{u}'(\mathbf{x}, t) = \mathbf{u}(\mathbf{x}, t) - \mathbf{U}(\mathbf{x}) = A(t)\mathbf{f}(\mathbf{x}) + A^*(t)\mathbf{f}^*(\mathbf{x}) \qquad (49.1)$$

for some complex amplitude $A \propto e^{st}$ with relative growth rate
$s = \sigma + i\omega$. When $R < R_c$ all disturbances are stable with $\sigma < 0$. (We
write here of the Reynolds number R although the ideas apply
equally to problems specified by other dimensionless parameters.)
When $R = R_c$ there is just one normal mode, with $s = s_1 = \sigma_1 + i\omega_1$,
say, which is marginally stable. As R increases above R_c, $\sigma_1 > 0$ but

$\sigma < 0$ for all the other modes. When R is sufficiently large these other modes also become unstable in turn. One usually finds that

$$\sigma_1 = k(R - R_c) + O\{(R - R_c)^2\} \quad \text{as } R \to R_c, \qquad (49.2)$$

where k is some positive constant. Thus if $0 < R - R_c \ll 1$ then the most unstable mode grows slowly but all other modes decay, linear theory showing that the most unstable mode becomes dominant very soon. It is now recognized that (after a long time if $R - R_c$ is small but quite soon in practice) the nonlinear self-interaction of the dominant mode generates harmonics and distorts the mean flow, thereby moderating the slow exponential growth. Landau described the instability by the equation

$$d|A|^2/dt = 2\sigma|A|^2 - \ell|A|^4 \qquad (49.3)$$

for the amplitude $|A|$ of the dominant mode, where ℓ is some constant, now called the *Landau constant*. Also equation (49.3) is called the *Landau equation*, although it is equivalent to the logistic equation in the theory of population growth. (We shall see that it is more properly regarded as a truncation of a system of ordinary differential equations whose other terms are often, but not always, negligible in hydrodynamic stability.) Of course if ℓ were zero equation (49.3) would reduce to the equation given by the linear theory. The second term on the right-hand side of equation (49.3) is due to the nonlinearity and may moderate or accelerate the exponential growth of the linear disturbance according to the signs of σ and ℓ.

Rewriting the Landau equation as a linear equation in $|A|^{-2}$, namely

$$\frac{d|A|^{-2}}{dt} + 2\sigma|A|^{-2} = \ell,$$

we find the explicit general solution

$$|A|^{-2} = \ell/2\sigma + (A_0^{-2} - \ell/2\sigma)\,e^{-2\sigma t}$$

if $\sigma \neq 0$, where A_0 is the initial value of $|A|$. Therefore

$$|A|^2 = A_0^2 \Big/ \Big\{ \frac{\ell}{2\sigma} A_0^2 + \Big(1 - \frac{\ell}{2\sigma} A_0^2\Big) e^{-2\sigma t} \Big\}. \qquad (49.4)$$

First take the case $\ell > 0$ and $R > R_c$, i.e. $\sigma > 0$. Then the solution (49.4) gives $|A| \sim A_0\, e^{\sigma t}$ as $t \to -\infty$ and $A_0 \to 0$, just as in the linear theory, but

$$|A| \to A_e \equiv (2\sigma/\ell)^{1/2} \quad \text{as } t \to +\infty, \tag{49.5}$$

whatever the value of A_0. This is called *supercritical stability*, the basic flow being linearly unstable for $R > R_c$ but settling down as a new laminar flow eventually. The new flow is, moreover, independent of the initial conditions except through the phase of the complex amplitude A of the dominant mode; it has period $2\pi/\omega_1$ if $\omega_1 \neq 0$ or is steady if $\omega_1 = 0$. The disturbance is said to *equilibrate*, because its amplitude tends to A_e after a long time. Further, equations (49.2) and (49.5) give

$$A_e \sim \{2k(R - R_c)/\ell\}^{1/2} \quad \text{as } R \downarrow R_c. \tag{49.6}$$

Thus, when $0 < R - R_c \ll 1$, A_e is small and even if higher-order terms, for example one in $|A|^6$, were included on the right-hand side of the Landau equation (49.3) they would remain small and the qualitative character of the solution would be unchanged (see Problem 7.1). The typical development of $|A|$ with time is sketched in Fig. 7.1(a) and the dependence of the amplitudes of the equilibrium solutions $|A| = 0$, $|A| = A_e$ upon R in Fig. 7.1(b). The branching of the curve of the equilibrium solutions at $R = R_c$, $|A| = 0$ is called a *bifurcation*. The Landau equation implies that the solution $|A| = 0$, which represents the steady basic flow, is stable for $R < R_c$ but unstable for $R > R_c$ and that $|A| = A_e$, which represents the new laminar flow, is stable where it exists, i.e. for $R > R_c$. In more complete models of hydrodynamic stability we shall see that there may be further bifurcations from the solution $|A| = 0$, e.g. where the next least stable normal mode of the basic flow becomes unstable, and from the solution $|A| = A_e$.

If $\ell > 0$ but $R < R_c$ then $\sigma < 0$ and equation (49.4) confirms that the disturbance decays in accord with the linear theory, i.e. $|A| \sim A_0\, e^{\sigma t}$ as $t \to \infty$ and $A_0 \to 0$. In this case the term $-\ell|A|^4$ of equation (49.3) due to the nonlinearity remains small for all time if it is initially small.

Landau suggested that $\ell > 0$ for 'ordinary' flow past bodies, but we must also consider the case $\ell < 0$, which in fact is characteristic of flows in channels. If, further, $R > R_c$, we have $\ell < 0$ and $\sigma > 0$ so

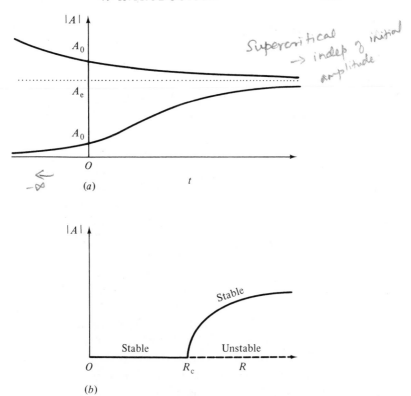

Supercritical → indep of initial amplitude

Fig. 7.1(a) Supercritical stability for $0 < R - R_c \ll 1$ and $\ell > 0$: the development of $|A|$ as a function of time for two initial values A_0. (b) The bifurcation curve: the amplitude of the equilibrium solution as a function of the Reynolds number for $\ell > 0$.

that both terms on the right-hand side of the Landau equation (49.3) are positive and therefore $|A|$ increases super-exponentially. Indeed, equation (49.4) shows that the solution breaks down after a finite time, $|A|$ becoming infinite at $t = (2\sigma)^{-1} \ln\{1 + 2\sigma/(-\ell A_0^2)\}$. Of course this kind of breakdown does not occur in practice; rather, the breakdown points to the need for terms in higher powers of $|A|$ on the right-hand side of equation (49.3). We plausibly anticipate that in this case none of the higher terms may be truncated and there is a fast transition to turbulence.

Finally, if $\ell < 0$ and $R < R_c$ then $\sigma < 0$ and the two terms on the right-hand side of equation (49.3) have opposite signs, the first or the second being the greater or less in magnitude according to whether $|A|$ is respectively less than or greater than $A_e \equiv \{(-2\sigma/(-\ell)\}^{1/2}$. In fact the solution (49.4) shows that if $A_0 < A_e$ then $|A| \sim A_0 A_e \, e^{\sigma t}/(A_e^2 - A_0^2)^{1/2}$ as $t \to \infty$ and if $A_0 > A_e$ then $|A| \to \infty$ as $t \to (-2\sigma)^{-1} \ln\{A_0^2/(A_0^2 - A_e^2)\}$. In this case A_e appears as a threshold value, because if $A_0 < A_e$ then $|A| \to 0$ as $t \to \infty$ and if $A_0 > A_e$ then the solution breaks down after a finite time. As we noted before, inclusion of higher-order terms on the right-hand side of equation (49.3) might make this breakdown a less-unrealistic rapid transition to turbulence. It can be seen that if the coefficients of the higher-order terms were all of order of magnitude one then there could be no steady small-amplitude solution as $R \uparrow R_c$ other than $|A| \sim A_e \propto (R_c - R)^{1/2}$. Then all the higher-order terms would become significant if $A_0 > A_e$ and $|A|$ grew to be of order of magnitude one, and one expects that *all* higher harmonics would be excited. This case is called *subcritical instability*, because instability may occur with finite amplitude $|A| > A_e$ when all infinitesimal disturbances are stable; it is commonly called *metastability* by physicists. The development of $|A|$ as a function of time and the equilibrium solutions $|A|$ as functions of R are shown in Fig. 7.2. Landau suggested that in this case there should be a 'lower' critical value of the Reynolds number, R_G (say), below which the bifurcated solution does not exist (so that the curve in Fig. 7.2(*b*) would turn back and $|A|$ would begin to increase with R). One expects that if $R < R_G$ then the basic flow is *globally asymptotically stable*, i.e. then all disturbances, however large initially, decay ultimately.

Landau (1944) asserted that Poiseuille flow in a pipe is an example of flows which are stable to infinitesimal disturbances at all values of the Reynolds number, i.e. for which $R_c = \infty$, yet which might develop into another steady or periodic flow for constant $|A| \neq 0$ when $R > R_G$. This value R_G might more properly be identified with the value of R (about 2000) below which Poiseuille flow is observed to be stable irrespective of conditions at the inlet to the pipe (see § 1).

Concentrating upon the case of supercritical stability, Landau (1944) considered the development of the flow as R increases slowly above R_c. He envisaged supercritical equilibration when $R - R_c$ ceases to be small. Then separation of the basic flow from

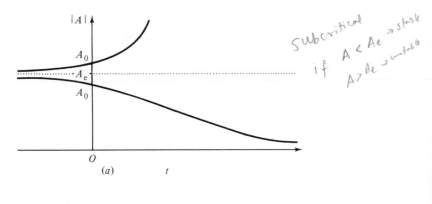

Subcritical
if $A < A_e \to$ stable
$A > A_e \to$ unstable

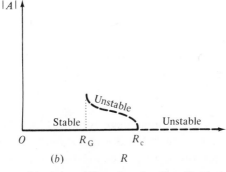

Fig. 7.2(a) Subcritical instability for $0 < R_c - R \ll 1$ and $\ell < 0$: the development of $|A|$ as a function of time for two initial values A_0. (b) The amplitude of the equilibrium solution as a function of the Reynolds number for $\ell < 0$.

the equilibrated disturbance has little meaning, because the strong nonlinear interactions of the disturbance modify the basic flow. The new flow is made up of all the harmonics of the fundamental with frequency ω_1 and some phase, β_1 say, so that the velocity has the form

$$\mathbf{u} = \sum_{p=-\infty}^{\infty} \mathbf{F}_p(\mathbf{x}) \exp\{ip(\omega_1 t + \beta_1)\}. \tag{49.7}$$

The phase β_1 is essentially arbitrary, depending on the particular initial conditions which happen to occur. Landau suggested that as R increases the periodic flow described by equation (49.7) changes until it becomes unstable itself (as we have discussed in § 48) and that its unstable disturbances will in turn equilibrate. This leads to yet another flow with two different periods, $2\pi/\omega_2$ (say) as well as $2\pi/\omega_1$, and two essentially arbitrary phases, β_2 (say) as well as β_1.

When the Reynolds number increases still further, more and more new periods appear in succession. The range of Reynolds numbers between successive appearances of new frequencies diminishes rapidly in size. The new flows themselves are on a smaller and smaller scale ... [and] the flow rapidly becomes complicated and confused

that is to say turbulence ensues. In short, turbulence may arise slowly through spectral evolution if the Landau constant is positive or suddenly if the Landau constant is negative. It is now, however, recognized that transition to turbulence is more complex. It can be seen that, in the limit of the sequence of transitions which Landau conjectured, the velocity becomes an almost periodic function of time. It follows that its autocorrelation does not decay after a long time; however, the autocorrelation of a turbulent velocity field is observed to decay rapidly.

49.2 *Discussion*

Landau's ideas had little immediate impact. Hopf (1948) advanced similar ideas of how turbulence might develop as the Reynolds number increases, through the repeated bifurcation of the solution representing the flow, but seems to have been unaware of Landau's work. Hopf gave a detailed mathematical treatment of a model problem, although his discussion of the physical character of nonlinear hydrodynamic stability was less rich. In 1958 Stuart sought to deduce the Landau equation from the energy equation for disturbances in Couette flow between rotating cylinders. Gor'kov (1957) and Malkus & Veronis (1958) independently calculated the *steady* nonlinear solution for slightly supercritical Bénard convection; they used an expansion similar to that of the Liapounov–Schmidt method to calculate the first few terms of the equilibrated solution $|A| = A_e$ in powers of $Ra - Ra_c$, where Ra is the Rayleigh number, but did not derive the Landau equation itself. The first to show how to derive the Landau equation from the system of partial differential equations governing a flow were Palm (1960) for a model of Bénard convection, and Stuart (1960a) and Watson (1960b) for plane parallel flows. Since 1960 the Landau equations appropriate to many flows have been derived by a number of authors.

Landau did not deal with the important practical point of how to derive the Landau equation from the partial differential system governing a flow. We shall illustrate the derivation in detail for a simple model problem in § 50. Here, following Eckhaus (1965), we note that one can in general expand the nonlinear velocity perturbation as

$$\mathbf{u}' = \mathbf{u} - \mathbf{U} = \sum_{j=1}^{\infty} \{A_j(t)\mathbf{f}_j(\mathbf{x}) + A_j^*(t)\mathbf{f}_j^*(\mathbf{x})\}, \qquad (49.8)$$

where $\{\mathbf{f}_j\}$ is any complete set of complex functions satisfying the appropriate boundary conditions. Various choices of $\{\mathbf{f}_j\}$ come to mind; we have already used some in various problems in §§ 17 and 30, but the set of eigenfunctions of the linear problem is a natural choice if the nonlinearity is weak. Similarly one can expand the perturbations of other quantities specifying the flow. Then the nonlinear partial differential equations may plausibly be reduced to an infinite set of ordinary differential equations by use of some method such as Galerkin's to give an *amplitude equation* (or *evolution equation*), say,

$$\frac{\mathrm{d}A_j}{\mathrm{d}t} = s_j A_j + N_j(A_k) \quad \text{for } j = 1, 2, \ldots, \qquad (49.9)$$

where the complex function N_j of the A_ks represents the nonlinear action of all the modes on the jth (including the self-interaction). The linear terms are in general simply $s_j A_j$ (with no summation) where $\mathbf{f}_j$ is the eigenfunction of the linear problem with relative growth rate s_j as its eigenvalue. This makes it plausible that the Landau equation arises as a truncated form of equation (49.9) when the problem has a symmetry such that $N_j(A_k)$ is the product of A_j and a function of the set of $|A_k|^2$.

It is found that the essence of the weakly nonlinear stability of a wide variety of flows is governed by a Landau equation. Derivation and truncation of the ordinary differential system leads to the form,

$$\frac{\mathrm{d}A}{\mathrm{d}t} = sA - \tfrac{1}{2}\ell|A|^2 A, \qquad (49.10)$$

where in general $\ell = \ell_r + i\ell_i$ as well as $s = \sigma + i\omega$ is complex. It readily follows that

$$\frac{\mathrm{d}|A|^2}{\mathrm{d}t} = 2\sigma|A|^2 - \ell_r|A|^4 \qquad (49.11)$$

and

$$\frac{d(phA)}{dt} = \omega - \tfrac{1}{2}\ell_i |A|^2. \qquad (49.12)$$

It has been found that ℓ is typically positive for models of Bénard convection and Taylor vortices, $\ell_r > 0$ typically for unbounded jets and shear layers, and $\ell_r < 0$ for boundary layers, channel flows and pipe flows.

An important addition to Landau's ideas is the theory of *resonant wave interactions*, whereby a wave of finite amplitude may become unstable. This has applications to oceanography and other fields. Resonant wave interactions, whereby three or more weakly nonlinear *stable* normal modes exchange energy with one another, are closely related both to parametric instability of a finite-amplitude wave (briefly mentioned in § 48) and to weakly nonlinear instability. They are introduced in § 51.

The ordinary differential system described by equation (49.9) can be used more generally to describe the development of the nonlinear disturbance $\mathbf{u}'(\mathbf{x}, t)$, whether it is small or not. This enables us to harness the powerful and well-developed theory of ordinary differential equations. A few ideas of this theory relevant to hydrodynamic instability are introduced in §§ 52.1 and 52.2. In particular, we see that the Landau equation is not the only form of truncated system suitable to describe nonlinear instability, and that turbulence need not arise as the slow transition due to successive bifurcations of quasi-periodic flows when the Landau constant is positive or as the fast transition when the Landau constant is negative. Rather, several types of transition are possible, one of which may represent the development of some flows and another of other flows.

Small imperfections, such as those due to end effects, may lead to small changes in the form of the Landau equation. These are introduced in § 52.3 by use of illustrations from catastrophe theory as well as from the theories of rigid-body dynamics and elasticity.

An important qualification to Landau's theory comes from the recognition of its implicit assumption that there is a unique and discrete most unstable mode. For flow in a bounded region we plausibly expect the normal modes of the linear eigenvalue problem to be both discrete and complete. But we already know that the normal modes are not discrete for flow in an unbounded region. For example, for Bénard convection between infinite horizontal planes and for plane Poiseuille flow we have shown that the normal modes depend continuously upon the wavenumber. Then the most unstable mode is only infinitesimally more unstable than its fellows. In the linear theory this gives rise to a dominant wavepacket of the most unstable modes rather than a single dominant mode (see § 47). In the nonlinear theory this means that the sum in equation (49.8) is more properly an integral or both a sum and an integral, and that the approximation by a single equation (49.3) is inadmissible. This idea was first examined by Newell & Whitehead and by Segel independently in 1969. We discuss it in § 52.4, showing how nonlinear instability may develop in space as well as time.

We also discuss nonlinear critical layers in § 52.5. The critical layer is so important in the linear theory of the stability of parallel flows that it is natural, albeit difficult, to enquire into the nonlinear development of a critical layer in an unstable flow.

Fruitful though Landau's work has been, by no means all of the current theory of nonlinear hydrodynamic stability is closely related to it. The energy method, using estimates based upon the equation for the rate of change of energy of the perturbation, has a very different and older origin. We used what is essentially the energy method in § 26 to find sufficient conditions for the linear stability of plane parallel flows. These conditions in fact apply also to nonlinear stability (because the nonlinear terms in the equations of motion may describe the convection of energy from one part of the flow to another, not the creation or destruction of energy). We describe the wider implications of this in § 53, considering the stability of more types of basic flow.

Finally, § 54 comprises a survey of the theory and observations of the nonlinear stability of some of the particular problems whose linear stability has been treated at length in previous chapters.

This chapter then is a description of the many substantial advances made in our understanding of nonlinear hydrodynamic stability during the past 40 years. These advances bring us nearer to the goal of understanding the physical mechanisms of transition from laminar to turbulent flow.

50 The derivation of ordinary differential systems governing stability

The ideas which Landau adumbrated in 1944 are described and elaborated in the previous section. He did not show how his equation could be derived for the stability of a given flow and did not identify the nonlinear interaction of the fundamental with its harmonics and the mean flow. These important steps were taken about 15 years later, when weakly nonlinear theories were developed, notably by Malkus & Veronis (1958), Stuart (1958, 1960a) and Watson (1960b). Malkus & Veronis were the first to apply successfully a weakly nonlinear theory to a problem of hydrodynamic stability. Shortly afterwards Davey (1962) deduced the first reliable nonlinear results in good quantitative agreement with experimental results.

Such applications of the theory to the stability of flows are long and technically complicated. So, to introduce methods of deriving an ordinary differential system from the partial differential system governing the stability of a given basic flow, we shall treat a simple mathematical model of flow in a channel. This model problem was suggested by the work of Matkowsky (1970). Then we suppose that

$$\frac{\partial u}{\partial t} - \sin u = \frac{1}{R} \frac{\partial^2 u}{\partial z^2}, \tag{50.1}$$

and

$$u = 0 \quad \text{at } z = 0 \quad \text{and} \quad \pi. \tag{50.2}$$

This may be regarded as a model of the flow of a fluid with velocity $u(z, t)$ along a channel between the parallel planes $z = 0$ and $z = \pi$, where R is the 'Reynolds' number.

The basic 'flow' is taken as the steady solution $u = U$, where $U \equiv 0$. Linearizing the perturbations of this trivial solution in the usual way, we find

and

$$\left. \begin{array}{l} \dfrac{\partial u'}{\partial t} - u' = \dfrac{1}{R} \dfrac{\partial^2 u'}{\partial z^2} \\[2mm] u' = 0 \quad \text{at } z = 0, \quad \pi. \end{array} \right\} \tag{50.3}$$

The solution of this linearized problem can easily be represented as a sum of the normal modes,

$$u' = \sum_{n=1}^{\infty} A_n \exp s_n t \sin nz, \tag{50.4}$$

where

$$s_n = 1 - n^2/R. \tag{50.5}$$

The nth mode is stable if and only if $R \leq n^2$. Therefore the 'flow' $U \equiv 0$ is stable if and only if all the modes are stable, i.e.

$$R \leq R_c \equiv \min_{n \geq 1} n^2 = 1. \tag{50.6}$$

Next we examine the nonlinear instability when the 'flow' is just unstable. If $0 < R - 1 \ll 1$ then all the normal modes except the first decay exponentially in time, so it is plausible to ignore the higher modes in the linearized initial-value problem. Accordingly, we approximate the linearized solution by

$$u' \cong A_1 \exp s_1 t \sin z. \tag{50.7}$$

This solution grows very slowly so that, however small the disturbance is initially, it will cease to be small only after a long time of the order of $-(\ln A_1)/s_1 \sim -(\ln A_1)/(R - 1)$ as $R \downarrow 1$. By this time nonlinearity will have modified the exponential growth and the solution given by equation (50.7) will have become invalid. To approximate the solution uniformly over so long a time, we anticipate that the nonlinear solution satisfies

$$u \sim u_1 \equiv A(t) \sin z \quad \text{as } R \to 1, \quad A \to 0 \tag{50.8}$$

for all time, where the amplitude equation is

$$\frac{dA}{dt} = a_1 A + a_2 A^2 + a_3 A^3 + \cdots. \tag{50.9}$$

Moreover, we assume that the exact solution can be expanded as

$$u = u_1 + u_2 + u_3 + \cdots, \tag{50.10}$$

where the fundamental mode at marginal stability is given by equation (50.8) and

$$u_r = O(A^r) \text{ as } A \to 0 \text{ for } r = 2, 3 \ldots. \tag{50.11}$$

To find the expansions (50.9) and (50.10) by iteration, we first transfer all the small terms of equation (50.1) to the right-hand side, writing

$$\frac{\partial^2 u}{\partial z^2} + u = \frac{\partial u}{\partial t} + (u - \sin u) + \frac{R-1}{R} \frac{\partial^2 u}{\partial z^2} \qquad (50.12)$$

without approximation. Note that each of the three terms on the right-hand side is small, the first because the disturbance varies slowly, the second because the nonlinearity is weak, and the third because the 'flow' is just supercritical. Note also that the linear operator $L \equiv \partial^2/\partial z^2 + 1$ associated with the left-hand side is such that $Lu_1 = 0$, where u_1 is the most unstable mode given by equation (50.8) at marginal stability.

Checking the first approximation, we equate all terms of order A in equation (50.12) to find

$$0 = \text{terms of order } A \text{ in } \left\{ \frac{dA}{dt} - \frac{(R-1)}{R} A \right\} \sin z$$

$$= \left(a_1 - \frac{R-1}{R} \right) A \sin z.$$

Therefore we identify $a_1 = s_1 \sim R - 1$ as $R \to 1$, in agreement with the linear theory.

For the next approximation, we equate terms of order A^2 in equation (50.12) to find

$$Lu_2 \equiv \frac{\partial^2 u_2}{\partial z^2} + u_2 = \text{terms of order } A^2 \text{ in } \left\{ \frac{\partial u_1}{\partial t} + \frac{R-1}{R} \frac{\partial^2 u_1}{\partial z^2} \right\}$$

$$= \text{terms of order } A^2 \text{ in } \left\{ \frac{dA}{dt} - \frac{R-1}{R} A \right\} \sin z$$

$$= a_2 A^2 \sin z. \qquad (50.13)$$

Similarly, the boundary conditions (50.2) give

$$u_2 = 0 \quad \text{at } z = 0, \pi. \qquad (50.14)$$

If the solution u_2 of the linear inhomogeneous problem given by equations (50.13) and (50.14) exists, we may multiply equation (50.13) by u_1, integrate from $z = 0$ to π, and deduce that

$$\int_0^\pi a_2 u_1 A^2 \sin z \, dz = \int_0^\pi (Lu_2) u_1 \, dz,$$

i.e.

$$a_2 A^3 \int_0^\pi \sin^2 z \, dz = \left[\frac{\partial u_2}{\partial z} A \sin z - u_2 A \cos z \right]_0^\pi$$

$$+ \int_0^\pi u_2 (L u_1) \, dz$$

$$= 0, \tag{50.15}$$

on integration by parts, and use of the boundary conditions (50.14) and of $L u_1 = 0$. This is called the *solvability condition* of equations (50.13) and (50.14), it being necessary for the existence of the solution u_2. Therefore the resonant term on the right-hand side of equation (50.13) must vanish in order that there is no secular term in u_2, i.e.

$$a_2 = 0. \tag{50.16}$$

We now go back to solve equations (50.13) and (50.14), seeing trivially that

$$u_2 \equiv 0, \tag{50.17}$$

i.e. that the second harmonic happens not to be excited. Of course any multiple of u_1 could be added to this solution u_2, but such an addition could be transferred to the fundamental solution u_1 by re-definition of the amplitude A. So we may take the solution (50.17) without loss of generality. This choice of normalization can be systematized by imposition of the orthogonality condition,

$$\int_0^\pi u_1 (u - u_1) \, dz = 0, \quad \text{say.} \tag{50.18}$$

For the next approximation, we equate terms of order A^3 in equations (50.12) and (50.2), finding in the limit as $R \to 1$ that

$$L u_3 = a_3 A^3 \sin z + \tfrac{1}{6} u_1^3$$

$$= (a_3 + \tfrac{1}{8}) A^3 \sin z - \tfrac{1}{24} A^3 \sin 3z, \tag{50.19}$$

and

$$u_3 = 0 \quad \text{at } z = 0, \pi. \tag{50.20}$$

Multiplying equation (50.19) by u_1, integrating from $z = 0$ to π, etc., we get the solvability condition

$$a_3 = -\tfrac{1}{8}. \tag{50.21}$$

Then one may go back to equations (50.19) and (50.20) and show that their solution is

$$u_3 = \tfrac{1}{192} A^3 \sin 3z. \tag{50.22}$$

Although one could go on to find a_4, u_4, a_5, etc. in turn, we stop the iteration here, having already found the Landau equation to the cubic approximation for $0 < R - 1 \ll 1$:

$$\frac{dA}{dt} = (R - 1)A - \tfrac{1}{8}A^3. \tag{50.23}$$

The Landau constant is positive (i.e. $a_3 < 0$) in this example, so there is supercritical stability with equilibration, etc.

This heuristic method can be made more systematic, though a formal presentation need not be a rigorous one. Matkowsky essentially defined $\varepsilon = (R - 1)^{1/2}$, $T = \varepsilon^2 t$ and $A_1 = A/\varepsilon$, and then showed formally that the method gives

$$\frac{dA_1}{dT} = A_1 - \tfrac{1}{8}A_1^3$$

in the limit as $\varepsilon \to 0$. This scaling of the variables ensures that A_1 and T are of order one when the three terms of the Landau equation are of the same order of magnitude, i.e. when the solution of the Landau equation develops substantially. Thus we require that $A^2/(R - R_c)$ is independent of R in the limit as $R \to R_c$ for fixed $(R - R_c)t$.

One may remark on another quite general property illustrated by this problem. It could have been anticipated that $a_2 = 0$ on the grounds of symmetry. For the problem is invariant if $z \to \pi - z$ or $u \to -u$, and so a similar invariance of the solution must follow. In particular, it follows that the Landau equation must be invariant if $A \to -A$. More generally, one finds that many hydrodynamic problems are invariant under any translation in some direction, and in particular to translation of one half wavelength of the fundamental, which corresponds to a change in the sign of the amplitude.

We have taken a simple model problem and solved it informally to illustrate the theory of weakly nonlinear stability. Realistic problems of hydrodynamic stability may be solved similarly, but the technical difficulties are usually more severe. These difficulties, associated with the inversion (or, rather, pseudo-inversion) of the operator L of the linear problem at marginal stability, may involve lengthy calculus, algebra or numerical work. They have prevented rigorous solution of all but a few problems, although many problems have been solved systematically by scaling the time and by taking formal expansions in the limits as $R \to R_c$ and $A \to 0$.

The very simplicity of the model problem has enabled us to avoid these difficulties and the tedium of a lot of calculus and algebra. At the same time it has led to the suppression of a few important points which often arise in more realistic problems. Firstly note that the method of deduction of the solvability condition (50.15) is appropriate because the operator L with the boundary conditions (50.14) happens to be self-adjoint and to have only one independent solution, namely u_1, of the homogeneous problem $Lu = 0$. More generally, in solving problems of weakly nonlinear hydrodynamic stability by this method, an inhomogeneous linear problem of the form of a partial differential equation $Lu_2 = G$ and boundary conditions $Bu_2 = H$ is likely to arise, where G and H are known in terms of u_1, etc. Then we would need to seek the adjoint operators $L^\dagger$ of L and $B^\dagger$ of B; to find them one usually chooses integration over the domain of the flow for an appropriate inner product and integrates by parts to get the adjointness identity. Next all the solutions $u^\dagger$ of the adjoint homogeneous problem $L^\dagger u^\dagger = 0$, $B^\dagger u^\dagger = 0$ are required. Finally we form the inner product $(G, u^\dagger)$ and deduce that $(G, u^\dagger) = (Lu_2, u^\dagger) = (u_2, L^\dagger u^\dagger) + F(H, u^\dagger) = F(H, u^\dagger)$, where F is some bilinear functional of H and of $u^\dagger$ evaluated on the boundaries. This gives the solvability conditions $(G, u^\dagger) = F(H, u^\dagger)$, which express the fact that G is in some sense orthogonal to all the solutions $u^\dagger$ of the adjoint problem in order that u_2 may exist. In the example above of the model problem, $u^\dagger$ is simply u_1 and H is zero.

Secondly, note that the Landau equation (50.23) is an appropriate description of the nonlinear self-interaction of the most unstable mode when slightly supercritical, but that we can with greater generality expand

$$u = \sum_{n=1}^{\infty} A_n(t) \sin nz \qquad (50.24)$$

in terms of the complete set of functions $\{\sin nz\}$ satisfying the boundary conditions (50.2). Substituting this expansion into equation (50.1) and equating coefficients of $\sin mz$ (or, more properly, multiplying by $(2/\pi) \sin mz$ and integrating from $z = 0$ to π), we deduce that

$$\frac{\mathrm{d}A_m}{\mathrm{d}t} = \frac{m^2}{2} A_m + \frac{2}{\pi} \int_0^\pi \sin mz \sin \left(\sum_{n=1}^\infty A_n \sin nz \right) \mathrm{d}z$$

$$= (1 - m^2/R)A_m + N_m(A_n), \tag{50.25}$$

say, where N_m is known in principle and may easily be expanded as a series in the odd powers of the A_ns, beginning with cubic terms. The system described by equation (50.25) is an example of equation (49.9).

Thirdly, note that the assumption of the expansion in equation (50.9) of $\mathrm{d}A/\mathrm{d}t$ in powers of A happens to be appropriate for the model problem, and is justifiable by use of equation (50.25). But the expansion in equation (50.9), and, indeed, a system of the form described by equation (49.9), is by no means applicable to *all* problems of hydrodynamic stability. A simple counter-example, where the term $\partial u/\partial t$ in equation (50.1) is replaced by $\partial^2 u/\partial t^2$, is given in Problem 7.5. A profound counter-example is given in Problem 7.10.

Fourthly, note that the method we have chosen to solve the model problem defined by equations (50.1) and (50.2) is not the only one used to solve weakly nonlinear problems. It was first used by Palm (1960), Stuart (1960a) and Watson (1960b) to treat problems of hydrodynamic stability. But the method has been varied somewhat by later authors and itself uses some ideas of bifurcation theory that have been traced back to Lindstedt's work on celestial mechanics in 1883. It is similar to the well-known method of Liapounov and Schmidt, which was first used at the beginning of this century to find equilibrium solutions near bifurcation (see, for example, Stakgold (1971)). Applying the Liapounov–Schmidt method to the model problem, one would assume that $\partial u/\partial t$ and therefore $\mathrm{d}A/\mathrm{d}t$ is zero and find $A = A_e$ as a power series in $R - R_c$; of course $A_e = O\{(R - R_c)^{1/2}\}$ as $R \downarrow R_c$, so square-roots would have to be included in the series. Another method, closely related to the Liapounov–Schmidt method, is to find the steady solution $A = A_e$ by expanding $R - R_c$ in powers of A_e rather than vice versa, and this has the advantage that only integral powers need be used; this method is essentially the one first used by Gor'kov (1957) and

Malkus & Veronis (1958) to study nonlinear Bénard convection. Again Eckhaus (1965) pioneered the use of more general Galerkin expansions of the form given by equation (49.8). In all these methods the state of linear marginal stability is perturbed, so similar inhomogeneous linear problems arise at successive orders of approximation whichever method is used.

51 Resonant wave interactions

We have just shown how both the new finite-amplitude equilibrium solution and its stability near the bifurcation at marginal stability of the original basic flow are determined by the Landau equation. If the Landau constant is positive then the new solution exists and is stable when the original basic flow is just supercritical; if the Landau constant is negative then the new solution exists and is unstable when the basic flow is just subcritical. Of course the stability as well as the structure of finite-amplitude disturbances is important in practice. Now a particularly important case of finite-amplitude disturbances comprises waves, such as surface waves on the ocean. So we shall describe next the weakly nonlinear stability of weakly nonlinear waves. The ideas to be used are somewhat similar to those discussed in the last two sections but have been developed more-or-less independently. Phillips (1960) was the first to consider weakly nonlinear interactions of waves, finding that interactions of surface gravity waves are enhanced when certain resonance conditions are satisfied. Since 1960 the theory has been developed and refined by Phillips and many other authors; also the theory has been applied to many kinds of waves in fluids. The most celebrated application of the theory is perhaps the demonstration of the instability of a Stokes wave by Benjamin & Feir (1967).

51.1 *Internal resonance of a double pendulum*

To describe the essence of the theory we shall again use a simple ordinary differential equation as an example. The example comes from rigid-body dynamics. Rott (1970) devised a double pendulum which provides a delightful visual and theoretical illustration of the essence of resonant interactions. (Such a pendulum may be easily constructed with the aid of the design details Rott gave.) Here we begin with his dynamical equations in the form

$$\frac{d^2\theta_1}{dt^2} + \omega_1^2\theta_1 = a\left\{ (\theta_1 - \theta_2)\frac{d^2\theta_2}{dt^2} - \left(\frac{d\theta_2}{dt}\right)^2 \right\}, \qquad (51.1)$$

and

$$\frac{d^2\theta_2}{dt^2} + \omega_2^2\theta_2 = b\left\{ (\theta_1 - \theta_2)\frac{d^2\theta_1}{dt^2} + \left(\frac{d\theta_1}{dt}\right)^2 \right\}, \qquad (51.2)$$

which describe approximately small but finite oscillations of the double pendulum. The angles of rotation of the two components of the pendulum are denoted by θ_1 and θ_2, and the frequencies of the corresponding normal modes of oscillation by ω_1 and ω_2 respectively. Rott showed that the terms on the right-hand sides represent the weak interactions of the modes, where a and b are certain coefficients determined by the construction of the double pendulum. First note the exact energy integral of these equations,

$$\frac{1}{2a}\left\{ \left(\frac{d\theta_1}{dt}\right)^2 + \omega_1^2\theta_1^2 \right\} + \frac{1}{2b}\left\{ \left(\frac{d\theta_2}{dt}\right)^2 + \omega_2^2\theta_2^2 \right\}$$

$$+ (\theta_2 - \theta_1)\frac{d\theta_1}{dt}\frac{d\theta_2}{dt} = \text{constant.} \qquad (51.3)$$

This problem may be solved intuitively as follows. Considering oscillations of the system described by equations (51.1) and (51.2) with small but finite amplitudes, we anticipate the occurrence of the normal modes $\theta_1 = \phi_1 \cos(\omega_1 t + \psi_1)$ and $\theta_2 = \phi_2 \cos(\omega_2 t + \psi_2)$ modified by their weakly nonlinear interactions. The quadratic terms on the right-hand sides involve products of the normal modes and hence trigonometric functions of $(\omega_1 \pm \omega_2)t$ because

$$\phi_1 \cos(\omega_1 t + \psi_1)\phi_2 \cos(\omega_2 t + \psi_2)$$
$$= \tfrac{1}{2}\phi_1\phi_2[\cos\{(\omega_1 - \omega_2)t + (\psi_1 - \psi_2)\}$$
$$+ \cos\{(\omega_1 + \omega_2)t + (\psi_1 + \psi_2)\}],$$

etc. Thus small terms of order $\phi_1\phi_2$ and periods $2\pi/|\omega_1 \pm \omega_2|$ will arise on the right-hand sides. Similarly terms of order ϕ_1^2 and period π/ω_1, and of order ϕ_2^2 and period π/ω_2, as well as constant terms of orders ϕ_1^2 and ϕ_2^2 will arise on the right-hand sides. The fundamental idea here is to regard all these small terms on the right-hand sides as terms forcing the linear oscillators represented by the left-hand sides. In general the response is of the same order as the forcing terms, but there is an exception, called *internal resonance*, when any of the forcing terms has the same period as one of the fundamental normal modes. This resonance occurs when $\omega_2 = 2\omega_1$ or $\omega_1 = 2\omega_2$. (It is also possible that there is resonance at third order if $\omega_1 = 3\omega_2$ or $\omega_2 = 3\omega_1$, because then the cubic terms in the expansions of the right-hand sides in powers of ϕ_1 and ϕ_2 may have period ω_1 or ω_2 respectively. There may also be resonance at higher orders similarly if $\omega_1 = 4\omega_2$ or $\omega_2 = 4\omega_1$, etc.) The resonance leads to an enhanced response which may change the amplitudes of the fundamental normal modes at first order after a long time, even though the interaction itself is at second or higher order.

To examine this, let us suppose that $\omega_2 = 2\omega_1$. (The case $\omega_1 = 2\omega_2$ is similar, so we need not discuss it.) Then we seek an approximate solution of the form

$$\left.\begin{aligned}
\theta_1 &= A_1 \exp{(i\omega_1 t)} + A_1^* \exp{(-i\omega_1 t)}, \\
\theta_2 &= A_2 \exp{(2i\omega_1 t)} + A_2^* \exp{(-2i\omega_1 t)},
\end{aligned}\right\} \tag{51.4}$$

where A_1 and A_2 are now taken as small, and hence slowly varying, complex functions of time. We shall not attempt to find higher approximations to the solution, and so we neglect quadratic terms of A_1 and A_2 in equation (51.4). Next we substitute equation (51.4) into equations (51.1) and (51.2), and neglect terms in accord with the approximation that A_1 and A_2 are both small. First we ignore the terms on the right-hand sides of equations (51.1) and (51.2) which do not resonate with the appropriate fundamentals, i.e. which do not vary like $\exp{(\pm i\omega_1 t)}$ in equation (51.1) or like $\exp{(\pm 2i\omega_1 t)}$ in equation (51.2); this is plausible because the response to the forcing by quadratic terms which vary like $\exp{(\pm 3i\omega_1 t)}$, $\exp{(\pm 4i\omega_1 t)}$, etc. is also of second order in A_1 and A_2. This effectively approximates equations (51.1) and (51.2) by

$$\frac{d^2\theta_1}{dt^2} + \omega_1^2\theta_1 = a\theta_1 \frac{d^2\theta_2}{dt^2}, \tag{51.5}$$

and

$$\frac{d^2\theta_2}{dt^2} + 4\omega_1^2\theta_2 = b\left\{\theta_1 \frac{d^2\theta_1}{dt^2} + \left(\frac{d\theta_1}{dt}\right)^2\right\}. \tag{51.6}$$

Secondly, we approximate $|\ddot{A}_1| \ll \omega_1 |\dot{A}_1| \ll \omega_1^2 |A_1|$ and $|\ddot{A}_2| \ll \omega_1 |\dot{A}_2| \ll \omega_1^2 |A_2|$; this approximation of slow variation will soon be found to be self-consistent by inspection of the resultant solution. Thus, equating coefficients of $\exp(\pm i\omega_1 t)$ in equation (51.5) and of $\exp(\pm 2i\omega_1 t)$ in equation (51.6), we deduce that

$$2i\omega_1 \frac{dA_1}{dt} = -4a\omega_1^2 A_1^* A_2 \tag{51.7}$$

and

$$4i\omega_1 \frac{dA_2}{dt} = -2b\omega_1^2 A_1^2 \tag{51.8}$$

approximately. We now see that $\dot{A}_1 = O(A_1 A_2)$ and $\dot{A}_2 = O(A_1^2)$ as $A_1, A_2 \to 0$, confirming that A_1 and A_2 do indeed vary slowly. It also follows by elimination of A_2 from equations (51.7) and (51.8) that

$$\frac{d}{dt}\left(\frac{1}{A_1^*}\frac{dA_1}{dt}\right) = -ab\omega_1^2 A_1^2. \tag{51.9}$$

Letting $\phi = |A_1|$ and $\psi = \mathrm{ph}\, A_1$ so that $A_1 = \phi e^{i\psi}$, we deduce that

$$\frac{1}{\phi}\frac{d^2\phi}{dt^2} - \frac{1}{\phi^2}\left(\frac{d\phi}{dt}\right)^2 - 2\left(\frac{d\psi}{dt}\right)^2 = -ab\omega_1^2\phi^2 \tag{51.10}$$

and

$$\frac{d^2\psi}{dt^2} + \frac{2}{\phi}\frac{d\phi}{dt}\frac{d\psi}{dt} = 0. \tag{51.11}$$

Therefore

$$\psi = k_1 \int_0^t \phi^{-2}dt + k_2 \tag{51.12}$$

for some constants k_1 and k_2 of integration of equation (51.11), and thence equation (51.10) becomes

$$\frac{d}{dt}\left(\frac{1}{\phi}\frac{d\phi}{dt}\right) = \frac{2k_1^2}{\phi^4} - ab\omega_1^2\phi^2.$$

Multiplying this equation by $\dot{\phi}/\phi$, integrating, and then multiplying by $8\phi^4$, we find

$$\left(\frac{d\phi^2}{dt}\right)^2 = k_3\phi^4 - 4ab\omega_1^2\phi^6 - 4k_1^2, \tag{51.13}$$

where k_3 is another constant of integration. It follows that ϕ^2 is an elliptic function, and therefore that the modulus, although not the phase, of A_1 is periodic in time. The solution can be found in explicit terms of ϕ, because equation (51.12) gives

$$A_1 = \phi \exp\left(ik_1 \int_0^t \phi^{-2}\,dt + k_2\right) \qquad (51.14)$$

and equation (51.7)

$$A_2 = -\frac{i}{2a\omega_1}\left(\frac{1}{\phi}\frac{d\phi}{dt} + \frac{ik_1}{\phi^2}\right)\exp\left(2ik_1\int_0^t \phi^{-2}\,dt + 2k_2\right). \qquad (51.15)$$

The solution can be found in this way for any given initial values of θ_1, $\dot\theta_1$, θ_2 and $\dot\theta_2$ because a fourth constant of integration is available on integration of equation (51.13). It can be seen that there is a slow periodic interchange of energy between the two fundamentals, consistent with equation (51.3) to the present order of approximation.

Rott (1970) gave this solution in greater detail, in fact solving the more general case of near resonance, i.e. the case when $\omega_2 = 2\omega_1 + \varepsilon$ and $|\varepsilon| \ll \omega_1$. He noted that after completion of a period of $\phi(t)$ the phases of the solution (51.14), (51.15) imply that the number of oscillations of the faster pendulum is exactly twice the number of the slower pendulum. The steady solution $A_1 = 0$, $A_2 = $ constant of equations (51.7) and (51.8) gives $\theta_1 = 0$, $\theta_2 = \phi_2 \cos(2\omega_1 t + \psi_2)$. (This is an exact solution of equations (51.5) and (51.6) but not of equations (51.1) and (51.2) because of the approximations made.) This solution, however, is unstable, so that if θ_1 and $\dot\theta_1$ are very small and θ_2 and $\dot\theta_2$ small initially then the first mode will slowly draw all the energy from the second mode, give it back again, and so on. Thus it is impossible in practice to generate and maintain a pure oscillation of the faster pendulum without very slowly exciting an oscillation of the slower pendulum. Similarly, a pure oscillation of the faster pendulum cannot be maintained, although the growth of the other normal mode is different in this case. Many more interesting properties of this double pendulum are discussed by Rott.

These general ideas of internal resonance have been attributed by Beth (1913) to work of Korteweg in 1897. A more modern treatment of an example of internal resonance, using the Hamiltonian and the method of averaging, is described in the textbook of Nayfeh (1973), § 5.5.3.

51.2 *Resonant wave interactions*

Moving from the introductory example to the substance of this section, we next examine the forms of resonance possible for interactions of waves in a fluid. For waves in an unbounded fluid there is not a finite number of normal modes, as for a dynamical system, but a continuum for all real values of the wavenumbers, so the possibilities for resonant wave interactions are richer. For isotropic waves in a fluid the quantities specifying the waves vary like $\exp\{i(\mathbf{k} \cdot \mathbf{x} - \omega t)\}$, where the real frequency ω is given in terms of the wavenumber vector $\mathbf{k}$ by some *dispersion relation*, $\omega = f(\mathbf{k})$ say. We usually treat waves in time-reversible and space-reversible systems, so that if ω is a frequency for a given $\mathbf{k}$, then so is $-\omega$ for the same $\mathbf{k}$ and ω for $-\mathbf{k}$.

The nonlinearity in the equations and boundary conditions governing any wave motion leads to wave interactions. Again by convention the linear terms may be put on the left-hand sides and the nonlinear terms on the right-hand sides of the equations and boundary conditions. Then the nonlinear terms may be regarded as small terms forcing the linear system. Two waves may interact at second order to excite a third wave by resonance now if not only their frequencies add to that of the third but also their wavenumbers, i.e. if

$$\omega_3 = \omega_1 + \omega_2 \quad \text{and} \quad \mathbf{k}_3 = \mathbf{k}_1 + \mathbf{k}_2,$$

where

$$\omega_n = f(\mathbf{k}_n) \quad \text{for } n = 1, 2, 3. \tag{51.16}$$

Thus resonance occurs at second order only if the dispersion relation is such that $f(\mathbf{k}_1 + \mathbf{k}_2) = f(\mathbf{k}_1) + f(\mathbf{k}_2)$ for some wavenumbers $\mathbf{k}_1$ and $\mathbf{k}_2$. In particular, one wave may interact with itself to excite another if $\omega_2 = 2\omega_1$ and $\mathbf{k}_2 = 2\mathbf{k}_1$; also it may interact with the other to excite itself under the same conditions if $-\omega_1 = f(-\mathbf{k}_1)$, because $\omega_1 = -\omega_1 + \omega_2$ and $\mathbf{k}_1 = -\mathbf{k}_1 + \mathbf{k}_2$. The resonance condition for wavenumbers as well as frequencies is necessary, of course, because a partial differential system represents the motion of waves in a fluid. The dispersion relation may be such that no triad of waves can satisfy all the resonance conditions. Indeed, $\omega^2 = g|\mathbf{k}|$ for surface waves on deep water, where $\mathbf{k}$ is the horizontal wavenumber vector, and in fact no resonance at second order is possible for these waves; there are, however, tetrads which resonate at third order with $\omega_1 + \omega_2 + \omega_3 + \omega_4 = 0$ and $\mathbf{k}_1 + \mathbf{k}_2 + \mathbf{k}_3 + \mathbf{k}_4 = 0$ (Phillips 1960). (Note that $\omega_1 = \omega_2 + \omega_3 + \omega_4$ is equivalent to $\omega_1 + \omega_2 + \omega_3 + \omega_4 = 0$ and $\mathbf{k}_1 = \mathbf{k}_2 + \mathbf{k}_3 + \mathbf{k}_4$ to $\mathbf{k}_1 + \mathbf{k}_2 + \mathbf{k}_3 + \mathbf{k}_4 = 0$ when the dispersion relation is symmetric in $\pm\omega$ and $\pm\mathbf{k}$.)

The solutions presented in the early work on resonant wave interactions contained secular terms proportional to the time and are not uniformly valid over an infinite time interval. Benney (1962) recognized the significance of the slow but substantial variation of the wave amplitudes and derived uniformly valid solutions.

To illustrate the theory we shall take the simple model equation

$$\frac{\partial^2 u}{\partial t^2} + \frac{\partial^4 u}{\partial x^4} + \frac{\partial^2 u}{\partial x^2} + u = u^2, \qquad (51.17)$$

after Bretherton (1964). The simplicity of this problem is largely due to the presence of only one space variable, and hence we need consider only a scalar wavenumber. The first things to do are to linearize the system, and find the normal modes and their dispersion relation. For equation (51.17), we neglect the right-hand side, take

$$u = A \exp\{i(kx - \omega t)\} + A^* \exp\{-i(kx - \omega t)\} \qquad (51.18)$$

for any constant complex amplitude A, and find

$$\omega^2 = 1 - k^2 + k^4. \qquad (51.19)$$

Secondly, the solution to the weakly nonlinear problem (51.17) may be expanded as

$$u = \sum_k A(k, t) \exp[i\{kx - \omega(k)t\}], \qquad (51.20)$$

where now $A(k, t)$ is expected to be a slowly varying function of t and ω satisfies equation (51.19). The summation is in general an integration over all wavenumbers, but often in practice only discrete waves are important. In order that u be real, we take

$A(-k, t) = A^*(k, t)$ and sum over equal and opposite values of k. Next substitution of equation (51.20) into equation (51.17) gives

$$\sum_{k_1} \left[\frac{d^2 A_1}{dt^2} - 2i\omega_1 \frac{dA_1}{dt} + \{-\omega_1^2 + (1 - k_1^2 + k_1^4)\}A_1 \right]$$

$$\times \exp\{i(k_1 x - \omega_1 t)\}$$

$$= \sum_{\substack{\omega_2 + \omega_3 = \omega_1 \\ k_2 + k_3 = k_1}} 2A_2 A_3 \exp\{i(k_1 x - \omega_1 t)\}. \tag{51.21}$$

(It can in fact be shown that the terms on the right-hand side can be grouped in this way; if the dispersion relation were such that this were not possible, then those terms which did not satisfy the resonance condition would be neglected.) Therefore

$$\sum_{k_1} \left(\frac{d^2 A_1}{dt^2} - 2i\omega_1 \frac{dA_1}{dt} \right) \exp\{i(k_1 x - \omega_1 t)\}$$

$$= \sum_{\substack{\omega_1 + \omega_2 + \omega_3 = 0 \\ k_1 + k_2 + k_3 = 0}} 2A_2^* A_3^* \exp\{i(k_1 x - \omega_1 t)\}. \tag{51.22}$$

The symmetric form $\omega_1 + \omega_2 + \omega_3 = 0$, $k_1 + k_2 + k_3 = 0$ of the resonance conditions shows at once that any two waves of a triad interact to excite the third. Also Bretherton (1964) showed that for the dispersion relation (51.19) there are two infinite families of resonant triads, such that for a given positive value of k_3 there exists a unique pair of values of k_1 and k_2, and therefore that each triad is essentially independent of the others. Therefore equation (51.22) gives

$$\frac{d^2 A_1}{dt^2} - 2i\omega_1 \frac{dA_1}{dt} = 2A_2^* A_3^*. \tag{51.23}$$

The equation for $-k_1$, $-\omega_1$ is the complex conjugate of this equation. Again we see that $\dot{A}_1 = O(A_2 A_3)$ and therefore we can neglect $\ddot{A}_1$ as $A_2, A_3 \to 0$. There are similar equations for $\dot{A}_2$ and $\dot{A}_3$, so finally we deduce the approximate equations

$$\left. \begin{array}{ll} \dfrac{dA_1}{dt} = i\omega_1^{-1} A_2^* A_3^*, & \dfrac{dA_2}{dt} = i\omega_2^{-1} A_3^* A_1^*, \\[2mm] \dfrac{dA_3}{dt} = i\omega_3^{-1} A_1^* A_2^*. \end{array} \right\} \tag{51.24}$$

Note that equations (51.24) give

$$\omega_2 A_2^* \frac{dA_2}{dt} = iA_1^* A_2^* A_3^* = \omega_3 A_3^* \frac{dA_3}{dt}.$$

Adding the complex conjugates of these equations, we deduce that

$$\frac{d}{dt}\omega_2|A_2|^2 = \frac{d}{dt}\omega_3|A_3|^2.$$

There is a similar equation for $|A_3|$ and $|A_1|$. Therefore

$$\omega_1(|A_1|^2 - A_{10}^2) = \omega_2(|A_2|^2 - A_{20}^2) = \omega_3(|A_3|^2 - A_{30}^2), \qquad (51.25)$$

where $|A_n| = A_{n0}$ at $t = 0$ for $n = 1, 2, 3$. These two independent integrals of equations (51.24) are consistent with the conservation of 'energy',

$$\tfrac{1}{2}(\omega_1^2|A_1|^2 + \omega_2^2|A_2|^2 + \omega_3^2|A_3|^2) = \text{constant}, \qquad (51.26)$$

because $\omega_1 + \omega_2 + \omega_3 = 0$.

To consider the stability of a single wave of frequency ω_1 and wavenumber k_1, we find the resonant triad and then suppose that A_2 and A_3 are very small. We may neglect all other modes because they do not excite the chosen wave. Thus we take $A_1 = \text{constant}$ and $A_2 = A_3 = 0$ as a basic solution and linearize equation (51.24), finding

$$\frac{dA_1}{dt} = 0, \quad \frac{d^2A_2}{dt^2} - \frac{|A_1|^2}{\omega_2\omega_3}A_2 = 0, \quad \frac{d^2A_3}{dt^2} - \frac{|A_1|^2}{\omega_2\omega_3}A_3 = 0 \quad (51.27)$$

after elimination of A_3 and A_2 respectively. Therefore A_1 is constant at the present order of approximation, but A_2 and A_3 may grow exponentially if $\omega_2\omega_3 > 0$ or oscillate sinusoidally if $\omega_2\omega_3 < 0$. Thus the given wave A_1 is unstable if $\omega_2\omega_3 > 0$ and stable if $\omega_2\omega_3 < 0$.

It is perhaps easier to examine this condition for stability by supposing that all the wavenumbers and frequencies are positive, because the dispersion relation (51.19) is even in ω and k. Then the resonance conditions may be rewritten in either the form

$$\omega_1 = \omega_2 + \omega_3 \quad \text{and} \quad k_1 = k_2 + k_3 \qquad (51.28)$$

or

$$\omega_1 = \omega_2 - \omega_3 \quad \text{and} \quad k_1 = k_2 - k_3. \qquad (51.29)$$

If condition (51.28) holds, then the wave A_1 is unstable, being subject to parasitic growth of A_2 and A_3. If condition (51.29) holds, then the wave A_1 is neutrally stable. This result of instability for the sum interaction and stability for the difference interaction holds, in fact, for all conservative coupled-mode systems (Hasselmann 1967).

Bretherton (1964) went on to note that the interaction equations (51.24) could be solved in exact terms of elliptic functions. One might add that equations (51.24) are similar to Euler's dynamical equations for the free motion of a rigid body about a fixed point (cf. Goldstein 1950, Chapter 5). The above results are now recognizable as analogues of some properties of the motion of a rigid body. In particular, the stability criterion is the analogue of the classic result that the steady rotation of a rigid body about its axis of greatest or least inertia is stable, and the rotation about the intermediate principal axis is unstable.

Bretherton (1964) analysed the approximation more carefully. He also treated nearly resonant interactions, where $k_1 + k_2 + k_3 = 0$ exactly but $\omega_1 + \omega_2 + \omega_3 = O(A)$ as $|A| \to 0$. This case of a triad that is not quite perfectly tuned is more realistic, because if the domain of 'flow' of equation (51.17) is unbounded then the sum (51.20) is more properly an integral, and if the domain is bounded then the sum is over discrete wave components which are unlikely to include the other two members of a triad which resonate exactly with a given wave.

Problems, such as Bretherton's, with only one space variable and therefore only a scalar wavenumber, usually give independent resonant triads. It should be noted that this is atypical of problems in more than one space variable, for which a given wave is usually a member of an infinite number of different triads.

We have used a direct intuitive method of finding the ordinary differential system which governs the slow variation of the amplitudes in order to make the nature of the wave interactions apparent. But any research worker solving a hydrodynamic problem should consider the use of a variational formulation of the problem and an averaged Lagrangian in order to systematize and shorten the detailed calculations. This lessens the likelihood of making elementary errors in the long calculations of the interaction coefficients of the amplitude equations. A variational formulation also leads more readily to the recognition of invariants which give conservation relations in the form of integrals of the amplitude equations, such as the energy integral. Whitham (1967), § 3 was the first to calculate resonant wave interactions by putting the equations for a problem (in fact the modified Korteweg–de Vries equation) into variational form and by averaging its Lagrangian. Simmons (1969) formulated this method quite generally, showing how to find conservation relations and many other useful results. Nayfeh (1973), §§ 5.8, 6.2.9 and 6.4.8 solved Bretherton's model problem to illustrate the method of multiple scales as well as the method of the averaged Lagrangian. Whitham (1974), § 15.6, however, in his book treats the theory of resonant wave interactions by the direct method.

In a book on hydrodynamic stability it is appropriate to emphasize the application of the theory of resonant wave interactions to the stability of finite-amplitude waves in a fluid. The approximations of the theory limit its application to weakly nonlinear waves. The development of the instability, however, may be followed beyond the initial stage of exponential growth of the parasitic modes to a full periodic interchange of the energies of the interacting modes. This may usually be done by use of elliptic functions. To apply the theory one chooses any weakly nonlinear wave, with amplitude $A_1(t)$, say, as the basic flow. To test its instability one considers its weakly nonlinear interactions with all the other waves. If a resonant triad can be found to include the basic wave, then the amplitude equation must be derived. Otherwise one must seek interactions of a tetrad and its amplitude equations at third order, etc. Finally one must examine the amplitude equations to find whether the basic wave is stable to small perturbations due to other members of the triad (or tetrad, etc.), much as we examined equations (51.27).

The theory of resonant wave interactions has been applied to capillary waves, internal gravity waves, Rossby waves, etc. as well as to surface gravity waves. Many of these applications have been reviewed recently by Phillips (1974).

An early and interesting application is to capillary waves. Their dispersion relation (a special case of Problem 1.4) is well known to be $\omega^2 = gk + \gamma k^3/\rho$, and McGoldrick (1965) found the second-order interactions of triads of waves. He (McGoldrick 1970) later developed the analysis and related it to some beautiful experiments.

Engineers who test models of ships in long tanks seek to generate regular trains of waves. For decades the engineers had been troubled by the generation of steep waves, which tend to break up into a series of wave groups, and discussed the malformation of those waves as being due to imperfections in their apparatus. Then Benjamin and Feir (Benjamin & Feir 1967, Feir 1967, Benjamin 1967) showed theoretically and experimentally that a Stokes wave is in fact unstable. Their analysis is essentially a treatment of the third-order interactions of a tetrad with wavenumbers k_1, k_1, $-k_1(1 + \varepsilon)$, $-k_1(1 - \varepsilon)$ for $0 < \varepsilon \ll 1$. Their frequencies are approximately ω_1, ω_1, $-\omega_1(1 + \frac{1}{2}\varepsilon)$ and $-\omega_1(1 - \frac{1}{2}\varepsilon)$ for deep-water waves, because $\omega^2 = g|\mathbf{k}|$, and so nearly add to zero. Benjamin and Feir called $-\omega_1(1 \pm \frac{1}{2}\varepsilon)$ *side-band frequencies* of the fundamental frequency of the basic wave, and the growth of the side bands leads to the instability of the fundamental wave. The mechanism of this instability has since been shown to be essentially equivalent to that of a side-band instability discovered earlier by Eckhaus (see § 52.4, Yuen & Lake (1975), Lake *et al.* (1977) and Stuart & DiPrima (1978)). The modulation of the amplitude of a water wave is governed by a nonlinear Schrödinger equation. This equation may be used to describe quite simply the nature of side-band instability and to show that weakly nonlinear water waves may recur owing to the properties of solitons. As a postscript we note the recent investigation by Longuet-Higgins (1978) of the linear stability of Stokes waves of all amplitudes, not just small ones. Longuet-Higgins found a mode of instability of steep waves quite near to breaking in addition to the Benjamin–Feir mode of instability of smallish waves (see also § 48.2).

52 Fundamental concepts of nonlinear stability

52.1 *Introduction to ordinary differential equations*

We have shared Landau's vision whereby hydrodynamic stability may be governed by a simple ordinary differential equation. With this vision we ignore the practical difficulties of deriving the appropriate differential equation. The severity of these difficulties is indicated in § 50. Yet if we do ignore them we can gain much insight into the qualitative character of nonlinear hydrodynamic stability.

Firstly, it should be remembered that the Landau equation is not the only ordinary differential equation to govern nonlinear stability. The Landau equation describes the essential behaviour of an infinite system of equations in the important, but special, case when a single weakly unstable mode and its lower harmonics dominate the flow. We have already shown that other amplitude equations may be needed instead of the Landau equation. More generally, we anticipate as many qualitatively different cases of nonlinear hydrodynamic instability as are met in the governing systems of ordinary differential equations.

Secondly, it should be remembered that the powerful and well-developed theory of ordinary differential equations is at our service. Here we can only summarize informally a few points of the classic theory, which is described in many good books, for example those by Coddington & Levinson (1955) and by Birkhoff & Rota (1978). Poincaré's theory of critical points and limit cycles of second-order equations at once suggests the possibility of a greater variety of phenomena than can be described by the Landau equation. The more recent discovery of generalizations of these critical solutions, namely the *strange attractors*, for third-order or higher-order systems suggests more-complicated phenomena, which might prove to be a manifestation of transition or even turbulence itself. Again, recent use of the centre manifold theorem suggests that the critical properties of an infinite system of ordinary differential equations governing hydrodynamic stability may be found in a system of finite, and indeed low, order. It is beyond the scope of this book to describe the rapidly developing mathematical theory of bifurcation for ordinary differential equations, but the modern books by Hartman (1964) and Hirsch & Smale (1974) are recommended. Also the application of the theory to hydrodynamic stability is described more fully by Ruelle & Takens (1971) and by Joseph (1976), Chapter II.

A system of ordinary differential equations may be put in the canonical form

$$\frac{d\mathbf{x}}{dt} = \mathbf{F}, \tag{52.1}$$

where $\mathbf{x}(t)$ and $\mathbf{F}(\mathbf{x}, t)$ are n- (or infinite-) dimensional vectors in the space $\mathbb{R}^n$ (or $\mathbb{R}^\infty$) with components x_i and F_i for $i = 1, 2, \ldots$. In hydrodynamic stability we often consider the stability of steady basic flows, and so are interested particularly in *autonomous* systems, i.e. those for which $\mathbf{F}$ does not depend explicitly upon t. We are also interested particularly in the steady solutions, or *equilibrium solutions*, for which $\mathbf{x} = \mathbf{X}$ (say), $\mathbf{F}(\mathbf{X}, t) = \mathbf{0}$ and $\mathbf{X} = \text{constant}$ for all t. Having found a solution $\mathbf{X}$, steady or unsteady, we may translate the system described by equation (52.1) and so suppose without loss of generality that $\mathbf{X} \equiv \mathbf{0}$; for writing $\mathbf{y} = \mathbf{x} - \mathbf{X}$ we get

$$\frac{d\mathbf{y}}{dt} = \mathbf{F}(\mathbf{y} + \mathbf{X}, t) - \frac{d\mathbf{X}}{dt} = \mathbf{G}(\mathbf{y}, t), \quad \text{say.} \qquad (52.2)$$

The stability of a solution $\mathbf{x} = \mathbf{X}$ is determined by the development as $t \to \infty$ of all the solutions $\mathbf{x}(t)$ according to their initial values $\mathbf{x}(0)$. The solution $\mathbf{X} \equiv \mathbf{0}$ is said to be *stable* (*in the sense of Liapounov*) if for any $\varepsilon > 0$ there exists a δ such that if $\|\mathbf{x}(0)\| < \delta$ then $\|\mathbf{x}(t)\| < \varepsilon$ for all $t > 0$. If, in addition, $\mathbf{x}(t) \to \mathbf{0}$ as $t \to \infty$, then $\mathbf{X} \equiv \mathbf{0}$ is said to be *asymptotically stable*. There is some freedom in defining the norm $\|\mathbf{x}(t)\|$, but we may choose it to be the greatest component of the vector $\mathbf{x}$ at time t. Conversely, the solution $\mathbf{X} \equiv \mathbf{0}$ is *unstable* if there exists some ε such that for any $\delta > 0$, however small, there exist values $\mathbf{x}_0$ and $t_0 > 0$ such that if $\mathbf{x}(0) = \mathbf{x}_0$ and $\|\mathbf{x}_0\| < \delta$ then $\|\mathbf{x}(t_0)\| > \varepsilon$. The *domain of attraction* of the solution $\mathbf{X} \equiv \mathbf{0}$ is the set of points $\mathbf{x}_0$ of $\mathbb{R}^n$ such that if $\mathbf{x}(0) = \mathbf{x}_0$ then $\mathbf{x}(t) \to \mathbf{0}$ as $t \to \infty$. So if the solution $\mathbf{X} \equiv \mathbf{0}$ is asymptotically stable then its domain of attraction includes some neighbourhood of $\mathbf{0}$ in $\mathbb{R}^n$. If $\mathbf{x}(t) \to \mathbf{0}$ as $t \to \infty$ whatever the initial values of $\mathbf{x}$, then the solution $\mathbf{X} \equiv \mathbf{0}$ is said to be *globally asymptotically stable*; it follows that the whole of $\mathbb{R}^n$ is the domain of attraction of $\mathbf{X} \equiv \mathbf{0}$.

It can be shown that a given steady or periodic basic solution $\mathbf{X}$ of a well-behaved nonlinear system described by equation (52.1) is asymptotically stable if all the solutions of the corresponding linearized system decay exponentially with time, and is unstable if one of them grows exponentially with time. This is why the linearized system is important. If all the solutions of the linearized system are neutrally stable, oscillating without exponential decay or growth, then the nonlinear terms may be used to determine the stability of the basic solution. (We have not hitherto emphasized the distinction between the stability of a basic flow and of the null solution of the corresponding linearized problem. It now seems plausible that the basic flow is asymptotically stable or unstable according as the null solution is asymptotically stable or unstable respectively. The stability of the basic flow may be doubtful when the null solution is neutrally stable. Even then it is plausible that the basic flow is stable unless the linearized system is marginally stable. Further circumstantial support of these conclusions may be drawn from experimental evidence in some cases. But a critical reader, seeking a more convincing mathematical demonstration, is not to be dismissed as a pedant.)

Poincaré found it fruitful to consider the *trajectories* or *orbits* in the *phase space* of an autonomous system, i.e. to consider the paths traced by the points $\mathbf{x}(t)$ in $\mathbb{R}^n$ as t increases. This is especially useful for second-order systems, because the phase space becomes the phase plane and has simple topological properties; for example, the plane is divided in two by a closed trajectory. The trajectories can meet only where the components of $\mathbf{F}$ vanish, i.e. at points of equilibrium or *critical points* of equation (52.1). These points are useful in determining the topological character of the trajectories of a given system, the domains of attraction, etc. Poincaré classified the critical points of second-order autonomous systems by examining the local properties of the trajectories in the phase plane. He found stable critical points (centres, spiral sinks, foci and nodes) and unstable ones (spiral sources, nodes and saddle points). There are also stable and unstable *limit cycles*, i.e. closed curves in the phase plane which represent periodic solutions which are approached by neighbouring solutions as $t \to \pm\infty$ respectively. Critical points and limit cycles are examples of *attractors*, i.e. sets of points which are approached by neighbouring solutions as $t \to \pm\infty$.

More recently interest has been aroused in attractors which are neither critical points nor closed curves, called *strange attractors*. They belong essentially to third-order and higher-order systems, and include the *horseshoe* and the *Lorenz attractor* (Problem 7.10). They are aptly called strange attractors because their properties are much more subtle than those of critical points and limit cycles. The solutions are not periodic. Although entirely deterministic, they share some properties of random systems. For example, several solutions which are initially arbitrarily close together may develop in substantially different ways as time increases. The trajectories originating in the appropriate domain of attraction tend to a subset of phase space in which they wander for ever in a way that may *appear* to be random. The correlation functions of quantities describing strange attractors, e.g. the second-order correlation

$$R_{ij}(t) = \lim_{T \to \infty} \frac{1}{2T} \int_{-T}^{T} x_i(s) x_j(s+t) \, ds,$$

decay rapidly as t increases from zero, as do similar correlations of quantities describing turbulence itself. The strange attractors have not been systematically classified in $\mathbb{R}^n$ or even $\mathbb{R}^3$ and it may be anticipated that much more about them will be discovered soon.

Ruelle & Takens (1971) used strange attractors to model transition, in marked contrast to Landau's model of spectral evolution. In Landau's model there is a succession of bifurcations as the Reynolds number increases, so that at each stage of transition the solution has components of the different periods $2\pi/\omega_1$, $2\pi/\omega_2$, etc., or, more precisely, the solution is a *quasi-periodic* † function of time. Spectral analysis of the solution gives resonance peaks at the frequencies ω_1, ω_2, etc.; yet spectral analysis of the aperiodic solution for a strange attractor gives a broad band of 'noise' rather than high peaks. So a strange attractor is suggestive of turbulence itself, and Landau's spectral evolution is suggestive of the more orderly transition seen in the cellular motion of Bénard convection or Taylor vortices.

52.2 *Introduction to bifurcation theory*

Bifurcation theory is the study of the changes of the qualitative character of the solutions, particularly equilibrium solutions, of nonlinear systems as a parameter varies. The nonlinear system may be algebraic, ordinary differential, or partial differential, etc., though we are interested in ordinary and partial differential systems

---

† A function $f(t)$ is said to be quasi-periodic if it may be expressed as $f(t) = g(\omega_1 t, \omega_2 t, \ldots, \omega_k t)$, where g has period 1 in each of its arguments separately and the frequencies ω_1, $\omega_2, \ldots, \omega_k$ are not rationally related. For example, $\cos t + \cos \sqrt{2} t \cos \pi t$ is a quasi-periodic function of t.

for their applications to hydrodynamic stability. We have already discussed the supercritical and subcritical bifurcations of Landau's theory in § 49. But there are other kinds of bifurcation relevant to hydrodynamic stability, and many others in general. Problems of bifurcation arise naturally from elementary particle dynamics. For example, when a particle is constrained to move along a smooth wire fixed in a steadily rotating vertical plane under the influence of gravity, the number and stability of the positions of equilibrium relative to the wire may change with the rate of rotation (see, for example, Problem 7.12). The modern theory of bifurcation, however, may be said to have arisen from the work of Poincaré, Liapounov and others on the figures of equilibrium of rotating self-gravitating masses of fluid; for a modern account of this work, based on the virial method, see Chandrasekhar (1969). The theory of bifurcation is now rapidly developing and multifarious, so there is no definitive textbook. For further reading, we recommend the lectures by Arnol'd (1972) and Sattinger (1973), who take rather different viewpoints. Marsden's (1978) survey is brief but broad, with many references. Also the books on ordinary differential equations which we have recommended cover some aspects of bifurcation.

Here we shall first describe heuristically the *Hopf bifurcation*, the most elementary bifurcation whereby a periodic, rather than a steady, solution may bifurcate at the margin of stability of a steady basic solution. The ideas of this kind of bifurcation in ordinary differential systems have been traced back to work of Poincaré, but were first fully treated by Hopf (1942). For the purpose of illustration, it will suffice to take as an example the simple system

$$\frac{dx}{dt} = (R - x^2 - y^2)x - y, \quad \frac{dy}{dt} = x + (R - x^2 - y^2)y, \quad (52.3)$$

where R is some real parameter.

The linear stability of the null solution $x, y \equiv 0$ is governed by

$$\frac{dx'}{dt} = Rx' - y', \quad \frac{dy'}{dt} = x' + Ry'. \quad (52.4)$$

Taking normal modes $x', y' \propto e^{st}$, we find

$$sx' = Rx' - y', \quad sy' = x' + Ry'. \quad (52.5)$$

Eliminating x' and y', we deduce the eigenvalue relation

$$s = R \pm \mathrm{i}. \tag{52.6}$$

Therefore the null solution is stable if $R \leq 0$ and unstable if $R > 0$. As R increases through $R_{\mathrm{c}} = 0$ the real part of the two complex conjugate eigenvalues increases through zero.

To examine the bifurcation it seems easiest to use the polar coordinates r and θ, where $x = r \cos \theta$ and $y = r \sin \theta$, because then the system may be solved explicitly. For equation (52.3) may be transformed to

$$\frac{\mathrm{d}r}{\mathrm{d}t} = r(R - r^2), \quad \frac{\mathrm{d}\theta}{\mathrm{d}t} = 1. \tag{52.7}$$

(The first of these equations is, of course, a Landau equation, and one may regard x as the real and y as the imaginary part of a complex amplitude $A = r \mathrm{e}^{\mathrm{i}\theta}$. It can be seen similarly that the complex equation (49.10) may describe a Hopf bifurcation.) The general solution of both equations is given by

$$r^2 = R r_0^2 / \{r_0^2 + (R - r_0^2) \, \mathrm{e}^{-2Rt}\}, \quad \theta = t + \theta_0, \tag{52.8}$$

where r_0 and θ_0 are the initial values. For $R \leq 0$ all solutions tend to the trivial one as $t \to \infty$, whatever the initial conditions are, i.e. the domain of attraction of the null solution is the whole of the xy-plane. For $R > 0$ there exist the stable solutions $x = R^{1/2} \cos(t + \theta_0)$, $y = R^{1/2} \sin(t + \theta_0)$ as well as the unstable null solution; note that the stable solutions are the same except for their phase θ_0. Also, for $R > 0$, (x, y) tends to one of the stable solutions as $t \to \infty$ if it initially is any point other than the origin; all trajectories of the phase plane of x and y spiral counter-clockwise into the circle $r = R^{1/2}$, i.e. the domain of attraction of this circle is the whole plane except the origin. This is an example of a limit cycle. It can be seen that the periodic solutions bifurcate from the null solution as R increases through zero.

Note that in the bifurcation of the Landau equation one real eigenvalue s passes through zero as R increases through R_{c}, whereas in the Hopf bifurcation the real part of two complex conjugate eigenvalues passes through zero. For this reason these are the two simplest canonical cases of bifurcation of real systems of ordinary differential equations. In each case the bifurcated solutions may exist either supercritically or subcritically.

Hopf bifurcations are often met in particle dynamics. Hopf (1942) himself speculated briefly that they are also met in hydrodynamic stability, and others have added to this speculation. In the spectral evolution envisaged by Landau one of the successive bifurcations as the Reynolds number increases could thereby introduce a periodic oscillation into what were previously steady flows. For example, the onset of the periodicity of a Kármán vortex street or of oscillatory Taylor vortices (Davey, DiPrima & Stuart 1968) may be describable by a Hopf bifurcation. Substantial work to show how periodic flows bifurcate from a steady one when the real part of two complex conjugate eigenvalues of the linearized problem passes through zero was also done by Iudovich (1971), Iooss (1972) and Joseph & Sattinger (1972). Again the bifurcated periodic solution is itself stable if it exists when the steady flow is supercritical and unstable when subcritical.

The Landau and Hopf bifurcations are important examples. But there are no general rules to determine the number and stability of the steady solutions which exist as the parameter varies, nor universal properties of bifurcation. Nonlinear phenomena are too diverse. Nonetheless, for fairly general well-behaved systems bifurcation may occur only where the real part of an eigenvalue of the linearized system passes through zero and does occur there if that eigenvalue has odd multiplicity (i.e. if it has an odd number of independent eigenfunctions). Also for many, but by no means all, systems if bifurcation occurs where the least stable mode of stability of a steady solution becomes unstable then the difference between the numbers of stable and unstable solutions is unchanged after bifurcation. (More specifically, it may be shown that in general the sum of the Leray–Schauder indices of all the possible steady flows for given boundary conditions is one, independently of the Reynolds number.) This is a property of the Landau equation $dA/dt = \sigma A - \ell A^3$ where $\sigma(R)$ changes sign if A is real but not necessarily positive. For this system we can represent the equilibrium solutions by the single variable A, so that bifurcation may be illustrated by a bifurcation curve in the RA-plane. If a system is of higher, or even infinite, order we can sometimes still choose a single variable, e.g.

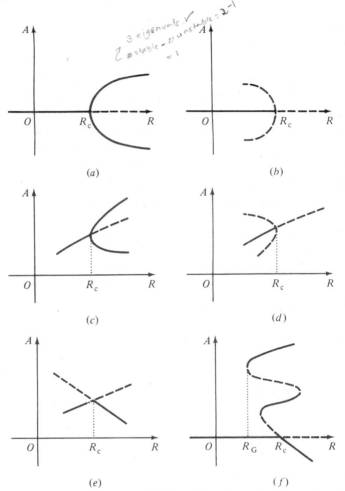

Fig. 7.3. Some common types of bifurcation curves. Stable solutions are denoted by continuous curves and unstable ones by dashed curves. (a) Supercritical bifurcation. (b) Subcritical bifurcation. (c) More general supercritical bifurcation. (d) More general subcritical bifurcation. (e) Transcritical bifurcation. (f) Another common type of bifurcation curve.

the amplitude of the least stable mode, to represent an informative bifurcation curve. Some bifurcation curves of the Landau equation described just above are shown in Figs. 7.3(a), (b); compare them with Figs. 7.1(b), 7.2(b), for which $|A| \geqslant 0$. More general forms of supercritical and subcritical bifurcation, where the basic equilibrium solution is some function of R and not the null function, are illustrated by bifurcation curves in Figs. 7.3(c), (d). The bifurcations of Figs. 7.3(*a*)–(*d*) are sometimes called *pitchfork bifurcations* because of the local shape of the curves. Another common type of bifurcation is sketched in Fig. 7.3(*e*); this is appropriate for dA/dt = $\{A - f(R)\}\{A - g(R)\}$, say. Yet another common type of bifurcation curve is sketched in Fig. 7.3(*f*); this is appropriate for dA/dt = $A\{R + f(A)\}$, say, where $R_c = -f(0)$. In all the examples of Fig. 7.3 the difference between the numbers of stable and unstable equilibrium solutions is unchanged after bifurcation.

We have described bifurcation heuristically. But many results on bifurcation have been rigorously proved by classical analysis, e.g. by use of the energy method or the Liapounov–Schmidt method, and many by functional analysis, e.g. by fixed-point theorems and by Leray–Schauder degree theory. Some of these results are discussed by Sattinger (1973). The first proof of bifurcation in hydrodynamics seems to be that of Ukhovskii & Iudovich (1963); they considered Bénard convection with free boundaries, investigated the multiplicity of the eigenvalues of the linearized problem, and applied a theorem of Krasnoselskii which is based upon degree theory. Perhaps the outstanding application of this method has been to the bifurcation of Taylor vortices from Couette flow at the critical Taylor number by Velte (1966), Iudovich (1966) and Ivanilov & Iakolev (1966) independently. The abstract and indirect nature of this method is, however, less informative than constructive methods. Kirchgässner & Sorger (1969), applying the Liapounov–Schmidt method in terms of functional analysis, also proved the bifurcation of Taylor vortices; showed that the amplitude of the radial velocity, etc. is proportional to $(T - T_c)^{1/2}$, where T_c is the critical value of the Taylor number T for disturbances of given axial wavelength; and related their results to observation.

52.3 *Structural stability*

Common sense has for a long time led scientists, when seeking to describe a natural phenomenon quantitatively, to choose a mathematical model which if changed a little does not have substantially different properties. It is desirable that a small change of the equations of a model should result in only a small change of ‖ definition each solution and not in any change of the qualitative character of the set of solutions. Mathematicians have recently formulated this intuitive idea by various definitions of *structural stability* (see, for example, Thom 1975; Hirsch & Smale 1974, Chapter 16). In the context of fluid dynamics these definitions mean essentially that a system is structurally stable if infinitesimal changes in the governing equations do not lead to any change in the qualitative character of the set of solutions, whereas stability in the usual sense means that infinitesimal changes in a given solution at some time lead to only infinitesimal changes of that solution for all time.

Frictionless systems in particle dynamics and in fluid dynamics provide examples of structurally unstable systems, because an infinitesimal amount of friction or viscosity may change the set of motions qualitatively. Of course this does not mean that the theory of inviscid fluids should be discarded but rather that the practical limitations of the theory should be recognized and understood. In the narrower field of hydrodynamic stability, the limitations of some of the models which we have used should be understood. From the beginning we have chosen simple flows with symmetric configurations in order that their solutions are tractable. For example, consider Bénard convection between horizontal rigid planes. It is at once clear that no experiment can be made with a Boussinesq fluid between two infinite horizontal planes at uniform and constant temperatures; rather a nearly pure real liquid or gas moves between nearly vertical side walls and nearly horizontal nearly flat plates maintained at temperatures which vary only slightly in space and time. What are the effects of these small 'imperfections' on the solution of Rayleigh's idealized model of Bénard convection? Even the most careful experimentalist cannot eliminate these imperfections, so we should examine whether they have qualitative or only small quantitative effects upon the flow. Further, we should examine whether the idealized model may be useful in practical conditions other than those in carefully controlled laboratory experiments. We have already discussed the effects of side walls upon Rayleigh's linear theory of convection in § 11, and shall go on to discuss the effects of these and other imperfections upon nonlinear convection. But first we shall consider more generally how imperfections may modify nonlinear stability.

The effect of imperfections upon elastic stability was studied long before that upon hydrodynamic stability, the fundamental ideas being put forward in 1945 by Koiter. Of the many later developments in the theory of solid mechanics, we note those of Sewell (1966) and Thompson & Hunt (1973) for providing examples of Thom's (1975) general mathematical theory of structural stability and morphogenesis. To introduce applications of these ideas to hydrodynamic stability, applications which have been initiated largely after the publication of Thom's book in 1975, we shall describe a simple statical model devised by Sewell (1966) to illustrate the buckling of an elastic column under a load.

Sewell supposed that a constant load P acts vertically on a point A of a rigid body which is free to rotate about a fixed smooth pivot O, as shown in Fig. 7.4(a). The load is displaced a distance $AB = \varepsilon l$ from the column OB of length l in order to simulate the asymmetry of a load, which is unavoidable for a column in practice. This system is said to be *perfect* if the column is symmetrically loaded (i.e. $\varepsilon = 0$) and otherwise *imperfect*. To model the resistance of an elastic column to bending, two linear springs of constant modulus E are taken to exert vertical restoring forces $Ecl \sin \theta$, where θ is the angle of rotation of the column OB from the vertical and $-\frac{1}{2}\pi < \theta < \frac{1}{2}\pi$. It follows that the potential energy of the system may be taken as

$$V(\theta) = c^2 l^2 E\{\sin^2 \theta - R(1 - \cos \theta + \varepsilon \sin \theta)\}, \qquad (52.9)$$

where (say) $R \equiv P/c^2 lE$. There is equilibrium when $dV/d\theta = 0$, i.e.

$$R(\sin \theta + \varepsilon \cos \theta) = \sin 2\theta. \qquad (52.10)$$

The equilibrium is stable or unstable according as $d^2V/d\theta^2$ is positive or negative respectively, and so the margin of stability of equilibrium is given by

$$R(\cos \theta - \varepsilon \sin \theta) = 2 \cos 2\theta. \qquad (52.11)$$

These results are illustrated by the sketch of the bifurcation curves in Fig. 7.4(b). It can be seen that the perfect system has subcritical instability but that an imperfect system (with $\varepsilon \neq 0$) has three equilibrium solutions of which one is unstable for all values of R and is entirely separate from the pair of stable and unstable solutions. Thus the perfect system is structurally unstable. To find the condition for marginal stability one may eliminate θ from equations (52.10) and (52.11), either analytically or numerically, to get a

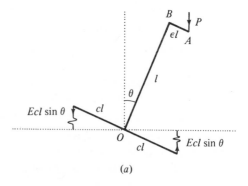

(a)

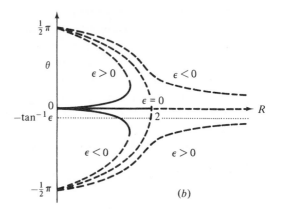

(b)

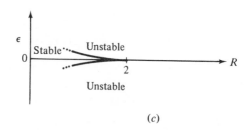

(c)

Fig. 7.4. A model of imperfect subcritical instability. (a) The loaded column with restoring springs. (b) Sketch of the bifurcation curves in the $R\theta$-plane for various values of ε. (c) Sketch of the cusp on the curve of marginal stability in the $R\varepsilon$-plane.

relationship between R and ε. This may be done explicitly for small θ, because then

$$V(\theta) = -c^2 l E\{\varepsilon R\theta + (\tfrac{1}{2}R - 1)\theta^2 + \tfrac{1}{3}\theta^4 + \cdots\} \qquad (52.12)$$

and $dV/d\theta$, $d^2V/d\theta^2 = 0$ can easily be shown to give $\varepsilon^2 \sim \tfrac{1}{36}(2-R)^3$ as $R \uparrow 2$. This is the equation of a cusp in the $R\varepsilon$-plane, as found by Sewell (1966) and shown in Fig. 7.4(c). The relationship between R and ε specifying marginal stability is now recognized as an example of a *catastrophe set* and its singularity is a cusp catastrophe (Thom 1975, p. 62). It can be seen that a small imperfection may lead to a substantial reduction in the critical load, because R decreases rapidly as ε increases at the cusp.

Sewell's problem serves as a model of the typical effects of an imperfection upon subcritical instability. Its wider implications can be seen more clearly in the light of Thom's catastrophe theory. The essential behaviour is found from the approximate form given by equation (52.12). So we may illustrate the effects of an imperfection upon supercritical bifurcation even more simply by choosing a mathematical model with

$$\frac{dA}{dt} = \varepsilon + (R - R_c)A - A^3, \qquad (52.13)$$

i.e. $dA/dt = -dV/dA$, where $V = \tfrac{1}{4}A^4 - \tfrac{1}{2}(R - R_c)A^2 - \varepsilon A$. This gives equilibrium if $dV/dA = 0$, i.e.

$$A^3 - (R - R_c)A - \varepsilon = 0, \qquad (52.14)$$

and marginal stability if in addition $d^2V/dA^2 = 0$, i.e.

$$A^2 = \tfrac{1}{3}(R - R_c). \qquad (52.15)$$

These results give the bifurcation curves sketched in Fig. 7.5(a). Also elimination of A from equations (52.14) and (52.15) gives the condition $\varepsilon^2 = \tfrac{4}{27}(R - R_c)^3$ for marginal stability; this is the equation of a cusp in the $R\varepsilon$-plane, as sketched in Fig. 7.5(b). Following Thom, we consider $dV/dA = 0$ as the equation of the equilibrium surface in the space of the *behaviour variable A* and the parameters, or *control variables*, R and ε. Accordingly, the $R\varepsilon$-plane is called the *control plane*. The equilibrium surface, of a type well known in catastrophe theory, is sketched in Fig. 7.5(c). The projection of

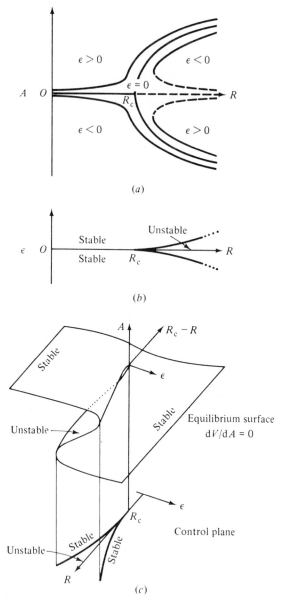

Fig. 7.5. A model of imperfect supercritical stability. (a) Sketch of the bifurcation curves in the RA-plane. (b) Sketch of the cusp on the curve of marginal stability in the $R\epsilon$-plane. (c) Sketch of the projection of the catastrophe set onto the control plane to give a curve of marginal stability.

points where there is marginal stability, i.e. where the tangent plane to the equilibrium surface is perpendicular to the control plane, onto the control plane gives the *catastrophe set* illustrated in Fig. 7.5(b). It should be noted that Fig. 7.5(a) shows that the perfect system ($\varepsilon = 0$) has the same supercritical stability as a Landau equation (see Fig. 7.3(a)) but is structurally unstable. For a given small non-zero value of ε there is one equilibrium, called the primary solution, which is stable for all values of R, and another separate pair of stable and unstable equilibrium solutions which exist only for $R > R_1(\varepsilon)$, say, where $R_1 = R_c + (9\varepsilon^2/4)^{1/3} + o(\varepsilon^{2/3})$ as $\varepsilon \to 0$. Thus the primary solution of an imperfect system does not become unstable at all; it changes continuously with R, slowly as R increases below R_c and more rapidly as R increases above R_c. In short, there is no abrupt change in the character of the primary solution as R increases. The other stable solution, called the secondary solution, which exists for $R \geqslant R_1$, can be realized only by a strong disturbance of the primary solution. If R decreases slowly after the secondary solution has been realized, then one expects a 'catastrophe', or sudden transition to the primary solution, as R decreases through R_1.

There is now ample evidence (cf. Thompson & Hunt 1973, 1975) that these ideas provide a basis for the qualitative description of the effects of imperfections upon elastic stability. Catastrophe theory provides a descriptive framework from which Koiter's (1945) fundamental ideas and their developments may be viewed. Of course the ideas have to be generalized to deal with more than one imperfection. It is found that for more than one imperfection the cusp catastrophe may be replaced by higher elementary catastrophes such as a swallow tail. Also a perfect system may be sensitive to some imperfections but not to others, i.e. the bifurcation curves of the perfect system may be qualitatively changed by some imperfections but not others. In fact different imperfections may change the bifurcation curves in different ways. These conclusions are supported by detailed treatment of many elastic systems, not just of simple finite-dimensional models such as we have used.

There is a small but growing body of evidence that these ideas provide a similar basis for hydrodynamic stability. Some of this evidence comes from the general mathematical results of Thom's

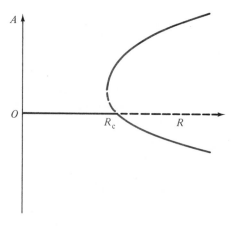

Fig. 7.6. Sketch of a bifurcation curve for imperfect supercritical bifurcation, with transcritical bifurcation at $A = 0$, $R = R_c$.

catastrophe theory, although the hypotheses of the theory, especially the hypothesis that there exists a potential function for the system, need to be justified for each application of the theory. Some of the evidence comes from case studies, both experimental and theoretical.

Bénard convection of a Boussinesq fluid had come in the 1960s to be recognized as an example of supercritical stability. Palm (1960), however, modelled convection in a fluid with temperature-dependent viscosity and found amplitude equations which give *transcritical bifurcation*. A simple example of transcritical bifurcation is illustrated in Fig. 7.6 and its local behaviour is given by the canonical equation

$$\frac{dA}{dt} = (R - R_c)A + \varepsilon A^2 - A^3. \qquad (52.16)$$

(We may scale t so that $\sigma = R - R_c$ and A so that $\ell = 1$.) It can be seen that this is an imperfect form of supercritical stability which gives subcritical instability as well. Liang, Vidal & Acrivos (1969) confirmed this by experimental and theoretical work on a fluid with temperature-dependent viscosity in a cylindrical container. In these investigations of nonlinear convection the boundary conditions were taken such that the state of rest, i.e. $A = 0$, is an equilibrium solution for all values of the Rayleigh number R. Daniels (1977),

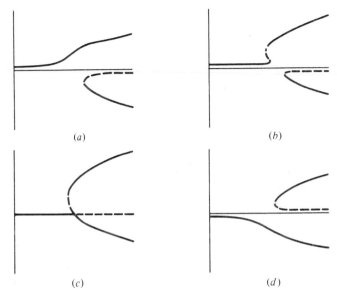

Fig. 7.7. Sketch of the morphogenesis of the bifurcation curves for imperfect supercritical stability as the imperfection changes. (After Benjamin 1978.)

Hall & Walton (1977) and Kelly & Pal (1978), however, investigated the effects of various imperfections upon Bénard convection of a uniform Boussinesq fluid due to irregular heating of the side walls. By applying weakly nonlinear theory to their various models, they found the imperfect supercritical stability as illustrated in Fig. 7.5. Their theoretical results are in accord with the very accurate observations of Ahlers (1974). Putting these two types of imperfection together, we find the canonical equation

$$\frac{\mathrm{d}A}{\mathrm{d}t} = \varepsilon_1 + (R - R_c)A + \varepsilon_2 A^2 - A^3 \qquad (52.17)$$

for small ε_1 and ε_2. Golubitsky & Schaeffer (1979) have classified the bifurcations associated with an equation of this form.

Benjamin (1978) has also applied the theories of structural stability and Leray–Schauder degree to describe qualitatively the end effects upon Taylor vortices. He argued that as the length of the rotating cylinders which confine the flow is increased the bifurcation curves in the RA-plane change continuously in the sense $(a) \to (d)$ of Fig. 7.7, and he supported this argument with experimental evidence.

52.4 *Spatial development of nonlinear stability*

An important limitation of Landau's theory is due to the assumption that the interaction of only one mode and its harmonics need be considered. This assumption is plausible if the eigenfunctions of the linearized problem are discrete and simple, so that when the flow is slightly unstable only one normal mode is unstable and all the others decay. When the flow is in an unbounded domain, however, the eigenfunctions depend continuously on a real wavenumber. Then a wavepacket of modes is unstable when the flow is slightly unstable. This in fact occurs for most of the cases we have treated. For example, Fig. 2.2(a) shows that, when a fluid at rest between infinite horizontal planes is heated from below and the Rayleigh number R is slightly supercritical, there is a small band of unstable waves, say $a_1(R) < a < a_2(R)$, where a is the wavenumber. The band width $a_2 - a_1$ is proportional to $(R - R_c)^{1/2}$ and centred on the wavenumber a_c as $R \downarrow R_c$. The linear theory determines an exponential growth rate for each wave in this band, the most unstable component with wavenumber a_c being only the first among equals. In the linear initial-value problem this leads to the development of a disturbance in space as well as time. So we anticipate that nonlinear disturbances similarly develop in space as well as time. In particular, the disturbance whose equilibrium is described by the Landau equation may become unstable in this way.

Side-band instability. In considering the stability of the bifurcated equilibrium solutions of the Landau equation, we have found subcritical instability and supercritical stability. But this finding was based upon solutions of only the Landau equation, *not* of the equations of motion of a fluid from which the Landau equation was derived. This effectively means that we considered the stability of the bifurcated solutions only to disturbances of the same periodicity as the least stable normal mode, not to *general* perturbations. Consider, for example, Couette flow between infinitely long rotating coaxial cylinders. For the nonlinear stability of this flow, Davey (1962) took axisymmetric modes of given axial wavenumber a as the Taylor number T increases through its critical value T_c, and computed the Landau constant ℓ, finding it to be positive. The Landau equation then represents the interactions of only the axisymmetric mode of given wavenumber a and its harmonics. A Taylor vortex when $0 < T - T_c \ll 1$ is accordingly stable to such disturbances but *may* be unstable to disturbances which do not have period $2\pi/a$ in the axial direction or which are not axisymmetric.

Eckhaus (1965), Chapter 8 investigated the stability of a periodic flow just after its supercritical bifurcation from an original basic flow. He took a fairly general model equation, found the bifurcated periodic solutions for $0 < R - R_c \ll 1$, and considered the linear stability of each of these periodic solutions to arbitrary small disturbances. The linear stability of the original basic 'flow' gives normal modes, the least stable of which have relative growth rates s such that $\sigma = \mathrm{Re}\, s$ has the form

$$\sigma = k(R - R_c) - m(a - a_c)^2$$
$$+ O\{(a - a_c)^3, (a - a_c)(R - R_c), (R - R_c)^2\} \qquad (52.18)$$

as $R \to R_c$ and $a \to a_c$ for some positive constants k and m. This gives the margin of stability,

$$R = R_c + m(a - a_c)^2/k + O\{(a - a_c)^3\} \quad \text{as } a \to a_c,$$

and a side band of unstable waves $a_1 < a < a_2$, where

$$a_2(R),\, a_1(R) = a_c \pm \{k(R - R_c)/m\}^{1/2} + O(R - R_c) \quad \text{as } R \downarrow R_c.$$
$$(52.19)$$

Eckhaus took the equilibrated solutions with amplitudes given by their Landau equations, and considered their linear stability to arbitrary small disturbances in the limit as $R \downarrow R_c$. He found that they were unstable if $a_1 < a < a_3$ or $a_4 < a < a_2$, where

$$a_c - a_3(R) \sim \{a_c - a_1(R)\}/\sqrt{3} \quad \text{and} \quad a_4(R) - a_c \sim \{a_2(R) - a_c\}/\sqrt{3}$$
$$\text{as } R \downarrow R_c \quad (52.20)$$

(see Fig. 7.8). Kogelman & DiPrima (1970) found the same result for more general systems of partial differential equations, including the system for axisymmetric disturbances of Couette flow. The mechanism of instability is a side-band instability analogous to the Benjamin–Feir instability of a Stokes wave.

Stuart & DiPrima (1978) identified this side-band instability clearly by using an amplitude equation, namely equation (52.27). They showed that the mechanism of Benjamin–Feir instability of a

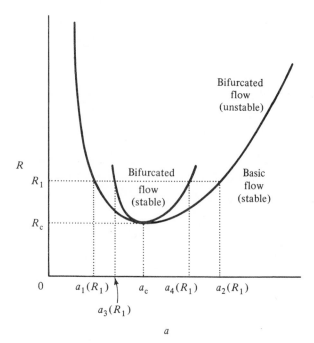

R

R_1

Bifurcated
flow
(unstable)

Bifurcated
flow
(stable)

Basic
flow
(stable)

R_c

0 $a_1(R_1)$ a_c $a_4(R_1)$ $a_2(R_1)$

$a_3(R_1)$

a

Fig. 7.8. The interval of instability of the basic flow and the subinterval of stability of the bifurcated flow. (Note that the bifurcated flow may also be unstable to three-dimensional disturbances not considered in the determination of the subinterval.)

Stokes water wave is similar to Eckhaus's mechanism, and also that Eckhaus's theory must be modified if the linearized problem or its eigenvalue is not real.

Nonlinear development of a wavepacket. Going on to consider more generally the development of weakly nonlinear disturbances in space as well as time, we recall the treatment of the linear theory in § 47. For a wide variety of linear problems, a single normal mode may be represented by a solution of the form

$$u' = A \exp\{i\alpha(x - ct)\} + A^* \exp\{-i\alpha(x - c^*t)\}, \quad (52.21)$$

where now α is the wavenumber. For many such problems the dispersion relation may be expanded about the wavenumber α_c of the most unstable mode to give the relative growth rate in the form

$$s = -i\alpha c = -i\omega_c - ic_g(\alpha - \alpha_c) + a_1(R - R_c) - a_2(\alpha - \alpha_c)^2$$
$$+ O\{(\alpha - \alpha_c)^3, (\alpha - \alpha_c)(R - R_c), (R - R_c)^2\} \quad (52.22)$$

as $\alpha \to \alpha_c$ and $R \to R_c$, where ω_c and the group velocity $c_g = [\mathrm{i}\,\mathrm{d}s/\mathrm{d}\alpha]_{\alpha=\alpha_c}$ are real constants, and where a_1 and a_2 are now some complex constants with positive real parts. The linear theory also gives the approximate solution for the packet of unstable waves in the form

$$u' = A(X, T) \exp\{\mathrm{i}(\alpha_c x - \omega_c t)\} + A^*(X, T) \exp\{-\mathrm{i}(\alpha_c x - \omega_c t)\},$$
$$(52.23)$$

where $X = \varepsilon x$, $T = \varepsilon t$ and $\varepsilon = |R - R_c|^{1/2} \to 0$. The slow modulation of A in space and time can represent the wavepacket. Corresponding to the expansion (52.22), it can be shown that A satisfies the equation

$$\varepsilon\left(\frac{\partial A}{\partial T} + c_g \frac{\partial A}{\partial X}\right) = \varepsilon^2\left(a_1 A + a_2 \frac{\partial^2 A}{\partial X^2}\right) + O(\varepsilon^3 A) \quad (52.24)$$

as $R \downarrow R_c$, because $\partial/\partial t = s$ and $\partial/\partial x = \mathrm{i}\alpha$ in the linearized system for a single mode. (The sign of the term $a_1 A$ changes if $R \uparrow R_c$.) This shows that, to the first approximation for small ε, A is a function of $X - c_g T$ alone, and therefore that the wavepacket travels downstream at the group velocity. The temporal growth of the wavepacket is determined by the first term on the right-hand side of equation (52.24) and its spatial spread by the second. If $c_g \neq 0$ it is informative to observe the wavepacket in a frame moving downstream at the group velocity; so we define

$$\xi = X - c_g T = \varepsilon(x - c_g t) \quad \text{and} \quad \tau = \varepsilon T = \varepsilon^2 t,$$

choosing the scales in order to balance all the terms of equation (52.24). This gives

$$\frac{\partial A}{\partial \tau} - a_2 \frac{\partial^2 A}{\partial \xi^2} = a_1 A. \qquad (52.25)$$

Although these ideas are applicable to wavepackets in many linear problems, it should be noted that for simplicity we have considered only two-dimensional disturbances of the form given in equation (52.23).

To incorporate weak nonlinearity into this theory, the ideas of the derivation of the Landau equation suggest that only the cubic term need be added to the linear equation (52.24).

Newell & Whitehead (1969) and Segel (1969) were the first to use this approach to nonlinear hydrodynamic stability. They derived equations of the form

$$\frac{\partial A_1}{\partial \tau} - a_2 \frac{\partial^2 A_1}{\partial X^2} = kA_1 - \tfrac{1}{2}\ell|A_1|^2 A_1 \qquad (52.26)$$

to describe the nonlinear modulation of the amplitude of a roll cell in two-dimensional Bénard convection. Here $A_1 = A/\varepsilon$ to balance the linear and nonlinear terms with A_1 of order of magnitude one as $\varepsilon \to 0$. For this problem $c_g = 0$ and the Landau constant ℓ is positive. Equation (52.26) reduces to equation (52.25) on linearization and to a Landau equation when A_1 is independent of X. These authors also considered three-dimensional modulation.

Stewartson & Stuart (1971) used this approach for plane Poiseuille flow, carefully deriving an equation of the form

$$\frac{\partial A_1}{\partial \tau} - a_2 \frac{\partial^2 A_1}{\partial \xi^2} = kA_1 - \tfrac{1}{2}\ell|A_1|^2 A_1. \qquad (52.27)$$

For this flow $c_g > 0$ and $\ell_r < 0$. Davey, Hocking & Stewartson (1974) have extended this work to accommodate three-dimensional disturbances. Also Newell & Whitehead (1971) and DiPrima, Eckhaus & Segel (1971) independently derived equation (52.27) for various basic flows.

Benney & Maslowe (1975) have presented the above theory very clearly and quite generally, and applied it to several examples of plane parallel flows.

52.5 *Critical layers in parallel flow*

Much of this book has been devoted to the mathematical details of critical layers in plane parallel flows, yet only the grossest properties, if any, of critical layers of homogeneous fluids have been directly observed in experiment or Nature. We have treated critical layers because of their importance in the linear theory of slightly unstable disturbances, in the determination of the stability criterion in particular, and because of their intrinsic interest; but to understand a critical layer in practice we note that it arises as a mathematical singularity for a single normal mode in an inviscid fluid. Thus different normal modes have critical layers in different planes, and a superposition of normal modes will in general smooth out the singularity. However, one mode may become dominant after some time. It follows that the singularity only arises in the limits as the amplitude of the disturbance tends to zero, as the viscosity tends to zero, and as time tends to infinity. The last limit is necessary for the transients due to an initial disturbance to die away and leave a single normal mode. The order of the three limits is not uniform, and subtle questions are involved in the relative magnitudes of the small effects of nonlinearity, viscosity and transience. Different practical applications of the theory may need different orders of these limits to represent different relative magnitudes, and thus need critical layers of different structures.

Lin (1958) was the first to recognize the importance of the relative magnitudes of the effects of viscosity to nonlinearity upon the structure of a critical layer; but R. E. Davis (1969) and, independently and in more detail, Benney & Bergeron (1969) discovered the substance of the current theory of nonlinear critical layers. The relative effects of viscosity and nonlinearity can in fact be represented by the dimensionless parameter

$$\lambda = 1/RA^{3/2},$$

where R is the Reynolds number and A is the dimensionless amplitude of the disturbance. Thus in the limit as $\lambda \to \infty$ we get the linear theory of the Orr–Sommerfeld equation for each normal mode, but other critical layers with $R \gg 1$ and $\lambda \approx 1$ or $\lambda \ll 1$ may arise in practice. Davis and Benney & Bergeron found the structure of critical layers in which nonlinearity dominates viscosity, i.e. $\lambda \ll 1$, and $R \gg 1$. The pattern of the streamlines in these layers resembles the cat's eyes of Kelvin (Fig. 4.3), although there are regions of uniform vorticity in the 'eyes'.

To understand what leads to the choice of the parameter λ, let us examine the nonlinear vorticity equation for two-dimensional instability of plane parallel flow of viscous fluid,

$$\left(\frac{\partial}{\partial t} + U\frac{\partial}{\partial x}\right)\Delta\psi' - \frac{d^2 U}{dz^2}\frac{\partial\psi'}{\partial x} - \frac{\partial(\psi', \Delta\psi')}{\partial(x, z)} = \frac{1}{R}\Delta^2\psi'. \quad (52.28)$$

Near a critical layer, of thickness δ, say, we estimate

$$\frac{\partial}{\partial t} + U\frac{\partial}{\partial x} \approx \alpha(U - c) \approx \alpha\frac{dU}{dz}\delta,$$

and

$$\partial/\partial x \ll \partial/\partial z \approx 1/\delta.$$

Thus

$$\left(\frac{\partial}{\partial t} + U\frac{\partial}{\partial x}\right)\Delta\psi' \approx A/\delta.$$

If we balance this vorticity convection term with the viscous term $R^{-1}\Delta^2\psi' \approx A/\delta^4 R$ then we find $\delta \approx R^{-1/3}$, as we have discussed at length in the linear theory of critical layers (Chapter 4). If, however, we consider a critical layer in which the convection term balances with the nonlinear term $\partial(\psi', \Delta\psi')/\partial(x, z) \approx A^2/\delta^3$, then we may find a nonlinear critical layer with thickness $\delta \approx A^{1/2}$. Thus λ is the cube of the ratio of the thickness of a linear viscous critical layer to that of a nonlinear inviscid one. For $\lambda \approx 1$, $R \gg 1$ we have a balance of the three terms in the vorticity equation, to give approximately

$$\left(\frac{\partial}{\partial t} + U\frac{\partial}{\partial x}\right)\frac{\partial^2\psi'}{\partial z^2} - \frac{\partial(\psi', \partial^2\psi'/\partial z^2)}{\partial(x, z)} = \frac{1}{R}\frac{\partial^4\psi'}{\partial z^4}. \quad (52.29)$$

Haberman (1972) showed the need for some modifications of the theory of Benney & Bergeron, finding that the vorticity and the velocity are continuous across the edge of the cat's eyes when $\lambda = 0$. Now the Rayleigh equation for the linear stability of plane parallel flow of an inviscid fluid is singular at the critical layer, and we stated in § 22 that this singularity of a marginally stable mode is interpreted by taking $\ln(z - z_c) = \ln|z - z_c|$ for $z > z_c$ but $\ln|z - z_c| + \theta i$ for $z < z_c$, where the phase change across the layer is given by $\theta = -\pi$ if $U_c' > 0$ (or π if $U_c' < 0$). This was used as a connexion formula. There is no singularity of the Orr–Sommerfeld equation, however, so we were able in § 27 to use the linear theory to justify this choice of θ by taking the limit as $R \to \infty$. Looking at the nonlinear critical layers, Haberman (1972) showed that this result is modified by nonlinearity, so that θ increases monotonically from $-\pi$ to zero as λ decreases from infinity to zero for fixed $R \gg 1$.

These theories of weakly nonlinear critical layers have been extended by Maslowe (1972, 1973) to apply to a stratified fluid, and some of his results bear an encouraging resemblance to the numerical results of Patnaik, Sherman & Corcos (1976) and to radar observations of the instability of atmospheric waves by, for example, Hardy, Glover & Ottersten (1969) and Browning (1971).

It is also desirable to show how a single nonlinear mode with a critical layer would arise and evolve in practice, for observational evidence is far from conclusive at present. One must remember that there is no singularity for a viscous fluid, even in the linear theory, although a normal mode may be nearly singular if $R_c \gg 1$ and $0 < R - R_c \ll 1$. Also normal modes of different wavelengths have different critical layers. So a general localized disturbance, being a superposition of these normal modes, does not have a critical layer. If, however, the flow is at a very high Reynolds number and is slightly unstable because the dimensionless number measuring some external force is nearly critical (for example, the Richardson number measuring buoyancy is slightly subcritical), then only a narrow waveband of modes is unstable. It follows that these may be expected to dominate any initial disturbance quite rapidly. Thus a wavepacket with a well-defined critical layer may dominate the disturbance by the time nonlinearity becomes significant.

Recently Brown & Stewartson (1978b) have carefully analysed some cases of the evolution of weakly nonlinear critical layers, and substantiated some results of the theory.

Strongly nonlinear critical layers present even more difficult problems, and their nature will surely take longer to be understood. Stuart (1967) has found a family of exact solutions for nonlinear waves in an inviscid fluid. In a certain limit they reduce to the linearized perturbation of a shear layer with a hyperbolic-tangent profile at marginal stability. Also Varley, Kazakia & Blythe (1977) have shown that, for basic parallel flows of an inviscid fluid with piecewise-linear profiles, the evolution of some strongly nonlinear long waves may be found by use of simple-wave solutions. These papers are intriguing, but cover rather special cases and are not yet closely related to the main body of the theory of nonlinear stability.

53 Additional fundamental concepts of nonlinear stability

The emphasis on the linear theory in this book in part reflects the importance of the linear theory in the determination of a criterion for stability and in part the inferior development of the nonlinear theory at the present time. The linear theory is also the pivot of the weakly nonlinear theory, which comprises a high proportion of the nonlinear theory which is now known. This weakly nonlinear theory has been introduced by the essays in the previous section. It was seen that, even when there is subcritical instability, linear phenomena are important in theory, if not in practice. There is, however, a growing body of knowledge of nonlinear stability not based directly upon the assumption that the disturbance is of small, albeit finite, amplitude and *a fortiori* not based upon the use of normal modes. Experiments have long contributed to this knowledge. Recent improvements in the performance of computers and in numerical analysis have led to the successful solution of problems of hydrodynamic stability by direct integration of the governing partial differential equations. These integrations may be regarded as very carefully controlled experiments and, like laboratory experiments, do not depend upon linearization of the problem. Also we have already mentioned the rapid growth of the global theory of differential equations. But there are a few specific methods in the strongly nonlinear theory of hydrodynamic stability, and these are introduced in this section.

53.1 *The energy method*

To obtain sufficient conditions for stability with respect to arbitrary disturbances we now wish to consider briefly the energy method which originated in the early work of Reynolds (1895) and Orr (1907). This aspect of the subject has been considered again more recently by Serrin (1959) and his work has since led to a considerable revival of interest in the subject.† He showed, for example, that a basic flow in a bounded region $\mathcal{V}$ is stable with respect to arbitrary disturbances provided the Reynolds number $R = VL/\nu$ is less than 5.71, where V is the maximum speed of the basic flow and L is the maximum diameter of $\mathcal{V}$. He also obtained similar stability criteria for flows in unbounded regions such as channels and pipes.

Consider then a basic flow of an incompressible viscous fluid with velocity vector $\mathbf{U}$, and let $\mathbf{u}'$ denote the velocity vector of the

---

† A fuller account of the energy method has been given recently by Joseph (1976).

perturbation flow. The total or altered flow $\mathbf{U}+\mathbf{u}'$ must satisfy the equations of motion and the same boundary conditions as $\mathbf{U}$ but the perturbation flow is otherwise arbitrary; in particular, it is not necessarily small. The kinetic energy of the perturbation flow is then given by

$$\mathcal{K} = \frac{1}{2} \int u_i'^2 \, d\mathbf{x}. \tag{53.1}$$

(We use dimensionless quantities with unit density, etc.)

If $\mathcal{V}$ is bounded then the integration here is over the whole of $\mathcal{V}$; if $\mathcal{V}$ is unbounded, however, we suppose that $\mathbf{u}'$ is spatially periodic in those directions in which $\mathcal{V}$ extends to infinity and the integration can then be taken over exactly one wavelength. If $\mathcal{K} \to 0$ as $t \to \infty$, then the basic flow is said to be (*asymptotically*) *stable in the mean*.

It can then be shown from the equations of motion that the rate of change of $\mathcal{K}$ is governed by the *Reynolds–Orr energy equation*

$$\frac{d\mathcal{K}}{dt} = -\int \{u_i' u_j' D_{ij} + R^{-1}(\partial u_i'/\partial x_j)^2\} \, d\mathbf{x}, \tag{53.2}$$

where D_{ij} is the rate-of-strain tensor of the basic flow

$$D_{ij} = \frac{1}{2}(\partial U_i/\partial x_j + \partial U_j/\partial x_i). \tag{53.3}$$

In the derivation of equation (53.2) it is found that all of the nonlinear terms in the equations of motion disappear in the process of integration and, as a result, equation (53.2) is quadratic in the perturbation velocity $\mathbf{u}'$. Now if the right-hand side of equation (53.2) is negative for arbitrary non-vanishing vectors $\mathbf{u}'$ satisfying $\nabla \cdot \mathbf{u}' = 0$, then $d\mathcal{K}/dt < 0$, and we have stability. The second term on the right of equation (53.2) is always negative and represents the viscous dissipation of kinetic energy. The first term, however, represents a transfer of energy between the basic flow and the perturbation flow. If it is positive so that energy is being transferred from the basic flow to the perturbation flow, then it is possible that the right-hand side of equation (53.2) may become positive, especially for large values of R, and this would lead to instability.

Before considering some of the general consequences of equation (53.2), it is worth while noting the form which this equation takes for two-dimensional, parallel flows. Thus, with $\mathbf{U} = \{U(z), 0, 0\}$ and $\mathbf{u}' = (u', 0, w')$ we have

$$\frac{d}{dt} \iint \frac{1}{2}(u'^2 + w'^2)\, dx\, dz$$

$$= -\iint \left\{ u'w' \frac{dU}{dz} + \frac{1}{R}\left(\frac{\partial w'}{\partial x} - \frac{\partial u'}{\partial z}\right)^2 \right\} dx\, dz. \quad (53.4)$$

This form of the energy equation is closely related to equation (26.10). To see this let

$$u' = \operatorname{Re}\{(D\phi)\, e^{i\alpha(x-ct)}\} \quad \text{and} \quad w' = -\operatorname{Re}\{i\alpha\phi\, e^{i\alpha(x-ct)}\}. \quad (53.5)$$

On averaging with respect to x, i.e. on integrating over one wavelength and dividing by the wavelength, we have

$$\langle u'^2 \rangle = \tfrac{1}{2}|D\phi|^2 \exp(2\alpha c_i t), \quad \langle w'^2 \rangle = \tfrac{1}{2}\alpha^2 |\phi|^2 \exp(2\alpha c_i t),$$

and

$$\langle u'w' \rangle = \tfrac{1}{4} i\alpha \{(D\phi)\phi^* - \phi(D\phi^*)\} \exp(2\alpha c_i t). \quad \left.\right\} (53.6)$$

Equation (53.4) then becomes

$$\tfrac{1}{2}\alpha c_i \int \{|D\phi|^2 + \alpha^2 |\phi|^2\}\, dz = -\tfrac{1}{4} i\alpha \int \{(D\phi)\phi^* - \phi(D\phi^*)\} \frac{dU}{dz}\, dz$$

$$- \tfrac{1}{2} R^{-1} \int \{|D^2\phi|^2 + 2\alpha^2 |D\phi|^2 + \alpha^4 |\phi|^2\}\, dz, \quad (53.7)$$

which is just the imaginary part of equation (26.10). Thus, so far as two-dimensional disturbances are concerned, the energy integral of the Orr–Sommerfeld equation is equivalent to the two-dimensional form of the Reynolds–Orr energy equation. When three-dimensional disturbances are considered, however, then the consequences of equation (53.2) are substantially different from those of equation (26.10).

A different form of the Reynolds–Orr energy equation is sometimes useful. On using the identity

$$u_i' u_j' D_{ij} = \partial(u_i' u_j' U_i)/\partial x_j - U_i u_j' \partial u_i'/\partial x_j - U_i u_i' \partial u_j'/\partial x_j,$$

Gauss's divergence theorem, and conditions (53.14), equation (53.2) may be rewritten as

$$\frac{d\mathcal{H}}{dt} = \int \{U_i u_j' \partial u_i'/\partial x_j - R^{-1}(\partial u_i'/\partial x_j)^2\}\, d\mathbf{x}. \quad (53.8)$$

Serrin (1959, §73) used this form to derive one of his various criteria for stability. By use of equation (53.8) and the inequality

$$0 \leqslant \tfrac{1}{2}(R^{-1/2}\partial u_i'/\partial x_j - R^{1/2}U_j u_i')^2,$$

he deduced that

$$\frac{d\mathscr{H}}{dt} \leqslant \frac{1}{2} \int \{RU_i^2 u_j'^2 - R^{-1}(\partial u_i'/\partial x_j)^2\}\, d\mathbf{x}$$

$$\leqslant \frac{1}{2} \int \{Ru_j'^2 - R^{-1}(\partial u_i'/\partial x_j)^2\}\, d\mathbf{x}, \qquad (53.9)$$

because the velocity scale V has been chosen so that $\max|U_i| = 1$. Further,

$$0 \leqslant (u_i' h_j + \partial u_i'/\partial x_j + \varepsilon \partial u_j'/\partial x_i)^2$$

for any real constant ε and continuously differentiable vector h_j. Whence it follows that

$$\int_{\mathscr{V}} \{(\partial h_j/\partial x_j - h_j^2)u_i'^2 + 2\varepsilon u_i' u_j' \partial h_i/\partial x_j\}\, d\mathbf{x}$$

$$\leqslant (1 + \varepsilon^2) \int_{\mathscr{V}} (\partial u_i'/x_j)^2\, d\mathbf{x}.$$

Serrin chose $h_i = \pi \tan(\pi x_i)$, with a suitable origin at the centre of $\mathscr{V}$ so that h_i is differentiable in the domain of integration. (Here the assumption that the length scale is the maximum diameter of $\mathscr{V}$ is used.) He then chose $\varepsilon = \tfrac{1}{2}(13^{1/2} - 3)$ to get the best estimate, and deduced that

$$b \int_{\mathscr{V}} u_i'^2\, d\mathbf{x} \leqslant \int_{\mathscr{V}} (\partial u_i'/\partial x_j)^2\, d\mathbf{x}, \qquad (53.10)$$

where $b = \tfrac{1}{2}(3 + 13^{1/2})\pi^2 = 32.6$. Now inequality (53.9) gives

$$\frac{d\mathscr{H}}{dt} \leqslant (R - b/R)\mathscr{H}, \qquad (53.11)$$

which, on integration, gives

$$\mathscr{H}(t) \leqslant \mathscr{H}(0) \exp\{(R - b/R)t\}, \qquad (53.12)$$

where $\mathscr{H}(0)$ is the initial kinetic energy of the perturbation flow. Therefore the basic flow is stable in the mean if $R < b^{1/2} = 5.71$.

Serrin's result shows that for any bounded flow there exists the critical value R_c of the Reynolds number such that the flow is stable if $R < R_c$ and is unstable for some R in the neighbourhood of R_c (or that the flow is stable for all values of R, i.e. $R_c = \infty$). However, in virtue of the very generality of the method of proof, the lower bound 5.71 is likely to be a poor estimate of R_c for any particular basic flow. The result, applying to any basic flow, whether it is steady or unsteady, in any domain, is not to be expected to give a good estimate of R_c. Yet it is valuable to be assured that there exists $R_c > 0$. The result also shows that if a flow is steady and $R < 5.71$ then it is the unique steady flow.

On sacrificing some simplicity, the energy equation may be used to give more information. According to equation (53.2), stability is assured if the right-hand side of that equation is negative, and the largest value of R, $\tilde{R}$ (say), satisfying this condition can then be determined from the variational problem

$$\frac{1}{\tilde{R}} = \max_{\mathbf{u}'} \frac{-\int u_i' u_j' D_{ij} \, \mathbf{dx}}{\int (\partial u_i'/\partial x_j)^2 \, \mathbf{dx}} \qquad (53.13)$$

for arbitrary vector fields $\mathbf{u}'$ satisfying

$$\boldsymbol{\nabla} \cdot \mathbf{u}' = 0 \quad \text{in} \quad \mathscr{V} \quad \text{and} \quad \mathbf{u}' = 0 \quad \text{on} \ \partial \mathscr{V}. \qquad (53.14)$$

With $\tilde{R}$ determined in this manner, equation (53.2) then gives

$$\frac{d\mathscr{K}}{dt} \leq -\left(\frac{1}{R} - \frac{1}{\tilde{R}}\right) \int (\partial u_i'/\partial x_j)^2 \, \mathbf{dx}, \qquad (53.15)$$

which shows that we have stability for $R < \tilde{R}$. To obtain the best possible estimate of the rate of decay of the kinetic energy we now let

$$\tilde{a} = \min_{\mathbf{u}'} \frac{\int (\partial u_i'/\partial x_j)^2 \, \mathbf{dx}}{\int u_i'^2 \, \mathbf{dx}}, \qquad (53.16)$$

where $\mathbf{u}'$ must again satisfy equations (53.14). Inequality (53.15) then becomes

$$\frac{d\mathscr{K}}{dt} \leq -2\tilde{a}\left(\frac{1}{R} - \frac{1}{\tilde{R}}\right) \mathscr{K}, \qquad (53.17)$$

which, on integration, gives

$$\mathscr{K}(t) \leq \mathscr{K}(0) \exp\left\{-2\tilde{a}\left(\frac{1}{R} - \frac{1}{\tilde{R}}\right) t\right\}. \qquad (53.18)$$

The Euler equation corresponding to equation (53.13) is

$$u'_j D_{ij} = -\frac{\partial \lambda'}{\partial x_i} + \rho^{-1} \Delta u'_i, \qquad (53.19)$$

where λ' is a Lagrange multiplier associated with the divergence constraint. This equation bears a striking similarity to the time-independent form of the Navier–Stokes equation. It must be solved subject to the conditions (53.14), and $\tilde{R}$ is then obtained as the least eigenvalue ρ of this eigenvalue problem. Similarly, the Euler equation corresponding to the variational principle (equation 53.16) is

$$\alpha^{-1} \Delta u'_i + u'_i = \partial \lambda' / \partial x_i, \qquad (53.20)$$

which has the same form as equation (53.19) with $D_{ij} = -\delta_{ij}$. This equation must also be solved subject to the conditions (53.14), and $\tilde{a}$ is then the least eigenvalue α.

We are primarily interested in the determination of $\tilde{R}$, and to illustrate some of the consequences of equation (53.19) we will now specialize to the case of parallel shear flows with $\mathbf{U} = \{U(z), 0, 0\}$. The rate-of-strain tensor D_{ij} then has only two non-vanishing components, $D_{13} = D_{31} = \frac{1}{2} U'(z)$, and in component form equation (53.19) becomes

$$\left.\begin{array}{l} \frac{1}{2} U' w' = -\lambda'_x + \rho^{-1} \Delta u', \\ 0 = -\lambda'_y + \rho^{-1} \Delta v', \\ \frac{1}{2} U' u' = -\lambda'_z + \rho^{-1} \Delta w', \end{array}\right\} \qquad (53.21)$$

where the subscripts now denote partial derivatives. If we consider two-dimensional disturbances in the xz-plane and let $u' = \psi'_z$, $v' = 0$, $w' = -\psi'_x$, then equations (53.21) reduce to

$$\rho^{-1} \Delta^2 \psi' + U' \psi'_{xz} + \frac{1}{2} U'' \psi'_x = 0, \qquad (53.22)$$

the normal-mode form of which is identical with equation (26.18). Thus, for disturbances of this type, the linear theory and the energy method lead to identical results. Orr (1907) further assumed that two-dimensional disturbances of this type would be less stable than three-dimensional disturbances, i.e. that the least eigenvalue of equations (53.21) can be obtained as the least eigenvalue of equation (53.22). Although Squire's theorem provides a justification for this assumption in the case of linear theory, it is not applicable to equations (53.21) and the assumption is false in general.

The determination of the least eigenvalue of equations (53.21) is clearly a formidable problem in general and results are known for only a few flows. In the case of plane Couette flow with $U(z) = z$ $(-1 \leqslant z \leqslant 1)$, however, Joseph (1966) has proved that the least eigenvalue of equations (53.21) is still associated with a two-dimensional disturbance but one which varies only in the yz-plane, i.e. the perturbed flow consists of rolls whose axes are in the direction of the basic flow. For this flow, with $\mathbf{u}'(\mathbf{x}) = \mathbf{u}(z)\,e^{i\beta y}$, and $\lambda'(\mathbf{x}) = \lambda(z)\,e^{i\beta y}$, equations (53.21) and the divergence constraint become

$$\tfrac{1}{2}w = \rho^{-1}(D^2 - \beta^2)u,$$
$$0 = -i\beta\lambda + \rho^{-1}(D^2 - \beta^2)v, \qquad (53.23)$$
$$\tfrac{1}{2}u = -D\lambda + \rho^{-1}(D^2 - \beta^2)w,$$
$$i\beta v + Dw = 0.$$

These equations can be combined to yield a single equation for w,

$$(D^2 - \beta^2)^3 w = -\tfrac{1}{4}\beta^2\rho^2 w, \qquad (53.24)$$

which must be solved subject to the boundary conditions

$$w = Dw = (D^2 - \beta^2)^2 w = 0 \quad \text{at } z = \pm 1. \qquad (53.25)$$

It will be recognized immediately that these equations are of exactly the same form as equations (10.9) and (10.10), which govern the Bénard problem with two rigid boundaries. Accordingly, the least eigenvalue is $\rho = \tfrac{1}{2}\sqrt{1708} \cong 20.7$ and plane Couette flow is therefore stable with respect to arbitrary disturbances provided $R < 20.7$.

The solutions of equation (53.19) have also been studied by Joseph & Carmi (1969) for Poiseuille flow in pipes, annuli and channels. For pipe flow with $U(r) = 1 - r^2$ $(0 \leqslant r \leqslant 1)$ they considered disturbances of the form $\mathbf{u}'(r, \theta, x) = \mathbf{u}(r)\exp\{i(\alpha x + n\theta)\}$ and found (numerically) that the least eigenvalue $(\rho = 81.49)$ is associated with a spiral mode for which $n = 1$ and $\alpha \cong 1$. Similarly, for plane Poiseuille flow with $U(z) = 1 - z^2$, Busse (1969) and Joseph & Carmi (1969) found (numerically) that the least eigenvalue $(\rho = 49.60)$ is associated with a two-dimensional disturbance in the yz-plane.

The energy method thus provides rigorous criteria for stability with respect to arbitrary disturbances whereas the linear theory can only provide criteria for instability. According to the linear theory, plane Couette flow and circular Poiseuille flow are both stable but plane Poiseuille flow is unstable if $R > 5772$. For parallel shear flows such as these, neither theory gives results which are close to the true physical situation as observed experimentally. There are, however, some important problems for which the two theories lead to identical results. These include the problem of thermal instability of a Boussinesq fluid, for which the energy method shows that all subcritical instabilities are impossible, i.e. we have stability for all values of the Rayleigh number below the critical value given by the linear theory (see Problem 7.13).

The direct method of Liapounov. The direct, or second, method of Liapounov is a natural extension of the energy method. Other positive definite functionals may be used to play the role of the energy integral in deducing sufficient conditions for stability by arguments similar to the energy method. Liapounov's direct method had been used for over 60 years to determine conditions for stability of systems of ordinary differential equations before Zubov (1957) and Movchan (1959) generalized the method in order to apply it to continuous systems. Some ingenuity in the choice of Liapounov functionals and in the use of inequalities is required to apply the direct method to hydrodynamic stability, and the applications have not proved very useful yet. Pritchard (1968), however, has elegantly derived some criteria for the nonlinear stability of Bénard convection and of Couette flow between rotating cylinders.

The use of bounds on flow quantities. The energy method is also closely related to the use of bounds on flow quantities in the theory of turbulence, since variational principles instead of the full equations of motion are applied in both methods.

Malkus (1954b) suggested that the solution realized in turbulent convection of a fluid between plates with prescribed temperatures is one for which the heat flux across the layer is a maximum. This led Howard (1963b) to treat the corresponding variational problem. Although current observational evidence does not support the universal validity of Malkus's suggestion, Howard's work has initiated the study of bounds on various flow quantities in order to apply them to various fully developed turbulent flows (see the review by Howard (1972)). It is doubtful whether any solution of these variational problems which has been found so far has any direct relation to turbulence, but some of the bounds themselves are encouragingly close to those observed in turbulent flows. Moreover, a mathematical demonstration of a bound on a flow quantity, such as the heat flux or viscous dissipation, gives definite information about this quantity, even if the bound is in fact too far from the value of the quantity realized in the observed flow to be of much practical use. Also this same bound applies to nonlinear stability as well as to turbulence.

53.2 *Maximum and minimum energy in vortex motion*

Kelvin (1887) recognized that the kinetic energy of a flow of an incompressible inviscid fluid has a stationary value when the flow is steady, and that the flow is stable if the stationary value is either a maximum or a minimum. Arnol'd (1966) proved these results by use of the calculus of variations and applied them to hydrodynamic stability of various flows. To simplify this theory, we present only Arnol'd's earlier work on two-dimensional flow.

Arnol'd (1965a) exploited the conservation of vorticity as well as energy in two-dimensional motion of an incompressible inviscid fluid, to find some quite general criteria for stability of steady basic flows. He considered first a general flow in a doubly connected plane domain D with a fixed boundary Γ which consists of two smooth closed curves Γ_1 and Γ_2. This flow is governed by the vorticity equation

$$\frac{\partial \Delta \psi}{\partial t} = \frac{\partial(\psi, \Delta \psi)}{\partial(x, z)} \tag{53.26}$$

in the xz-plane, where $u = \partial\psi/\partial z$ and $w = -\partial\psi/\partial x$. The boundary conditions on the streamfunction may be expressed so that $\psi = 0$ on Γ_1 and $\psi = q(t)$ on Γ_2 for some function q. It follows that the energy integral $\mathcal{K} = \frac{1}{2} \iint_D (\nabla\psi)^2 \, dx \, dz$ is a constant of the motion. Also the vorticity $\Delta\psi$ of each incompressible fluid particle is constant, so the integral $\mathcal{F} = \iint_D f(\Delta\psi) \, dx \, dz$ is another constant of the motion for *any* integrable function f.

Arnol'd considered next the extrema of the functional $\mathcal{H} = \mathcal{F} + \mathcal{K}$ with respect to variations of smooth functions ψ which satisfy the boundary conditions and also $\oint_{\Gamma_j} \partial \psi / \partial n \, ds = C_j$ for $j = 1, 2$, where C_1 and C_2 are some constants. The first variation is given by

$$\delta \mathcal{H} = -\iint_D \{\nabla \psi \cdot \nabla \delta \psi + f'(\Delta \psi) \Delta \delta \psi\} \, dx \, dz$$

$$= -\iint_D \Delta h \, \delta \psi \, dx \, dz + \oint_\Gamma \left(\frac{\partial h}{\partial n} \delta \psi + f' \frac{\partial \delta \psi}{\partial n} \right) ds,$$

on integrating by parts, where $h = \psi - f'(\Delta \psi)$. Therefore

$$\delta \mathcal{H} = -\iint_D \Delta h \delta \psi \, dx \, dz. \tag{53.27}$$

Similarly, the second variation can be shown to be

$$\delta^2 \mathcal{H} = \tfrac{1}{2} \iint_D \{(\nabla \delta \psi)^2 + f''(\Delta \psi)(\Delta \delta \psi)^2\} \, dx \, dz. \tag{53.28}$$

Now the stream function Ψ of any steady basic flow satisfies the equation

$$\Psi = g(\Delta \Psi) \tag{53.29}$$

for some function g. This follows from the vorticity equation (53.26) or from the variational equation (53.27) if $f' = g$ so that $h \equiv 0$. Therefore $\mathcal{H}$ is a local minimum or maximum where $\psi = \Psi$ if the integrand of equation (53.28) is positive or negative definite respectively at each point of D. Arnol'd deduced, by appeal to a generalization of a well-known result for dynamical systems of finite dimension, that the basic flow is stable to disturbances of finite (but possibly small) amplitude if the integrand of equation (53.28) is either positive or negative definite in some frame of coordinates.

(The essence of Arnol'd's deduction can be understood by use of a three-dimensional analogy. Suppose then that $\mathcal{H} = \text{constant}$ is represented by some surface in $\mathbb{R}^3$ and $\mathcal{K}$ by a Cartesian coordinate. Then each fluid motion is represented by the trajectory where a surface $\mathcal{H} = \text{constant}$ is cut by a plane $\mathcal{K} = \text{constant}$. If $\mathcal{H}$ has a maximum or a minimum when $\psi = \Psi$, then the surface $\mathcal{H} = \mathcal{H}_0$ is locally a paraboloid touching the plane $\mathcal{K} = \mathcal{K}_0$ at the 'point' Ψ, where $\mathcal{H}_0 = \mathcal{H}(\Psi)$ and $\mathcal{K}_0 = \mathcal{K}(\Psi)$. Therefore if ψ takes any value ψ_1 near to Ψ at some instant the ensuing motion will be represented by the elliptic trajectory where the plane $\mathcal{K} = \mathcal{K}(\psi_1)$ cuts the paraboloid $\mathcal{H} = \mathcal{H}(\psi_1)$. It follows that $\mathcal{K}$ will remain near $\mathcal{K}_0$ and therefore that ψ will remain near Ψ for all time. The method depends upon finding a local maximum or minimum of $\mathcal{H}$, and so applies to disturbances of finite amplitude, but these disturbances may be of small amplitude if another extremum happens to be nearby.)

Arnol'd (1965a) applied this result first to parallel flow between the planes Γ_1, with equation $z = z_1$, and Γ_2, with $z = z_2$. In this example, the basic velocity is given by $U(z) = \Psi'(z) = dg(U')/dz = g'(U')U''$ and the basic vorticity by $\Delta\Psi = U'(z)$, so that $f''(\Delta\psi) = U/U''$. He deduced that the flow is stable if $U/U'' > 0$ for $z_1 < z < z_2$. This result follows also if $(U - U_0)/U'' > 0$ for any constant U_0, and so gives Fjørtoft's theorem (§ 22). Indeed Arnol'd's use of the energy and vorticity equations is somewhat similar to the method Fjørtoft (1950) used earlier, although Fjørtoft used the linearized equations rather than a variational principle. Arnol'd also used his method to show that the swirling flow in an irregular annular domain is stable to two-dimensional perturbations if the velocity profile is concave.

Dikii (1965a,b), Blumen (1968) and Dikii & Kurganskii (1971) applied Arnol'd's method to flows relative to a rotating frame in order to find various criteria of stability. Arnol'd (1965b) himself examined further the properties of a flow whose kinetic energy and vorticity are conserved. Also Drazin & Howard (1966), § 2.3 found results reminiscent of Arnol'd's by direct use of the equations of energy and vorticity, rather than by use of a variational principle.

53.3 *Application of boundary-layer theory to cellular instability*

When steady Bénard convection sets in at the critical value of the Rayleigh number, the flow is very slow, but, as the Rayleigh number increases far above its critical value, a strong cellular flow develops. This experimental observation led Pillow (1952) to assume that there is a strong steady cellular flow and to neglect viscosity and heat conduction in the interior of the cells. Thus he deduced that the fluid in the interior has uniform temperature and vorticity, and he analysed the boundary layers at the rigid plates and at the edges of the cells. This led to the result that the heat transfer is proportional to $(T_0 - T_1)^{5/4}$, where T_0 is the temperature of the lower plate and T_1 of the upper plate, i.e. that the Nusselt number $Nu \propto Ra^{1/4}$. This agrees quite well with the values measured in many experiments when Ra is large enough for the convection to be strong yet not so large that the flow is unsteady. For example, Threlfall (1975) found that $Nu = 0.173Ra^{0.28}$ very accurately when $4 \times 10^5 < Ra < 2 \times 10^9$ in some experiments on liquid helium (for which $Pr \approx 0.8$).

Batchelor (Donnelly & Simon 1960, Appendix) applied similar ideas to Taylor vortices, and deduced that the torque between the rotating cylinders over a length H is approximately proportional to $H\rho\Omega_1^2 R_1^4 (\nu/\Omega_1 R_1 d)^{1/2} (d/R_1)^{1/4}$ when the outer cylinder is at rest, where the gap width $d = R_2 - R_1$, etc. Again there is fair agreement with observed values of the torque when the Taylor number is a few times its critical value (Donnelly & Simon 1960).

54 Some applications of the nonlinear theory

In developing the linear theory of hydrodynamic stability, we have described the major physical mechanisms and theoretical methods, and used them to understand the instability of a few flows. These flows were chosen because their instability is both intrinsically important and illustrates well the mechanisms and methods. Similarly, we have described the major mechanisms and methods of nonlinear hydrodynamic stability, and next shall apply them to those flows whose linear stability we have previously treated. This last section cannot be expected to contain the last word on nonlinear hydrodynamic stability, because research is still progressing rapidly. Nonetheless, it will be seen that many parts of the theory are already well understood and supported by experimental work.

54.1 *Bénard convection*

Bénard convection of a Boussinesq fluid provides the simplest case for which bifurcation and transition to turbulence may be studied theoretically, yet a case for which it is not easy to compare many theoretical and experimental results, because of the complexity of the properties of real fluids. In reviewing the nonlinear theory here, we first discuss convection of a Boussinesq fluid between two infinite horizontal planes, and then consider the modifications due to non-Boussinesq properties of fluids and to the presence of side walls. We shall not emphasize the various boundary conditions, but often in the history of the development of the theory a particular nonlinear phenomenon has been analysed first for the simplest case of free boundaries and then for rigid boundaries, only quantitative differences between the effects of the boundary conditions being found.

Gor'kov (1957) and, independently but more completely, Malkus & Veronis (1958), found weakly nonlinear *steady* solutions with various plan-forms of the cells for slightly supercritical values of the Rayleigh number R. Their work implies that the Landau constant is positive, and gives the magnitude and structure of the disturbance, and hence the heat transfer, as a function of R for each assumed plan-form, although it does not distinguish flows for which there is ascent at the centres of the cells from those for which there is descent.

In the classical linear theory of Bénard convection the most unstable mode is degenerate, in the sense that the horizontal plan-form of its cells is indeterminate, although its vertical variation and horizontal wavenumber are determinate; the ultimate plan-form of a given disturbance is thus determined by the initial horizontal distribution of the most unstable mode. Yet in experiments with a given apparatus, the plan-form of the cells may consistently be observed and appears to be largely independent of the initial conditions. Malkus & Veronis (1958) suggested that the observed plan-form should correspond to that one of all the possible nonlinear solutions which is itself stable with respect to infinitesimal disturbances of the form of the other solutions.

Schlüter, Lortz & Busse (1965) went on to consider the general steady solution of a given horizontal wavenumber a for small positive values of $R - R_c$. They found an infinity of such solutions, and showed that of these the only stable ones correspond to two-dimensional roll cells. Of these roll cells, some with $a > a_c$ are stable but none with $a < a_c$ is. Schlüter *et al.* also calculated the heat transfer for small values of $R - R_c$, showing that for convection

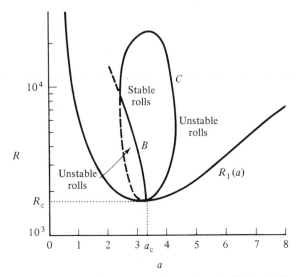

Fig. 7.9. The region of stability of two-dimensional rolls in a fluid of infinite Prandtl number with respect to non-oscillatory perturbations. (After Busse 1967a.)

between rigid boundaries the Nusselt number Nu satisfies

$$R_c(Nu - 1)/(R - R_c)$$

$$\rightarrow \begin{cases} (0.69942 - 0.00472Pr^{-1} + 0.00832Pr^{-2})^{-1} \text{ for rolls} \\ (0.77890 + 0.03996Pr^{-1} + 0.06363Pr^{-2})^{-1} \text{ for square cells} \\ (0.89360 + 0.04959Pr^{-1} + 0.06787Pr^{-2})^{-1} \text{ for hexagonal cells} \end{cases}$$

$$\text{as } R \downarrow R_c \text{ for fixed } Pr \neq 0. \tag{54.1}$$

These results are consistent with Malkus's (1954a,b) hypothesis of maximal heat transport, namely that, of all the possible motions, the one realized will be that for which the heat transport is a maximum, because the Nusselt number is largest for rolls.

Using a Galerkin numerical method, Busse (1967a) extended these results of Schlüter *et al.* for values of R up to 30 000 for the special case of rigid boundaries and large values of the Prandtl number. His results are illustrated in Fig. 7.9. There are no steady solutions of horizontal wavenumber a for $R < R_1(a)$, where R_1 here denotes the least eigenvalue of the appropriate linear problem.

For $R > R_1$, there is an infinity of steady finite-amplitude solutions, of which the only stable ones are the two-dimensional rolls for values of a and R within the region bounded by the curves (B) and (C). The curve (B) is the margin outside of which each roll becomes unstable to disturbances of the form of oblique rolls, which grow without oscillations. This is called *zigzag instability* on account of the pattern of the ensuing steady convection, which Busse & Whitehead (1971) observed in some experiments with a silicone oil (of Prandtl number 100). The curve (C) of Fig. 7.9 is the margin outside of which each two-dimensional roll becomes unstable to disturbances of the form of rolls perpendicular to itself. This is called *cross-roll instability* (Busse & Whitehead 1971), and it also sets in as a cellular motion with exchange of stabilities. This represents a limit of current theoretical knowledge. If, however, R is above the value (about 23 000) of the topmost point of the curve (C) in Fig. 7.9, then another form of instability has been observed experimentally by Busse & Whitehead (1971) to set in; this instability leads to a steady three-dimensional motion, as if composed of two perpendicular rolls of different wavenumbers, and is called *bimodal instability*. At even higher values of the Rayleigh number a further instability is observed, the motion ceases to be steady, and an oscillatory flow arises (see also Whitehead & Parsons (1978)).

These theoretical and experimental results differ somewhat if the value of the Prandtl number is not large; the quantitative results depend upon its value, though not strongly. Also when the Prandtl number is sufficiently low (less than about 1.1 for rigid and 3.5 for free boundaries) the curve (C) is the margin outside of which each roll becomes unstable to disturbances of the form of rolls parallel to itself; this is a special case of the side-band instability discovered by Eckhaus (see § 52.4, Clever & Busse (1974)). Also, Busse & Clever (1979) showed both theoretically and experimentally that two further types of instability of rolls, called *knot instability* and *skewed varicose instability*, occur when the value of the Prandtl number lies in a certain finite positive interval.

The instability of two-dimensional steady rolls in convection of a fluid of small Prandtl number was investigated theoretically by Busse (1972) and Clever & Busse (1974). They found that when the amplitude of a roll exceeds a certain critical value there is shear (rather than gravitational) instability of the roll. Thus an oscillatory three-dimensional flow sets in at much lower values of the Rayleigh number than it does for a fluid of large Prandtl number.

The occurrence of hexagonal cells is now attributed to non-Boussinesq properties of the fluid. Palm (1960) was the first to consider theoretically the effects of the variation of viscosity with temperature upon nonlinear cellular convection. Of the many subsequent theoretical papers those by Busse (1967b) and Palm, Ellingsen & Gjevik (1967) may be chosen to give the current view. The kinematic viscosity is assumed to vary slightly with temperature. It is then shown at length that for $R_c < R < R_1$ only hexagonal cells are stable, for $R_1 < R < R_2$ both hexagonal cells and two-dimensional rolls are stable, and for $R_2 < R$ only two-dimensional rolls are stable, where here $R_1 - R_c$ and $R_2 - R_c$ are certain small functions of the parameters which are proportional to the derivative $d\nu/d\theta$ of the kinematic viscosity of the fluid with respect to the temperature. In addition, subcritical instability occurs; the steady hexagonal cells are in fact stable also for the small range $R_{-1} < R < R_c$, and there is transcritical bifurcation at $R = R_c$, where $R_c - R_{-1}$ is a certain function proportional to $d\nu/d\theta$ (see also § 52.3). For each hexagonal convection cell the flow ascends at the centre if $d\nu/d\theta < 0$ and descends if $d\nu/d\theta > 0$. These theoretical results are confirmed qualitatively by the experiments of Hoard, Robertson & Acrivos (1970) which were done in a cylinder containing a liquid hydrocarbon (namely Aroclor 1248) whose viscosity varies strongly with temperature.

The side walls also have important effects, which must be borne in mind when comparing theoretical with experimental results. Davis (1968) and Segel (1969) were the first to apply the weakly nonlinear theory to convection in a container with side walls. Of the subsequent theoretical papers which have treated this difficult problem, we have in § 52.3 mentioned some on two-dimensional convection and add the one by Brown & Stewartson (1978a) on axisymmetric convection in a cylindrical container. The effects of side walls are particularly important for convection in deep layers, experiments showing that concentric circular rolls arise in a container with a circular cross-section and square cells in one with a square cross-section (see, for example, the review by Koschmieder (1974)).

The early experimentalists (see § 12) carefully measured the heat transfer as a function of the Rayleigh number for various fluids. This determined the critical Rayleigh number in agreement with the linear theory. Now the heat transfer also depends, albeit less strongly, upon the plan-form, the Prandtl number and other factors. So it is difficult to compare theoretical and experimental results in detail. Nonetheless, there is in general satisfactory agreement between them. In particular, Koschmieder & Pallas (1974) measured the heat transfer across convecting layers (of depth $d \approx 5$ mm) of various silicone oils (with $500 \le Pr \le 1700$) in a cylindrical container (of diameter $D \approx 13$ cm) for $0 < R \le 170R_c$. For small positive values of $R - R_c$ they saw concentric circular rolls and their measurements give $R(Nu - 1)/(R - R_c) \cong 1.48$; this compares well with the theoretical result given by equations (54.1), which gives $R_c(Nu - 1)/(R - R_c) \to 1.43$ for two-dimensional rolls as $R \downarrow R_c$ for any large value of Pr. Koschmieder & Pallas's measurements of Nu when R was a few times greater than its critical value agree quite well with the numerical results of Lipps & Somerville (1971). Ahlers (1974) made some exceptionally accurate measurements on convection in layers ($d \approx 1$ mm) of liquid and gaseous helium ($0.6 \le Pr \le 1.5$) in a cylindrical container (diameter $D \approx 10$ mm). The experimental techniques available at low temperatures allowed him to work with very small temperature differences across the layers of fluid, conditions for which the Boussinesq approximation is very good. He found that Nu increased from the value one smoothly as R increased through R_c, not abruptly at the onset of instability as given by the usual theory. This result is in accord with the theory of imperfections (see § 52.3), and may be due to side-wall effects. For slightly larger values of R, Ahlers fitted his data with the formula

$$Nu = 1 + 1.034 \, \varepsilon + 0.981 \, \varepsilon^3 - 0.866 \, \varepsilon^5$$

for $1.07 \, R_c \le R \le 2.5 \, R_c$ and $Pr = 1.17$, where $\varepsilon = (R - R_c)/R$. This is in qualitative agreement with the theoretical results given by equations (54.1), although Ahlers did not observe the plan-form. He also found that $Nu - 1 = 0.77\{(R/R_c) - 1\}^{0.334}$ for $30 \, R_c \le R \le$

150 R_c. Threlfall (1975), in further experiments on convection in helium at low temperatures, found that $Nu = 0.173\,R^{0.28}$ for $230\,R_c \leqslant R \leqslant 10^6\,R_c$, in fair agreement with Pillow's theoretical result (see § 53.3).

Schmidt & Saunders (1938) seem to be the first to have reported a discontinuity in the rate of increase of Nu with R. Their measurements of the heat transfer through a layer of water give a sudden change in the slope of the graph of the function $Nu(R)$ at about $R = 45\,000$, when the convection became unsteady. In further experiments on a few liquids, Malkus (1954a) found several such discrete changes, which he attributed to changes in the mode of convection. This is a plausible interpretation in view of the theories of the various nonlinear modes and stabilities which we have described, yet the existence of the changes has been affirmed by some experimentalists and denied by others. For example, Threlfall (1975) observed a sudden drop of the heat transfer at about $R \approx 3.2 \times 10^5$ as well as changes of the slope of the function $Nu(R)$ at lower values of R, changes which Ahlers (1974) sought and failed to find. One may speculate that these differences are due to different initial conditions, side effects or small imperfections in the experimental designs.

Observation of steady convection cells shows that their size tends to increase, and therefore their horizontal wavenumber to decrease, as the Rayleigh number increases above its critical value. This seems to be in disagreement with the theoretical results illustrated in Fig. 7.9, because the stable two-dimensional rolls are found to have wavenumbers greater than those on the boundary (B). This analytical conclusion is supported by numerical work (for example, Roberts (1966)). Resolution of this disagreement is awaited, but again it must be remembered that the experiments are not exactly modelled by the ideal assumptions of the theories without side effects, etc.

Krishnamurti (1968) considered both theoretically and experimentally how convection is initiated by slow heating of the lower plate or cooling of the upper plate. She assumed that the average basic temperature of a Boussinesq fluid increases or decreases slowly at a constant rate. Her results are similar to those for a steady basic state of a fluid whose viscosity depends upon the temperature.

In particular, she found that there is subcritical instability and that hexagonal rather than roll cells arise. The fluid ascends or descends at the centres of the cells according to whether the average basic temperature decreases or increases respectively with time.

The nature of the onset of turbulence itself remains obscure, and may indeed depend upon the fluid, the apparatus, and how the convection is initiated. Modern experimental techniques, however, make it possible to measure and analyse the frequency spectra of unsteady flows, and offer some evidence of aperiodicity like that of a strange attractor as well as of the periodicity and quasi-periodicity of the flows which we have described above. Ahlers (1974) observed that small-amplitude oscillations of the heat transfer about a mean value arose with a broad frequency band as R increases through a critical value R_t equal to about twice R_c. In later experiments Ahlers & Behringer (1978) found that R_t decreases with aspect ratio $D/2d$ of the cylindrical container and they suggested that $R_t \cong R_c$ when $D/2d = 57$. The aperiodic component of the flow had not been observed before, perhaps because its time scale is slow. These experimental results are somewhat similar to the theoretical results of McLaughlin & Martin (1975), who derived a system of ordinary differential equations after truncating a spectral representation of the convection and found solutions resembling a strange attractor. It is, however, difficult to justify the truncation when it is used in this way to describe what is evidently a strongly nonlinear phenomenon involving many components.

The nonlinear phenomena of only classical Bénard convection, rather than its many variations and applications, have been described for brevity. Some of the variations and applications are mentioned in § 13, and more in Hopfinger, Atten & Busse (1979), where several modern references may be found.

54.2 *Couette flow*

The theory of nonlinear stability of Couette flow is another good case study of transition to turbulence as well as an intrinsically important topic. It is easier to compare with experimental results than is the theory of nonlinear Bénard convection, because there are no non-Boussinesq effects on Taylor vortices. Chapters 2 and 3 show, however, that the theory of linear stability of Couette flow has more structure and is more difficult than that of Bénard convection of a Boussinesq fluid between horizontal planes. The theory of nonlinear stability of Couette flow is therefore correspondingly more difficult than that of Bénard convection, although the difficulties are of a technical nature and will be suppressed in the following discussion.

Stuart (1958) considered theoretically axisymmetric disturbances of given axial wavelength and used the narrow-gap approximation. He took the Reynolds–Orr energy equation, rather than the complete nonlinear equations of motion and boundary conditions, and substituted a solution with the same spatial form as the solution of the *linearized* problem. Thus he made what is called the *shape assumption*, namely that the nonlinear disturbance has the same spatial structure as the linear one although their amplitudes $A(t)$ may differ. This approximation serves to give a simplified derivation of a Landau equation by neglect of the harmonics and of the distortion of the fundamental. It is a good approximation only if the total nonlinear effect is nearly the same as that due only to the distortion of the mean flow. Stuart thereby calculated the Landau constant, finding it to be positive, and deduced the torque between the cylinders. There is quite good agreement with Taylor's (1936) measurements of the torque.

Using the later method of Stuart (1960a) and Watson (1960b), Davey (1962) treated slightly unstable disturbances more systematically. He considered only axisymmetric disturbances of given axial wavelength, but did not confine his results to those for the narrow-gap approximation. Davey also analysed the separate effects due to the distortion of the mean flow, to the distortion of the fundamental, and to the generation of the second harmonic, and found that the shape assumption happens to give a very good approximation when the gap between the cylinders is narrow. The Landau constant is positive (for the cases with $0 \leqslant \Omega_2/\Omega_1 < 1$ which were analysed) and the equilibration amplitude is therefore proportional to $(T - T_c)^{1/2}$ as the Taylor number $T \downarrow T_c$. The detailed results depend upon the ratios of the radii and of the angular velocities of the cylinders. Davey found good agreement between this theory and various measurements of the torque in slightly supercritical conditions.

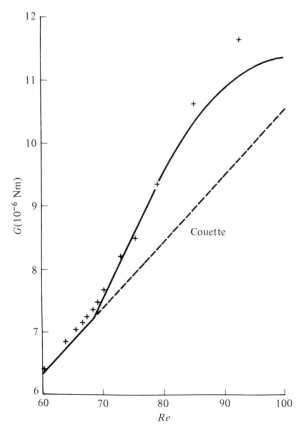

Fig. 7.10. The variation of the torque G with the Reynolds number $Re = R_1(R_2 - R_1)\Omega_1/\nu$ for $\Omega_2 = 0$ and $R_1/R_2 = \frac{1}{2}$: $+$ experimental measurements (Donnelly & Simon 1960, Table 2), —— theoretical values (DiPrima & Eagles 1977, Table IV).

Kirchgässner & Sorger (1969) confirmed these theoretical results by a rigorous application of the Liapounov–Schmidt method, and also calculated the next higher term in the expansion of the torque as $T \downarrow T_c$. Also DiPrima & Eagles (1977) used the method of Stuart and Watson to find this higher term, but there are some unresolved discrepancies between their results and those of Kirchgässner & Sorger. The measurements of the torque in experiments with $R_1/R_2 = \frac{1}{2}$ (Donnelly & Simon 1960) are compared with the theoretical results (DiPrima & Eagles 1977) in Fig. 7.10.

Kogelman & DiPrima (1970) showed that Taylor vortices them-selves are unstable to a side band of axisymmetric disturbances if

$$a_1 - a_c < a - a_c < (a_1 - a_c)/\sqrt{3}$$

or

$$(a_2 - a_c)/\sqrt{3} < a - a_c < a_2 - a_c \quad \text{as } T \downarrow T_c,$$

where the linear theory gives marginal stability of Couette flow for the wavenumbers $a = a_1(T)$ and $a_2(T)$. This implies that steady Taylor vortices should not be observed with wavenumbers in these intervals, and this is consistent with the experimental results of Snyder (1969) and Burkhalter & Koschmieder (1973).

Davey, DiPrima & Stuart (1968) considered the stability of Taylor vortices to disturbances which have the same axial wavenumber but are not axisymmetric. They assumed that the gap between the cylinders is narrow, selected four fundamental modes of azimuthal wavenumber n, and considered their weakly nonlinear interaction. Thereby they found that Taylor vortices are unstable to disturbances which differ in axial phase when the Taylor number is greater than some value, $T_c'(n)$ say. The disturbances are waves which travel around the Taylor vortices in the azimuthal direction. Davey et al. calculated T_c' only for $n = 1$, 2 and 4, and found that $T_c'(4) > T_c'(2) > T_c'(1) \cong 1.08\, T_c$. This suggests that the waviness should set in first with $n = 1$ as T increases to $T_c'(1)$. Experiments, however, show that (see, for example, Coles (1965)) the waviness sets in with $n = 4$. Eagles (1971) made more extensive calculations, not only for the case of a narrow gap, showing that $T_c'(n)$ depends upon R_1/R_2, as well as confirming the results of Davey et al. Eagles (1974) also calculated the torque for the wavy vortices, and found that if one selects $n = 4$ then the theoretical results agree well with the measurements of the torque, although the most unstable mode in fact has wavenumber $n = 1$. This gives a sudden decrease in the slope of the torque as a function of the Taylor number at the transition. Thus the wavy mode gives a lower value of the torque than would a Taylor vortex at the same value of T.

Batchelor's 'inviscid-core-and-boundary-layer' model of strong Taylor vortices gives the torque

$$G \propto \rho H \Omega_1^2 R_1^4 (\nu/\Omega_1 R_1 d)^{1/2} (d/R_1)^{1/4}$$

when $\Omega_2 = 0$ and Ω_1 is much greater than its critical value for the onset of Taylor instability yet not so great that the Taylor vortices break down (Donnelly & Simon 1960, Appendix). This compares well with experimental results, for Donnelly & Simon found that

$$G \propto \rho H \Omega_1^2 R_1^4 (\nu/\Omega_1 R_1 d)^l (d/R_1)^m$$

where $l \cong 0.5$ and $m \cong 0.3$ for speeds well beyond critical. (They also fitted the formula

$$G = a\Omega_1^{-1} + b\Omega_1^{1.36}$$

to the results of various experiments over ranges of values of Ω_1 up to about ten times the critical value, where a and b are constants which depend on the radii of the cylinders, etc.) It may be noted, however, that Batchelor's result was derived for steady axisymmetric vortices, not vortices with travelling waves.

Pai (1943) was the first to note the non-uniqueness of flow between rotating cylinders. In experiments with the outer cylinder at rest, he observed that there were six vortices from the onset of instability but that when the rate of rotation of the inner cylinder was slowly and substantially increased the number of vortices would suddenly decrease to four. There was hysteresis, so that either four or six vortices might exist under given steady conditions in an intermediate range of values of the angular velocity according to whether the flow had been attained by slowly decreasing or increasing the angular velocity. It is difficult to interpret the results of Pai, which were mostly at such large values of the angular velocity that the flow was turbulent, but they show clearly that the flow under given steady conditions depends upon the history of the state as well as the steady conditions themselves. Koschmieder (1979) made some more refined experiments at very large values of the Taylor number, and found that turbulent vortices with all wavelengths in a wide range may be realized by different acceleration procedures. These experiments also show that the *mean* flow, up to and beyond the onset of turbulence, has a structure somewhat similar to that of Taylor vortices.

The most comprehensive set of experiments on Taylor vortices and their secondary instabilities is the one reported by Coles (1965). He observed many different kinds of doubly periodic flows for wide ranges of the angular velocities of the two cylinders. He counted the number m of vortices in the axial direction between the ends of the cylinders as well as the number n of waves in the azimuthal direction around the cylinders. The number pair m/n changes in a complicated but repeatable succession of transitions as the angular velocities change slowly. If the outer cylinder is at rest and the angular velocity Ω_1 of the inner cylinder is slowly increased from rest then waves $m/4$ begin to travel around when a flow with m Taylor vortices breaks down. As the rate of rotation increases further, m tends to decrease and n to increase slowly and somewhat erratically. Coles observed, however, that if the slow increase in the angular velocity is reversed then a different succession of transitions follows in general, not the original succession in reverse order. Further, by successively increasing and decreasing the angular velocity in various ways, Coles found that any of several values of m/n may occur at the same fixed value of Ω_1; this again shows clearly that the flow is not unique even under steady conditions. Some of Coles's experiments on unstable and turbulent Couette flow are illustrated by educational film loops (Coles FL 1963a,b).

The modern development of the laser Doppler technique to measure the velocity of a fluid enabled Gollub & Swinney (1975) to adopt a different experimental approach. They measured the radial component of the velocity at essentially a fixed point half way between the cylinders, and then analysed the variation of the velocity in time. This led to the identification of a succession of transitions when they fixed $\Omega_2 = 0$ and slowly increased Ω_1 from zero. (1) They observed the onset of Taylor instability when $Re \equiv R_1(R_2 - R_1)\Omega_1/\nu = 128$ (a Reynolds number is often used instead of a Taylor number), and confirmed that Taylor vortices are steady. (2) They found the onset of the secondary instability to wavy travelling disturbances with azimuthal wavenumber $n = 4$ at $Re = 160$, the power spectrum of the velocity having dominant peaks at a fundamental frequency, ω_1 say, and at its harmonics, $2\omega_1, 3\omega_1, \ldots$. (3) They saw that the spatial form of the waves persisted as Ω_1 increased further, but they noted a transition to another time-dependent motion with a new component when Re increased to a critical value of about 1350. This component coexists with, but is weaker and of lower frequency, ω_2 (say), than the wavy disturbance.

On increasing Ω_1 further, ω_1 remains constant, but the background noise increases further and the frequency ω_2 decreases, until finally the new component is no longer observable when Re reaches a critical value of about 2000. (4) Next a component with a fundamental frequency ω_3, of about $\frac{2}{3}\omega_1$, and its harmonics appear. (5) As Re increases further to yet another critical value, of about 2500, the sharp peaks in the power spectrum at integral multiples of ω_1 and ω_3 suddenly disappear, leaving a broad doublet. This transition is reversible, and seems to mark the onset of aperiodicity in the flow. A further increase of Ω_1 leads to a broadening of the doublet but no further qualitative change in the spectrum. Similar results have been reported subsequently by these and other experimentalists (see, for example, Fenstermacher, Swinney & Gollub (1979)).

The final transition observed by Gollub & Swinney has some resemblance to that due to a strange attractor, and has accordingly been taken to agree with the hypotheses of Ruelle & Takens (1971) about transition to turbulence. Yet the theoretical problem of finding a particular strange attractor which is manifestly a good model of the experiments and whose solutions agree well with the detailed experimental results remains to be solved.

Emphasis has historically been placed upon the case when the outer cylinder is at rest. This makes observations and measurements of the torque somewhat easier. Nonetheless, many experiments, by Taylor (1923) and others, have been done with co-rotation and counter-rotation. For example, Coles (1965), p. 390 noted the rapid onset of turbulence, characteristic of subcritical instability, when the outer cylinder counter-rotates much more rapidly than the inner rotates. Then he observed what he called spiral turbulence.

In a typical flow . . . a turbulent strip and a laminar strip are wrapped around the cylinders in a spiral pattern much like the pattern of stripes on a barber's pole. This spiral pattern rotates at very nearly the mean angular velocity of the two cylinders, without any significant change in size or shape.

Coles observed spiral turbulence predominantly when there was counter-rotation, but sometimes when the inner cylinder was at rest and even when both cylinders rotated in the same direction.

Also the intermittency of the turbulence varied according to conditions.

End effects are usually insignificant in the determination of the torque and the critical value of the Taylor number when the aspect ratio H/d of the length H of the cylinders to the gap width d is large. End effects, however, may be important in various ways, and should be borne in mind when comparing the results of experiments on pairs of cylinders with different aspect ratios. There is some evidence (Coles 1965, p. 399; Cole 1976, p. 7) that the onset of Taylor vortices is not abrupt; also small weak pairs of vortices may be seen at the ends of the cylinders before the onset of the Taylor vortices all along the cylinders (Burkhalter & Koschmieder 1973, p. 552; Cole 1976, p. 7). Indeed Jackson, Robati & Mobbs (1977) saw weak cellular flows for T at least as low as $0.3\ T_c$. These experimental results are consistent with a slightly imperfect supercritical bifurcation, as illustrated in Fig. 7.5(a) for small ε. It has also been observed that the onset of the secondary instability of Taylor vortices depends strongly upon the aspect ratio (Snyder 1969, § 7).

Benjamin (1978) examined these end effects both theoretically and experimentally. He used an apparatus with a stationary outer cylinder, and with aspect ratios H/d between three and four to emphasize end effects. He carefully observed a primary mode of a steady flow with two weak toroidal vortices for all values of the angular velocity Ω_1 of the inner cylinder which are not too large. The mode intensifies smoothly as Ω_1 increases, in accord with the theory of an imperfect supercritical bifurcation. The mode developed when the angular velocity was held steady after a sudden start from rest for values of the Reynolds number $Re \equiv \Omega_1 R_1^2/\nu$ from zero to 318 in a series of experiments with $H/d = 3.25$. For values of Re between 318 and 322, however, a secondary mode with four steady toroidal vortices developed. Yet if Ω_1 is increased very slowly from rest then the primary mode of two vortices remained until Re reached about 900, when there was transition to a flow with wave disturbances travelling in the azimuthal direction. Benjamin used his experimental results to show when these (and other) steady modes may exist for various values of the aspect ratio, and plotted and analysed the bifurcation set. This shows clearly the non-uniqueness and the hysteresis of steady flows.

54.3 *Parallel shear flows*

The current state of knowledge of the nonlinear stability of parallel flows is much less complete than that of Bénard convection or of Couette flow. This is in part because the theory of the linear stability of parallel flows, as seen in Chapter 4, is considerably more difficult. Further, nearly parallel flows bring complications into the theory. Indeed, it is not a question of whether most nearly parallel flows, for example the Blasius boundary layer on a flat plate, are unstable so much as where they are unstable. Also comparison of theory with experiments is not easy. The finiteness of a laboratory makes it impossible to suppress all end and side effects, even on exactly parallel flows such as plane Poiseuille flow or plane Couette flow.

Plane Poiseuille flow. This flow is useful as a prototype to illustrate the fundamentals of the theory, because it is an exact solution of the Navier–Stokes equations with a finite value of the critical Reynolds number. It is, however, not easily approximated in the laboratory.

Meksyn & Stuart (1951) treated theoretically the interaction through the Reynolds stress of a linearized disturbance and a mean flow driven down a channel by a prescribed pressure gradient, i.e. the perturbation is governed by the Orr–Sommerfeld equation and the exact equation of mean longitudinal momentum (x-momentum). Meksyn & Stuart solved both equations together to determine the mean flow as well as the perturbation. In this *mean-field approximation*, the interaction of the mean flow and the fundamental, whereby the variation of each with the transverse coordinate z is modified, is represented but the generation of harmonics is neglected. This leads to an equilibrium, or 'threshold', amplitude A_e of the fundamental above which the disturbances are unstable. Meksyn & Stuart found this subcritical instability to occur for Reynolds numbers R (based upon the maximum value of the mean velocity and the half-width of the channel) greater than a value of about 2900.

Stuart (1960a) and Watson (1960b) devised a more rational approach, specifying a method to find the Landau equation for weakly nonlinear disturbances when R is close to R_c. Detailed calculations of the Landau constant for plane Poiseuille flow were subsequently made by Reynolds & Potter (1967) and Pekeris & Shkoller (1967). They found that the real part ℓ_r of the complex

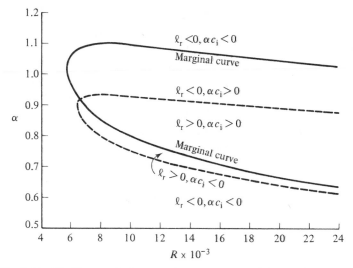

Fig. 7.11. The linear relative growth rate $\sigma = \alpha c_i$ and the Landau constant ℓ_r for plane Poiseuille flow $U = 1 - z^2$ $(-1 \leqslant z \leqslant 1)$, where $\mathrm{d}|A|^2/\mathrm{d}t = 2\alpha c_i|A|^2 - \ell_r|A|^4$. (After Pekeris & Shkoller 1967.)

Landau constant is negative for the critical mode (with wavenumber $\alpha = \alpha_c = 1.02$ and $R = R_c = 5772$). It follows that a time-periodic two-dimensional wave bifurcates at $R = R_c$ and that there is subcritical instability for $0 < R_c - R \ll 1$. They also found, however, that ℓ_r is positive on other parts of the marginal curve. Their results are summarized in Fig. 7.11. Chen & Joseph (1973) reached similar conclusions by more rigorous methods. Higher approximations to the nonlinear wave solutions of given wavelengths have been calculated (Herbert 1976, 1977a,b) by numerical solution of the nonlinear modal equations formulated by Heisenberg (1924). Itoh (1974b) treated weakly nonlinear spatially growing modes of prescribed frequencies.

Stewartson & Stuart (1971) and DiPrima, Eckhaus & Segel (1971) considered the spatial as well as the temporal development of weakly nonlinear slightly unstable modes. They showed that the variation of the amplitude A of the most unstable modes is governed by an equation of the form

$$\frac{\partial A_1}{\partial \tau} - a_2 \frac{\partial^2 A_1}{\partial \xi^2} = kA_1 - \tfrac{1}{2}\ell|A_1|^2 A_1, \qquad (54.2)$$

where $A_1 = A/\varepsilon$, $\xi = \varepsilon(x - c_g t)$, $\tau = \varepsilon^2 t$ and $\varepsilon^2 = R - R_c$ (see § 52.4). Hocking & Stewartson (1972) solved equation (54.2) in various cases. They found that if $R > R_c$ and $\ell_r < 0$ then A_1 may tend to infinity as τ tends to a finite value for fixed ξ. This rapid increase of A (until the equation becomes invalid) or 'burst' is reminiscent of sudden transition to turbulence. Also Davey, Hocking & Stewartson (1974) derived a coupled pair of partial differential equations which govern the modulation of a three-dimensional wavepacket of most unstable modes.

Following the experimental work of Reynolds (1883) and others on transition of flow in a circular pipe, Davies & White (1928) observed the flow in channels of various rectangular sections with width/depth ratios from about 40 to 160. Their inlet pipe generated quite strong eddies at the upstream end of the channel. From their observations of the flows downstream under a wide range of conditions they recognized 'three distinct types of flow: (a) one in which eddies cannot exist', for values of R less than a critical value of about 100; '(b) one in which eddies may exist, due to an initial disturbance, but cannot be sustained ..., the initial eddies therefore ultimately disappearing' for a range of about $100 < R < 1000$; 'and (c) one in which eddies once generated will be maintained without decrement, ... corresponding to truly turbulent flow' for R greater than the critical value of about 1000. Meksyn & Stuart (1951) suggested that their theory is qualitatively consistent with flow of type (c), there being instability to sufficiently large disturbances when $R_G < R < R_c$, where $R_G \cong 1000$ in the experiments of Davies & White but 2900 in the theory. Joseph (1976), vol. I, p. 88 suggested further that plane Poiseuille flow of type (a) is globally and monotonically stable, i.e. the total kinetic energy of any disturbance decays monotonically for $R < R_\mathscr{E}$; and of type (c) is globally but not monotonically stable for $R_\mathscr{E} < R < R_G$. It may be noted, however, that the inlet conditions of the experiment of Davies & White differ considerably from the theoretical ideal of an infinite channel.

Nishioka, Iida & Ichikawa (1975) used a long channel with a width/depth ratio of 27.4, chiefly to test the nonlinear theory. By reducing the level of background turbulence at the inlet they were able to maintain laminar flow for values of R up to 8000, which is substantially higher than the theoretical value of $R_c = 5772$. Inserting a vibrating ribbon at various specified frequencies, they confirmed the theoretical results of Itoh (1974a,b) for the spatial growth of both linear and weakly nonlinear modes. The experimental relation between the threshold amplitude and frequency supports the theoretical relation of Itoh (1974b). They further noted a rapid growth of harmonics but little distortion of the mean flow in the initial stage of the development of subcritical disturbances. The eventual transition to turbulence seems to occur suddenly as a secondary instability, the velocity profile developing a point of inflexion with strong shear during each cycle of the fundamental oscillation.

Recently there have been a few numerical integrations of the Navier–Stokes equations to determine the evolution of various disturbances (Orszag & Kells 1980). These are particularly valuable to study strongly nonlinear disturbances and to bridge some of the gaps between analytic and experimental results.

Plane Couette flow. Although, as seen in § 31, this flow is stable to infinitesimal disturbances for all values of the Reynolds number R (which we choose to be based upon half the basic-velocity difference across the channel and the half-width of the channel), the decay rates of the least stable modes vanish with $c_i = O(R^{-1/3})$ as $R \to \infty$. So Ellingsen, Gjevik & Palm (1970) proceeded formally to calculate the Landau constant for several pairs of values of R and wavenumber α, and found that $\ell_r < 0$. This suggests that there may be subcritical instability for disturbances of small finite amplitudes, even though there is no marginally stable normal mode for any finite value of R. Ellingsen *et al.* noted, however, that $\ell_r = O(R^{-1/3})$, the same order of magnitude as c_i, so that the threshold amplitude A_e appears not to vanish as $R \to \infty$. Moreover, the coefficients of the higher terms in $|A|^4$, $|A|^6$, etc. in the power series of $d|A|^2/dt$ are also of order $R^{-1/3}$ as $R \to \infty$, so that the truncation of the higher terms may at best be justified as a numerical approximation.

It was noted in § 31 that the linear theory of disturbances of plane Couette flow is somewhat similar to the linear theory of *axisymmetric* disturbances of Poiseuille flow in a pipe, although non-axisymmetric disturbances may be less stable and therefore more important in practice. Some disturbances are, to the first approximation for large values of R, confined to a thin layer near the wall in each problem, and therefore the curvature of the pipe is negligible. So Davey & Nguyen's (1971) weakly nonlinear theory of axisymmetric disturbances of Poiseuille flow in a pipe is relevant here. Their calculations suggest that there is subcritical instability of Poiseuille flow, much as do those of Ellingsen *et al.* for plane Couette flow. In addition, Davey & Nguyen treated the stability of plane Couette flow directly. They considered larger values of α than did Ellingsen *et al.* because the least stable modes are very short waves, the wall modes having $\alpha = O(R^{1/2})$ as $R \to \infty$. Their numerical results suggest that for these wall modes of plane Couette flow $A_e = O(R^{-1})$ as $R \to \infty$.

Also experimental work on the stability of plane Couette flow is difficult and incomplete. Reichardt (1956) found laminar flow only for R less than about 750, in a way to suggest that $R_G \approx 750$.

The Blasius boundary layer. This flow is important in its own right and also useful as a prototype for other nearly parallel flows. Chapter 4 indicates the difficulty in the linear theory of treating the effects of the divergence of the basic flow. When in addition treating weakly nonlinear effects, care should be taken to ensure that they are not less significant than any effect of the divergence which is neglected. The combined effects have yet to be treated satisfactorily.

Nonetheless Itoh (1974c) has carefully treated the weakly nonlinear growth of spatial modes, and his calculations plausibly give results qualitatively similar to those for plane Poiseuille flow. In particular, ℓ_r is positive on only part of the marginal curve. Herbert (1975) has reached a similar conclusion. Also Smith (1979a, b) has used a more systematic method, considering asymptotic approximations to the basic flow as well as the disturbance for large values of the Reynolds number.

Benney & Lin (1960) suggested a different approach to explain the observed three-dimensional nature of the instability of a boundary layer on a flat plate. Certain normal modes which vary periodically in the spanwise y-direction as well as the longitudinal x-direction are selected, and their weakly nonlinear interaction is calculated. It is found that they generate longitudinal vortices (with axes parallel to the x-axis and separated by the y-wavelength) which redistribute the momentum in the yz-planes perpendicular to the mean flow. This leads to a periodic intensification and reduction of the mean shear. Greenspan & Benney (1963) suggested that this variation of the mean flow, periodic in space as well as time, leads to local turbulent spots because of a secondary instability. The secondary instability is a strong one, like that of an unsteady parallel flow of an inviscid fluid, occurring rapidly in bursts where and when the shear is strong. Benney (1964) applied this method in detail to a boundary-layer profile, and his results bear an encouraging resemblance to the experimental observations by Klebanoff, Tidstrom & Sargent (1962) of the development of longitudinal vortices in boundary layers at transition. Yet in the Benney–Lin theory it is assumed that the instability is dominated by the interaction of a two-dimensional and a three-dimensional mode with the same longitudinal wavenumber α and the same phase velocity c_r, and two such fundamental modes satisfy the eigenvalue relation for the Blasius boundary-layer profile no better than approximately (Stuart 1960b, p. 91). Taking into account the difference of the phase velocities of the interacting modes, Antar & Collins (1975) showed that there is a slow modulation of the mean secondary flow.

There have also been a few numerical integrations of the Navier–Stokes equations for the Blasius boundary layer to determine the evolution of various disturbances (Fasel 1976, 1977). These overcome some of the theoretical difficulties of the divergence of the flow and of strong nonlinearity, and soon may be able to deal with three-dimensional disturbances adequately.

Reviewing earlier experiments at the National Bureau of Standards, Klebanoff *et al.* (1962) noted three consecutive stages in the development of boundary-layer instability downstream: (1) two-dimensional Tollmien–Schlichting waves consistent with linear theory; (2) strongly three-dimensional nonlinear waves; and (3) the birth of turbulent spots. Seeking to understand the three-dimensional nature of the instability they used a vibrating ribbon with a controlled spanwise variation to generate the instability. They then found longitudinal vortices distributed spanwise and the breakdown into turbulence as a consequence of a secondary instability, somewhat similar to the theoretical results described in the paragraph above.

These are only the main established features of the complicated three-dimensional process of transition. More details of recent research are given in some surveys edited by Eppler & Fasel (1980).

Unbounded shear layers. It is seen in Chapter 4 that use of a profile for an unbounded shear layer in the Orr–Sommerfeld equation gives $R_c = 0$, although shear layers are far from parallel where the local Reynolds number is small. There is therefore no theory available now to give a good approximation to R_c for a real shear layer.

Nonetheless Schade (1964) took a hyperbolic-tangent profile and formally calculated the Landau constant ℓ for a point on the marginal curve in the limit as $R \to \infty$, and found that ℓ is positive. Maslowe (1977) also calculated ℓ for this antisymmetric profile at several points on the marginal curve in the $R\alpha$-plane, and found that ℓ is positive at each point. The results vary little as R decreases from infinity to about 100.

Experimental work, however, has largely been based upon the use of vibrating ribbons and loudspeakers to force unstable waves of prescribed frequency, and so is better adapted to the theory of spatially than of temporally growing modes. The observed growth rates (see, for example, Freymuth (1966), Browand (1966) and Miksad (1972)) then agree very well with the theory of linear stability of a hyperbolic-tangent profile in an inviscid fluid (Michalke 1965).

Sato (1959) observed some oscillations whose frequency is half that of the initially dominant oscillation. This led Kelly (1967) to suggest that the fastest-growing normal mode first grows to dominate the other modes, develops into a finite-amplitude oscillation by self-interaction, and then by parametric instability generates a subharmonic wave of half the frequency and half the wavenumber of the fundamental dominant mode. Investigating in detail this interaction of the mean flow, the fundamental and the subharmonic, Kelly deduced that the growth rate of the subharmonic exceeds the growth rate of the fundamental when the amplitude of the velocity of the fundamental exceeds 12 per cent of the velocity difference across the shear layer. This result is confirmed by the experiments of Browand (1966).

Miksad (1972) has reviewed the results of many good experiments on the instability of free shear layers, and also reported his own. He classified six successive types of flow, as the instability develops downstream. (1) The instability is initiated in a region close to the origin of the shear layer. A disturbance of small amplitude has exponential growth rate, speed and wavelength well described by the linear theory of spatially growing modes in an inviscid fluid. (2) In a region downstream, further exponential growth leads to nonlinearity, and to the generation of subharmonics as well as harmonics. (3) The fundamental, harmonics and subharmonics then equilibrate in unison. (4) A second region of subharmonic generation follows next, in accord with the theory of Kelly (1967). The fundamental remains little changed, but the harmonics begin to decay. (5) Then three-dimensional distortion of the fundamental begins; the formation of longitudinal vortices resembles that proposed by Benney & Lin (1960). (6) Finally there is a region of breakdown into turbulence.

Jets and wakes. The theoretical and experimental difficulties for jets and wakes, as well as the results, are somewhat similar to those for unbounded shear layers. Liu (1969) showed how to calculate the Landau constant for a wake at $R = \infty$ after neglect of the divergence of the flow, and reported that $\ell_r > 0$. Sato & Kuriki's (1961) experimental results on the wake of a flat plate agree quite well with the growth rates given by the linear theory for temporal modes in an inviscid fluid, and, one might presume, agree better for spatial modes. The results also suggest a succession of regions downstream similar to those observed in an unbounded shear layer, except that no substantial subharmonic component is observed. Later experiments by Sato (1970) give more information about the nonlinear development of the instability.

Zabusky & Deem (1971) integrated the Euler and Navier–Stokes equations directly in order to simulate transition in a wake. They took the Bickley jet and the Gaussian wake as basic flows in a series of numerical experiments in which the development of various two-dimensional strongly nonlinear perturbations was computed. This reveals much information of the detailed features of the flow. This is a valuable complement to analytical and experimental methods, although further work remains to be done to relate theoretical to experimental results.

In summary, one may say that the transition from a laminar to a turbulent flow comprises a complex sequence of mechanisms. These mechanisms are not the same for all flows. Cellular instabilities, wave instabilities and turbulent spots are forerunners of well-developed turbulence. The onset of turbulence is characterized by the sudden development of a strongly three-dimensional strongly rotational aperiodic flow which varies rapidly and irregularly in space and time. The early stages of nonlinear instability of some simple flows may be described quite well now by the use of analytic theory and numerical experiments. This enables us to understand, and thereby to predict or to control, the breakdown of these flows with some success, although we know little about the onset of turbulence itself.

Problems for chapter 7

7.1. *A Landau equation with a cubic term.* If

$$\frac{\mathrm{d}A}{\mathrm{d}t} = aA + 2bA^2 + cA^3$$

for $A > 0$, $a > 0$, $b < 0$, and $b^2 > ac$, show that $A \uparrow A_e$ as $t \to \infty$ provided that A is sufficiently small at $t = 0$, where $A_e = -a/2b + O(a^2)$ as $a \downarrow 0$ for fixed values of b and c.

7.2. *The Landau equation: a cautionary example.* If

$$\frac{\mathrm{d}A}{\mathrm{d}t} = \left(\frac{\sigma^3}{3\ell}\right)^{1/2} \sin\left\{\left(\frac{3\ell}{\sigma}\right)^{1/2} A\right\}$$

and A is sufficiently small at $t = 0$, show that $A \to \pi(\sigma/3\ell)^{1/2}$ as $t \to \infty$. $[\sigma t = \ln[\tan\{\frac{1}{2}(3\ell/\sigma)^{1/2}A\}] + \text{constant}$. Note that $\mathrm{d}A^2/\mathrm{d}t = 2\sigma A^2 - \ell A^4 + (3\ell^2/20\sigma)A^6 + \cdots$ but that $A \nrightarrow A_e \equiv (2\sigma/\ell)^{1/2}$ as $t \to \infty$ and $\sigma \downarrow 0$, because then the coefficients of the terms in the sixth and higher powers of A become large.]

7.3. *Nonlinear instability for a modified form of the Burgers equation.* Take the model system,

$$\frac{\partial u}{\partial t} + \frac{\partial u^2}{\partial z} + R\left(\int_0^1 u^2 \mathrm{d}z\right) u - u = \frac{1}{R}\frac{\partial^2 u}{\partial z^2}$$

and

$$u = 0 \quad \text{at } z = 0 \quad \text{and} \quad 1,$$

and show that the basic 'flow' $u(z, t) = U \equiv 0$ gives rise to a linearized solution $u = U + u'$ with normal modes of the form $u' = \exp(s_n t)u_n(z)$, where

$$u_n = \sin\{(n+1)\pi z\}, \quad s_n = 1 - (n+1)^2\pi^2/R \quad \text{for } n = 0, 1, 2, \ldots$$

Deduce that there is stability if and only if $R \leqslant R_c = \pi^2$.

Considering the Fourier expansion

$$u(z, t) = \sum_{n=0}^{\infty} A_n(t) u_n(z)$$

of the nonlinear solution, show that

$$\frac{dA_n}{dt} = s_n A_n - \frac{1}{2} R \left(\sum_{k=0}^{\infty} A_k^2 \right) A_n$$

$$- (n+1)\pi \left(\frac{1}{2} \sum_{k=0}^{n-1} A_k A_{n-k-1} - \sum_{k=0}^{\infty} A_k A_{k+n+1} \right).$$

Hence deduce that there is an equilibrium solution $A_n = A_{en}$ for $R > R_c$, where

$$A_{e0} \sim \pm \{6(R - R_c)/5\pi^4\}^{1/2}, \quad A_{e1} \sim 2(R - R_c)/5\pi^3,$$

$$A_{en} = O\{(R - R_c)^{(n+1)/2}\} \quad \text{for } n = 2, 3, \ldots \quad \text{as } R \downarrow R_c.$$

[Burgers (1948), pp. 173–174, Eckhaus (1965), pp. 45–47.]

7.4. *Nonlinear instability for the Eckhaus model problem.* Take the system

$$\frac{\partial u}{\partial t} + \frac{\partial u}{\partial z} \frac{\partial^2 u}{\partial x^2} = \frac{1}{R} \left(\frac{\partial^2 u}{\partial x^2} + \frac{\partial^2 u}{\partial z^2} \right) - \frac{\partial^4 u}{\partial x^4}$$

and

$$u(x, 0, t) = 0, \ u(x, 1, t) = 1 \quad \text{for } -\infty < x < \infty, \ t \geqslant 0,$$

and show that the basic 'flow' $u(x, z, t) = U \equiv z$ gives rise to a linearized solution $u = U + u'$ with normal modes of the form

$$u'(x, z, t) = e^{i\alpha x + st} \phi(z),$$

where

$$\phi = \sin\{(n+1)\pi z\}, \quad s = f(n, \alpha) \equiv \alpha^2(1 - \alpha^2) - \{(n+1)^2\pi^2 + \alpha^2\}/R$$

for $n = 0, 1, \ldots$ and any real wavenumber α. Deduce that the (n, α) mode is stable if and only if $R \leqslant R_n(\alpha) = \{(n+1)^2\pi^2 + \alpha^2\}/\alpha^2(1 - \alpha^2)$. Hence show that the flow is stable if and only if $R \leqslant R_c = R_0(\alpha_c) = 41.5$ where $\alpha_c^2 = \pi^2\{(1 + \pi^{-2})^{1/2} - 1\} = 0.488$.

Show that the Landau equation for the amplitude $A(t)$ of the most unstable mode is

$$\frac{dA}{dt} = f(0, \alpha_c)A - \ell A^3 \quad \text{as } R \downarrow R_c,$$

where

$$\ell = -\frac{1}{4}\pi^2 \alpha_c^4 \left[\frac{1}{f(1, 0)} - \frac{1}{2f(1, 2\alpha_c)} \right]_{R = R_c} > 0.$$

[Eckhaus (1965), pp. 73–76.]

7.5. *A modified form of Matkowsky's model problem.* Consider the system

$$\frac{\partial^2 u}{\partial t^2} + f(u) = \frac{1}{R}\frac{\partial^2 u}{\partial z^2} \quad \text{for } 0 < z < \pi$$

and

$$u = 0 \quad \text{at } z = 0, \pi,$$

where

$$f(0) = 0, \quad f'(0) < 0, \quad f''(0) = 0, \quad f'''(0) > 0.$$

Show that the solution $u = U \equiv 0$ is linearly stable if $R \le R_c = -1/f'(0)$.

Solve the weakly nonlinear problem for $0 < R - R_c \ll 1$, showing that $u \sim A(t) \sin z$, where

$$\frac{d^2 A}{dt^2} = (R_c^{-1} - R^{-1})A - \tfrac{1}{8}f'''(0)A^3.$$

Find the 'energy' integral of this equation, and hence discuss the qualitative character of its solutions.

What is the equation for $d^2 A/dt^2$ when $f''(0) \ne 0$?

$[u' = \exp(s_n t)\sin nz$ for $n = 1, 2, \dots$, where $s_n^2 = -f'(0) - n^2/R$. $u_2 \equiv 0$. $\tfrac{1}{2}(dA/dt)^2 = \tfrac{1}{2}(R_c^{-1} - R^{-1})A^2 - \tfrac{1}{32}f'''(0)A^4 + \text{constant}$. If $f''(0) \ne 0$, then $a_2 = 0$, $u_2 = \tfrac{1}{4}R_c f''(0)(1 + \tfrac{1}{3}\cos 2z)$ and $a_3 = -\tfrac{1}{8}f'''(0) - \tfrac{5}{24}(f''(0))^2$.]

7.6. *Segel's model of nonlinear Bénard convection.* Consider the system

$$\frac{\partial w}{\partial t} + \frac{\partial^3 (\tfrac{1}{2}w^2)}{\partial z^3} = \Delta^3 w - R\Delta_1 w \quad \text{for } -\infty < x, y < \infty, \quad 0 < z < 1,$$

and

$$w = \partial^2 w/\partial z^2 = \partial^4 w/\partial z^4 = 0 \quad \text{at } z = 0, 1$$

as a model of nonlinear convection between horizontal stress-free planes at $z = 0, 1$, where R is the 'Rayleigh number'. Taking normal modes of the form $w' = W(z)f(x, y)\,e^{st}$ to solve the linearized problem, show that

$$W = W_j = A_j \sin j\pi z \quad \text{for } j = 1, 2, \dots, \quad \text{and} \quad \Delta_1 f + a^2 f = 0,$$

where $s = a^2 R - (j^2\pi^2 + a^2)^3$ and a is any real wavenumber. For a given value of a, deduce that the first mode is stable if and only if $R \le R_c(a) = (\pi^2 + a^2)^3/a^2$.

Taking the 'roll cell' $w_1 = A \cos ax \sin \pi z$ as the fundamental in the weakly nonlinear problem, deduce that its Landau equation is

$$\frac{\mathrm{d}A}{\mathrm{d}t} = a^2(R - R_c)A - \tfrac{1}{4}\pi^3(2d_0 + d_1)A^3,$$

where $d_0 = 1/64\pi^3$ and $d_1 = \pi^3/60(\pi^2 + a^2)^3$. [Segel (1965), Segel (1966), pp. 166–167, 174–175.]

7.7. *The amplitude equations for resonant interactions of a triad of waves.* A triad of waves which resonate at second order gives rise to amplitude equations of the form

$$\mathrm{d}A_1/\mathrm{d}t = C_1 A_2^* A_3^*, \qquad \mathrm{d}A_2/\mathrm{d}t = C_2 A_3^* A_1^*, \qquad \mathrm{d}A_3/\mathrm{d}t = C_3 A_1^* A_2^*,$$

where the C_n are complex interaction coefficients which may depend upon the frequencies and wavenumbers of the triad as well as the parameters of the particular problem being treated, and where $\omega_1 + \omega_2 + \omega_3 = 0$ and $k_1 + k_2 + k_3 = 0$. Show that $C_3|A_2|^2 = C_2|A_3|^2 + \text{constant}$ if C_2 and C_3 are both either real or purely imaginary. If, moreover, $C_1\omega_1^2 + C_2\omega_2^2 + C_3\omega_3^2 = 0$, deduce the 'energy' integral

$$\tfrac{1}{2}(\omega_1^2|A_1|^2 + \omega_2^2|A_2|^2 + \omega_3^2|A_3|^2) = \text{constant}.$$

Suppose that $C_1 < 0 < C_2, C_3$ and that $A_1 = 0$, $A_2 = A_{20}$ and $A_3 = A_{30}$ at $t = 0$, where $A_{20}, A_{30} > 0$ and $C_3 A_{20}^2 = C_2 A_{30}^2$. Then show that

$$A_1 = -(-C_1 A_{20}^2/C_2)^{1/2} \tanh at, \quad A_2 = A_{20} \operatorname{sech} at, \quad A_3 = A_{30} \operatorname{sech} at$$

for $t > 0$, where $a = (-C_3 C_1 A_{20}^2)^{1/2}$. Verify that this solution satisfies the 'energy' integral above.

7.8. *A model of resonant wave interactions.* Linearize the equation

$$\frac{\partial^2 \phi}{\partial t^2} + \left(\frac{\partial^2}{\partial x^2} + 1\right)^2 \phi = 2\frac{\partial \phi}{\partial x}\frac{\partial^2 \phi}{\partial x^2},$$

and show that the normal modes have the form $\phi = A \exp\{\mathrm{i}(kx - \omega t)\}$, where $\omega^2 = (k^2 - 1)^2$.

Examining the dispersion relation, show that there exist triads such that $\omega_1 = \omega_2 + \omega_3$ and $k_1 = k_2 + k_3$. Verify that a limiting form of these triads is the dyad with $k_1 = \sqrt{2}$, $\omega_1 = 1$, $k_2 = k_3 = 1/\sqrt{2}$, $\omega_2 = \omega_3 = \tfrac{1}{2}$.

Taking $\phi = \tfrac{1}{2}\sum_k A \exp\{\mathrm{i}(kx - \omega t)\}$ for real $A(k, t)$, deduce that resonant interactions of the dyad are governed by

$$\mathrm{d}A_1/\mathrm{d}t = 2^{-5/2}A_2^2, \quad \mathrm{d}A_2/\mathrm{d}t = -A_1 A_2/\sqrt{2}$$

approximately when A_1 and A_2 are small. Hence show that $4A_1^2 + A_2^2$ is a constant of the motion, and that if $A_1 = 0$ and $A_2 = a$ at $t = 0$ then $A_1 = \tfrac{1}{2}a \tanh(2^{-3/2}at)$, $A_2 = a \operatorname{sech}(2^{-3/2}at)$ for $t > 0$.

7.9. *Side-band instability for a nonlinear Klein–Gordon equation.* Show that the normal modes for the equation

$$\frac{\partial^2 \phi}{\partial t^2} - \frac{\partial^2 \phi}{\partial x^2} + \phi = -\sigma \phi^3$$

are given by $\phi = A \exp\{i(kx - \omega t)\}$, where $\omega^2 = k^2 + 1$.

Let $\phi = \sum_k A(k, t) \exp\{i(kx - \omega t)\}$, where $A(-k, t) = A^*(k, t)$ and $\omega(-k) = -\omega(k)$ in order that ϕ is real. Deduce that

$$\frac{dA_1}{dt} = -\frac{i\sigma}{2\omega_1} \sum_{k_2+k_3+k_4=k_1} A_2 A_3 A_4 \exp\{-i(\omega_2 + \omega_3 + \omega_4 - \omega_1)t\}$$

approximately when the amplitudes are small.

Take now a basic wave with k_0, $\omega_0 > 0$ and make up a particular tetrad with $-k_0$ and ω_0, $k_+ = k_0(1 + \varepsilon)$ and $\omega_+ > 0$, and $k_- = k_0(1 - \varepsilon)$ and $\omega_- > 0$ as wavenumbers and frequencies respectively, where $0 < \varepsilon \ll 1$. Let $\Omega = \omega_+ + \omega_- - 2\omega_0$. Deduce that $\Omega \sim \varepsilon^2 k_0^2/\omega_0^3$ as $\varepsilon \to 0$. Show that

$$\frac{dA_0}{dt} = -\frac{i\sigma}{2\omega_0}(3A_0^2 A_0^* + 6A_0 A_+ A_+^* + 6A_0 A_- A_-^* + 6A_0^* A_+ A_- \, e^{-i\Omega t}),$$

$$\frac{dA_+}{dt} = -\frac{i\sigma}{2\omega_+} (6A_0 A_0^* A_+ + 3A_0^2 A_-^* \, e^{i\Omega t} + 6A_- A_-^* A_+ + 3A_+^2 A_+^*)$$

and that the equation for dA_-/dt is analogous to the one for dA_+/dt.

To investigate the stability of the basic wave, suppose that $|A_+|$, $|A_-| \ll |A_0|$ and linearize the amplitude equations. Hence show that

$$\frac{dA_0}{dt} = -\frac{3i\sigma}{2\omega_0}|A_0|^2 A_0,$$

$$\frac{dA_+}{dt} = -\frac{i\sigma}{2\omega_+}(6|A_0|^2 A_+ + 3A_0^2 A_-^* \, e^{i\Omega t}).$$

Show that the solution of these equations and the analogous one for dA_-/dt is given by $A_0 = a_0 \, e^{-i\rho t}$ and $A_\pm = a_\pm \, e^{i\lambda_\pm t}$ approximately, where a_0 and $a_\pm$ are real constants, $\rho = \frac{3}{2}a_0^2\sigma/\omega_0$ and $\lambda_\pm$ are the roots of $\lambda^2 + (2\rho - \Omega)\lambda + \rho(\rho - 2\Omega) = 0$. Hence show that the basic wave is unstable if $6\sigma a_0^2 + \varepsilon^2 k_0^2/\omega_0^2 < 0$. [Whitham (1974), § 15.6.]

*7.10. *The Lorenz attractor.* In a model of nonlinear Bénard convection, the interaction of modes is described by the dimensionless system

$$\frac{dx}{dt} = -Pr \, x + Pr \, y,$$

$$\frac{dy}{dt} = kx - y - zx,$$

$$\frac{dz}{dt} = -bz + xy,$$

where x, y and z are the amplitudes of the modes, Pr is the Prandtl number, $k = Ra/Ra_c$ and $b = 4/(1+a^2)$. As in Chapter 2, Ra is the Rayleigh number, $Ra_c = (\pi^2 + a^2)^3/a^2$ its critical value for free–free boundaries, and a is a wavenumber. Show that there is a steady solution $x = y = z = 0$ of this system which is stable if and only if $k \leq 1$. Show further that there exist steady solutions $x = y = \pm\{b(k-1)\}^{1/2}$, $z = k - 1$ if $k > 1$, and that these are stable if $Pr \leq b + 1$ or if $Pr > b + 1$ and $k \leq Pr(Pr + b + 3)/(Pr - b - 1)$.

Prove that if x, y or z is sufficiently large then $\frac{1}{2}\{x^2 + y^2 + (z - k - Pr)^2\}$ decreases as time increases. Prove also that $\nabla \cdot \mathbf{u} < 0$, where $\mathbf{u} = (dx/dt, dy/dt, dz/dt)$, and interpret this result. How might the unsteady solutions behave when all three equilibrium solutions are unstable? Discuss in general the development of the solutions of this system with time, using a computer to get some numerical results to illustrate your discussion. [Lorenz (1963), Marsden & McCracken (1976), pp. 141–149.]

7.11. *Bifurcation for an ordinary differential system.* Given that

$$dx/dt = \{1 - f(x, y)/R\}x - y, \quad dy/dt = x + \{1 - f(x, y)/R\}y,$$

show that $dr/dt = \{1 - f(x, y)/R\}r$, where $r = (x^2 + y^2)^{1/2}$.

Deduce that if $f(0, 0) > 0$ the null solution is stable when $0 < R < f(0, 0)$ and unstable when $R > f(0, 0)$. Further, show that the null solution is globally stable if $f(x, y) > R > 0$ for all x, y.

If $f(x, y) = g(r)$ and $R = g(r_0)$ for some positive function g and some positive constant r_0, verify that there exist solutions of the form $x = r_0 \cos(t + \theta_0)$, $y = r_0 \sin(t + \theta_0)$ in addition to the null solution. Prove that these periodic solutions are themselves stable if $g'(r_0) > 0$ and unstable if $g'(r_0) < 0$.

Further, given that $g(r) = 1 + \{1 + (r-1)^2\}^{-1}$, sketch the bifurcation curve in the first quadrant of the r_0R-plane. Is the Hopf bifurcation of the periodic solutions from the null solution at $(0, 1)$ supercritical or subcritical? What is the value of R_G such that the null solution is globally stable for all $R < R_G$ but not globally stable for some R in each neighbourhood of R_G? Discuss the domains of attraction of the periodic solutions in the xy-plane for various fixed values of R.

7.12. *A simple problem of particle dynamics which exhibits structurally unstable bifurcation.* A particle moves along a smooth wire which is fixed in a vertical plane. Gravity acts on the particle, and the plane rotates with constant angular velocity ω about a vertical axis in the plane. Take Cartesian coordinates (x, z) in the plane, where Oz is the upward vertical, so that the wire has an equation of the form $z = f(x)$ and the axis of rotation has equation $x = \varepsilon a > 0$. You are given that the equation of motion of the particle is then

$$(1 + f'^2)\ddot{x} + f'f''\dot{x}^2 - \omega^2(x - \varepsilon a) + gf' = 0.$$

Show that there may be equilibrium at $x = X$ where X is any root of $gf''(X) = \omega^2(X - \varepsilon a)$, and that this equilibrium is stable if $\omega^2 < gf''(X)$ and unstable if $\omega^2 > gf''(X)$.

If $f(x) = \frac{1}{2}a(x/a)^2\{1 + (x/a)^2\}$, show that there is supercritical bifurcation at $R \equiv a\omega^2/g = 1$ and $X = 0$ when $\varepsilon = 0$, but that the bifurcation has a different character when $0 < \varepsilon \ll 1$. Sketch the bifurcation curves in the RX-plane for both cases. Further, show that for any real value of ε there is marginal stability if at bifurcation $27\varepsilon^2 R^2 = 4(R - 1)^3$, the equation of a curve with a cusp at $(0, 1)$ in the εR-plane.

7.13. *The energy method for Bénard convection.* It can be seen from Chapter 2 that the nonlinear equations governing Bénard convection of a Boussinesq fluid may be expressed as

$$\partial u_i/\partial t + u_j \partial u_i/\partial x_j = -\partial p/\partial x_i + Ra\, Pr\, \theta\delta_{i3} + Pr\, \Delta u_i,$$

$$\partial\theta/\partial t + u_j \partial\theta/\partial x_j - w = \Delta\theta,$$

$$\partial u_i/\partial x_i = 0,$$

where the steady basic state of rest with heat conduction is represented by the solution $u_i, \theta \equiv 0$.

If the flow has wavelength $2\pi/a_x$ in x and $2\pi/a_y$ in y, and satisfies any of the usual boundary conditions at $z = 0$ and 1 (e.g. $\theta, u_i = 0$ at a rigid boundary), deduce the 'energy' equation

$$\frac{d(\mathcal{K} + Pr\, Ra\, \mathcal{H})}{dt} = 2Pr \int \{Ra\, \theta w - \frac{1}{2}(\partial u_i/\partial x_j)^2 - \frac{1}{2}Ra(\partial\theta/\partial x_j)^2\}\, d\mathbf{x},$$

where

$$\mathcal{K} = \int \frac{1}{2}u_i^2\, d\mathbf{x}, \quad \mathcal{H} = \int \frac{1}{2}\theta^2\, d\mathbf{x},$$

and the domain of the integrations is the cuboid $0 \le x \le 2\pi/a_x,\ 0 \le y \le 2\pi/a_y,\ 0 \le z \le 1$.

Using the constraint $\partial u_i/\partial x_i = 0$ with the Lagrange multiplier P, show that the variational principle to minimize the integral on the right-hand side of the 'energy' equation above gives the Euler equations,

$$0 = -\nabla P + Ra\, \theta\mathbf{k} + \Delta\mathbf{u},$$

and

$$-w = \Delta\theta.$$

Hence show that all disturbances, whether large or small, are stable if $Ra < Ra_c$, where Ra_c is the critical value of the Rayleigh number determined by the linear theory with the appropriate boundary conditions. [Joseph (1965).]

A CLASS OF GENERALIZED AIRY
FUNCTIONS

In § 27 we found that the local turning-point approximations to the solutions of the Orr–Sommerfeld equation could be expressed in terms of double integrals of Airy functions. These approximations are simply the first terms in the inner expansions of the solutions. To infer the general structure of the uniform expansions, however, it is necessary to characterize the class of functions in terms of which the inner expansions of the solutions can be expressed to *all* orders. It is natural therefore to begin with a discussion of Airy functions and the required class of functions then emerges in two stages by a systematic generalization of the Airy functions. The present discussion is a modified version of the account given by Reid (1972).

A1 The Airy functions $A_k(z)$

Consider first the solutions of Airy's equation $u'' - zu = 0$ which are often taken as Ai (z) and Bi (z). When z is real, these solutions are *numerically satisfactory* in the sense of J. C. P. Miller (cf. Olver 1974, pp. 154–155). When z is complex, however, Ai (z) is recessive in the sector $|\mathrm{ph}\, z| < \frac{1}{3}\pi$, but there is no sector in which Bi (z) is recessive. Furthermore, since Airy's equation is invariant with respect to the transformations $z \to z\mathrm{e}^{\pm 2\pi\mathrm{i}/3}$, this suggests that we consider three solutions defined by

$$\left.\begin{aligned}
A_1(z) &= \mathrm{Ai}(z), \\
A_2(z) &= \mathrm{e}^{2\pi\mathrm{i}/3}\mathrm{Ai}\,(z\,\mathrm{e}^{2\pi\mathrm{i}/3}), \\
A_3(z) &= \mathrm{e}^{-2\pi\mathrm{i}/3}\mathrm{Ai}\,(z\,\mathrm{e}^{-2\pi\mathrm{i}/3}).
\end{aligned}\right\} \tag{A1}$$

and

The basic idea here is due to Olver (1974), p. 413, but the present notation differs slightly from his. The solutions $A_k(z)$ ($k = 1, 2, 3$) are thus recessive in the sectors $\mathbf{S}_k$ shown in Fig. A1. They are not,

Fig. A1. The Stokes lines (left) and the anti-Stokes lines (right) for the Airy functions in the z-plane.

of course, linearly independent but must be related by the connexion formula,

$$\sum_{k=1}^{3} A_k(z) = 0, \qquad (A2)$$

which follows immediately from the integral representation of Ai(z). Any two of these functions form a pair of linearly independent solutions of Airy's equation and their Wronskians are

$$\mathcal{W}(A_1, A_2) = \mathcal{W}(A_2, A_3) = \mathcal{W}(A_3, A_1) = -1/2\pi i. \qquad (A3)$$

A2 The functions $A_k(z, p)$, $B_0(z, p)$ and $B_k(z, p)$

The first generalization we wish to consider was motivated initially by the observation that the first three terms in the inner expansions of the solutions of dominant-recessive type can be expressed explicitly in terms of the Airy functions $A_k(z)$, their first derivatives, and their first integrals together with polynomial coefficients (cf. Problem 4.8), and this observation eventually led to a consideration of the generalized Airy functions defined by

$$A_k(z, p) = \frac{1}{2\pi i} \int_{L_k} t^{-p} \exp\left(zt - \tfrac{1}{3}t^3\right) dt, \qquad (A4)$$

where $p = 0, \pm1, \pm2, \ldots$ and L_k ($k = 1, 2, 3$) are the contours shown in Fig. A2. This class of functions has also been discussed to some extent by Grohne (1954) in connexion with his study of the stability of plane Couette flow. For some purposes, however, it is

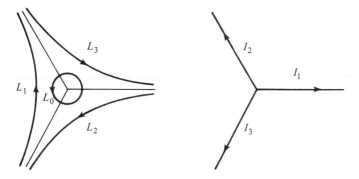

Fig. A2. The paths of integration in the t-plane.

convenient to allow p to be an unrestricted complex variable and in that case we will suppose that a cut has been introduced in the t-plane running from the origin to infinity along the positive real axis so that $0 \leqslant \mathrm{ph}\, t < 2\pi$.

It is easily seen that the functions $A_k(z, p)$ are solutions of the differential equation

$$(AD + p - 1)u = 0, \tag{A5}$$

where $D = d/dz$ and $A = D^2 - z$, and this equation could have been used as the starting point for the present discussion. On differentiating equation (A5) and noting that $DAD = (AD - 1)D$ we have

$$AD^2 A_k(z, p+1) = -(p-1)A_k(z, p). \tag{A6}$$

By using this equation we can immediately obtain the particular integrals of the inhomogeneous equation $AD^2 u = A_k(z, p)$ for all values of p except $p = 1$, and this result provides a simple algorithm for deriving the first three terms of the inner expansions of the solutions of dominant-recessive type. It also shows, however, that these are the only terms which can be expressed in terms of this class of functions and that to proceed to higher orders we must consider a more general class of functions.

The solutions associated with the paths L_2 and L_3 can be expressed in terms of $A_1(z, p)$ by means of the rotation formulae

and

$$A_2(z, p) = e^{-2(p-1)\pi i/3} A_1(z\, e^{2\pi i/3}, p) \left.\right\}$$
$$A_3(z, p) = e^{2(p-1)\pi i/3} A_1(z\, e^{-2\pi i/3}, p). \left.\right\} \tag{A7}$$

The derivatives of these solutions satisfy the relation

$$D^n A_k(z, p) = A_k(z, p - n) \tag{A8}$$

and from equation (A5) we then have the recursion formula

$$A_k(z, p - 3) - z A_k(z, p - 1) + (p - 1) A_k(z, p) = 0. \tag{A9}$$

The following special cases may be noted:

and

$$\left.\begin{aligned}
A_k(z, -1) &= A_k'(z), \\
A_k(z, 0) &= A_k(z), \\
A_k(z, 1) &= \int_{\infty_k}^{z} A_k(t)\, dt,
\end{aligned}\right\} \tag{A10}$$

where ∞_k denotes a path of integration that tends to infinity in the sector $\mathbf{S}_k$. Thus, for other values of p, $A_k(z, p)$ can, by use of the recursion formula (A9), be expressed as a linear combination of $A_k(z, 0)$ and $A_k(z, \pm 1)$ with polynomial coefficients. It should also be noted that $A_k(z, -p)$ denotes the pth derivative of $A_k(z)$ and that $A_k(z, p)$ denotes the p-fold integral of $A_k(z)$ from ∞_k to z. For numerical purposes it is convenient to have the initial values

$$A_1(0, p) = e^{-p\pi i} 3^{-(p+2)/3} / \Gamma(\tfrac{1}{3}(p + 2)) \quad (p \in \mathbb{C}); \tag{A11}$$

the corresponding values of $A_2(0, p)$ and $A_3(0, p)$ then follow directly from the rotation formulae (A7).

In discussing the asymptotic expansions of these functions it is convenient to adopt the convention that ph z lies in the range $-\tfrac{4}{3}\pi < \text{ph } z < \tfrac{2}{3}\pi$. An application of the method of steepest descents to $A_1(z, p)$ and subsequent use of the rotation formulae (A7) then shows that when $p = 0, \pm 1, \pm 2, \ldots$, the required expansions can all be expressed in terms of the quantities

$$A_\pm(z, p) = \tfrac{1}{2} \pi^{-1/2} (\pm 1)^p z^{-(2p+1)/4} e^{\pm \xi} \sum_{s=0}^{\infty} (\pm 1)^s a_s(p) \xi^{-s}, \tag{A12}$$

where $\xi = \tfrac{2}{3} z^{3/2}$ and

$$\left.\begin{aligned}
a_0(p) &= 1, \\
a_1(p) &= \frac{1}{2^3 3^2}(12p^2 + 24p + 5), \\
a_2(p) &= \frac{1}{2^7 3^4}(144p^4 + 1344p^3 + 3864p^2 + 3504p + 385), \\
\cdots\,;
\end{aligned}\right\} \tag{A13}$$

more generally, $a_s(p)$ is a polynomial in p of degree $2s$. In terms of these quantities we have

$$
\left.
\begin{aligned}
A_1(z,p) &\sim A_-(z,p) && (z \in \mathbf{T}_2 \cup \mathbf{T}_3), \\
A_2(z,p) &\sim iA_+(z,p) && (z \in \mathbf{T}_3 \cup \mathbf{T}_1), \\
A_3(z,p) &\sim
\begin{cases}
-A_-(z,p) & (z \in \mathbf{T}_1) \\
-iA_+(z,p) & (z \in \mathbf{T}_2),
\end{cases}
\end{aligned}
\right\}
\tag{A14}
$$

where $\mathbf{T}_k$ are the sectors shown in Fig. A1. In equations (A14) the sectors of validity have been restricted to insure completeness.

To obtain the asymptotic expansions of $A_k(z,p)$ in the sectors $\mathbf{T}_k$, it is first necessary to consider the well-balanced solution of equation (A5) which is associated with the closed path L_0 shown in Fig. A2. Thus, we let

$$
B_0(z,p) = \frac{1}{2\pi i} \int_{L_0} t^{-p} \exp\left(zt - \tfrac{1}{3}t^3\right) dt. \tag{A15}
$$

If $p \leq 0$ then $B_0(z,p) \equiv 0$; otherwise it is a polynomial in z of degree $p-1$ which, by the residue theorem, is simply the coefficient of t^{p-1} in the expansion of $\exp\left(zt - \tfrac{1}{3}t^3\right)$. The first few of these polynomials are

$$
\left.
\begin{aligned}
&B_0(z,1) = 1, && B_0(z,4) = \frac{1}{3!}z^3 - \frac{1}{3}, \\
&B_0(z,2) = z, && B_0(z,5) = \frac{1}{4!}z^4 - \frac{1}{3}z, \\
&B_0(z,3) = \frac{1}{2!}z^2, && B_0(z,6) = \frac{1}{5!}z^5 - \frac{1}{6}z^2.
\end{aligned}
\right\}
\tag{A16}
$$

We may also note the initial values

$$
\left.
\begin{aligned}
& B_0(0, 3m+1) = (-1)^m/3^m m! \\
& B_0(0, 3m+2) = B_0(0, 3m+3) = 0 \quad (m = 0, 1, 2, \ldots).
\end{aligned}
\right\}
\tag{A17}
$$

and

It should be emphasized that $B_0(z,p)$ not only satisfies the same differential equation as $A_k(z,p)$ but also satisfies the same recursion formula. Thus, given $B_0(z,1)$ and the fact that $B_0(z,p) \equiv 0$ for $p \leq 0$, we can obtain $B_0(z,p)$ for $p \geq 2$ recursively from equation (A9). The polynomials $B_0(z,p)$ also satisfy equation (A6); hence we can immediately obtain the particular integrals of the equation $AD^2 u = B_0(z,p)$ for all values of p except $p = 1$.

We have thus obtained four solutions of equation (A5). They cannot, of course, be linearly independent but must be related by the connexion formula

$$\sum_{k=1}^{3} A_k(z, p) = -B_0(z, p). \tag{A18}$$

This formula reduces, as it must, to equation (A2) when $p = 0$. By using this connexion formula we then find that the asymptotic behaviour of $A_k(z, p)$ in $\mathbf{T}_k$ is given by

$$\begin{bmatrix} A_1(z, p) \\ A_2(z, p) \\ A_3(z, p) \end{bmatrix} \sim \begin{bmatrix} -i & -1 & 1 \\ i & -1 & -1 \\ -i & -1 & -1 \end{bmatrix} \begin{bmatrix} A_+(z, p) \\ B_0(z, p) \\ A_-(z, p) \end{bmatrix} \quad (p \in \mathbb{Z}). \tag{A19}$$

When $p = 0, -1, -2, \ldots$, equation (A5) also has solutions of balanced type which can conveniently be taken in the form

$$B_k(z, p) = \int_{I_k} t^{-p} \exp\left(zt - \tfrac{1}{3}t^3\right) dt, \tag{A20}$$

where the paths I_k are shown in Fig. A2. These solutions also satisfy equations (A7)–(A9) with $A_k(z, p)$ replaced by $B_k(z, p)$. It is easy to see that the functions $B_k(z, 0)$ are solutions of the inhomogeneous Airy equation $Au = 1$ and that $B_1(z, 0) = \pi \text{Hi}(z)$, where $\text{Hi}(z)$ is one of Scorer's functions (see, for example, Abramowitz & Stegun (1964), p. 448). We also note the initial values

$$B_1(0, p) = 3^{-(p+2)/3} \Gamma(\tfrac{1}{3}(1 - p)). \tag{A21}$$

The asymptotic expansion of $B_1(z, p)$ in $\mathbf{T}_1$ can easily be obtained by the method of integration by parts and the corresponding expansion of $B_2(z, p)$ and $B_3(z, p)$ then follow from the rotation formulae (A7). In this way we obtain

$$B_k(z, p) \sim (-1)^{1-p}(-p)! z^{p-1}\{1 - \tfrac{1}{3}(p-1)(p-2)(p-3)z^{-3} + \cdots\}$$
$$(z \in \mathbf{T}_k), \tag{A22}$$

where the sectors of validity have again been restricted to insure completeness. The six solutions $A_k(z, p)$ and $B_k(z, p)$ are not, of course, linearly independent but must satisfy the three connexion formulae

and

$$
\left.
\begin{aligned}
B_1(z, p) - B_2(z, p) &= 2\pi i A_3(z, p), \\
B_2(z, p) - B_3(z, p) &= 2\pi i A_1(z, p) \\
B_3(z, p) - B_1(z, p) &= 2\pi i A_2(z, p).
\end{aligned}
\right\} \qquad \text{(A23)}
$$

These equations further imply that

$$
\sum_{k=1}^{3} A_k(z, p) = 0 \quad (p = 0, -1, -2, \dots) \qquad \text{(A24)}
$$

in agreement with equation (A18). The connexion formulae (A23) provide a simple method of obtaining the complete asymptotic expansions of $B_k(z, p)$ in the complementary sectors $\mathbf{I} \backslash \mathbf{T}_k$; in particular they show that $B_k(z, p)$ are dominant in $\mathbf{S}_k$.

The Wronskian of any linearly independent set of solutions of equation (A5) is necessarily a constant. In most hydrodynamical applications, however, a numerically satisfactory set of solutions would include $A_1(z, p)$, $A_2(z, p)$, and either $B_0(z, p)$ or $B_3(z, p)$, and their Wronskians are given by

$$
\left.
\begin{aligned}
\mathcal{W}(A_1, A_2, B_0)(z, p) &= \frac{1}{(p-1)!}(-1)^p \frac{1}{2\pi i} \quad (p = 1, 2, \dots) \\
\mathcal{W}(A_1, A_2, B_3)(z, p) &= -(-p)! \frac{1}{2\pi i} \quad (p = 0, -1, -2, \dots).
\end{aligned}
\right\} \qquad \text{(A25)}
$$

It is also of some interest to consider the second-order Wronskians of the solutions. For this purpose let u and v be any two solutions of equation (A5). Then it is not difficult to show that their Wronskian satisfies the equation

$$
(AD - p)\mathcal{W}(u, v) = 0, \qquad \text{(A26)}
$$

the solutions of which are $A_k(z, 1-p)$ and either $B_0(z, 1-p)$ ($p = 0, -1, -2, \dots$) or $B_k(z, 1-p)$ ($p = 1, 2, \dots$). Consider, for example, the Wronskian of $A_1(z, p)$ and $A_2(z, p)$. It is clearly balanced in $\mathbf{T}_3$ and must, therefore, be a constant multiple of either $B_0(z, 1-p)$ or $B_3(z, 1-p)$. The values of the constants can be determined either by evaluation at $z = 0$ or, more easily, by letting $z \to \infty$ in $\mathbf{T}_3$.

Repetition of this argument then gives

$\mathscr{W}(A_k, A_{k+1})(z, p)$

$$= \begin{cases} -(-p)!(-1)^{-p}\dfrac{1}{2\pi i}B_0(z, 1-p) & (p = 0, -1, -2, \ldots) \\[2mm] -\dfrac{1}{(p-1)!}\dfrac{1}{2\pi i}B_{k+2}(z, 1-p) & (p = 1, 2, \ldots), \end{cases} \qquad \text{(A27)}$$

where the suffix k is enumerated modulo 3. Similarly we obtain

$$\left.\begin{array}{l} \mathscr{W}(A_k, B_0)(z, p) = -\dfrac{1}{(p-1)!}A_k(z, 1-p) \quad (p = 1, 2, \ldots) \\[3mm] \text{and} \\[3mm] \mathscr{W}(A_k, B_l)(z, p) = (-p)!(-1)^{-p}A_k(z, 1-p) \\[2mm] \qquad\qquad\qquad (k \neq l, p = 0, -1, -2, \ldots). \end{array}\right\} \qquad \text{(A28)}$$

A3 The functions $A_k(z, p, q)$ and $B_k(z, p, q)$

To obtain the inner expansions of the solutions of dominant-recessive type of order ε^3 it is necessary to solve the inhomogeneous equation $AD^2u = A_k(z, 1)$. From equation (A6) it follows that a solution of this equation is given formally by $-[\partial A_k(z, p)/\partial p]_{p=2}$. Since differentiation of equation (A4) with respect to p introduces a factor $-\ln t$ into the integrand of that equation, this suggests that we consider a further generalization of the Airy functions defined by

$$A_k(z, p, q) = \frac{1}{2\pi i}\int_{L_k} t^{-p}(\ln t)^q \exp\left(zt - \tfrac{1}{3}t^3\right) dt, \qquad \text{(A29)}$$

where $p = 0, \pm 1, \pm 2, \ldots$, $q = 0, 1, 2, \ldots$, and a branch cut has been placed along the positive real axis in the t-plane so that $0 \leqslant \text{ph } t < 2\pi$. We note first that $A_k(z, p, 0) \equiv A_k(z, p)$ and that the derivatives satisfy the relation

$$D^n A_k(z, p, q) = A_k(z, p-n, q). \qquad \text{(A30)}$$

The functions $A_k(z, p, q)$ are solutions of the inhomogeneous equation

$$(AD + p - 1)A_k(z, p, q) = qA_k(z, p, q - 1), \qquad \text{(A31)}$$

from which we have the recursion formula

$$A_k(z, p-3, q) - zA_k(z, p-1, q) + (p-1)A_k(z, p, q)$$
$$= qA_k(z, p, q-1). \quad (A32)$$

Thus, it is again sufficient to let $p = 0$ and ± 1 (say). On differentiating equation (A31) we obtain

$$AD^2 A_k(z, p+1, q) = -(p-1)A_k(z, p, q) + qA_k(z, p, q-1), \quad (A33)$$

which shows that the inner expansions of the solutions of dominant-recessive type can be obtained to all orders in terms of the functions $A_k(z, p, q)$. For example, the particular integrals of the equation $AD^2 u = A_k(z, 1)$ are simply $A_k(z, 2, 1)$. Corresponding to equations (A7) we now have the rotation formulae

$$A_2(z, p, q) = e^{-2(p-1)\pi i/3} \sum_{n=0}^{q} \binom{q}{n}\left(\frac{2}{3}\pi i\right)^n A_1(z\, e^{2\pi i/3}, p, q-n)$$

and

$$A_3(z, p, q) = e^{2(p-1)\pi i/3} \sum_{n=0}^{q} \binom{q}{n}\left(-\frac{2}{3}\pi i\right)^n A_1(z\, e^{-2\pi i/3}, p, q-n).$$

$$(A34)$$

The values of $A_1(0, p, q)$ can be obtained by repeated differentiation of equation (A11) and the corresponding values of $A_2(0, p, q)$ and $A_3(0, p, q)$ then follow by use of the rotation formulae (A34). Thus, for example, we have

$$A_1(0, p, 1) = \tfrac{1}{3}[\,\psi(\tfrac{2}{3} + \tfrac{1}{3}p) + \ln 3 + 3\pi i]A_1(0, p) \quad (p \in \mathbb{C}). \quad (A35)$$

The asymptotic expansion of $A_1(z, p, q)$ can be obtained either by the method of steepest descents or, more easily, by repeated differentiation of $A_-(z, p)$ with respect to p. In this latter approach, the factor $(-1)^p$ which appears in equation (A12) must first be replaced by $e^{-p\pi i}$, and in this way we find, for example, that

$$A_1(z, p, 1) \sim \tfrac{1}{2}\pi^{-1/2}(-1)^p z^{-(2p+1)/4}\, e^{-\xi}$$

$$\times \sum_{s=0}^{\infty} (-1)^s[(\tfrac{1}{2}\ln z + \pi i)a_s(p) - a_s'(p)]\xi^{-s} \quad (z \in \mathbf{T}_2 \cup \mathbf{T}_3).$$

$$(A36)$$

The corresponding expansions for $A_2(z, p, 1)$ and $A_3(z, p, 1)$ then follow from the rotation formulae (A34).

To deal with the inner expansions of the solutions of balanced type we define the functions

$$B_k(z, p, q) = \frac{1}{2\pi i} \int_{\infty \exp [2(k-1)\pi i/3]}^{(0+)} t^{-p} (\ln t)^q \exp\left(zt - \tfrac{1}{3}t^3\right) dt,$$

(A37)

the asymptotic expansions of which are purely balanced in the sectors $\mathbf{T}_k$. To simplify the connexion formulae which will be given later, it is convenient to consider a fourth function of this type with $k = 4$ in equation (A37); this function is also purely balanced in $\mathbf{T}_1$. We may note that when p is an integer $B_k(z, p, 0) \equiv B_0(z, p)$ for all values of k and that $B_k(z, 0, 1) \equiv B_k(z, 0)$.

These functions also satisfy equations (A30)–(A33) with $A_k(z, p, q)$ replaced by $B_k(z, p, q)$. Thus, for example, the particular integrals of the equation $AD^2u = 1$ are simply $B_k(z, 2, 1)$. The rotation formulae for these functions are

$$B_2(z, p, q) = e^{-2(p-1)\pi i/3} \sum_{n=0}^{q} \binom{q}{n} \left(\frac{2}{3}\pi i\right)^n B_1(z\, e^{2\pi i/3}, p, q, -n),$$

$$B_3(z, p, q) = e^{-4(p-1)\pi i/3} \sum_{n=0}^{q} \binom{q}{n} \left(\frac{4}{3}\pi i\right)^n B_1(z\, e^{4\pi i/3}, p, q-n),$$

and

$$B_4(z, p, q) = \sum_{n=0}^{q} \binom{q}{n} (2\pi i)^n B_1(z, p, q-n).$$

(A38)

The determination of the initial values of these functions is somewhat tricky but the following three values will suffice for subsequent purposes:

$$B_1(0, 1, 1) = \tfrac{1}{3}(\ln 3 - \gamma) + \pi i,$$

$$B_1(0, 0, 1) = 3^{-2/3} \Gamma(\tfrac{1}{3}),$$

and

$$B_1(0, -1, 1) = 3^{-1} \Gamma(\tfrac{2}{3}).$$

(A39)

To obtain the asymptotic expansion of $B_1(z, p, q)$ we first consider $B_1(z, p, 0)$ for unrestricted (complex) values of p. By using a formula given by Erdélyi *et al.* (1953), p. 14 we obtain

$$B_1(z, p, 0) \sim \sum_{s=0}^{\infty} \frac{(-1)^s}{3^s s!} \frac{z^{p-3s-1}}{\Gamma(p-3s)} \quad (p \in \mathbb{C}, z \in \mathbf{T}_1). \quad (A40)$$

If p is an integer then this expression vanishes identically for all $p \leq 0$; otherwise the series terminates, and we recover $B_0(z, p)$. On differentiation of this result with respect to p we find, for example, that

$$B_1(z, 1, 1) \sim -\ln z - \gamma + \sum_{s=1}^{\infty} \frac{(3s-1)!}{3^s s!} z^{-3s} \quad (z \in \mathbf{T}_1).$$
$$(A41)$$

The functions $A_k(z, p, q)$ and $B_k(z, p, q)$ are related by the connexion formulae

$$\sum_{k=1}^{3} A_k(z, p, q) = -B_1(z, p, q), \quad (A42)$$

$$\left.\begin{aligned}
B_2(z, p, q) - B_3(z, p, q) &= \sum_{n=1}^{q} \binom{q}{n}(2\pi i)^n A_1(z, p, q-n), \\
B_3(z, p, q) - B_4(z, p, q) &= \sum_{n=1}^{q} \binom{q}{n}(2\pi i)^n A_2(z, p, q-n), \\
\text{and} \qquad & \\
B_1(z, p, q) - B_2(z, p, q) &= \sum_{n=1}^{q} \binom{q}{n}(2\pi i)^n A_3(z, p, q-n).
\end{aligned}\right\} \quad (A43)$$

Equation (A42) also follows from addition of equations (A43) together with the last of equations (A38).

In discussing the eigenvalue relation for the Orr–Sommerfeld problem in § 41 it was found necessary to evaluate the second-order Wronskians of $A_1(z, p)$ and $B_3(z, p, 1)$ for $p = 0$ and 1. This can be done quite simply by first showing that these Wronskians satisfy inhomogeneous equations of the form given by equation (A31) with $q = 1$, the solutions of which can then be expressed in terms of $A_1(z, p)$ and $A_1(z, p, 1)$. Alternatively, we will describe a method which has the great advantage that it can easily be generalized so as to apply to the Orr–Sommerfeld equation itself. For this purpose let

$$u = A_1(z, 1) \quad \text{and} \quad v = B_3(z, 1, 1). \quad (A44)$$

We also let $u = u_1$, $\mathrm{D}u = u_2$, $\mathrm{D}^2 u = u_3$ and similarly for v. Then u_i and v_i ($i = 1, 2, 3$) satisfy the equations

$$\mathrm{D}u_i = A_{ij}u_j \quad \text{and} \quad \mathrm{D}v_i = A_{ij}v_j + \delta_{i3}, \qquad (A45)$$

where

$$\mathbf{A} = \begin{bmatrix} 0 & 1 & 0 \\ 0 & 0 & 1 \\ 0 & z & 0 \end{bmatrix}. \qquad (A46)$$

If we now let $w_i^1 = u_i$ and $w_i^2 = v_i$ then equations (A45) become

$$\mathrm{D}w_i^r = A_{ij}w_j^r + \delta_{i3}\delta_{r2} \quad (r = 1, 2). \qquad (A47)$$

The 2×2 minors of the matrix w_i^r ($i = 1, 2, 3$; $r = 1, 2$) are given by

$$W_{ij} = \varepsilon_{rs}w_i^r w_j^s, \qquad (A48)$$

where

$$\varepsilon_{rs} = \begin{cases} 1 & \text{if } r = 1 \quad \text{and} \quad s = 2 \\ -1 & \text{if } r = 2 \quad \text{and} \quad s = 1 \\ 0 & \text{otherwise,} \end{cases} \qquad (A49)$$

and a simple calculation shows that they satisfy the equation

$$\mathrm{D}W_{ik} = A_{ij}W_{jk} - A_{kj}W_{ji} + w_i^1\delta_{k3} - w_k^1\delta_{i3}. \qquad (A50)$$

The three independent components of this equation are

$$\left.\begin{aligned} \mathrm{D}W_{12} &= W_{13}, \\ \mathrm{D}W_{13} &= zW_{12} + W_{23} + w_1^1, \\ \mathrm{D}W_{23} &= w_2^1. \end{aligned}\right\} \qquad (A51)$$

and

On eliminating W_{13} and W_{23} we have

$$(\mathrm{AD} - 1)W_{12} = 2A_1(z). \qquad (A52)$$

The general solution of this equation then follows from equations (A5) and (A31) in the form

$$W_{12} = 2A_1(z, 0, 1) + CA_1(z). \qquad (A53)$$

The arbitrary constant in this equation can easily be evaluated by setting $z = 0$ and using the initial values given by equations (A11) and (A35). In this way we obtain

$$W_{12} \equiv \mathscr{W}\{A_1(z, 1), B_3(z, 1, 1)\}$$
$$= 2A_1(z, 0, 1) + (\gamma - 4\pi i)A_1(z). \qquad (A54)$$

Similarly, from the first and third of equations (A51) we have

$$W_{13} \equiv \frac{\mathrm{d}}{\mathrm{d}z} \mathscr{W}\{A_1(z, 1), B_3(z, 1, 1)\}$$
$$= 2A_1(z, -1, 1) + (\gamma - 4\pi i)A_1(z, -1) \qquad (A55)$$

and

$$W_{23} \equiv \mathscr{W}\{A_1(z), B_3(z, 0, 1)\} = A_1(z, 1). \qquad (A56)$$

A4 The zeros of $A_1(z, p)$

In many stability problems for parallel shear flows it is found that the limiting behaviour of the damped modes as $\alpha R \to \infty$ can be expressed in terms of the zeros of $A_1(z, p)$ and it is of some interest, therefore, to consider briefly some of the properties of these zeros. Throughout this discussion we will suppose that p is real but not necessarily an integer.

Wasow (1953a) has proved that $A_1(z, 1)$ has no real zeros and his proof can easily be extended to show that $A_1(z, p)$ has no real zeros when $p > \frac{1}{2}$. It has not been proved, however, that the zeros are all real when $p \leq \frac{1}{2}$ but the existing evidence, both analytical and numerical, strongly suggests that they are, and we shall proceed on that assumption.

When $p \leq \frac{1}{2}$ it is convenient to denote the zeros of $A_1(z, p)$ by $a_s(p)$ ($s = 1, 2, \ldots$). Thus, if a_s and a'_s denote, as usual, the zeros of $\mathrm{Ai}(z)$ and $\mathrm{Ai}'(z)$, respectively, then we have

$$a_s(0) = a_s \quad \text{and} \quad a_s(-1) = a'_s. \qquad (A57)$$

Furthermore, for half-integral values of p it is not difficult to show that $A_1(z, p)$ can be expressed in terms of products of $\mathrm{Ai}(x)$ and $\mathrm{Ai}'(x)$, where $x = 2^{-2/3}z$. Thus, for example, we have

and
$$\left. \begin{array}{l} A_1(z, \tfrac{1}{2}) = -\mathrm{i}2^{2/3}\pi^{1/2}\{\mathrm{Ai}(x)\}^2 \\ A_1(z, -\tfrac{1}{2}) = -\mathrm{i}2\pi^{1/2}\mathrm{Ai}(x)\mathrm{Ai}'(x). \end{array} \right\} \qquad (A58)$$

Accordingly, for $p = \pm \frac{1}{2}$ we have

$$a_{2s-1}(\tfrac{1}{2}) = a_{2s}(\tfrac{1}{2}) = 2^{2/3} a_s$$

and

$$a_{2s-1}(-\tfrac{1}{2}) = 2^{2/3} a_s', \quad a_{2s}(-\tfrac{1}{2}) = 2^{2/3} a_s. \tag{A59}$$

The values of the first two zeros of $A_1(z, p)$ have been tabulated by Davey & Reid (1977a) for $p = -0.5(0.05)1.0$.

When $p > \frac{1}{2}$ the zeros of $A_1(z, p)$ occur in complex conjugate pairs, and it is then more convenient to denote them by $\alpha_s(p)$ and $\alpha_s^*(p)(s = 1, 2, \ldots)$ where, by convention, $\mathrm{Im}\{-\alpha_s(p)\} > 0$. Davey & Reid (1977a) have also tabulated $\alpha_1(p)$ for $p = 0.5(0.05)1.0$.

It is also useful to have asymptotic approximations to these zeros. From the results given above in §§ A2 and A3 it can be shown that

$$A_1(-z, p) \sim \frac{1}{\Gamma(p)} e^{-p\pi i} z^{p-1}$$
$$+ \pi^{-1/2} e^{-p\pi i} z^{-(2p+1)/4} \sin \left(\xi - \tfrac{1}{2} p \pi + \tfrac{1}{4}\pi\right) \tag{A60}$$

and this approximation is valid in the complete sense as $z \to \infty$ in the sector $|\mathrm{ph}\, z| < \frac{1}{3}\pi$. When $p < \frac{1}{2}$, the algebraic term in equation (A60) can be neglected, and we then obtain

$$\{-a_s(p)\}^{2/3} \sim \tfrac{3}{8}\pi(4s + 2p - 1) \quad (p < \tfrac{1}{2}). \tag{A61}$$

This approximation is not uniformly valid, however, as $p \uparrow \frac{1}{2}$, the lack of uniformity being due to the fact that the zeros of $A_1(z, \frac{1}{2})$ are of multiplicity two. When $p \geq \frac{1}{2}$, the algebraic term in equation (A60) cannot be neglected, and it is then not difficult to generalize the result given by Zondek & Thomas (1953) for $p = 1$ to obtain

$$\{-\alpha_s(p)\}^{3/2} \sim \tfrac{3}{8}\pi(8s + 2p - 3)$$
$$+ i\tfrac{3}{2} \cosh^{-1}\{\frac{\sqrt{\pi}}{\Gamma(p)}[\tfrac{3}{8}\pi(8s + 2p - 3)]^{p-1/2}\} \quad (p \geq \tfrac{1}{2}). \tag{A62}$$

This approximation does remain uniformly valid as $p \downarrow \frac{1}{2}$.

WEAKLY NON-PARALLEL THEORIES FOR THE BLASIUS BOUNDARY LAYER

The discussion on pp. 240–2 of the weakly non-parallel theories for the Blasius boundary layer requires some revision and in this Addendum, therefore, we will attempt to clarify the current status of the non-parallel theories and their relationship to the experimental data. The theories due to Bouthier (1972, 1973) and Saric & Nayfeh (1975) are based on the method of multiple scales whereas the theory due to Gaster (1974) is based on an iterative method. These theories seek to include all non-parallel effects correctly to order $R_x^{-1/2}$, where $R_x = U_* x_*/\nu$. An interesting feature of the non-parallel theories is that the stability characteristics of the flow cease to be independent of z and exhibit different rates of amplification depending not only upon which flow quantity is observed but also on the particular path in the (R_x, z)-plane along which the observations are made. In an attempt to remove the ambiguity associated with the dependence on the path of observation, Saric & Nayfeh defined the condition for marginal stability by suppressing the effect due to the streamwise change in the shape of the disturbance velocity profile. This led to the curve labelled 'Non-parallel' in Fig. 4.30. The experimental results pre-date the present level of understanding of the dependence of the results on the path of observation and of the possible subcritical behaviour associated with the finite disturbance caused by the vibrating ribbon which leads to a distortion of the basic flow. Furthermore, the effect of the mean wake of the ribbon on the delicate determination of the local marginal location has not yet been established experimentally. Without some reasonably convincing justification for the neglect of the profile distortion effect, the agreement with the experimental data shown in Fig. 4.30 must therefore be regarded as largely fortuitous. When all of the non-parallel terms are included in a consistent fashion, the theory of Saric & Nayfeh leads[†] to results which are close to those obtained

† W. S. Saric, private communication, 1982.

by Bouthier and Gaster. Although these theories differ in some details, it now appears that they all lead to similar results and in particular they predict curves of marginal stability which are close to the parallel flow results as shown in Gaster's Figs. 1 and 2.

The preceding discussion suggests that extreme care is required to obtain meaningful comparisons between theoretical and experimental results. On the basis of the existing evidence, however, it is perhaps fair to conclude that non-parallel effects are much smaller for the Blasius boundary layer than they are for the asymptotic suction boundary layer and that they are not sufficient, as shown in Gaster's Figs. 3 and 4, to obtain a marginal curve in agreement with the experimental data where the frequency is high and the Reynolds number low. It should be emphasized, however, that the experiments are very difficult to perform when the frequency is high and the Reynolds number is low. Perhaps other factors will have to be considered to improve the agreement between the theoretical and experimental results but those are matters for the future.

We are grateful for the generous help we have received from M. Gaster, L. M. Mack, M. V. Morkovin and W. S. Saric in this attempt to clarify some of the main issues involved in this difficult but important aspect of the subject.

March 1982

BIBLIOGRAPHY AND AUTHOR INDEX

Numbers in square brackets refer to the pages on which the reference is cited.

Abramowitz, M. & Stegun, I. A. (Editors) (1964). *Handbook of mathematical functions.* Appl. Math. Ser. No. 55. Washington, D.C.: U.S. Govt. Printing Office. Also New York: Dover (1965). [pp. 356, 357, 470.]

Ahlers, G. (1974). Low-temperature studies of the Rayleigh–Bénard instability and turbulence. *Phys. Rev. Lett.* **33**, 1185–8. [pp. 415, 440, 441, 442.]

Ahlers, G. & Behringer, R. P. (1978). Evolution of turbulence from the Rayleigh–Bénard instability. *Phys. Rev. Lett.* **40**, 712–16. [p. 442.]

al-Amir, Z. (1968). Hydrodynamic stability of parallel and non-parallel jets. Ph.D. dissertation. Leeds University. [p. 237.]

Alterman, Z. (1961). Kelvin–Helmholtz instability in media of variable density. *Phys. Fluids* **4**, 1177–9. [p. 364.]

Antar, B. N. & Benek, J. A. (1978). Temporal eigenvalue spectrum of the Orr–Sommerfeld equation for the Blasius boundary layer. *Phys. Fluids* **21**, 183–9. [p. 156.]

Antar, B. N. & Collins, F. G. (1975). Numerical calculation of finite amplitude effects in unstable laminar boundary layers. *Phys. Fluids* **18**, 289–97. [p. 455.]

Arnol'd, V. I. (1965a). Conditions for nonlinear stability of stationary plane curvilinear flows of an ideal fluid. *Dokl. Akad. Nauk SSSR* **162**, 975–8. Translated in *Soviet Math. Dokl.* **6**, 773–7 (1965). [pp. 432, 434.]

Arnol'd, V. I. (1965b). Variational principle for three-dimensional steady-state flows of an ideal fluid. *Prikl. Mat. Mekh.* **29**, 846–51. Translated in *J. Appl. Math. Mech.* **29**, 1002–8 (1965). [p. 434.]

Arnol'd, V. I. (1966). Sur un principe variationnel pour les écoulements stationnaires des liquides parfaits et ses applications aux problèmes de stabilité non linéaires. *J. Mécanique* **5**, 29–43. [p. 432.]

Arnol'd, V. I. (1972). Lectures on bifurcation and versal families. *Uspekhi Mat. Nauk* **27**, No. 5, 119–84. Translated in *Russian Math. Surveys* **27**, No. 5, 54–123. [p. 403.]

Bahl, S. K. (1970). Stability of viscous flow between two concentric rotating porous cylinders. *Defense Sci. J.* **20**, 89–96. [p. 123.]

Baines, P. G. & Gill, A. E. (1969). On thermohaline convection with linear gradients. *J. Fluid Mech.* **37**, 289–306. [p. 65.]

Banks, W. H. H. & Drazin, P. G. (1973). Perturbation methods in boundary-layer theory. *J. Fluid Mech.* **58**, 763–75. [p. 331.]

Banks, W. H. H., Drazin, P. G. & Zaturska, M. B. (1976). On the normal modes of parallel flow of inviscid stratified fluid. *J. Fluid Mech.* **75**, 149–71. [p. 331.]

Barcilon, V. (1964). Role of the Ekman layers in the stability of the symmetric regime obtained in a rotating annulus. *J. Atmos. Sci.* **21**, 291–9. [p. 365.]

Barry, M. D. J. & Ross, M. A. S. (1970). The flat plate boundary layer. Part 2. The effect of increasing thickness on stability. *J. Fluid Mech.* **43**, 813–18. [p. 240.]

Batchelor, G. K. (1967). *An introduction to fluid dynamics.* Cambridge University Press. [pp. 14, 15, 34.]

Bénard, H. (1900). Les tourbillons cellulaires dans une nappe liquide. *Revue Gén. Sci. Pur. Appl.* **11**, 1261–71 and 1309–28. [p. 32.]

Benjamin, T. B. (1961). The development of three-dimensional disturbances in an unstable film of liquid flowing down an inclined plane. *J. Fluid Mech.* **10**, 401–19. [pp. 346, 348.]

Benjamin, T. B. (1967). Instability of periodic wave trains in nonlinear dispersive systems. *Proc. Roy. Soc.* A **299**, 59–75. [p. 398.]

Benjamin, T. B. (1972). The stability of solitary waves. *Proc. Roy. Soc.* A **328**, 153–83. [p. 363.]

Benjamin, T. B. (1978). Bifurcation phenomena in steady flows of a viscous fluid. *Proc. Roy. Soc.* A **359**, 1–26 and 27–43. [pp. 415, 449.]

Benjamin, T. B. & Feir, J. E. (1967). The disintegration of wave trains on deep water. Part 1. Theory. *J. Fluid Mech.* **27**, 417–30. [pp. 360, 387, 398.]

Benjamin, T. B. & Ursell, F. (1954). The stability of the plane free surface of a liquid in vertical periodic motion. *Proc. Roy. Soc.* A **225**, 505–15. [pp. 354, 355, 358, 359.]

Benney, D. J. (1962). Non-linear gravity wave interactions. *J. Fluid Mech.* **14**, 577–84. [p. 393.]

Benney, D. J. (1964). Finite amplitude effects in an unstable laminar boundary layer. *Phys. Fluids* **7**, 319–26. [p. 455.]

Benney, D. J. & Bergeron, R. F., Jr. (1969). A new class of nonlinear waves in parallel flows. *Stud. Appl. Math.* **48**, 181–204. [p. 421.]

Benney, D. J. & Lin, C. C. (1960). On the secondary motion induced by oscillations in a shear flow. *Phys. Fluids* **3**, 656–7. [pp. 454, 457.]

Benney, D. J. & Maslowe, S. A. (1975). The evolution in space and time of nonlinear waves in parallel shear flows. *Stud. Appl. Math.* **54**, 181–205. [p. 420.]

Bernstein, I. B., Frieman, E. A., Kruskal, M. D. & Kulsrud, R. M. (1958). An energy principle for hydromagnetic stability problems. *Proc. Roy. Soc.* A **244**, 17–40. [p. 368.]

Betchov, R. & Criminale, W. O., Jr. (1967). *Stability of parallel flows.* New York: Academic Press. [p. 203.]

Betchov, R. & Szewczyk, A. (1963). Stability of a shear layer between parallel streams. *Phys. Fluids* **6**, 1391–6. [pp. 238, 239.]

Beth, H. J. E. (1913). The oscillations about a position of equilibrium where a simple linear relation exists between the frequencies of the principal vibrations. *Phil. Mag.* (6) **26**, 268–324. [p. 391.]

Bickley, W. G. (1937). The plane jet. *Phil. Mag.* (7) **23**, 727–31. [p. 196.]

Bippes, H. & Görtler, H. (1972). Dreidimensionale Störungen in der Grenzschicht an einer konkaven Wand. *Acta Mech.* **14**, 251–67. [p. 118.]

Birikh, R. V. & Gershuni, G. Z. & Zhukhovitskii, E. M. (1965). On the spectrum of perturbations of plane parallel flows at low Reynolds numbers. *Prikl. Mat. Mekh.* **29**, 88–98. Translated in *J. Appl. Math. Mech.* **29**, 93–104 (1966). [p. 159.]

Birkhoff, G. & Rota, G.-C. (1978). *Ordinary differential equations,* 3rd edn, New York: John Wiley & Sons. [p. 399.]

Bisshopp, F. E. (1963). Asymmetric inviscid modes of instability in Couette flow. *Phys. Fluids* **6**, 212–17. [p. 88.]

Bjerknes, V., Bjerknes, J., Bergeron, T. & Solberg, H. (1933). *Physikalische hydrodynamik.* Berlin: Springer. Translated into French as *Hydrodynamique physique* (1934). Paris: Presses Universitaires de France. [p. 364.]

Bleistein, N. & Handelsman, R. A. (1975). *Asymptotic expansions of integrals.* New York: Holt, Rinehart and Winston. [p. 262.]

Block, M. J. (1956). Surface tension as the cause of Bénard cells and surface deformation in a liquid film. *Nature* **178**, 650–1. [p. 34.]

Blumen, W. (1968). On the stability of quasi-geostrophic flow. *J. Atmos. Sci.* **25**, 929–31. [p. 434.]

Bogy, D. B. (1979). Drop formation in a circular liquid jet. *Ann. Rev. Fluid Mech.* **11**, 207–28. [p. 27.]

Bohr, N. (1909). Determination of the surface-tension of water by the method of jet vibration. *Phil. Trans. Roy. Soc.* A **209**, 281–317. Also *Collected works* (1972), vol. 1, pp. 29–65. Amsterdam: North-Holland. [pp. 25, 370.]

Boussinesq, J. (1903). *Théorie analytique de la chaleur,* vol. 2, p. 172. Paris: Gauthier–Villars. [p. 35.]

Bouthier, M. (1972). Stabilité linéaire des écoulements presque parallèles. *J. Mécanique* **11**, 599–621. [pp. 242, 363, 479.]

Bouthier, M. (1973). Stabilité linéaire des écoulements presque parallèles. Partie II. La couche limite de Blasius. *J. Mécanique* **12**, 75–95. [pp. 242, 479.]

Bretherton, F. P. (1964). Resonant interactions between waves. The case of discrete oscillations. *J. Fluid Mech.* **20**, 457–79. [pp. 393, 394, 396.]

Brewster, D. B., Grosberg, P. & Nissan, A. H. (1959). The stability of viscous flow between horizontal concentric cylinders. *Proc. Roy. Soc.* A **251**, 76–91. [pp. 110, 113.]

Brewster, D. B. & Nissan, A. H. (1958). Hydrodynamics of flow between horizontal concentric cylinders. I. Flow due to rotation of cylinder. *Chem. Eng. Sci.* **7**, 215–21. [p. 113.]

Browand, F. K. (1966). An experimental investigation of the instability of an incompressible, separated shear layer. *J. Fluid Mech.* **26**, 281–307. [p. 456.]

Brown, S. N. & Stewartson, K. (1978a). On finite amplitude Bénard convection in a cylindrical container. *Proc. Roy. Soc.* A **360**, 455–69. [p. 439.]

Brown, S. N. & Stewartson, K. (1978b). The evolution of a small inviscid disturbance to a marginally unstable stratified shear flow; stage two. *Proc. Roy. Soc.* A **363**, 175–94. [p. 423.]

Browning, K. A. (1971). Structure of the atmosphere in the vicinity of large-amplitude Kelvin–Helmholtz billows. *Quart. J. Roy. Meteor. Soc.* **97**, 283–99. [p. 423.]

Burgers, J. M. (1948). A mathematical model illustrating the theory of turbulence. In *Advances in applied mechanics*, vol. 1, eds. R. von Mises and T. von Kármán, pp. 171–99. New York: Academic Press. [p. 459.]

Burkhalter, J. E. & Koschmieder, E. L. (1973). Steady supercritical Taylor vortex flow. *J. Fluid Mech.* **58**, 547–60. [pp. 108, 445, 449.]

Busse, F. H. (1967a). On the stability of two-dimensional convection in a layer heated from below. *J. Math. and Phys.* **46**, 140–50. [p. 437.]

Busse, F. H. (1967b). The stability of finite amplitude cellular convection and its relation to an extremum principle. *J. Fluid Mech.* **30**, 625–49. [p. 439.]

Busse, F. H. (1968). Shear flow instabilities in rotating systems. *J. Fluid Mech.* **33**, 577–89. [p. 122.]

Busse, F. H. (1969). Bounds on the transport of mass and momentum by turbulent flow between parallel plates. *Z. Angew. Math. Phys.* **20**, 1–14. [p. 430.]

Busse, F. H. (1972). The oscillatory instability of convection rolls in a low Prandtl number fluid. *J. Fluid Mech.* **52**, 97–112. [p. 438.]

Busse, F. H. & Clever, R. M. (1979). Instabilities of convection rolls in a fluid of moderate Prandtl number. *J. Fluid Mech.* **91**, 319–35. [p. 438.]

Busse, F. H. & Whitehead, J. A. (1971). Instabilities of convection rolls in a high Prandtl number fluid. *J. Fluid Mech.* **47**, 305–20. [p. 438.]

Case, K. M. (1960a). Stability of inviscid plane Couette flow. *Phys. Fluids* **3**, 143–8. [pp. 80, 149, 150, 151.]

Case, K. M. (1960b). Stability of an idealized atmosphere. I. Discussion of results. *Phys. Fluids* **3**, 149–54. [pp. 149, 329.]

Chandrasekhar, S. (1954a). On characteristic value problems in high order differential equations which arise in studies of hydrodynamic and hydromagnetic stability. *Amer. Math. Monthly* **61**, 32–45. [p. 95.]

Chandrasekhar, S. (1954b). The stability of viscous flow between rotating cylinders. *Mathematika* **1**, 5–13. [p. 99.]

Chandrasekhar, S. (1961). *Hydrodynamic and hydromagnetic stability*. Oxford: Clarendon Press. [pp. 25, 28, 49, 62, 64, 66, 67, 79, 93, 98, 99, 103, 110, 121, 205, 327, 340.]

Chandrasekhar, S. (1969). *Ellipsoidal figures of equilibrium*. New Haven: Yale University Press. [pp. 75, 403.]

Chang, T. S. & Sartory, W. K. (1967). Hydromagnetic stability of dissipative flow between rotating permeable cylinders. Part 1. Stationary critical modes. *J. Fluid Mech.* **27**, 65–79. [p. 123.]

Charney, J. G. (1947). The dynamics of long waves in a baroclinic westerly current. *J. Meteor.* **4**, 135–62. [p. 335.]

Charney, J. G. & Stern, M. E. (1962). On the stability of internal baroclinic jets in a rotating atmosphere. *J. Atmos. Sci.* **19**, 159–72. [p. 366.]

Chen, T. S. & Joseph, D. D. (1973). Subcritical bifurcation of plane Poiseuille flow. *J. Fluid Mech.* **58**, 337–51. [pp. 244, 451.]

Chen, T. S., Sparrow, E. M. & Tsou, F. K. (1971). The effect of mainflow transverse velocities in linear stability theory. *J. Fluid Mech.* **50**, 741–50. [p. 240.]

Chester, C., Friedman, B. & Ursell, F. (1957). An extension of the method of steepest descents. *Proc. Camb. Phil. Soc.* **53**, 599–611. [pp. 256, 262]

Christopherson, D. G. (1940). Note on the vibration of membranes. *Quart. J. Math.* **11**, 63–5. [p. 56.]

Clemmow, P. C. & Dougherty, J. P. (1969). *Electrodynamics of particles and plasmas*. Reading, Mass.: Addison–Wesley. [pp. 152, 153, 351.]

Clenshaw, C. W. & Elliott, D. (1960). A numerical treatment of the Orr–Sommerfeld equation in the case of a laminar jet. *Quart. J. Mech. Appl. Math.* **13**, 300–13. [pp. 205, 234, 235.]

Clever, R. M. & Busse, F. H. (1974). Transition to time-dependent convection. *J. Fluid Mech.* **65**, 625–45. [p. 438.]

Coddington, E. A. & Levinson, N. (1955). *Theory of ordinary differential equations*. New York: McGraw–Hill. [pp. 359, 399.]

Cole, J. A. (1976). Taylor-vortex instability and annulus-length effects. *J. Fluid Mech.* **75**, 1–15. [p. 449.]

Coles, D. (1965). Transition in circular Couette flow. *J. Fluid Mech.* **21**, 385–425. [pp. 107, 108, 445, 446, 448, 449.]

Colson, D. (1954). Wave-cloud formation at Denver. *Weatherwise* **7**, 34–5. [p. 22.]

Conrad, P. W. & Criminale, W. O. (1965a). The stability of time-dependent laminar flow: parallel flows. *Z. Angew. Math. Phys.* **16**, 233–54. [p. 363.]

Conrad, P. W. & Criminale, W. O. (1965b). The stability of time-dependent laminar flow: flow with curved streamlines. *Z. Angew. Math. Phys.* **16**, 569–82. [p. 363.]

Conte, S. D. (1966). The numerical solution of linear boundary value problems. *SIAM Rev.* **8**, 309–21. [p. 208.]

Conte, S. D. & Miles, J. W. (1959). On the numerical integration of the Orr–Sommerfeld equation. *J. Soc. Ind. Appl. Math.* **7**, 361–6. [p. 138.]

Corcos, G. M. & Sellars, J. R. (1959). On the stability of fully developed flow in a pipe. *J. Fluid Mech.* **5**, 97–112. [p. 219.]

Couette, M. (1890). Études sur le frottement des liquides. *Ann. Chim. Phys.* (6) **21**, 433–510. [p. 70.]

Craik, A. D. D. (1981). The development of wavepackets in unstable flows. *Proc. Roy. Soc.* A **373**, 457–76. [p. 349.]

Criminale, W. O., Jr. & Kovasznay, L. S. G. (1962). The growth of localized disturbances in a laminar boundary layer. *J. Fluid Mech.* **14**, 59–80. [p. 348.]

Curle, N. (1957). On hydrodynamic stability in unlimited fields of viscous flow. *Proc. Roy. Soc.* A **238**, 489–501. [p. 234.]

Daniels, P. G. (1977). The effect of distant sidewalls on the transition to finite amplitude Bénard convection. *Proc. Roy. Soc.* A **358**, 173–97. [p. 414.]

D'Arcy, D. F. (1951). Studies in vortical flow. M.S. thesis. Massachusetts Institute of Technology. [p. 123.]

Davey, A. (1962). The growth of Taylor vortices in flow between rotating cylinders. *J. Fluid Mech.* **14**, 336–68. [pp. 380, 416, 443.]

Davey, A. (1973a). On the stability of plane Couette flow to infinitesimal disturbances. *J. Fluid Mech.* **57**, 369–80. [p. 212.]

Davey, A. (1973b). A simple numerical method for solving Orr–Sommerfeld problems. *Quart. J. Mech. Appl. Math.* **26**, 401–11. [pp. 208, 209.]

Davey, A. (1977). On the numerical solution of difficult eigenvalue problems. *J. Comput. Phys.* **24**, 331–8. [p. 209, 210, 211.]

Davey, A. (1979). On the removal of the singularities from the Riccati method. *J. Comput. Phys.* **30**, 137–44. [p. 319.]

Davey, A., DiPrima, R. C. & Stuart, J. T. (1968). On the instability of Taylor vortices. *J. Fluid Mech.* **31**, 17–52. [pp. 405, 445.]

Davey, A. & Drazin, P. G. (1969). The stability of Poiseuille flow in a pipe. *J. Fluid Mech.* **36**, 209–18. [pp. 219, 220.]

Davey, A., Hocking, L. M. & Stewartson, K. (1974). On the nonlinear evolution of three-dimensional disturbances in plane Poiseuille flow. *J. Fluid Mech.* **63**, 529–36. [pp. 420, 452.]

Davey, A. & Nguyen, H. P. F. (1971). Finite-amplitude stability of pipe flow. *J. Fluid Mech.* **45**, 701–20. [p. 454.]

Davey, A. & Reid, W. H. (1977a). On the stability of stratified viscous plane Couette flow. Part 1. Constant buoyancy frequency. *J. Fluid Mech.* **80**, 509–25. [pp. 216, 478.]

Davey, A. & Reid, W. H. (1977b). On the stability of stratified viscous plane Couette flow. Part 2. Variable buoyancy frequency. *J. Fluid Mech.* **80**, 527–34. [p. 333.]

Davies, S. J. & White, C. M. (1928). An experimental study of the flow of water in pipes of rectangular section. *Proc. Roy. Soc.* A **119**, 92–107. [p. 452.]

Davis, R. E. (1969). On the high Reynolds number flow over a wavy boundary. *J. Fluid Mech.* **36**, 337–46. [p. 421.]

Davis, S. H. (1967). Convection in a box: linear theory. *J. Fluid Mech.* **30**, 465–78. [pp. 59, 61.]

Davis, S. H. (1968). Convection in a box: on the dependence of preferred wave-number upon the Rayleigh number at finite amplitude. *J. Fluid Mech.* **32**, 619–24. [p. 439.]

Davis, S. H. (1969). On the principle of exchange of stabilities. *Proc. Roy. Soc.* A **310**, 341–58. [p. 94.]

Davis, S. H. (1976). The stability of time-periodic flows. *Ann. Rev. Fluid Mech.* **8**, 57–74. [p. 360.]

Davis, S. H. & Segel, L. A. (1968). Effects of curvature and property variation on cellular convection. *Phys. Fluids* **11**, 470–6. [p. 41.]

Dean, W. R. (1928). Fluid motion in a curved channel. *Proc. Roy. Soc.* A **121**, 402–20. [p. 109.]

Deardorff, J. W. (1963). On the stability of viscous plane Couette flow. *J. Fluid Mech.* **15**, 623–31. [p. 212.]

De Villiers, J. M. (1975). Asymptotic solutions of the Orr–Sommerfeld equation. *Phil. Trans. Roy. Soc.* A **280**, 271–316. [pp. 253, 279.]

Dikii, L. A. (1960a). On the stability of plane parallel flows of an inhomogeneous fluid. *Prikl. Mat. Mekh.* **24**, 249–59. Translated in *J. Appl. Math. Mech.* **24**, 357–69 (1960). [p. 149.]

Dikii, L. A. (1960b). The stability of plane-parallel flows of an ideal fluid. *Dokl. Akad. Nauk SSSR* **135**, 1068–71. Translated in *Sov. Phys. Doklady* **5**, 1179–82 (1960). [p. 80.]

Dikii, L. A. (1964). On the stability of plane-parallel Couette flow. *Prikl. Mat. Mekh.* **28**, 389–92. Translated in *J. Appl. Math. Mech.* **28**, 479–83 (1964). [pp. 212, 246.]

Dikii, L. A. (1965a). On the nonlinear theory of hydrodynamic stability. *Prikl. Mat. Mekh.* **29**, 852–5. Translated in *J. Appl. Math. Mech.* **29**, 1009–12 (1965). [p. 434.]

Dikii, L. A. (1965b). On the nonlinear theory of the stability of zonal flows. *Fiz. Atmosfery i Okeana* **1**, 1117–22. Translated in *Izv. Atmos. Oceanic Phys.* **1**, 653–5 (1965). [p. 434.]

Dikii, L. A. & Kurganskii, M. N. (1971). Integral conservation law for perturbations of zonal flow, and its application to stability studies. *Fiz. Atmosfery i Okeana* **7**, 939–45. Translated in *Izv. Atmos. Oceanic Phys.* **7**, 623–6 (1971). [p. 434.]

DiPrima, R. C. (1954). A note on the asymptotic solutions of the equation of hydrodynamic stability. *J. Math. Phys.* **33**, 249–57. [p. 177.]

DiPrima, R. C. (1955). Application of the Galerkin method to problems in hydrodynamic stability. *Quart. Appl. Math.* **13**, 55–62. [p. 99.]

DiPrima, R. C. (1959). The stability of viscous flow between rotating concentric cylinders with a pressure gradient acting round the cylinders. *J. Fluid Mech.* **6**, 462–8. [p. 113.]

DiPrima, R. C. (1961). Stability of nonrotationally symmetric disturbances for viscous flow between rotating cylinders. *Phys. Fluids* **4**, 751–5. [p. 104.]

DiPrima, R. C. & Eagles, P. M. (1977). Amplification rates and torques for Taylor-vortex flows between rotating cylinders. *Phys. Fluids* **20**, 171–5. [p. 444.]

DiPrima, R. C., Eckhaus, W. & Segel, L. A. (1971). Non-linear wave-number interaction in near-critical two-dimensional flows. *J. Fluid Mech.* **49**, 705–44. [pp. 420, 451.]

DiPrima, R. C. & Habetler, G. J. (1969). A completeness theorem for non-selfadjoint eigenvalue problems in hydrodynamic stability. *Arch. Rat. Mech. Anal.* **34**, 218–27. [p. 158.]

DiPrima, R. C. & Sani, R. (1965). The convergence of the Galerkin method for the Taylor–Dean stability problem. *Quart. Appl. Math.* **23**, 183–7. [p. 96.]

Dolph, C. L. & Lewis, D. C. (1958). On the application of infinite systems of ordinary differential equations to perturbations of plane Poiseuille flow. *Quart. Appl. Math.* **16**, 97–110. [p. 204.]

Donnelly, R. J. (1958). Experiments on the stability of viscous flow between rotating cylinders. I. Torque measurements. *Proc. Roy. Soc.* A **246**, 312–25. [p. 107.]

Donnelly, R. J. (1964). Experiments on the stability of viscous flow between rotating cylinders. III. Enhancement of stability by modulation. *Proc. Roy. Soc.* A **281**, 130–9. [p. 360.]

Donnelly, R. J. & Fultz, D. (1960). Experiments on the stability of spiral flow between rotating cylinders. *Proc. Nat. Acad. Sci., Wash.* **46**, 1150–4. [p. 107.]

Donnelly, R. J. & Glaberson, W. (1966). Experiments on the capillary instability of a liquid jet. *Proc. Roy. Soc.* A **290**, 547–56. [pp. 25, 27.]

Donnelly, R. J. & Simon, N. J. (1960). An empirical torque relation for supercritical flow between rotating cylinders. *J. Fluid Mech.* **7**, 401–18. [pp. 107, 108, 435, 444, 446.]

Drazin, P. G. (1961). Discontinuous velocity profiles for the Orr–Sommerfeld equation. *J. Fluid Mech.* **10**, 571–83. [pp. 198, 201, 202, 235.]

Drazin, P. G. (1977a). On the stability of cnoidal waves. *Quart. J. Mech. Appl. Math.* **30**, 91–105. [p. 351.]

Drazin, P. G. (1977b). On the instability of an internal gravity wave. *Proc. Roy. Soc.* A **356**, 411–32. [p. 360.]

Drazin, P. G. (1978). Variations on a theme of Eady. In *Rotating fluids in geophysics*, eds. P. H. Roberts and A. M. Soward, pp. 139–69. London: Academic Press. [p. 335.]

Drazin, P. G. & Howard, L. N. (1962). The instability to long waves of unbounded parallel inviscid flow. *J. Fluid Mech.* **14**, 257–83. [pp. 142, 250.]

Drazin, P. G. & Howard, L. N. (1966). Hydrodynamic stability of parallel flow of inviscid fluid. In *Advances in applied mechanics*, vol. 7, ed. G. Kuerti, pp. 1–89. New York: Academic Press. [pp. 125, 133, 134, 233, 234, 333, 364, 365, 434.]

Duty, R. L. & Reid, W. H. (1964). On the stability of viscous flow between rotating cylinders. Part 1. Asymptotic analysis. *J. Fluid Mech.* **20**, 81–94. [pp. 99, 101, 102.]

Eady, E. A. (1949). Long waves and cyclone waves. *Tellus* **1**, 33–52. [p. 335.]

Eagles, P. M. (1966). The stability of a family of Jeffery–Hamel solutions for divergent channel flow. *J. Fluid Mech.* **24**, 191–207. [p. 164.]

Eagles, P. M. (1969). Composite series in the Orr–Sommerfeld problem for symmetric channel flow. *Quart. J. Mech. Appl. Math.* **22**, 129–82. [pp. 248, 250, 253, 268, 279.]

Eagles, P. M. (1971). On stability of Taylor vortices by fifth-order amplitude expansions. *J. Fluid Mech.* **49**, 529–50. [p. 445.]

Eagles, P. M. (1974). On the torque of wavy vortices. *J. Fluid Mech.* **62**, 1–9. [p. 445.]

Eckhaus, W. (1965). *Studies in non-linear stability theory.* Springer Tracts in Natural Philosophy, vol. 6. Berlin: Springer-Verlag. [pp. 377, 387, 398, 417, 438, 459.]

Eddington, A. S. (1926). *The internal constitution of the stars.* Cambridge University Press. [p. 12.]

Eliassen, A., Høiland, E. & Riis, E. (1953). Two-dimensional perturbations of a flow with constant shear of a stratified fluid. *Inst. Weather Climate Res., Oslo,* Publ. No. 1. [pp. 149, 150.]

Ellingsen, T., Gjevik, B. & Palm, E. (1970). On the non-linear stability of plane Couette flow. *J. Fluid Mech.* **40**, 97–112. [pp. 453, 454.]

Eppler, R. & Fasel, H. (Editors) (1980). *Laminar-turbulent transition.* Berlin: Springer-Verlag. [p. 456.]

Erdélyi, A., Magnus, W., Oberhettinger, F. & Tricomi, F. G. (1953). *Higher transcendental functions*, vol. 1. New York: McGraw-Hill. [p. 475.]

Esch, R. E. (1957). The instability of a shear layer between two parallel streams. *J. Fluid Mech.* **3**, 289–303. [pp. 199, 201.]

Faraday, M. (1831). On the forms and states assumed by fluids in contact with vibrating elastic surfaces. *Phil. Trans. Roy. Soc.* **121**, 319–40. Also *Experimental researches in chemistry and physics* (1859), pp. 335–58. London:

Richard Taylor and William Francis. [pp. 354, 358.]

Fasel, H. (1976). Investigation of the stability of boundary layers by a finite-difference model of the Navier–Stokes equations. *J. Fluid Mech.* **78**, 355–83. [p. 455.]

Fasel, H. (1977). Reaktion von zweidimensionalen, laminaren, inkompressiblen Grenzschichten auf periodische Störungen in der Außenströmung. *Z. Angew. Math. Mech.* **57**, T180–3. [p. 455.]

Feir, J. E. (1967). Some results from wave pulse experiments (discussion). *Proc. Roy. Soc.* A **299**, 54–8. [p. 398.]

Fenstermacher, P. R., Swinney, H. L. & Gollub, J. (1979). Dynamical instabilities and the transition to chaotic Taylor vortex flow. *J. Fluid Mech.* **94**, 103–28. [p. 448.]

Fjørtoft, R. (1950). Application of integral theorems in deriving criteria of stability for laminar flows and for the baroclinic circular vortex. *Geofys. Publ.*, Oslo **17**, No. 6, pp. 1–52. [pp. 132, 434.]

Foote, J. R. & Lin, C. C. (1950). Some recent investigations in the theory of hydrodynamic stability. *Quart. Appl. Math.* **8**, 265–80. [p. 140.]

Fowkes, N. D. (1968). A singular perturbation method. Part I. *Quart. Appl. Math.* **26**, 57–69. [p. 254.]

Fraenkel, L. E. (1969). On the method of matched asymptotic expansions. Part I. A matching principle. *Proc. Camb. Phil. Soc.* **65**, 209–31. [pp. 253, 276.]

Freymuth, P. (1966). On transition in a separated laminar boundary layer. *J. Fluid Mech.* **25**, 683–704. [p. 456.]

Gage, K. S. (1971). The effect of stable thermal stratification on the stability of viscous parallel flows. *J. Fluid Mech.* **47**, 1–20. [p. 333.]

Gage, K. S. & Reid, W. H. (1968). The stability of thermally stratified plane Poiseuille flow. *J. Fluid Mech.* **33**, 21–32. [p. 333.]

Gallagher, A. P. (1974). On the behaviour of small disturbances in plane Couette flow. Part 3. The phenomenon of mode-pairing. *J. Fluid Mech.* **65**, 29–32. [pp. 212, 216.]

Gallagher, A. P. & Mercer, A. McD. (1962). On the behaviour of small disturbances in plane Couette flow. *J. Fluid Mech.* **13**, 91–100. [p. 212.]

Gallagher, A. P. & Mercer, A. McD. (1964). On the behaviour of small disturbances in plane Couette flow. Part 2. The higher eigenvalues. *J. Fluid Mech.* **18**, 350–2. [p. 212.]

Gary, J. & Helgason, R. (1970). A matrix method for ordinary differential eigenvalue problems. *J. Comput. Phys.* **5**, 169–87. [pp. 204, 207.]

Gaster, M. (1962). A note on the relation between temporally-increasing and spatially-increasing disturbances in hydrodynamic stability. *J. Fluid Mech.* **14**, 222–4. [pp. 152, 243, 350, 352.]

Gaster, M. (1965). On the generation of spatially growing waves in a boundary layer. *J. Fluid Mech.* **22**, 433–41. [pp. 153, 349, 350.]

Gaster, M. (1968). The development of three-dimensional wave packets in a boundary layer. *J. Fluid Mech.* **32**, 173–84. [p. 349.]

Gaster, M. (1974). On the effects of boundary-layer growth on flow stability. *J. Fluid Mech.* **66**, 465–80. [pp. 241, 479.]

Gaster, M. (1975). A theoretical model of a wave packet in the boundary layer on a flat plate. *Proc. Roy. Soc.* A **347**, 271–89. [pp. 151, 349.]

Gaster, M. & Davey, A. (1968). The development of three-dimensional wave-packets in unbounded parallel flows. *J. Fluid Mech.* **32**, 801–8. [p. 349.]

Gaster, M. & Grant, I. (1975). An experimental investigation of the formation and development of a wave packet in a laminar boundary layer. *Proc. Roy. Soc.* A **347**, 253–69. [p. 151.]

Georgescu, A. (1970). Note on Joseph's inequalities in stability theory. *Z. Angew. Math. Phys.* **21**, 258–60. [p. 162.]

Gershuni, G. Z. & Zhukhovitskii, E. M. (1976). *Convective stability of incompressible fluids.* (Translated from the Russian edition of 1972 by D. Louvish.) Jerusalem: Israel Program for Scientific Translations. [p. 62.]

Gersting, J. M., Jr. & Jankowski, D. F. (1972). Numerical methods for Orr–Sommerfeld problems. *Int. J. Numer. Meth. Engin.* **4**, 195–206. [p. 209.]

Gibson, R. D. & Cook, A. E. (1974). The stability of curved channel flow. *Quart. J. Mech. Appl. Math.* **27**, 149–60. [pp. 111, 112.]

Gilbert, F. & Backus, G. E. (1966). Propagator matrices in elastic wave and vibration problems. *Geophysics* **31**, 326–32. [p. 311.]

Gill, A. E. (1965). On the behaviour of small disturbances to Poiseuille flow in a circular pipe. *J. Fluid Mech.* **21**, 145–72. [p. 219.]

Godunov, S. K. (1961). On the numerical solution of boundary-value problems for systems of linear ordinary differential equations. *Uspekhi Mat. Nauk* **16**, 171–4. [p. 208.]

Goldstein, H. (1950). *Classical mechanics.* Cambridge, Mass.: Addison–Wesley. [p. 396.]

Goldstein, S. (1931). On the stability of superposed streams of fluids of different densities. *Proc. Roy. Soc.* A **132**, 524–48. [pp. 324, 331.]

Goldstein, S. (Editor) (1938). *Modern developments in fluid dynamics.* vol. 1. Oxford: Clarendon Press. [p. 230.]

Gollub, J. P. & Swinney, H. L. (1975). Onset of turbulence in a rotating fluid. *Phys. Rev. Lett.* **35**, 927–30. [pp. 447, 448.]

Golubitsky, M. & Schaeffer, D. (1979). A theory for imperfect bifurcation via singularity theory. *Comm. Pure. Appl. Math.* **32**, 21–98. [p. 415.]

Gor'kov, L. P. (1957). Stationary convection in a plane liquid layer near the critical heat transfer point. *Zh. Eksp. Teor. Fiz.* **33**, 402–7. Translated in *Sov. Phys.—*

JETP **6**, 311–15 (1958). [pp. 376, 386, 436.]

Görtler, H. (1940). Über eine dreidimensionale Instabilität laminarer Grenzschichten an konkaven Wänden. *Nachr. Ges. Wiss. Göttingen, N.F.* **2**, No. 1, pp. 1–26. (Translated as 'On-the three-dimensional instability of laminar boundary layers on concave walls', *Tech. Memor. Nat. Adv. Comm. Aero., Wash.* No. 1375 (1954).) [pp. 116, 117, 118.]

Gossard, E. E. & Hooke, W. H. (1975). *Waves in the atmosphere.* Amsterdam: Elsevier. [p. 333.]

Gotoh, K. (1965). The damping of the large wave-number disturbances in a free boundary layer flow. *J. Phys. Soc. Japan* **20**, 164–9. [pp. 237, 238.]

Graaf, J. G. A. de & Held, E. F. M. van der (1953). The relation between the heat transfer and the convection phenomena in enclosed plane air layers. *Appl. Sci. Res.* A **3**, 393–409. [p. 60.]

Graebel, W. P. (1966). On determination of the characteristic equations for the stability of parallel flows. *J. Fluid Mech.* **24**, 497–508. [pp. 252, 267.]

Graham, A. (1933). Shear patterns in an unstable layer of air. *Phil. Trans. Roy. Soc.* A **232**, 285–96. [p. 61.]

Granoff, B. (1972). Asymptotic solutions of a 6th order differential equation with two turning points. Part II: Derivation by reduction to a first order system. *SIAM J. Math. Anal.* **3**, 93–104. [pp. 95, 101.]

Granoff, B. & Bleistein, N. (1972). Asymptotic solutions of a 6th order differential equation with two turning points. Part 1: Derivation by method of steepest descent. *SIAM J. Math. Anal.* **3**, 45–57. [pp. 95, 101.]

Greenspan, H. P. & Benney, D. J. (1963). On shear-layer instability, breakdown and transition. *J. Fluid Mech.* **15**, 133–53. [p. 455.]

Gregory, N. & Walker, W. S. (1950). The effect on transition of isolated surface excrescences in the boundary layer. *Rep. Memor. Aero. Res. Coun., Lond.* No. 2779 (1951 Vol.). [p. 118.]

Grohne, D. (1954). Über das Spektrum bei Eigenschwingungen ebener Laminarströmungen. *Z. Angew. Math. Mech.* **34**, 344–57. (Translated as 'On the spectrum of natural oscillations of two-dimensional laminar flows,' *Tech. Memor. Nat. Adv. Comm. Aero., Wash.* No. 1417 (1957).) [pp. 212, 213, 214, 221, 466.]

Grosch, C. E. & Salwen, H. (1968). The stability of steady and time-dependent plane Poiseuille flow. *J. Fluid Mech.* **34**, 177–205. [p. 205.]

Grosch, C. E. & Salwen, H. (1978). The continuous spectrum of the Orr–Sommerfeld equation. Part 1. The spectrum and the eigenfunctions. *J. Fluid Mech.* **87**, 33–54. [p. 156.]

Haberman, R. (1972). Critical layers in parallel flows. *Studies in Appl. Math.* **51**, 139–61. [p. 422.]

Hains, F. D. (1967). Stability of plane Couette–Poiseuille flow.

Phys. Fluids **10**, 2079–80. [p. 223.]

Hales, A. L. (1937). Convection currents in geysers. *Mon. Not. Roy. Astron. Soc., Geophys. Suppl.* **4**, 122–31. [p. 68.]

Hall, P. & Walton, I. C. (1977). The smooth transition to a convective régime in a two-dimensional box. *Proc. Roy. Soc.* A **358**, 199–221. [p. 415.]

Hämmerlin, G. (1955). Über das Eigenwertproblem der dreidimensionalen Instabilität laminarer Grenzschichten an konkaven Wänden. *J. Rat. Mech. Anal.* **4**, 279–321. [pp. 111, 116, 119.]

Hämmerlin, G. (1958). Die Stabilität der Strömung in einem gekrümmten Kanal. *Arch. Rat. Mech. Anal.* **1**, 212–24. [pp. 109, 111.]

Hardy, K., Glover, K. M. & Ottersten, H. (1969). Radar investigations of atmospheric structure and CAT in the 3 to 20-km region. In *Clear air turbulence and its detection*, eds. Y.-H. Pao & A. Goldburg, pp. 402–15. New York: Plenum Press. [p. 423.]

Harris, D. L. & Reid, W. H. (1964). On the stability of viscous flow between rotating cylinders. Part 2. Numerical analysis. *J. Fluid Mech.* **20**, 95–101. [pp. 99, 100.]

Hart, J. E. (1979). Finite amplitude baroclinic instability. *Ann. Rev. Fluid Mech.* **11**, 147–72. [p. 335.]

Hartman, P. (1964). *Ordinary differential equations.* New York: John Wiley & Sons. [p. 399.]

Hasselmann, K. (1967). A criterion for nonlinear wave stability. *J.*

Fluid Mech. **30**, 737–9. [p. 396.]

Haurwitz, B. (1931). Zur Theorie der Wellenbewegungen in Luft und Wasser. *Veröff. Geophys. Inst. Univ. Leipzig* **6**, No. 1. [pp. 324, 364.]

Hazel, P. (1972). Numerical studies of the stability of inviscid stratified shear flows. *J. Fluid Mech.* **51**, 39–61. [pp. 329, 330, 364, 365.]

Heading, J. (1962). *An introduction to phase-integral methods.* London: Methuen & Co. Ltd. [p. 170.]

Heisenberg, W. (1924). Über Stabilität und Turbulenz von Flussigkeitsströmen. *Ann. Phys., Lpz.* (4) **74**, 577–627. (Translated as 'On stability and turbulence of fluid flows', *Tech. Memor. Nat. Adv. Comm. Aero., Wash.* No. 1291 (1951).) [pp. 126, 139, 164, 165, 202, 370, 451.]

Helmholtz, H. von (1868). Über discontinuirliche Flüssigkeitsbewegungen. *Monats. Königl. Preuss. Akad. Wiss. Berlin* **23**, 215–28. Also *Wissenschaftliche Abhandlungen* (1882), vol. I, pp. 146–57. Leipzig: J. A. Barth. Translated into English by F. Guthrie as 'On discontinuous movements of fluids,' *Phil. Mag.* (4) **36**, 337–46 (1868). [pp. 14, 125.]

Helmholtz, H. von (1890). Die Energie der Wogen und des Windes. *Ann. Phys.* **41**, 641–62. Also *Sitz. Akad. Wiss. Berlin* **7**, 853–72, and *Wissenschaftliche Abhandlungen* (1895), vol. III, pp. 333–55. Leipzig: J. A. Barth. [p. 20.]

Herbert, T. (1975). On finite amplitudes of periodic disturbances of the boundary layer along a flat plate. In *Proceedings of the Fourth International Conference on Numerical Methods in Fluid Dynamics*, ed. R. D. Richtmyer, pp. 212–17. Lecture Notes in Physics, no. 35. Berlin: Springer-Verlag. [p. 454.]

Herbert, T. (1976). Periodic secondary motions in a plane channel. In *Proceedings of the Fifth International Conference on Numerical Methods in Fluid Dynamics*, eds. A. I. van de Vooren & P. J. Zandbergen, pp. 235–40. Lecture Notes in Physics, no. 59. Berlin: Springer-Verlag. [p. 451.]

Herbert, T. (1977a). Zur nichtlinearen Stabilität ebener Parallelströmungen. *Z. Angew. Math. Mech.* **57**, T187–9. [p. 451.]

Herbert, T. (1977b). Finite amplitude stability of plane parallel flows. In *Fluid dynamics panel symposium* (Lyngby, Denmark, 2–4 May 1977), CP224, pp. 3/1–10. Neuilly sur Seine, France: AGARD. [p. 451.]

Hirsch, M. W. & Smale, S. (1974). *Differential equations, dynamical systems, and linear algebra.* New York: Academic Press. [pp. 399, 408.]

Hoard, C. Q., Robertson, C. R. & Acrivos, A. (1970). Experiments on the cellular structure in Bénard convection. *Int. J. Heat Mass Transfer* **13**, 849–56. [p. 439.]

Hocking, L. M. (1975). Non-linear instability of the asymptotic suction velocity profile. *Quart. J. Mech. Appl. Math.* **28**, 341–53. [p. 229.]

Hocking, L. M. & Stewartson, K. (1972). On the nonlinear response of a marginally unstable plane parallel flow to a two-dimensional disturbance. *Proc. Roy. Soc.* A **326**, 289–313. [p. 452.]

Holstein, H. (1950). Über die äussere und innere Reibungsschicht bei Störungen laminarer Strömungen. *Z. Angew. Math. Mech.* **30**, 25–49. [pp. 172, 177, 178.]

Hopf, E. (1942). Abzweigung einer periodischen Lösung von einer stationären Lösung eines Differentialsystems. *Ber. Verh. Sächs. Akad. Wiss. Leipzig. Math.-phys. Kl.* **94**, 1–22. Translated in *The Hopf bifurcation and its applications* (1976), pp. 163–93, eds. J. E. Marsden & M. McCracken. New York: Springer-Verlag. [pp. 403, 405.]

Hopf, E. (1948). A mathematical example displaying features of turbulence. *Comm. Appl. Math.* **1**, 303–22. [p. 376.]

Hopf, L. (1914). Der Verlauf kleiner Schwingungen auf einer Strömung reibender Flüssigkeit. *Ann. Phys., Lpz.* (4) **44**, 1–60. [pp. 212, 213, 214.]

Hopfinger, E. J., Atten, P. & Busse, F. H. (1979). Instability and convection in fluid layers: a report on Euromech 106. *J. Fluid Mech.* **92**, 217–40. [p. 442.]

Howard, L. N. (1959). Hydro-dynamic stability of a jet. *J. Math. Phys.* **37**, 283–98. [pp. 164, 197, 198.]

Howard, L. N. (1961). Note on a paper of John W. Miles. *J. Fluid Mech.* **10**, 509–12. [pp. 142, 328.]

Howard, L. N. (1962). Review of *Hydrodynamic and hydromagnetic stability* by S. Chandrasekhar. *J. Fluid Mech.* **13**, 158–60. [p. 77.]

Howard, L. N. (1963a). Neutral curves and stability boundaries in stratified flow. *J. Fluid Mech.* **16**, 333–42. [p. 330.]

Howard, L. N. (1963b). Heat transport by turbulent convection. *J. Fluid Mech.* **17**, 405–32. [p. 431.]

Howard, L. N. (1972). Bounds on flow quantities. *Ann. Rev. Fluid Mech.* **4**, 473–94. [p. 431.]

Howard, L. N. & Gupta, A. S. (1962). On the hydrodynamic and hydromagnetic stability of swirling flows. *J. Fluid Mech.* **14**, 463–76. [p. 77.]

Howard, L. N. & Maslowe, S. A. (1973). Stability of stratified shear flow. *Boundary-Layer Meteorol.* **4**, 511–23. [pp. 326, 332, 333.]

Hughes, T. H. (1972). Variable mesh numerical method for solving the Orr–Sommerfeld equation. *Phys. Fluids* **15**, 725–8. [pp. 190, 207.]

Hughes, T. H. & Reid, W. H. (1964). The effect of a transverse pressure gradient on the stability of Couette flow. *Z. Angew. Math.*

Phys. **15**, 573–81. [pp. 111, 113, 114, 115, 116.]

Hughes, T. H. & Reid, W. H. (1965a). On the stability of the asymptotic suction boundary-layer profile. *J. Fluid Mech.* **23**, 715–35. [pp. 228, 246, 249.]

Hughes, T. H. & Reid, W. H. (1965b). The stability of laminar boundary layers at separation. *J. Fluid Mech.* **23**, 737–47. [pp. 134, 136, 164, 230, 250.]

Hughes, T. H. & Reid, W. H. (1968). The stability of spiral flow between rotating cylinders. *Phil. Trans. Roy. Soc.* A **263**, 57–91. [pp. 189, 250, 299, 304.]

Huppert, H. E. (1973). On Howard's technique for perturbing neutral solutions of the Taylor–Goldstein equation. *J. Fluid Mech.* **57**, 361–8. [pp. 331, 333.]

Ince, E. L. (1927). *Ordinary differential equations.* London: Longmans, Green & Co. Ltd. [p. 78.]

Iooss, G. (1972). Existence et stabilité de la solution périodique secondaire intervenant dans les problèmes d'évolution du type Navier–Stokes. *Arch. Rat. Mech. Anal.* **47**, 301–29. [p. 405.]

Itoh, N. (1974a). A power series method for the numerical treatment of the Orr–Sommerfeld equation. *Trans. Japan Soc. Aero. Space Sci.* **17**, 65–75. [pp. 244, 316, 453.]

Itoh, N. (1974b). Spatial growth of finite wave disturbances in parallel and nearly parallel flows. Part 1. The theoretical analysis and the numerical results for plane Poiseuille flow. *Trans. Japan Soc. Aero. Space Sci.* **17**, 160–74. [pp. 451, 453.]

Itoh, N. (1974c). Spatial growth of finite wave disturbances in parallel and nearly parallel flows. Part 2. The numerical results for the flat plate boundary layer. *Trans. Japan Soc. Aero. Space Sci.* **17**, 175–86. [p. 454.]

Iudovich, V. I. (1966). Secondary flows and fluid instability between rotating cylinders. *Prikl. Mat. Mekh.* **30**, 688–98. Translated in *J. Appl. Math. Mech.* **30**, 822–33 (1966). [p. 407.]

Iudovich, V. I. (1971). The onset of auto-oscillations in a fluid. *Prikl. Mat. Mekh.* **35**, 638–55. Translated in *J. Appl. Math. Mech.* **35**, 587–603 (1971). [p. 405.]

Ivanilov, Iu. P. & Iakolev, G. N. (1966). The bifurcation of fluid flow between rotating cylinders. *Prikl. Mat. Mekh.* **30**, 768–73. Translated in *J. Appl. Math. Mech.* **30**, 910–16 (1966). [p. 407.]

Jackson, P. A., Robati, B. & Mobbs, F. R. (1977). Secondary flows between eccentric rotating cylinders at sub-critical Taylor numbers. In *Superlaminar flow in bearings*, eds. D. Dowson, M. Godet, and C. M. Taylor, pp. 9–14. London: Mech. Engineering Publ. Ltd. [p. 449.]

Jeffreys, H. (1926). The stability of a layer of fluid heated below. *Phil. Mag.* (7) **2**, 833–44. Also *Collected papers* (1975), vol. 4, pp. 457–68. London: Gordon and Breach. Also reprinted by Saltzman (1962). [p. 12.]

Jeffreys, H. (1928). Some cases of instability in fluid motion. *Proc. Roy. Soc.* A **118**, 195–208. Also *Collected papers* (1975), vol. 4, pp. 469–84. London: Gordon and Breach. Also reprinted by Saltzman (1962). [pp. 42, 51, 97.]

Jeffreys, H. (1942). Asymptotic solutions of linear differential equations. *Phil. Mag.* (7) **33**, 451–6. Also *Collected papers* (1977), vol. 6, pp. 117–22. London: Gordon and Breach. [p. 171.]

Jordinson, R. (1970). The flat plate boundary layer. Part 1. Numerical integration of the Orr–Sommerfeld equation. *J. Fluid Mech.* **43**, 801–11. [pp. 225, 226, 227, 240, 243, 244.]

Jordinson, R. (1971). Spectrum of eigenvalues of the Orr–Sommerfeld equation for Blasius flow. *Phys. Fluids* **14**, 2535–7. [p. 156.]

Joseph, D. D. (1965). On the stability of the Boussinesq equations. *Arch. Rat. Mech. Anal.* **20**, 59–71. [p. 464.]

Joseph, D. D. (1966). Nonlinear stability of the Boussinesq equations by the method of energy. *Arch. Rat. Mech. Anal.* **22**, 163–84. [p. 430.]

Joseph, D. D. (1968). Eigenvalue bounds for the Orr–Sommerfeld equation. *J. Fluid Mech.* **33**, 617–21. [pp. 161, 163.]

Joseph, D. D. (1969). Eigenvalue bounds for the Orr–Sommerfeld equation. Part 2. *J. Fluid Mech.* **36**, 721–34. [pp. 161, 162, 164.]

Joseph, D. D. (1971). Stability of convection in containers of arbitrary shape. *J. Fluid Mech.* **47**, 257–82. [p. 68.]

Joseph, D. D. (1976). *Stability of fluid motions* (2 vols.). Springer Tracts in Natural Philosophy, vols. 27 and 28. Berlin: Springer-Verlag. [pp. 399, 424, 452.]

Joseph, D. D. & Carmi, S. (1969). Stability of Poiseuille flow in pipes, annuli and channels. *Quart. Appl. Math.* **26**, 575–99. [pp. 221, 430.]

Joseph, D. D. & Sattinger, D. H. (1972). Bifurcating time periodic solutions and their stability. *Arch. Rat. Mech. Anal.* **45**, 79–109. [p. 405.]

Kaplan, R. E. (1964). The stability of laminar incompressible boundary layers in the presence of compliant boundaries. Aeroelastic and Structures Research Laboratory, Report No. ASRL-TR 166-1, Massachusetts Institute of Technology. [p. 208.]

Kármán, T. von (1934). Some aspects of the turbulence problem. *Proceedings 4th International Congress for Applied Mechanics*, Cambridge, England, pp. 54–91. Also *Collected works* (1956), vol. III, pp. 120–55. London: Butterworths Scientific Publications. [p. 73.]

Keller, H. B. (1968). *Numerical methods for two-point boundary-value problems.* Waltham, Mass.: Cinn-Blaisdell. [p. 208.]

Keller, H. B. (1976). *Numerical solution of two point boundary value problems.* SIAM Regional Conference Series in Applied Mathematics. Philadelphia:

Society for Industrial and Applied Mathematics. [p. 317.]

Kelly, R. E. (1965). The stability of an unsteady Kelvin–Helmholtz flow. *J. Fluid Mech.* **22**, 547–60. [p. 369.]

Kelly, R. E. (1967). On the stability of an inviscid shear layer which is periodic in space and time. *J. Fluid Mech.* **27**, 657–89. [pp. 456, 457.]

Kelly, R. E. & Pal, D. (1978). Thermal convection with spatially periodic boundary conditions: resonant wavelength excitation. *J. Fluid Mech.* **86**, 433–56. [p. 415.]

Kelvin, Lord (1871). Hydrokinetic solutions and observations. *Phil. Mag.* (4), **42**, 362–77. Also *Mathematical and physical papers* (1910), vol. IV, pp. 69–85. Cambridge University Press. [pp. 14, 20, 28, 125.]

Kelvin, Lord (1880a). On the vibrations of a columnar vortex. *Phil. Mag.* (5) **10**, 155–68. Also *Mathematical and physical papers* (1910), vol. IV, pp. 152–65. Cambridge University Press. [pp. 77, 122.]

Kelvin, Lord (1880b). On a disturbing infinity in Lord Rayleigh's solution for waves in a plane vortex stratum. *Nature* **23**, 45–6. Also *Mathematical and physical papers* (1910), vol. IV, pp. 186–7. Cambridge University Press. [p. 141.]

Kelvin, Lord (1887). On the stability of steady and of periodic fluid motion. *Phil. Mag.* (5) **23**, 459–64, 529–39. Also *Mathematical and physical papers* (1910), vol.

IV, pp. 166–83. Cambridge University Press. [p. 432.]

Kirchgässner, K. (1961). Die Instabilität der Strömung zwischen zwei rotierenden Zylindern gegenüber Taylor–Wirbeln für beliebige Spaltbreiten. Z. Angew. Math. Phys. 12, 14–30. [p. 103.]

Kirchgässner, K. & Sorger, P. (1969). Branching analysis for the Taylor problem. Quart. J. Mech. Appl. Math. 22, 183–209. [pp. 407, 444.]

Klebanoff, P. S., Tidstrom, K. D. & Sargent, L. M. (1962). The three-dimensional nature of boundary-layer instability. J. Fluid Mech. 12, 1–34. [p. 455.]

Kobayashi, R. (1972). Note on the stability of a boundary layer on a concave wall with suction. J. Fluid Mech. 52, 269–72. [pp. 120, 123.]

Kogelman, S. & DiPrima, R. C. (1970). Stability of spatially periodic supercritical flows in hydrodynamics. Phys. Fluids 13, 1–11. [pp. 417, 445.]

Koiter, W. T. (1945). On the stability of elastic equilibrium. Dissertation, Delft, Holland. Translated in NASA Tech. Trans. F-10,833 (1967). [pp. 408, 413.]

Koschmieder, E. L. (1966). On convection on a uniformly heated plane. Beiträge Phys. Atmos. 39, 1–11. [p. 60.]

Koschmieder, E. L. (1974). Bénard convection. In Advances in chemical physics, vol. 26, eds. I. Prigogine & S. A. Rice, pp. 177–212. New York: Wiley–Interscience. [pp. 62, 440.]

Koschmieder, E. L. (1979). Turbulent Taylor vortex flow. J. Fluid Mech. 93, 515–27 and 800. [p. 446.]

Koschmieder, E. L. & Pallas, S. G. (1974). Heat transfer through a shallow, horizontal convecting fluid layer. Int. J. Heat Mass Transfer 17, 991–1002. [pp. 33, 440.]

Krauss, W. (1966). Methoden und Ergebnisse der Theoretischen Ozeanographie. II. Interne Wellen. Berlin: Gebrüder Borntraeger. [p. 325.]

Krishnamurti, R. (1968). Finite amplitude convection with changing mean temperature. J. Fluid Mech. 33, 445–55 and 457–63. [p. 441.]

Krueger, E. R. & DiPrima, R. C. (1962). Stability of nonrotationally symmetric disturbances for inviscid flow between rotating cylinders. Phys. Fluids 5, 1362–7. [p. 88.]

Krueger, E. R., Gross, A. & DiPrima, R. C. (1966). On the relative importance of Taylor-vortex and non-axisymmetric modes in flow between rotating cylinders. J. Fluid Mech. 24, 521–38. [pp. 104, 107.]

Kurtz, E. F., Jr. & Crandall, S. H. (1962). Computer-aided analysis of hydrodynamic stability. J. Math. Phys. 41, 264–79. [p. 207.]

Lagerstrom, P. A. (1964). Laminar flow theory. In Theory of laminar flows, ed. F. K. Moore, pp. 20–285. Princeton University Press. [pp. 120, 231.]

Lake, B. M., Yuen, H. C., Rungaldier, H. & Ferguson, W. E.

(1977). Nonlinear deep-water waves: theory and experiment. Part 2. Evolution of a continuous wave train. *J. Fluid Mech.* **83**, 49–74. [p. 398.]

Lakin, W. D., Ng, B. S. & Reid, W. H. (1978). Approximations to the eigenvalue relation for the Orr–Sommerfeld problem. *Phil. Trans. Roy. Soc.* A **289**, 347–71. [pp. 292, 295, 296, 301, 302, 303, 307, 310, 311.]

Lakin, W. D. & Reid, W. H. (1970). Stokes multipliers for the Orr–Sommerfeld equation. *Phil. Trans. Roy. Soc.* A **268**, 325–49. [pp. 189, 248, 252, 297, 319.]

Lamb, H. (1932). *Hydrodynamics*, 6th edn. Cambridge University Press. [pp. 17, 27, 28, 30.]

Landau, L. D. (1944). On the problem of turbulence. *C.R. Acad. Sci. U.R.S.S.* **44**, 311–14. Also *Collected papers* (1965), pp. 387–91. Oxford: Pergamon Press; New York: Gordon and Breach. [pp. 370, 374, 380.]

Landau, L. D. (1946). On the vibrations of the electronic plasma. *J. Phys. U.S.S.R.* **10**, 23–34. Also *Collected papers* (1965), pp. 445–60. Oxford: Pergamon Press; New York: Gordon and Breach. [pp. 152, 349.]

Landau, L. D. & Lifshitz, E. M. (1959). *Fluid mechanics*. Vol. 6 of *Course of theoretical physics*. (Translated from the Russian by J. B. Sykes and W. H. Reid.) London: Pergamon Press. [pp. 1, 349, 370.]

Langer, R. E. (1931). On the asymptotic solutions of ordinary differential equations, with an application to the Bessel functions of large order. *Trans. Amer. Math. Soc.* **33**, 23–64. [p. 251.]

Langer, R. E. (1932). On the asymptotic solutions of differential equations, with an application to the Bessel functions of large complex order. *Trans. Amer. Math. Soc.* **34**, 447–80. [p. 251.]

Langer, R. E. (1949). The asymptotic solutions of ordinary linear differential equations of the second order, with special reference to a turning point. *Trans. Amer. Math. Soc.* **67**, 461–90. [p. 251.]

Langer, R. E. (1957). On the asymptotic solutions of a class of ordinary differential equations of the fourth order, with special reference to an equation of hydrodynamics. *Trans. Amer. Math. Soc.* **84**, 144–91. [p. 251.]

Langer, R. E. (1959). Formal solutions and a related equation for a class of fourth order differential equations of hydrodynamic type. *Trans. Amer. Math. Soc.* **92**, 371–410. [p. 251.]

Lee, L. H. & Reynolds, W. C. (1967). On the approximate and numerical solution of Orr–Sommerfeld problems. *Quart. J. Mech. Appl. Math.* **20**, 1–22. [p. 209.]

Lees, L. & Reshotko, E. (1962). Stability of the compressible laminar boundary layer. *J. Fluid Mech.* **12**, 555–90. [p. 203.]

Lewis, D. J. (1950). The instability of liquid surfaces when accelerated in a direction perpendicular to their planes. II.

Proc. Roy. Soc. A **117**, 81–96. [p. 20.]

Liang, S. F., Vidal, A. & Acrivos, A. (1969). Buoyancy-driven convection in cylindrical geometries. *J. Fluid Mech.* **36**, 239–56. [p. 414.]

Lin, C. C. (1945). On the stability of two-dimensional parallel flows. *Quart. Appl. Math.* **3**, 117–42, 218–34 and 277–301. [pp. 126, 134, 157, 164, 202, 225, 226, 227, 250.]

Lin, C. C. (1955). *The theory of hydrodynamic stability.* Cambridge University Press. [pp. 102, 126, 134, 136, 164, 231, 292.]

Lin, C. C. (1957a). On uniformly valid asymptotic solutions of the Orr–Sommerfeld equation. *Proceedings 9th International Congress for Applied Mechanics (Brussels)*, vol. I, pp. 136–48. [pp. 251, 252, 285.]

Lin, C. C. (1957b). On the stability of the laminar boundary layer. *Proceedings of the Symposium on Naval Hydrodynamics*, pp. 353–71. (National Research Council Publication 515), Washington, D.C. [pp. 251, 252, 282, 285, 286.]

Lin, C. C. (1958). On the instability of laminar flow and its transition to turbulence. *Proceedings of the Symposium on Boundary Layer Research (Freiburg)*, ed. H. Görtler, pp. 144–60. Berlin: Springer-Verlag. [pp. 251, 285, 421.]

Lin, C. C. (1961). Some mathematical problems in the theory of the stability of parallel flows. *J. Fluid Mech.* **10**, 430–8. [p. 156.]

Lin, C. C. (1964). Some examples of asymptotic problems in mathematical physics. In *Asymptotic solutions of differential equations and their applications*, ed. C. H. Wilcox, pp. 129–43. New York: John Wiley & Sons. [pp. 251, 253.]

Lin, C. C. & Rabenstein, A. L. (1960). On the asymptotic solutions of a class of ordinary differential equations of the fourth order. I. Existence of regular formal solutions. *Trans. Amer. Math. Soc.* **94**, 24–57. [pp. 251, 266.]

Lin, C. C. & Rabenstein, A. L. (1969). On the asymptotic theory of a class of ordinary differential equations of fourth order. II. Existence of solutions which are approximated by the formal solutions. *Studies in Appl. Math.* **48**, 311–40. [pp. 251, 267.]

Lindstedt, A. (1883). Beitrag zur Integration der Differential gleichungen der Störungstheorie. *Mém. Acad. Imp. Sci. St.-Pétersbourg* (7) **31**, no. 4. [p. 386.]

Ling, C.-H. & Reynolds, W. C. (1973). Non-parallel flow corrections for the stability of shear flows. *J. Fluid Mech.* **59**, 571–91. [p. 241.]

Lipps, F. B. & Somerville, R. C. J. (1971). Dynamics of variable wavelength in finite-amplitude Bénard convection. *Phys. Fluids* **14**, 759–65. [p. 440.]

Liu, J. T. C. (1969). Finite-amplitude instability of the compressible laminar wake. Weakly nonlinear theory. *Phys. Fluids* **12**, 1763–74. [p. 457.]

Lock, R. C. (1955). The stability of the flow of an electrically conducting fluid between parallel planes under a transverse magnetic field. *Proc. Roy. Soc.* A **233**, 105–25. [p. 248.]

Longuet-Higgins, M. S. (1978). The instability of gravity waves of finite amplitude in deep water. II. Subharmonics. *Proc. Roy. Soc.* A **360**, 489–505. [pp. 360, 398.]

Lorenz, E. N. (1963). Deterministic nonperiodic flow. *J. Atmos. Sci.* **20**, 130–41. [p. 463.]

Low, A. R. (1929). On the criterion for stability of a layer of viscous fluid heated from below. *Proc. Roy. Soc.* A **125**, 180–95. Reprinted by Saltzman (1962). [p. 66.]

MacCreadie, W. T. (1931). On the stability of the motion of a viscous fluid. *Proc. Nat. Acad. Sci.*, *Wash.* **17**, 381–8. [p. 163.]

McGoldrick, L. F. (1965). Resonant interactions among capillary-gravity waves. *J. Fluid Mech.* **21**, 305–31. [p. 397.]

McGoldrick, L. F. (1970). An experiment on second-order capillary gravity resonant wave interactions. *J. Fluid Mech.* **40**, 251–71. [p. 397.]

Mack, L. M. (1965). Computation of the stability of the laminar compressible boundary layer. In *Methods in computational physics*, vol. 4, ed. B. Alder, pp. 247–99. New York: Academic Press. [p. 203.]

Mack, L. M. (1976). A numerical study of the temporal eigenvalue spectrum of the Blasius boundary layer. *J. Fluid Mech.* **73**, 497–520. [pp. 156, 221, 222.]

Mackrodt, P.-A. (1976). Stability of Hagen–Poiseuille flow with superimposed rigid rotation. *J. Fluid Mech.* **73**, 153–64. [p. 220.]

McLaughlin, J. B. & Martin, P. C. (1975). Transition to turbulence in a statically stressed fluid system. *Phys. Rev.* A (3) **12**, 186–203. [p. 442.]

Magnus, W. & Winkler, S. (1966). *Hill's equation.* New York: Wiley–Interscience. [pp. 359, 369.]

Malkus, W. V. R. (1954a). Discrete transitions in turbulent convection. *Proc. Roy. Soc.* A **225**, 185–95. [pp. 437, 441.]

Malkus, W. V. R. (1954b). The heat transport and spectrum of thermal turbulence. *Proc. Roy. Soc.* A **225**, 196–212. [pp. 431, 437.]

Malkus, W. V. R. & Veronis, G. (1958). Finite amplitude cellular convection. *J. Fluid Mech.* **4**, 225–60. [pp. 376, 380, 387, 436.]

Mallock, A. (1896). Experiments on fluid viscosity. *Phil. Trans. Roy. Soc.* A **187**, 41–56. [pp. 70, 105.]

Malurkar, S. L. & Srivastava, M. P. (1937). On the differential equation of the instability of a thin layer of fluid heated from below. *Proc. Indian Acad. Sci.* A **5**, 34–6. [p. 95.]

Marsden, J. E. (1978). Qualitative methods in bifurcation theory. *Bull. Amer. Math. Soc.* **84**, 1125–48. [p. 403.]

Marsden, J. E. & McCracken, M. (1976). *The Hopf bifurcation and its applications.* Applied

Mathematical Sciences, vol. 19. New York: Springer-Verlag. [p. 463.]

Maslowe, S. A. (1972). The generation of clear air turbulence by nonlinear waves. *Studies Appl. Math.* **51**, 1–16. [p. 422.]

Maslowe, S. A. (1973). Finite-amplitude Kelvin–Helmholtz billows. *Boundary-Layer Meteor.* **5**, 43–52. [p. 422.]

Maslowe, S. A. (1977). Weakly nonlinear stability of a viscous free shear layer. *J. Fluid Mech.* **79**, 689–702. [p. 456.]

Maslowe, S. A. & Thompson, J. M. (1971). Stability of a stratified free shear layer. *Phys. Fluids* **14**, 453–8. [p. 332.]

Matkowsky, B. J. (1970). A simple nonlinear dynamic stability problem. *Bull. Amer. Math. Soc.* **76**, 620–5. [p. 380.]

Matthiessen, L. (1868). Akustiche Versuche, die kleinsten Transversalwellen der Flüssigkeiten betressend. *Ann. Phys. Chem.* (2) **134**, 107–17. [pp. 355, 358.]

Matthiessen, L. (1870). Ueber die Transversalschwingungen tönender tropfbarer und elastischer Flüssigkeiten. *Ann. Phys. Chem.* (2) **141**, 375–93. [pp. 355, 358.]

Meister, B. (1962). Das Taylor–Deansche Stabilitätsproblem für beliebige Spaltbreiten. *Z. Angew. Math. Phys.* **13**, 83–91. [p. 113.]

Meksyn, D. (1946a). Stability of viscous flow between rotating cylinders. I. *Proc. Roy. Soc.* A **187**, 115–28. [p. 98.]

Meksyn, D. (1946b). Stability of viscous flow between rotating cylinders. II. Cylinders rotating in opposite directions. *Proc. Roy. Soc.* A **187**, 480–91. [pp. 99, 101.]

Meksyn, D. (1946c). Stability of viscous flow between rotating cylinders. III. Integration of a sixth order linear equation. *Proc. Roy. Soc.* A **187**, 492–504. [p. 101.]

Meksyn, D. (1950). Stability of viscous flow over concave cylindrical surfaces. *Proc. Roy. Soc.* A **203**, 253–65. [pp. 116, 118.]

Meksyn, D. (1961). *New methods in laminar boundary-layer theory.* Oxford: Pergamon Press. [pp. 98, 99, 101.]

Meksyn, D. & Stuart, J. T. (1951). Stability of viscous motion between parallel planes for finite disturbances. *Proc. Roy. Soc.* A **208**, 517–26. [pp. 450, 452.]

Michalke, A. (1964). On the inviscid instability of the hyperbolic-tangent velocity profile. *J. Fluid Mech.* **19**, 543–56. [p. 238.]

Michalke, A. (1965). On spatially growing disturbances in an inviscid shear layer. *J. Fluid Mech.* **23**, 521–44. [p. 456.]

Michalke, A. & Timme, A. (1967). On the inviscid instability of certain two-dimensional vortex-type flows. *J. Fluid Mech.* **29**, 647–66. [p. 122.]

Mihaljan, J. M. (1962). A rigorous exposition of the Boussinesq approximations applicable to a thin layer of fluid. *Astrophys. J.* **136**, 1126–33. [p. 36.]

Miksad, R. W. (1972). Experiments on the nonlinear stages of free-shear-layer transition. *J. Fluid Mech.* **56**, 695–719. [p. 456.]

Miles, J. W. (1960). The hydrodynamic stability of a thin film of liquid in uniform shearing motion. *J. Fluid Mech.* **8**, 593–610. [p. 188.]

Miles, J. W. (1961). On the stability of heterogeneous shear flows. *J. Fluid Mech.* **10**, 496–508. [p. 328.]

Miles, J. W. (1963). On the stability of heterogeneous shear flows. Part 2. *J. Fluid Mech.* **16**, 209–27. [p. 329.]

Miles, J. W. (1967). Surface-wave damping in closed basins. *Proc. Roy. Soc.* A **297**, 459–75. [p. 358.]

Mises, R. von (1912a). Beitrag zum Oszillationsproblem. In *Festschrift H. Weber*, pp. 252–82. Leipzig: Teubner. [p. 213.]

Mises, R. von (1912b). Kleine Schwingungen und Turbulenz. *Jber. Deutsch. Math.-Verein* **21**, 241–8. [p. 213.]

Mises, R. von & Friedrichs, K. O. (1942). *Fluid dynamics* (mimeographed lecture notes, Brown University, Providence, Rhode Island), chap. IV, pp. 200–10. Also *Fluid dynamics, Applied Mathematical Sciences*, Vol. 5 (1971), pp. 285–97. New York: Springer-Verlag. [p. 133.]

Movchan, A. A. (1959). The direct method of Liapunov in stability problems of elastic systems. *Prikl. Mat. Mekh.* **23**, 483–93. Translated in *J. Appl. Math. Mech.* **23**, 686–700 (1959). [p. 431.]

Muller, D. E. (1956). A method for solving algebraic equations using an automatic computer. *Math. Tables Aids Comput.* **10**, 208–15. [p. 207.]

Murdock, J. W. & Stewartson, K. (1977). Spectra of the Orr–Sommerfeld equation. *Phys. Fluids* **20**, 1404–11. [p. 156.]

Nachtsheim, P. R. (1964). An initial value method for the numerical treatment of the Orr–Sommerfeld equation for the case of plane Poiseuille flow. *Nat. Aero. Space Admin. Tech. Note* No. D-2414. [p. 208.]

Nayfeh, A. H. (1973). *Perturbation methods*. New York: John Wiley & Sons. [pp. 392, 397.]

Newell, A. C. & Whitehead, J. A. (1969). Finite bandwidth, finite amplitude convection. *J. Fluid Mech.* **38**, 279–303. [pp. 379, 420.]

Newell, A. C. & Whitehead, J. A. (1971). Review of the finite bandwidth concept. In *Instability of continuous systems*, ed. H. Leipholz, pp. 284–9. Berlin: Springer-Verlag. [p. 420.]

Ng, B. S. (1977). Approximations to the eigenvalue relation for the Orr–Sommerfeld problem: Numerical results for plane Poiseuille flow. Indiana University–Purdue University at Indianapolis, Indianapolis Center for Advanced Research Inc., Report No. FDL-77-011. [p. 304.]

Ng, B. S. & Reid, W. H. (1979). An initial value method for eigenvalue problems using compound matrices. *J. Comput. Phys.* **30**, 125–36. [pp. 311, 315, 316.]

Nield, D. A. (1972). On the inviscid solutions of the Orr–Sommerfeld equation. *Math. Chronicle* **2**, 43–52. [pp. 139, 182, 184, 296.]

Nishioka, M., Iida, S. & Ichikawa, Y. (1975). An experimental investigation of the stability of plane Poiseuille flow. *J. Fluid Mech.* **72**, 731–51. [pp. 244, 245, 452.]

Noether, F. (1921). Das Turbulenzproblem. *Z. Angew. Math. Mech.* **1**, 125–38 and 218–19. [p. 370.]

Oberbeck, A. (1879). Ueber die Wärmleitung der Flüssigkeiten bei Berücksichtigung der Strömungen infolge von Temperaturdifferenzen. *Ann. Phys. Chem.* **7**, 271–92. [p. 35.]

Obremski, H. J., Morkovin, M. V. & Landahl, M. (1969). A portfolio of stability characteristics of incompressible boundary layers. AGARDograph No. 134, NATO, Paris. [p. 232.]

Olver, F. W. J. (1974). *Asymptotics and special functions.* New York: Academic Press. [pp. 172, 175, 214, 253, 255, 294, 295, 346, 465.]

Orr, W. M'F. (1907). The stability or instability of the steady motions of a perfect liquid and of a viscous liquid. *Proc. Roy. Irish Acad.* A **27**, 9–68 and 69–138. [pp. 80, 125, 149, 163, 424, 429.]

Orszag, S. A. (1971). Accurate solution of the Orr–Sommerfeld stability equation. *J. Fluid Mech.* **50**, 689–703. [pp. 191, 204, 205, 221, 304, 306, 307.]

Orszag, S. A. & Kells, L. C. (1980). Transition to turbulence in plane

Poiseuille and plane Couette flow. *J. Fluid Mech.* **96**, 161–205. [p. 453.]

Osborne, M. R. (1967). Numerical methods for hydrodynamic stability problems. *SIAM J. Appl. Math.* **15**, 539–57. [p. 207.]

Pai, S. I. (1943). Turbulent flow between rotating cylinders. *Tech. Notes Nat. Adv. Comm. Aero., Wash.* No. 892. [p. 446.]

Palm, E. (1960). On the tendency towards hexagonal cells in steady convection. *J. Fluid Mech.* **8**, 183–92. [pp. 376, 386, 414, 439.]

Palm, E., Ellingsen, T. & Gjevik, B. (1967). On the occurrence of cellular motion in Bénard convection. *J. Fluid Mech.* **30**, 651–61. [p. 439.]

Patnaik, P. C., Sherman, F. S. & Corcos, G. M. (1976). A numerical simulation of Kelvin–Helmholtz waves of finite amplitude. *J. Fluid Mech.* **73**, 215–40. [p. 422.]

Pearson, J. R. A. (1958). On convection cells induced by surface tension. *J. Fluid Mech.* **4**, 489–500. [p. 34.]

Pedley, T. J. (1969). On the instability of viscous flow in a rapidly rotating pipe. *J. Fluid Mech.* **35**, 97–115. [p. 221.]

Pekeris, C. L. (1936). On the stability problem in hydrodynamics. *Proc. Camb. Phil. Soc.* **32**, 55–66. [p. 159.]

Pekeris, C. L. (1948). Stability of the laminar flow through a straight pipe of circular cross-section to infinitesimal disturbances which are symmetrical about the axis of the pipe. *Proc.*

Nat. Acad. Sci., Wash. **34**, 285–95. [p. 218.]

Pekeris, C. L. & Shkoller, B. (1967). Stability of plane Poiseuille flow to periodic disturbances of finite amplitude in the vicinity of the neutral curve. *J. Fluid Mech.* **29**, 31–8. [pp. 244, 450, 451.]

Pellew, A. & Southwell, R. V. (1940). On maintained convective motion in a fluid heated from below. *Proc. Roy. Soc.* A **176**, 312–43. Reprinted by Saltzman (1962). [pp. 45, 49, 57, 66, 67.]

Phillips, O. M. (1960). On the dynamics of unsteady gravity waves of finite amplitude. Part 1. The elementary interactions. *J. Fluid Mech.* **9**, 193–217. [pp. 387, 393.]

Phillips, O. M. (1974). Nonlinear dispersive waves. *Ann. Rev. Fluid Mech.* **6**, 93–110. [p. 397.]

Pillow, A. F. (1952). The free convection cell in two dimensions. *Rep. Aero. Res. Lab., Melbourne* A 79. [pp. 434, 441.]

Poincaré, H. (1885). Sur l'équilibre d'une masse fluide animée d'un mouvement de rotation. *Acta Math.* **7**, 259–380. Also *Oeuvres* (1952), Tome VII, pp. 32–140. Paris: Gauthier–Villars. [p. 12.]

Potter, M. C. (1966). Stability of plane Couette–Poiseuille flow. *J. Fluid Mech.* **24**, 609–19. [p. 223.]

Pritchard, A. J. (1968). A study of the classical problems of hydrodynamic stability by the Liapunov method. *J. Inst. Math. Applics* **4**, 78–93. [p. 431.]

Rabenstein, A. L. (1958). Asymptotic solutions of $u^{iv} + \lambda^2(zu'' + \alpha u' + \beta u) = 0$ for large $|\lambda|$. *Arch. Rat. Mech. Anal.* **1**, 418–35. [pp. 252, 286.]

Raney, D. C. & Chang, T. S. (1971). Oscillatory modes of instability for flow between rotating cylinders with a transverse pressure gradient. *Z. Angew. Math. Phys.* **22**, 680–90. [pp. 113, 116].

Rayleigh, Lord (1879). On the instability of jets. *Proc. London Math. Soc.* **10**, 4–13. Also *Scientific papers* (1899), vol. I, pp. 361–71. Cambridge University Press. [pp. 22, 25.]

Rayleigh, Lord (1880). On the stability, or instability, of certain fluid motions. *Proc. London Math. Soc.* **11**, 57–70. Also *Scientific papers* (1899), vol. I, pp. 474–87. Cambridge University Press. [pp. 69, 81, 121, 125, 131.]

Rayleigh, Lord (1883a). On maintained vibrations. *Phil. Mag.* (5) **15**, 229–35. Also *Scientific papers* (1900), vol. II, pp. 188–93. Cambridge University Press. [p. 355.]

Rayleigh, Lord (1883b). Investigation of the character of the equilibrium of an incompressible heavy fluid of variable density. *Proc. London Math. Soc.* **14**, 170–7. Also *Scientific papers* (1900), vol. II, pp. 200–7. Cambridge University Press. [pp. 19, 63, 324, 325, 363.]

Rayleigh, Lord (1883c). On the crispations of fluid resting upon a vibrating support. *Phil. Mag.* (5) **16**, 50–8. Also *Scientific papers* (1900), vol. II, pp. 212–19.

Cambridge University Press. [p. 355.]

Rayleigh, Lord (1884). Presidential address. *Brit. Assoc. Report*, pp. 1–23, Montreal. Also *Science* (o.s.) **4**, 179–84 (1884); *Scientific papers* (1900), vol. II, pp. 333–54. Cambridge University Press. [p. 32.]

Rayleigh, Lord (1892a). On the question of the stability of the flow of fluids. *Phil. Mag.* (5) **34**, 59–70. Also *Scientific papers* (1902), vol. III, pp. 575–84. Cambridge University Press. [pp. 158, 160.]

Rayleigh, Lord (1892b). On the instability of cylindrical fluid surfaces. *Phil. Mag.* (5) **34**, 177–80. Also *Scientific papers* (1902), vol. III, pp. 594–6. Cambridge University Press. [p. 30.]

Rayleigh, Lord (1894). *The theory of sound*, 2nd edn, London: Macmillan. [pp. 22, 30, 146, 147, 246.]

Rayleigh, Lord (1913). On the motion of a viscous fluid. *Phil. Mag.* (6) **26**, 776–86. Also *Scientific papers* (1920), vol. VI, pp. 187–96. Cambridge University Press. [p. 71.]

Rayleigh, Lord (1914). Further remarks on the stability of viscous fluid motion. *Phil. Mag.* (6) **28**, 609–19. Also *Scientific papers* (1920), vol. VI, pp. 266–75. Cambridge University Press. [p. 212.]

Rayleigh, Lord (1916a). On convection currents in a horizontal layer of fluid, when the higher temperature is on the under side. *Phil. Mag.* (6) **32**, 529–46. Also *Scientific papers* (1920), vol. VI, pp. 432–46. Cambridge University Press. Reprinted by Saltzman (1962). [pp. 32, 33, 35, 37, 41, 42, 50.]

Rayleigh, Lord (1916b). On the dynamics of revolving fluids. *Proc. Roy. Soc.* A **93**, 148–54. Also *Scientific papers* (1920), vol. VI, pp. 447–53. Cambridge University Press. [pp. 69, 72.]

Reichardt, H. (1956). Über die Geschwindigkeitsverteilung in einer geradlinigen turbulenten Couetteströmung. *Z. Angew. Math. Mech.* **36**, S26–9. [p. 454.]

Reid, W. H. (1958). On the stability of viscous flow in a curved channel. *Proc. Roy. Soc.* A **244**, 186–98. [pp. 109, 111, 112.]

Reid, W. H. (1960). Inviscid modes of instability in Couette flow. *J. Math. Anal. Applics* **1**, 411–22. [pp. 83, 84, 85, 86, 87, 88.]

Reid, W. H. (1961). Inviscid modes of instability in flow over a concave wall. *J. Math. Anal. Applics* **2**, 419–27. [p. 123.]

Reid, W. H. (1965). The stability of parallel flows. In *Basic developments in fluid dynamics*, vol. 1, ed. M. Holt, pp. 249–307. New York: Academic Press. [pp. 175, 177, 179, 187, 189, 190, 249, 250.]

Reid, W. H. (1972). Composite approximations to the solutions of the Orr–Sommerfeld equation. *Studies in Appl. Math.* **51**, 341–68. [pp. 248, 253, 268, 276, 279, 282, 465.]

Reid, W. H. (1974a). Uniform asymptotic approximations to the solutions of the Orr–Sommerfeld equation. Part 1. Plane Couette flow. *Studies in Appl. Math.* **53**, 91–110. [pp. 214, 215, 318.]

Reid, W. H. (1974b). Uniform asymptotic approximations to the solutions of the Orr–Sommerfeld equation. Part 2. The general theory. *Studies in Appl. Math.* **53**, 217–24. [pp. 248, 280.]

Reid, W. H. (1979). An exact solution of the Orr–Sommerfeld equation for plane Couette flow. *Studies in Appl. Math.* **61**, 83–91. [p. 319.]

Reid, W. H. & Harris, D. L. (1958). Some further results on the Bénard problem. *Phys. Fluids* **1**, 102–10. [pp. 51, 53, 66.]

Reid, W. H. & Harris, D. L. (1959). Streamlines in Bénard convection cells. *Phys. Fluids* **2**, 716–17. [pp. 57, 58.]

Reynolds, O. (1883). An experimental investigation of the circumstances which determine whether the motion of water shall be direct or sinuous, and of the law of resistance in parallel channels. *Phil. Trans. Roy. Soc.* **174**, 935–82. Also *Scientific papers* (1901), vol. II, pp. 51–105. Cambridge University Press. [pp. 1, 2, 3, 20, 124, 125, 370, 452.]

Reynolds, O. (1895). On the dynamical theory of incompressible viscous fluids and the determination of the criterion. *Phil. Trans. Roy. Soc.* A **186**, 123–64. Also *Scientific papers* (1901), vol. II, pp. 535–77. Cambridge University Press. [p. 424.]

Reynolds, W. C. & Potter, M. C. (1967). Finite-amplitude instability of parallel shear flows. *J. Fluid Mech.* **27**, 465–92. [pp. 194, 195, 223, 244, 304, 307, 450.]

Roberts, P. H. (1965). Appendix (to paper by R. J. Donnelly and K. W. Schwarz). The solution of the characteristic value problems. *Proc. Roy. Soc.* A **283**, 550–5. [p. 104.]

Roberts, P. H. (1966). On non-linear Bénard convection. In *Non-equilibrium thermodynamics, variational techniques, and stability*, eds. R. J. Donnelly, R. Herman & I. Prigogine, pp. 125–62. University of Chicago Press. [p. 441.]

Roberts, P. H. (1967). *An introduction to magnetohydrodynamics*. London: Longmans, Green & Co. Ltd. [pp. 340, 341.]

Romanov, V. A. (1973). Stability of plane-parallel Couette flow. *Funkcional Anal. i Proložen.* **7**, no. 2, 62–73. Translated in *Functional Anal. & Its Applics* **7**, 137–46 (1973). [p. 213.]

Rosenblat, S. & Herbert, D. M. (1970). Low-frequency modulation of thermal instability. *J. Fluid Mech.* **43**, 385–98. [p. 359.]

Ross, J. A., Barnes, F. H., Burns, J. G. & Ross, M. A. S. (1970). The flat plate boundary layer. Part 3. Comparison of theory with experiment. *J. Fluid Mech.* **43**, 819–32. [pp. 241, 242, 243, 244.]

Ross, M. A. S. (1973). Numerical treatment of fluid dynamical stability problems. In *Advances in numerical fluid dynamics,*

AGARD Lecture Series No. 64, pp. 7/1–21. [p. 209.]

Rossby, H. T. (1969). A study of Bénard convection with and without rotation. *J. Fluid Mech.* **36**, 309–35. [p. 60.]

Rott, N. (1970). A multiple pendulum for the demonstration of non-linear coupling. *Z. Angew. Math. Phys.* **21**, 570–82. [pp. 387, 391.]

Rotunno, R. (1978). A note on the stability of a cylindrical vortex sheet. *J. Fluid Mech.* **87**, 761–71. [p. 122.]

Ruelle, D. & Takens, F. (1971). On the nature of turbulence. *Comm. Math. Phys.* **20**, 167–92 and **23**, 343–4. [pp. 399, 402, 448.]

Saffman, P. G. & Taylor, G. I. (1958). The penetration of a fluid into a porous medium or Hele–Shaw cell containing a more viscous liquid. *Proc. Roy. Soc. A* **245**, 312–29. [p. 31.]

Saltzman, B. (Editor) (1962). *Selected papers on the theory of thermal convection with special application to the Earth's planetary atmosphere.* New York: Dover. [pp. 34, 62.]

Salwen, H. & Grosch, C. E. (1972). The stability of Poiseuille flow in a pipe of circular cross-section. *J. Fluid Mech.* **54**, 93–112. [p. 219.]

Sani, R. (1968). An extension and convergence proof of an approximate method due to Pellew and Southwell. *Z. Angew. Math. Mech.* **48**, 65–6. [p. 96.]

Saric, W. S. & Nayfeh, A. H. (1975). Nonparallel stability of boundary-layer flows. *Phys. Fluids* **18**, 945–50. [pp. 242, 479.]

Sato, H. (1959). Further investigation on the transition of two-dimensional separated layer at subsonic speeds. *J. Phys. Soc. Japan* **14**, 1797–810. [p. 456.]

Sato, H. (1970). An experimental study of non-linear interaction of velocity fluctuations in the transition region of a two-dimensional wake. *J. Fluid Mech.* **44**, 741–65. [p. 457.]

Sato, H. & Kuriki, K. (1961). The mechanism of transition in the wake of a thin flat plate placed parallel to a uniform flow. *J. Fluid Mech.* **11**, 321–52. [p. 457.]

Sattinger, D. H. (1973). *Topics in stability and bifurcation theory.* Lecture Notes in Mathematics, vol. 309. Berlin: Springer-Verlag. [pp. 403, 407.]

Schade, H. (1964). Contribution to the nonlinear stability theory of inviscid shear layers. *Phys. Fluids* **7**, 623–8. [p. 456.]

Schensted, I. V. (1960). Contributions to the theory of hydrodynamic stability. Ph.D. thesis. University of Michigan. [p. 158.]

Schlichting, H. (1933). Zur Entstehung der Turbulenz bei der Plattenströmung. *Nachr. Ges. Wiss. Göttingen, Math.-phys. Kl.,* pp. 181–208. [pp. 225, 240, 243, 352.]

Schlichting, H. (1968). *Boundary layer theory,* 6th edn. New York: McGraw-Hill. [p. 232.]

Schlüter, A., Lortz, D. & Busse, F. (1965). On the stability of steady finite amplitude convection. *J. Fluid Mech.* **23**, 129–44. [p. 436.]

Schmidt, R. J. & Saunders, O. A. (1938). On the motion of a fluid heated from below. *Proc. Roy. Soc.* A **165**, 216–28. [pp. 60, 441.]

Schubauer, G. B. & Skramstad, H. K. (1947). Laminar boundary-layer oscillations and transition on a flat plate. *J. Res. Nat. Bur. Stand.* **38**, 251–92; also *Rep. Nat. Adv. comm. Aero., Wash.* No. 909 (1948). [pp. 226, 240, 241, 242.]

Scott, M. R. (1973). An initial value method for the eigenvalue problem for systems of ordinary differential equations. *J. Comput. Phys.* **12**, 334–47. [pp. 209, 210.]

Scotti, R. S. & Corcos, G. M. (1972). An experiment on the stability of small disturbances in a stratified free shear layer. *J. Fluid Mech.* **52**, 499–528. [p. 333.]

Segel, L. A. (1965). The structure of non-linear cellular solutions to the Boussinesq equations. *J. Fluid Mech.* **21**, 345–58. [p. 461.]

Segel, L. A. (1966). Non-linear hydrodynamic stability theory and its applications to thermal convection and curved flows. In *Non-equilibrium thermodynamics, variational techniques, and stability*, eds. R. J. Donnelly, R. Herman & I. Prigogine, pp. 165–97. University of Chicago Press. [p. 461.]

Segel, L. A. (1969). Distant side-walls cause slow amplitude modulation of cellular convection. *J. Fluid Mech.* **38**, 203–24. [pp. 379, 420, 439.]

Serrin, J. (1959). On the stability of viscous fluid motions. *Arch. Rat. Mech. Anal.* **3**, 1–13. [pp. 424, 427.]

Sewell, M. J. (1966). On the connexion between stability and the shape of the equilibrium surface. *J. Mech. Phys. Solids* **14**, 203–30. [pp. 409, 411.]

Sexl, T. (1927). Zur Stabilitätsfrage der Poiseuilleschen und Couetteschen strömung. *Ann. Phys., Lpz.* (4) **83**, 835–48. [p. 217.]

Shen, S. F. (1954). Calculated amplified oscillations in the plane Poiseuille and Blasius flows. *J. Aero. Sci.* **21**, 62–4. [pp. 240, 241, 243.]

Shen, S. F. (1961). Some considerations of the laminar stability of incompressible time-dependent basic flows. *J. Aerospace Sci.* **28**, 397–404 and 417. [p. 361.]

Shen, S. F. (1964). Stability of laminar flows. In *Theory of laminar flows*, ed. F. K. Moore, pp. 719–853. Princeton University Press. [pp. 175, 221.]

Silcock, G. (1975). On the stability of parallel stratified shear flows. Ph.D. dissertation. University of Bristol. [p. 235.]

Silveston, P. L. (1958). Wärmedurchgang in waagerechten Flüssigkeitsschichten. *Forsch. Gebiete Ingenieurwes.* **24**, 29–32 and 59–69. [p. 60.]

Simmons, W. F. (1969). A variational method for weak resonant wave interactions. *Proc. Roy. Soc.* A **309**, 551–75. [p. 397.]

Singh, K., Lumley, J. L. & Betchov, R. (1963). Modified Hankel functions and their integrals to argument 10. *Engineering Res. Bull.* No. B-87, Pennsylvania State University. [p. 173.]

Sloan, D. M. (1977). Eigenfunctions of systems of linear ordinary differential equations with separated boundary conditions using Riccati transformations. *J. Comput. Phys.* **24**, 320–30. [p. 209.]

Sloan, D. M. & Wilks, G. (1976). Riccati transformations for eigenvalues of systems of linear ordinary differential equations with separated boundary conditions. *J. Inst. Math. Applics* **18** 117–27. [p. 209, 313.]

Sloan, D. M. & Wilks, G. (1977). The Riccati transformation method and the computation of eigenvalues of complex linear differential systems. *J. Comput. Appl. Math.* **3**, 195–9. [p. 209.]

Smith, A. M. O. (1955). On the growth of Taylor–Görtler vortices along highly concave walls. *Quart. Appl. Math.* **13**, 233–62. [p. 120.]

Smith, F. T. (1979a). On the non-parallel flow stability of the Blasius boundary layer. *Proc. Roy. Soc.* A **366**, 91–109. [p. 454.]

Smith, F. T. (1979b). Nonlinear stability of boundary layers for disturbances of various sizes. *Proc. Roy. Soc.* A **368**, 573–89. [p. 454.]

Snyder, H. A. (1968a). Stability of rotating Couette flow. II. Comparison with numerical results. *Phys. Fluids* **11**, 1599–605. [p. 107.]

Snyder, H. A. (1968b). Experiments on rotating flows between noncircular cylinders. *Phys. Fluids* **11**, 1606–11. [p. 107.]

Snyder, H. A. (1969). Wavenumber selection at finite amplitude in rotating Couette flow. *J. Fluid Mech.* **35**, 273–98. [pp. 445, 449.]

Snyder, H. A. & Lambert, R. B. (1966). Harmonic generation in Taylor vortices between rotating cylinders. *J. Fluid Mech.* **26**, 545–62. [p. 108.]

Sommerfeld, A. (1908). Ein Beitrag zur hydrodynamischen Erklaerung der turbulenten Fluessigkeitsbewegungen. *Proceedings 4th International Congress of Mathematicians, Rome*, vol. III, pp. 116–24. [p. 125.]

Sorokin, V. S. (1961). Nonlinear phenomena in closed flows near critical Reynolds numbers. *Prikl. Mat. Mekh.* **25**, 248–58. Translated in *J. Appl. Math. Mech.* **25**, 366–81 (1961). [p. 28.]

Southwell, R. V. & Chitty, L. (1930). On the problem of hydrodynamic stability.—I. Uniform shearing motion in a viscous liquid. *Phil. Trans. Roy. Soc.* A **229**, 205–53. [pp. 158, 160, 212.]

Sparrow, E. M., Goldstein, R. J. & Jonsson, V. K. (1964). Thermal instability in a horizontal fluid layer: effect of boundary conditions and non-linear temperature profile. *J. Fluid Mech.* **18**, 513–28. [pp. 42, 63.]

Spiegel, E. A. & Veronis, G. (1960). On the Boussinesq approximation for a compressible fluid. *Astrophys. J.* **131**, 442–7. [p. 36.]

Squire, H. B. (1933). On the stability of three-dimensional disturbances of viscous flow between parallel walls. *Proc. Roy. Soc. A* **142**, 621–8. [pp. 129, 155.]

Stakgold, I. (1971). Branching of solutions of nonlinear equations. *SIAM Rev.* **13**, 289–332. [p. 386.]

Stern, M. E. (1960). The 'salt fountain' and thermohaline convection. *Tellus* **12**, 172–5. [pp. 63, 65.]

Stewartson, K. & Stuart, J. T. (1971). A non-linear instability theory for a wave system in plane Poiseuille flow. *J. Fluid Mech.* **48**, 529–45. [pp. 420, 451.]

Stoker, J. J. (1950). *Nonlinear vibrations in mechanical and electrical systems.* New York: Interscience. [p. 354.]

Stork, K. & Müller, U. (1972). Convection in boxes: experiments. *J. Fluid Mech.* **54**, 599–611. [p. 61.]

Stuart, J. T. (1958). On the non-linear mechanics of hydrodynamic stability. *J. Fluid Mech.* **4**, 1–21. [pp. 376, 380, 443.]

Stuart, J. T. (1960a). On the non-linear mechanics of wave disturbances in stable and unstable parallel flows. Part 1. The basic behaviour in plane Poiseuille flow. *J. Fluid Mech.* **9**, 353–70. [pp. 376, 380, 386, 443, 450.]

Stuart, J. T. (1960b). Non-linear effects in hydrodynamic stability. *Proceedings 10th International Congress of Applied Mechanics,* Stresa, Italy, eds. F. Rolla & W. T. Koiter, pp. 63–97. Amsterdam: Elsevier. [p. 455.]

Stuart, J. T. (1963). Hydrodynamic stability. In *Laminar boundary layers,* ed. L. Rosenhead, pp. 492–579. Oxford: Clarendon Press. [pp. 172, 178, 223, 231.]

Stuart, J. T. (1964). On the cellular patterns in thermal convection. *J. Fluid Mech.* **18**, 481–98. [pp. 55, 56, 58.]

Stuart, J. T. (1967). On finite amplitude oscillations in laminar mixing layers. *J. Fluid Mech.* **29**, 417–40. [p. 423.]

Stuart, J. T. (1977). Bifurcation theory in non-linear hydrodynamic stability. In *Applications of bifurcation theory,* ed. P. H. Rabinowitz, pp. 127–47. New York: Academic Press. [p. 370.]

Stuart, J. T. & DiPrima, R. C. (1978). The Eckhaus and Benjamin–Feir resonance mechanisms. *Proc. Roy. Soc. A* **362**, 27–41. [pp. 398, 417.]

Synge, J. L. (1933). The stability of heterogeneous liquids. *Trans. Roy. Soc. Can.* (3) **27**, 1–18. [pp. 78, 329.]

Synge, J. L. (1938). Hydrodynamical stability. *Semicentenn. Publ. Amer. Math. Soc.* **2**, 227–69. [pp. 94, 161, 214, 247, 320.]

Tam, K. K. (1968). On the asymptotic solution of the Orr–Sommerfeld equation by the method of multiple-scales. *J. Fluid Mech.* **34**, 145–58. Corrigenda in **35**, p. 829 (1969). [p. 253.]

Tatsumi, T. (1952). Stability of the laminar inlet-flow prior to the formation of Poiseuille régime, II. *J. Phys. Soc. Japan* **7**, 495–502. [p. 219.]

Tatsumi, T. & Gotoh, K. (1960). The stability of free boundary layers between two uniform streams. *J. Fluid Mech.* **7**, 433–41. [p. 199.]

Tatsumi, T., Gotoh, K. & Ayukawa, K. (1964). The stability of a free boundary layer at large Reynolds numbers. *J. Phys. Soc. Japan* **19**, 1966–80. [pp. 237, 250.]

Tatsumi, T. & Kakutani, T. (1958). The stability of a two-dimensional laminar jet. *J. Fluid Mech.* **4**, 261–75. [pp. 197, 198, 199.]

Tayler, R. J. (1957a). The influence of an axial magnetic field on the stability of a constricted gas discharge. *Proc. Phys. Soc.* B **70**, 1049–63. [pp. 340, 345.]

Tayler, R. J. (1957b). Hydromagnetic instabilities of a cylindrical gas discharge. 5. Influence of alternating magnetic fields. *A.E.R.E. Theor. Rept* 2263. [p. 369.]

Taylor, G. I. (1921). Experiments with rotating fluids. *Proc. Camb. Phil. Soc.* **20**, 326–9. [pp. 71, 90.]

Taylor, G. I. (1923). Stability of a viscous liquid contained between two rotating cylinders. *Phil. Trans. Roy. Soc.* A **223**, 289–343. Also *Scientific papers* (1971), vol. IV, pp. 34–85. Cambridge University Press. [pp. 69, 71, 90, 97, 102, 104, 105, 106, 448.]

Taylor, G. I. (1931). Effect of variation in density on the stability of superposed streams of fluid. *Proc. Roy. Soc.* A **132**, 499–523. Also *Scientific papers* (1960), vol. II, pp. 219–39. Cambridge University Press. [pp. 324, 331, 333.]

Taylor, G. I. (1936). Fluid friction between rotating cylinders. *Proc. Roy. Soc.* A **157**, 546–64 and

565–78. Also *Scientific papers* (1960), vol. II, pp. 380–96 and 397–408. Cambridge University Press. [pp. 106, 212, 443.]

Taylor, G. I. (1950). The instability of liquid surfaces when accelerated in a direction perpendicular to their planes. I. *Proc. Roy. Soc.* A **201**, 192–6. Also *Scientific papers* (1963), vol. III, pp. 532–6. Cambridge University Press. [p. 19.]

Thom, R. (1975). *Structural stability and morphogenesis.* Reading, Mass.: Benjamin. [pp. 408, 409, 411.]

Thomas, L. H. (1953). The stability of plane Poiseuille flow. *Phys. Rev.* (2) **91**, 780–3. [pp. 202, 206, 221, 222.]

Thompson, J. M. T. & Hunt, G. W. (1973). *A general theory of elastic stability.* London: John Wiley & Sons. [pp. 409, 413.]

Thompson, J. M. T. & Hunt, G. W. (1975). Towards a unified bifurcation theory. *Z. Angew. Math. Phys.* **26**, 581–603. [p. 413.]

Thomson, J. (1882). On a changing tesselated structure in certain liquids. *Proc. Phil. Soc. Glasgow.* **13**, 464–8. [p. 32.]

Thorpe, S. A. (1968). A method of producing a shear flow in a stratified fluid. *J. Fluid Mech.* **32**, 693–704. [p. 22.]

Thorpe, S. A. (1969). Experiments on the instability of stratified shear flows: immiscible fluids. *J. Fluid Mech.* **39**, 25–48. [p. 20.]

Thorpe, S. A. (1971). Experiments on the instability of stratified shear flows: miscible fluids. *J.*

Fluid Mech. **46**, 299–319. [p. 333.]

Threlfall, D. C. (1975). Free convection in low-temperature gaseous helium. *J. Fluid Mech.* **67**, 17–28. [pp. 435, 441.]

Tietjens, O. (1925). Beiträge zur Entstehung der Turbulenz. *Z. Angew. Math. Mech.* **5**, 200–17. [p. 188.]

Tippelskirch, H. von (1956). Über Konvektionszellen, insbesondere im flüssigen Schwefel. *Beiträge Phys. Atmos.* **20**, 37–54. [p. 61.]

Tollmien, W. (1929). Über die Entstehung der Turbulenz. *Nachr. Ges. Wiss. Göttingen, Math.-phys. Kl.*, pp. 21–44. Translated as 'The production of turbulence', *Tech. Memor. Nat. Adv. Comm. Aero., Wash.* No. 609 (1931). [pp. 126, 137, 164, 165, 166, 177, 224, 225.]

Tollmien, W. (1935). Ein allgemeines Kriterium der Instabilität laminarer Geschwindigkeitsverteilungen. *Nachr. Wiss. Fachgruppe, Göttingen, Math.-phys. Kl.* **1**, 79–114. Translated as 'General instability criterion of laminar velocity distributions', *Tech. Memor. Nat. Adv. Comm. Aero., Wash.* No. 792 (1936). [pp. 124, 132, 134.]

Tollmien, W. (1947). Asymptotische Integration der Störungsdifferentialgleichung ebener laminarer Strömungen bei hohen Reynoldschen Zahlen. *Z. Angew. Math. Mech.* **25/27**, 33–50 and 70–83. [pp. 126, 164, 175, 177, 179, 299.]

Turner, J. S. (1973). *Buoyancy effects in fluids.* Cambridge University Press. [p. 333.]

Ukhovskii, M. R. & Iudovich, V. I. (1963). On the equations of steady-state convection. *Prikl. Mat. Mekh.* **27**, 295–300. Translated in *J. Appl. Math. Mech.* **27**, 432–40 (1963). [p. 407.]

Van Dyke, M. (1975). *Perturbation methods in fluid mechanics,* annotated edition. Stanford, Calif.: The Parabolic Press. [pp. 123, 174, 178, 180, 276, 278.]

Varley, E., Kazakia, J. Y. & Blythe, P. A. (1977). The interaction of large amplitude barotropic waves with an ambient shear flow: critical flows. *Phil. Trans. Roy. Soc.* A **287**, 189–236. [p. 423.]

Velte, W. (1966). Stabilität und Verzweigung stationärer Lösungen der Navier–Stokesschen Gleichungen beim Taylorproblem. *Arch. Rat. Mech. Anal.* **22**, 1–14. [p. 407.]

Walowit, J., Tsao, S. & DiPrima, R. C. (1964). Stability of flow between arbitrarily spaced concentric cylindrical surfaces including the effect of a radial temperature gradient. *J. Appl. Mech.* **31**, *Trans. A.S.M.E.* **86**, 585–93. [pp. 92, 103, 109.]

Wang, D. P. (1968). Finite amplitude effect in the stability of a jet of circular cross-section. *J. Fluid Mech.* **34**, 299–313. [p. 25.]

Wasow, W. (1953a). On small disturbances of plane Couette flow. *J. Res. Nat. Bur. Stand.* **51**, 195–202. [pp. 257, 477.]

Wasow, W. (1953b). Asymptotic solution of the differential

equation of hydrodynamic stability in a domain containing a transition point. *Annals Math.* **58**, 222–52. [p. 251.]

Watson, J. (1960a). Three-dimensional disturbances in flow between parallel planes. *Proc. Roy. Soc.* A **254**, 562–9. [p. 346.]

Watson, J. (1960b). On the nonlinear mechanics of wave disturbances in stable and unstable parallel flows. Part 2. The development of a solution for plane Poiseuille flow and for plane Couette flow. *J. Fluid Mech.* **9**, 371–89. [pp. 376, 380, 386, 443, 450.]

Watson, J. (1962). On spatially-growing finite disturbances in plane Poiseuille flow. *J. Fluid Mech.* **14**, 211–21. [pp. 152, 350.]

Wazzan, A. R., Okamura, T. T. & Smith, A. M. O. (1967). Stability of laminar boundary layers at separation. *Phys. Fluids* **10**, 2540–5. [pp. 230, 231.]

Wendt, F. (1933). Turbulente Strömungen zwischen zwei rotierenden koaxialen Zylindern. *Ingen.-Archiv* **4**, 577–95. [p. 107.]

Whitehead, J. A., Jr. & Parsons, B. (1978). Observations of convection at Rayleigh numbers up to 760,000 in a fluid with large Prandtl number. *Geophys. Astrophys. Fluid Dyn.* **9**, 201–17. [p. 438.]

Whitham, G. B. (1967). Variational methods and applications to water waves. *Proc. Roy. Soc.* A **299**, 6–25. [p. 397.]

Whitham, G. B. (1974). *Linear and nonlinear waves.* New York: John Wiley & Sons. [pp. 397, 462.]

Wilkinson, J. H. (1965). *The algebraic eigenvalue problem.* Oxford: Clarendon Press. [p. 204.]

Witting, H. (1958). Über den Einfluss der Stromlinienkrümmung auf die Stabilität laminarer Strömungen. *Arch. Rat. Mech. Anal.* **2**, 243–83. [pp. 102, 116, 119, 120.]

Wood, W. W. (1964). Stability of viscous flow between rotating cylinders. *Z. Angew. Math. Phys.* **15**, 313–14. [p. 94.]

Yih, C.-S. (1955). Stability of two-dimensional parallel flows for three-dimensional disturbances. *Quart. Appl. Math.* **12**, 434–5. [p. 322.]

Yih, C.-S. (1965). *Dynamics of nonhomogeneous fluids.* New York: Macmillan. [p. 325.]

Yih, C.-S. (1972). Spectral theory of Taylor vortices. *Arch. Rat. Mech. Anal.* **46**, 218–40 and 288–300. [p. 94.]

Yuen, H. C. & Lake, B. M. (1975). Nonlinear deep water waves: theory and experiment. *Phys. Fluids* **18**, 956–60. [p. 398.]

Yuen, M.-C. (1968). Nonlinear capillary instability of a liquid jet. *J. Fluid Mech.* **33**, 151–63. [p. 25.]

Zabusky, N. J. & Deem, G. S. (1971). Dynamical evolution of two-dimensional unstable shear flows. *J. Fluid Mech.* **47**, 353–79. [p. 457.]

Zierep, J. (1961). Thermokonvektive Zellularströmungen bei inkonstanter Erwärmung der

Grundfläche. *Z. Angew. Math. Mech.* **41**, 114–25. [p. 67.]

Zondek, B. & Thomas, L. H. (1953). Stability of a limiting case of plane Couette flow. *Phys. Rev.* (2) **90**, 738–43. [pp. 216, 478.]

Zubov, V. I. (1957). *Metody Liapunova i ikh Primenenie.* Leningrad: Gos. Univ. English translation by L. F. Boron: *Methods of A. M. Lyapunov and their application.* Groningen: P. Noordhoff Ltd. (1964). [p. 431.]

MOTION PICTURE INDEX

In the 1960s the Educational Development Center produced several films for the National Committee for Fluid Mechanics Films. It made 16mm sound films, each one taking about 30 minutes to project, and 8mm (super, standard or both) film loops in cassettes, each taking typically four minutes. Most of the loops contain edited excerpts from the long films. Many of the films are of demonstrations and experiments of hydrodynamic instability, and most of these are excellent. The long films are for hire or sale, and the loops for sale. They are distributed in the United States and many other countries by the Encyclopaedia Britannica Educational Corporation, 425 North Michigan Avenue, Chicago, Illinois 60611, and elsewhere by some special distributors.

16mm Films

Bryson, A. E. (F1967). *Waves in fluids.* B/W–No. 21611. [p. 20.]

Long, R. R. (F1968). *Stratified flow.* Color–No. 21618. [p. 22.]

Mollo-Christensen, E. L. (F1968). *Flow instabilities.* Color–No. 21619. [pp. 14, 31.]

Stewart, R. W. (F1968). *Turbulence.* Color–No. 21626. [p. 4.]

Trefethen, L. M. (F1965). *Surface tension in fluid mechanics.* Color–No. 21610. [p. 27.]

8mm Film Loops

Brown, F. N. M. (FL1964). *Stages of boundary-layer instability and transition.* B/W–No. S–FM092. [p. 239.]

Bryson, A. E. (FL1967). *Small-amplitude gravity waves in an open channel.* B/W–No. S–FM139. [p. 20.]

Coles, D. (FL1963a). *Instabilities in circular Couette flow.* B/W–No. S–FM031. [pp. 108, 447.]

Coles, D. (FL1963b). *Examples of turbulent flow between concentric rotating cylinders.* B/W–No. S–FM032. [pp. 108, 447.]

Dattner, A. (FL1963). *Current-induced instabilities of a mercury jet.* B/W–No. S–FM046. [p. 340.]

Fultz, D. (FL1964). *Inertia oscillations in rotating fluid.* B/W–No. S–FM044. [pp. 77, 121.]

Fultz, D., Kaylor, R. & Hide, R. (FL1965). *Buoyancy-induced waves in rotating fluid.* Color–No. S–FM043. [p. 334.]

Lippisch, A. M. (FL1964). *Tollmien–Schlichting waves.* B/W–No. S–FM023. [p. 239.]

Mollo-Christensen, E. L. (FL1968a). *Examples of flow instability* (Part I). Color–No. S–FM146. [p. 33.]

Mollo-Christensen, E. L. (FL1968b). *Experimental study of a flow instability.* B/W–No. S–FM148. [p. 14.]

Mollo-Christensen, E. L. & Wille, R. (FL1968). *Examples of flow instability* (Part II). Color–No. S–FM147. [p. 22.]

Stewart, R. W. (FL1968). *Laminar and turbulent pipe flow.* Color–No. S–FM134. [p. 4.]

Trefethen, L. M. (FL1965). *Breakup of liquid into drops.* Color–No. S–FM076. [p. 27.]

SUBJECT INDEX